The
Chemical Engineering
Guide to Heat Transfer
Volume I
Plant Principles

Edited by

Kenneth J. McNaughton
and the Staff of Chemical Engineering

HEMISPHERE PUBLISHING CORPORATION
Washington   New York   London

CHEMICAL ENGINEERING

McGraw-Hill Publications Co., New York, N.Y.

DISTRIBUTION OUTSIDE NORTH AMERICA
SPRINGER-VERLAG

Berlin   Heidelberg   New York   Tokyo

Copyright © 1986 by Chemical Engineering McGraw-Hill Pub. Co.
1221 Avenue of the Americas, New York, New York 10020

Library of Congress Cataloging-in-Publication Data
Main entry under title:

The Chemical engineering guide to heat transfer.

  Includes bibliographies and index.
  Contents: v. 1. Plant principles—v. 2. Equipment.
  1. Heat—Transmission. 2. Heat exchangers.
I. McNaughton, Kenneth J. II. Chemical engineering.
TJ260.C426 1985   660.2'8427   85-26987
ISBN 0-07-606939-7 (v. 1) Chemical Engineering
ISBN 0-07-606940-0 (v. 2) Chemical Engineering
ISBN 0-08116-465-0 (v. 1). Hemisphere
ISBN 0-89116-466-9 (v. 2) Hemisphere
DISTRIBUTED OUTSIDE NORTH AMERICA:
ISBN 3-540-16177-5 (v. 1) Springer-Verlag Berlin
ISBN 3-540-16178-3 (v. 2) Springer-Verlag Berlin

CONTENTS

Section IV Heat Recovery

Section V Steam

Section VI Cost

PREFACE

When I was studying chemical engineering at college, I loved heat transfer. It was neat. Q = CAT. That's all I needed to know. And heat-in equals heat-out. Boy! If only the rest of life was so simple!

Some people say that engineers are drawn to science because they lack the skills to deal with people. That scientific types take comfort in being able to handle a field that responds according to inviolate laws, unlike their lawless and unpredictable fellow beings. Other observers, perhaps more charitable, suggest that those who inherit the skills to deal with a scientific universe may not give themselves the time to come to grips with the more elusive rules that attempt to explain human behavior.

I think both are fascinating fields worth pursuing. And who can say that the two won't come together? What with all the exciting developments in our understanding of molecular biology, surely we are overdue for some breakthroughs in psychology as well.

As it turns out, heat transfer isn't so simple anyway. But it is neat. And we are very fortunate to have developed our communication skills so well. In this book, for instance, we have accumulated ten years of practical wisdom from the pages of *Chemical Engineering,* on the subject of heat transfer—basic plant principles.

Here, assembled in one volume, are the writings of fellow chemical engineers who also graduated with the basic understanding that Q = CAT and that heat-in equals heat-out. But they went on to specialize, and now they share with us their knowledge about how the theory is applied in the plants of today's chemical process industries.

All the different types of heat exchangers and how to select the right one. Shell-and-tube equipment, how it works. Design—calculator programs and modeling. Heat recovery—optimizing, conserving, saving, networks, efficiency. Steam—the conveyor of heat. And of course, cost—the bottom line.

Life may not be so simple, but this book is going to make your life easier when it comes to heat transfer applications in the plant. And so will the companion volume, which covers all the different sorts of heat transfer equipment.

Section I
Heat Exchangers

Types, performance and applications
Specifying and selecting
Materials of construction

Types, performance and applications

Engineers today must evaluate a large variety of heat exchangers in order to select the most suitable and economical one for a service. Many configurations are discussed and their salient features analyzed.

C. Scaccia and *G. Theoclitus*, *C-E Air Preheater*

☐ Performance objectives influence the selection of the type of heat exchanger, the flow arrangement and the materials of construction. If, for example, maximum heat recovery is paramount, the flows should be countercurrent. If, on the other hand, heat transfer between two streams is to be limited, other flow arrangements may be used. If sulfur oxide vapors are present, the problem of acid dewpoint must be considered.

Information on flow quantities, temperatures, pressures and compositions is fundamental to a rational design. Accurate process flows are indispensable for sizing exchangers to meet the required thermal duties at the smallest pressure drops and to minimize capital cost.

Changing process conditions or reduced ambient temperatures may result in material temperatures higher or lower than desired. Corrective measures are available, including hot-gas bypassing or gas cooling, cold-air bypassing, air preheating, or hot-air recirculation. When lower efficiency can be tolerated, the cold-end metal temperature may be controlled via a parallel-flow arrangement.

Both gas-phase and liquid-phase corrosion (whether by oxidation or acidity) must be carefully considered, although the latter usually presents the more severe problem, at the specification and design stages. Fouling of surfaces can lower performance and raise pressure drop, which can lead to operating problems. If fouling is anticipated, the means must be provided for cleaning the heat-transfer surfaces, such as by soot blowing or water washing.

Some heat exchangers are less prone to fouling. For example, in the rotary regenerative exchanger, the counterflow provides cleaning action, and surface expansion and contraction loosen soot, which the high-velocity air stream then removes.

Heat exchangers are generally named according to the way that the hot and cold fluids enter and leave. Usually, this also describes the relationship of the flows to each other.

Counterflow recovers maximum heat

In the counterflow exchanger, fluids flow in opposite directions, usually after entering at opposite ends. This arrangement takes maximum advantage of the heat-transfer surfaces, requiring the least surface for a given duty and achieving maximum heat recovery and temperature effectiveness.

Some critical terms defined

Temperature (or recuperative) effectiveness, ε, is the ratio of the difference between the inlet and outlet temperatures of one of the two fluids (the one whose temperature difference is greater) to the inlet temperature differences of the two fluids:

$$\varepsilon = (t_{in} - t_{out})_{max}/(t_{h,in} - t_{c,in})$$

Here, t_{in} represents the fluid inlet temperatures, and t_{out} the outlet temperatures.

C_{min}/C_{max} is the ratio of the heat capabilities ($\dot{m} c_p$) of the fluids. The numerator and denominator refer, respectively, to the fluid having the least and the greatest heat capacity. A C_{min}/C_{max} approaching 1 describes a gas/gas or liquid/liquid heat exchanger, when the specific heats of the two streams are approximately the same, whereas a value approaching 0 describes a gas/liquid one. Gas/gas exchangers handling high-specific-

heat gases, such as hydrogen, could also have values of C_{min}/C_{max} approaching 0.

Temperature effectiveness is always taken as positive, and the fluid having the greatest difference between its inlet and outlet temperatures has the minimum heat capacity.

NTU_{max} is a dimensionless quantity that expresses the heat exchanger's size. It is the product of an overall heat-transfer coefficient (U) and heat-transfer surface (A), divided by the value of C_{min}. The overall coefficient defines the thermal conductance of the exchanger and is obtained from the heat-transfer coefficients of the two fluids and the thermal conductivity of the boundary wall. Exchanger effectiveness rises with increasing NTU values. However, a higher NTU, and consequently greater effectiveness, implies a larger exchanger surface or pressure drop or both.

In Fig. 1, it can be seen that in the counterflow exchanger the limiting temperature for the cold fluid is the hot-fluid inlet temperature; and for the hot fluid, it is the cold-fluid inlet temperature. These temperature levels represent the ultimate transfer of heat in a counterflow exchanger, which would require an infinite heat-transfer surface.

The high heat-recovery and temperature effectiveness of this exchanger makes it attractive when the first consideration is process economy.

In some cases, however, high heat recovery must be avoided, as when, for example, low temperatures at the cold end could lead to acid dewpoint corrosion, or high temperatures at the hot end could expose the heat-transfer surfaces to too-high temperatures and cause structural failure.

Metal temperatures in the counterflow exchanger are shown in Fig. 1. The exchanger's theoretical performance is depicted in Fig. 2. (See the box above for definitions of effectiveness, C_{min}/C_{max} and NTU_{max}.)

Crossflow exchanger saves space

In the crossflow exchanger, the two fluids usually enter at right angles to each other, and pass over each other as they flow through the exchanger (Fig. 3). Design of the inlet and outlet headers economizes space, which can be important when an exchanger must be fitted into a process system.

Crossflow-exchanger tubes sometimes have perpendicular radial fins. Round tubes and finned tubes, as well as various types of flat-plate envelopes, are also used in this exchanger. These plates may be embossed to enhance heat-transfer coefficients.

Indicated in Fig. 3a are the fluid-temperature profiles of a crossflow exchanger in which the fluids pass through without mixing. The manner in which the fluids pass by each other fixes the temperature constraints that prevent the exchanger from reaching high-temperature effectiveness, and introduces a cold corner, where vapor condensation and corrosion can take place.

Fig. 3b represents the case in which the fluid flowing

over the tubes can be considered completely mixed, with the flow inside the tubes unmixed. The mixing of one of the fluids increases the metal temperature in the cold corner, because the mixed fluid will be at a higher temperature. However, the mixing lowers the maximum value of the temperature effectiveness.

Theoretical curves describing the performance of the crossflow exchanger are shown in Fig. 4. Performances that can be achieved when both fluids are unmixed and when only one is mixed are compared for the case in which $C_{min}/C_{max} = 1.0$ (the behavior depicted is similar for other ratios). The Fig. 4 curves show that the crossflow exchanger does not perform as well as the counterflow exchanger.

When higher temperature effectiveness is desired from the crossflow exchanger, a multipass crossflow system may be considered. Fig. 5 depicts a two-pass crossflow arrangement. It shows how the Fig. 3b single-pass crossflow exchanger can be upgraded to achieve a higher heat recovery. Multipass crossflows can be arranged in many different ways [1,2,3].

Parallel flow limits effectiveness

In the parallel-flow exchanger, the hot and cold fluids flow in the same direction, parallel to each other. Both fluids enter at the same end and exit at the opposite end. Fig. 6 shows the fluid temperatures as vectors whose lengths indicate their relative magnitude.

Temperature effectiveness is limited by the fluids flowing in the same direction. The outlet temperature of the cold fluid cannot exceed that of the hot fluid. This temperature constraint holds for all parallel-flow exchangers, and little can be gained by making them larger. When the two outlet temperatures are nearly equal, the parallel-flow exchanger is performing at its maximum.

Mainly because of its limited temperature effectiveness, the parallel-flow exchanger is frequently not seriously considered for heat recovery. However, it is useful when system flows are oriented in parallel, which enable it to be installed with minimum work or cost.

An additional advantage of this exchanger is that it

Nomenclature

A	Area	**Subscripts**	
C_i	Correction factors	c	Carbon dioxide
C_p	Specific heat	g	Gas
L	Mean beam length	S	Surface
$\dot{m}$	Mass flowrate	W	Water
NTU	Number of transfer units		
Q_{gs}	Radiative interchange rate		
T	Temperature		
α	Absorptivity, the ratio of radiation absorbed by a surface (or gas volume) to the incident radiation		
ε	Emissivity, the ratio of radiation emitted by a surface to the radiation emitted by a black body at the same temperature		
ρ	Reflectivity, the ratio of reflected to incident radiation ($\rho = 1 - \alpha - \tau$)		
τ	Transmissivity, the ratio of radiation that passes through a surface (or gas volume) to the incident radiation		

lessens the hazards of corrosive conditions, which can be destructive to the heat-transfer surface, particularly when the metal temperatures reach the dewpoint of the gases. If acid vapors are present in the gases, operating at the dewpoint can lead to acid attack and eventual destruction of the metal heat-transfer surface. Operation at the dewpoint can also lead to fouling and plug-

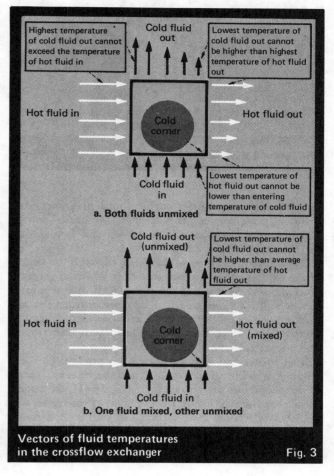

a. Both fluids unmixed

b. One fluid mixed, other unmixed

Vectors of fluid temperatures in the crossflow exchanger   Fig. 3

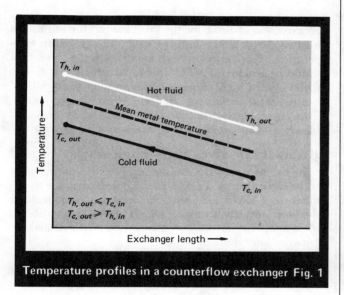

Temperature profiles in a counterflow exchanger Fig. 1

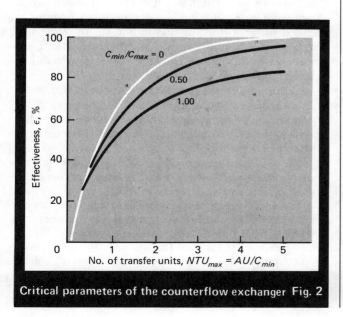

Critical parameters of the counterflow exchanger Fig. 2

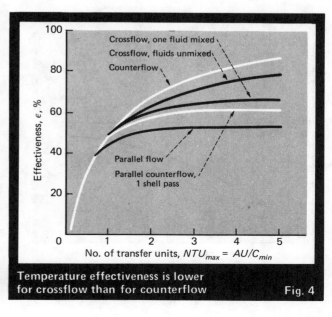

Temperature effectiveness is lower for crossflow than for counterflow   Fig. 4

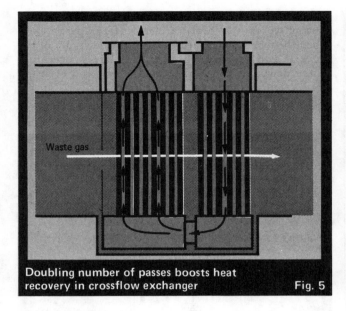

Doubling number of passes boosts heat recovery in crossflow exchanger   Fig. 5

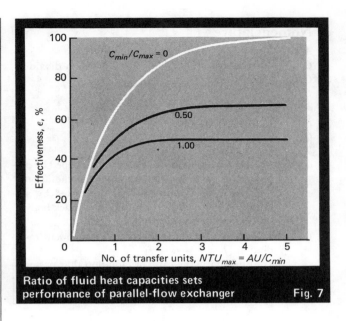

Ratio of fluid heat capacities sets performance of parallel-flow exchanger   Fig. 7

ging of the heat-transfer surface, and to lower heat-transfer performance and higher pumping power-requirements. Ultimately, a shutdown may be required to clean the exchanger.

Typical temperatures of the fluids through the length of the exchanger are shown in Fig. 6, which also represents the temperature of the metal wall that separates the two fluids. The mean metal temperature lies approximately midway between the curves that define the temperature drop through the fluid boundary layers and the metal wall.

The theoretical performance of a parallel-flow exchanger is depicted in Fig. 7. When the heat capacities of the two fluids are equal ($C_{min}/C_{max} = 1$), the curves show that the maximum temperature effectiveness will be 50%. When the heat capacity of one fluid is much greater than the other (which occurs particularly in a condenser), C_{min}/C_{max} approaches zero, and the curve for $C_{min}/C_{max} = 0$ shows an achievable temperature

effectiveness of 100% for the fluid having the minimum heat capacity. In the limiting case, the maximum temperature effectiveness in both instances is the value that would be reached if NTU were to become infinite. The high value of NTU corresponds to an infinite amount of heat-transfer surface or an infinite exchanger length. This is, of course, impractical. Such limits are used mostly as guidelines by heat-exchanger designers.

Recuperator gives high efficiency

In the recuperative exchanger, heat is continuously transferred from a fluid at a higher temperature to another at a lower temperature through a heat-transfer medium (usually a metal), which permits operation at high temperatures and pressures with minimal leakage between the unmixed fluids.

The geometry of the heat-transfer medium is a major factor in promoting the exchange of heat between the two unmixed fluids. The most elementary geometry is the smooth-wall circular tube. Heat transfer can be improved by modifying the geometry to include extended surfaces (fins) or undulations in the tube wall. Such modifications increase the heat-transfer coefficients on both sides of the tube, boosting overall exchanger efficiency.

An example of the fuel saving in million Btu/h that can be achieved in the operation of a direct-flame thermal oxidizer by the incorporation of a recuperative heat exchanger having undulated flat tubes is presented in the table on p. 126.

Fig. 8 illustrates the effect of heat-exchanger efficiency on annual fuel costs and savings. In this example, the process-gas flowrate is 20,000 std ft³/m, and ΔT is the difference between the incineration temperature (1,500°F) and the process-gas temperature (400°F), or 1,100°F. The fuel rate without the exchanger is 27.0 million Btu/h, and the fuel cost is $5.00/million Btu. At an annual operation of 8,000 h, the fuel cost is $1.080 million per year. At heat exchanger efficiencies of 40%, 48%, 57%, 65%, 80% and 90%, the annual fuel savings are shown to be $340,000,

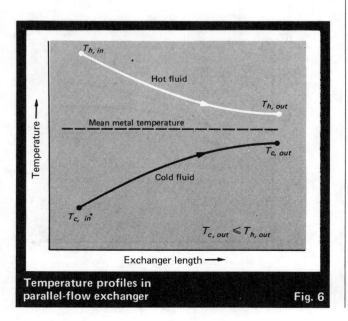

Temperature profiles in parallel-flow exchanger   Fig. 6

$430,000, $540,000, $640,000, $825,000 and $950,000, respectively.

High-temperature-exchanger design

One of the assumptions used in analyzing the performance of a low-temperature gas-to-gas heat ex-changer is that all fluid-to-surface heat transfer occurs only by convection. In the case of many high-temperature gas-to-gas exchangers, however, this assumption is invalid.

Although heat transfer by convection does occur in these exchangers, radiative heat transfer represents a

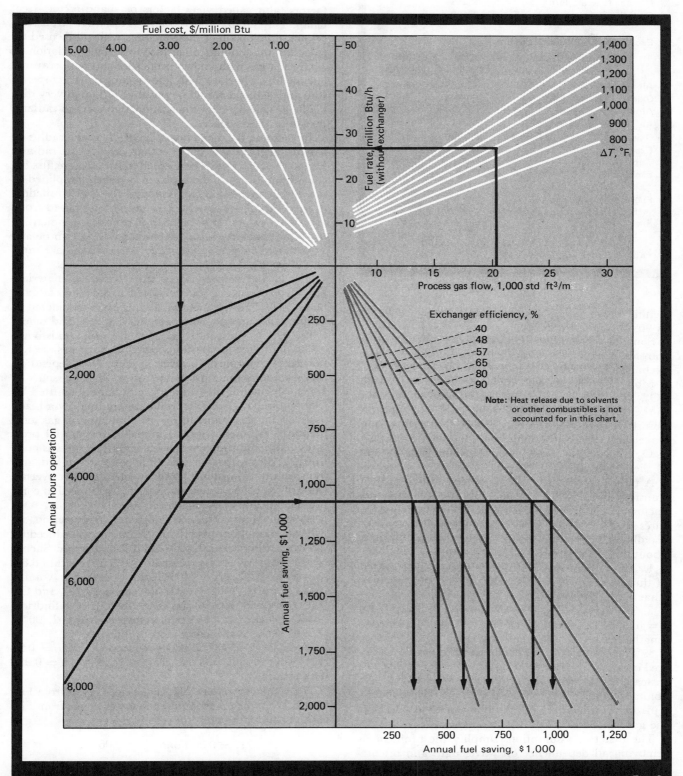

Heat exchangers of higher efficiency increase incinerator fuel savings   Fig. 8

Fuel saving from application of recuperative exchanger

Oxidizer process conditions:

Air-side flowrate, lb/h	78,732
Air-side inlet temperature, °F	80
Gas-side flowrate, lb/h	79,243
Gas-side inlet temperature, °F	1,400

Exchanger process conditions:

	No exchanger	With exchanger*
Air-side flow, lb/h	78,732	78,732
Oxidized-gas flow, lb/h	80,303	79,243
Air inlet temperature, °F	80	80
Air preheat temperature, °F	–	993
Oxidized-gas inlet temperature, °F	1,400	1,400
Exchanger air side pressure drop, in. H$_2$O	–	6.5
Exchanger gas-side pressure drop, in. H$_2$O	–	5.0
Fuel required (gross), Btu/h $\times$ 10^6	30.0	9.8
Fuel saving (gross), Btu/h $\times$ 10^6	–	20.2

*Air-side effectiveness equals 69%.

significant fraction of the total heat transfer—as much as 90% in some instances.

Some of the fundamental aspects of radiation heat-transfer calculations and their application to heat-exchanger performance analyses are now presented.

Most triatomic gaseous molecules (CO_2, H_2O, CO, SO_2, NO, NO_2, etc.) possess spectral absorption bands in the thermal radiation spectrum. These are known as absorbing gases because they absorb and emit certain wavelengths in the thermal radiation band. Gases such as O_2 and N_2, however, are not absorbing gases because of their molecular symmetry.

A typical problem involving gaseous-radiation heat transfer would be one in which an absorbing gas was enclosed by a number of surfaces, S_i, having temperatures T_i. When solving such a problem, the engineer usually makes a number of reasonable assumptions about the surfaces and gas. These assumptions include:

1. The radiant flux leaving a surface is independent of direction, i.e., the surfaces are diffuse, not specular (that is, mirrorlike).

2. The surfaces are gray and opaque, so that $\tau = 0$, $\varepsilon = \alpha$, $\rho = 1 - \varepsilon$, and ε and α are independent of wavelength.

3. The geometry of the enclosure need not be considered in detail when calculating gas-to-surface radiation. Instead, the average photon-path length from the gas volume to the enclosure (called the mean beam length, L) is used to determine the amount of radiation from the gas to its enclosure.

The first two assumptions simplify the problem by removing all dependence on wavelength and orientation of the surfaces with respect to either the incident or emitted flux. The last assumption is probably the most restrictive of the three because it implies that the gas temperature is uniform. Although this introduces some error, it is preferable to performing a tedious volume integration and is an acceptable approximation when temperature gradients in the gas are small.

It should be noted that none of these assumptions removes from the problem the orientation of each surface with respect to all other surfaces. Configuration factors, also called view factors or shape factors (see Chap. 2 of Hottel and Sarafin, "Radiative Transfer" [4], for a discussion of these factors), will still be used to calculate the amount of surface-to-surface radiation.

Given a gas with known properties and temperature, enclosed by a surface with known properties, temperature and shape, it would be possible, although very difficult, to calculate by integration the average radiant flux onto that surface.

By defining the mean beam length, L, as the radius of a hemispherical gas volume that produces a radiant flux at the center of its base equal to the average flux on the original surface, the problem is greatly simplified—if a simple expression can be found for L as a function of geometry. Fortunately, L is easily determined from tabulated values for a number of common geometries; or, if the geometry is an uncommon one, L can be approximated by multiplying the volume-to-surface ratio of the gas by 3.5 [4].

The surface emissivities of many different materials have been measured and tabulated as functions of temperature [4,5]; consequently, if a surface is considered to be gray and opaque, its absorptivity, $\alpha = \varepsilon$, and reflectivity, $\rho = 1 - \varepsilon$, can be calculated from its emissivity.

The radiative properties of an absorbing gas are not so easily determined, however, because they depend on more variables than just temperature. The emissivity of a gas—defined as the ratio of the radiation emitted by that volume of gas to the radiation emitted by a black body at the gas temperature—is a function of gas temperature, T_G, mean beam length, L, the partial pressures of all absorbing gas species, and the total gas pressure [4]. Also, when two absorbing gases are mixed, and the spectral absorption bands of one of them overlap those of the other, a correction term, $\Delta\varepsilon$, is used in computing the emissivity of the mixture.

As an example of a gas emissivity computation, we will calculate the emissivity of the flue gas produced by a burner firing natural gas with 10% excess air. Such a gas would have a total pressure of 1 atm, contain 8.8% by volume CO_2 ($p_c = 0.088$ atm), 17.1% by volume H_2O ($p_w = 0.171$ atm), with the remainder N_2 and O_2. If it is assumed that the gas is enclosed in a cylindrical 1-ft-dia. stack, and that the gas temperature is 2,200°R (1,740°F), ε_g can be computed as follows:

From Table 7.3 of "Radiative Transfer" [4], $L = 0.94$ ft. Therefore, $p_w L = 0.161$ (ft) (atm), and $p_c L = 0.083$ (ft) (atm).

The emissivity due to CO_2 alone, $\varepsilon_c^!$, is read from Fig. 6-9, and $\varepsilon_w^!$, the emissivity due to water alone, from Fig. 6-11 of the same book; thus:

$$\varepsilon_c^! = 0.066 \quad \text{and} \quad \varepsilon_w^! = 0.067$$

To find the emissivity of the gas, ε_g, use the equation:

$$\varepsilon_g = C_c \varepsilon_c^! + C_w \varepsilon_w^! - \Delta\varepsilon \qquad (1)$$

Here, C_c, C_w and $\Delta\varepsilon$ are correction terms that can be read from Fig. 7.7, 7.8, and 7.9, respectively, of McAdams' "Heat Transmission" [3] as $C_c = 1.0$, $C_w = 1.11$, and $\Delta\varepsilon = 0.003$.

Thus, $\varepsilon_g = 0.137$.

It is evident from the figures used in computing ε_g and Fig. 6.25 of Hottel and Sarafin [4] that ε_g for a real gas will always rise as pL increases and approach an asymptotic value of less than 1.0 at large values of pL. As the gas temperature gains, ε_g may rise or fall, but the total radiation emitted by a gas volume of fixed dimensions will always go up with temperature because any decline in ε_g with the temperature is overwhelmed by the hike in T^4.

If a simplistic model of gas-to-surface radiation is adequate, the assumption could be made that the gas is gray, so that $\varepsilon_g = \alpha_g$. Then, if the gas were surrounded by a *single* surface at temperature T_s and had an emissivity of ε_s, the net radiative-interchange rate between gas and surface would be:

$$Q_{gs} = \frac{\sigma A_s \ (T_g^{\,4} - T_s^{\,4})}{\dfrac{1}{\varepsilon_g} + \dfrac{1}{\varepsilon_s} - 1} \qquad (2)$$

Here, σ = Stefan-Boltzmann constant, 0.1713×10^{-8} Btu/(h) (ft^2) ($^\circ$R^4), and A_s = surface area of the gas's enclosure.

Applying this formula to the previous example, with an assumed wall temperature of $1,600\,^\circ$R ($1,140\,^\circ$F) and an assumed wall emissivity of 0.8, we obtain a radiative flux, Q_{gs}/A_s, of 3,823 Btu/(h) (ft^2), and a linearized radiation coefficient, $(Q_{gs}/A_s)\ (T_g - T_s)$, of 6.38 Btu/(h) (ft^2) ($^\circ$R). Because Eq. (2) is identical to the equation for radiant interchange between two infinite parallel gray plates, it implies that the gas has been treated as if it were a surface with temperature T_g, emissivity ε_g, and area A_s.

Since real gases are not actually gray, the gas radiation model can be made more realistic by assuming the gas is not gray, i.e., $\alpha_g \neq \varepsilon_g$. This means the gas absorptivity, α_g (defined to be the fraction of blackbody radiation emitted by a surface at temperature T_s that is absorbed by the gas), must be calculated. Gas absorptivity is a function of T_g, T_s, the partial pressures of all absorbing gas molecules, total gas pressure, and mean beam length.

The following empirical procedure describes the calculation of α_g:

1. Evaluate $\alpha_c^!$ and $\alpha_w^!$ from Hottel and Sarafin's Fig. 6-9 and 6-11 [4] as functions of pLT_s/T_g and T_s in the same manner that $\varepsilon_c^!$ and $\varepsilon_c^!$ were evaluated as functions of pL and T_g.

2. $\alpha_c = C_c \alpha_c^! (T_g/T_s)^{0.65}$ and $\alpha_w = C_w \alpha_w^! (T_g/T_s)^{0.45}$

3. $\Delta\alpha = \Delta\varepsilon$, evaluated at T_s

4. $\alpha_g = \alpha_c + \alpha_w - \Delta\alpha$

Using this procedure to calculate α_g for the same gas for which ε_g was calculated previously, it is found that α_g is 0.178 for an assumed wall temperature of $1,600\,^\circ$R ($1,140\,^\circ$F).

Now that ε_g and α_g have been calculated, a formula similar to Eq. (2) is required to predict the heat radiated

Continuous operation may depend on servicing arrangements

J. H. Ferguson, C-E Lummus

Exchangers should be engineered to facilitate repair or routine maintenance by providing for removal of channel covers, channels, shell covers, tube bundles and even the complete exchanger. Such an arrangement might also include a monorail and its supporting structure, and space for mobile equipment.

Exchangers must be designed to operate continuously for a specified period. If an exchanger may require cleanings sooner than during scheduled shutdowns, it should be designed for ease of service, and perhaps paralleled with another exchanger, with both valved so that one can be cleaned while the other is operating. Occasionally, an exchanger may be temporarily bypassed during servicing.

Usually, it is important to avoid shutting down a plant because of exchanger problems. Downtime will almost always be much more costly than any exchanger, and special provisions necessary to keep it on line can be justified.

Exchangers located at grade level are easier to service. When shells are stacked, their height should be limited to what can be reached by the available service equipment.

A high-pressure-drop allowance for an exchanger can be beneficial: better heat transfer, which lowers initial cost; and faster fluid velocity, which can reduce fouling, thereby cutting maintenance cost and lengthening on-stream time. Viscous-fluid and dry-gas streams are especially affected by the pressure drop allowance.

Exchanger types and ease of servicing

The removable floating-head tube bundle abets cleaning of both the shellside and tubeside, and facilitates retubing and repairs. For severe services, a spare tube bundle may be stored.

The removable tube bundle is the most difficult to service, and requires special handling equipment. A mechanical extractor can make the removal simpler and safer.

The fixed-tubesheet exchanger is the most economical, and easy to service. It should be selected if the shellside can be cleaned chemically and there are no solids to settle out. Differential expansion between the shell and tubes may make an expansion joint in the shell necessary.

The U-tube exchanger, which has the advantage of a removable bundle, is more economical than the floating-head type. Having fewer flanges and gaskets, it is easier to service and maintain.

The double-pipe exchanger is often more economical and flexible, and the easiest to service and maintain. Because of its standardized design, it affords maximum interchangeability, requiring storage of a minimum quantity of spare parts. To suit changes in process conditions, it can be easily rearranged or enlarged by duplicating its elements.

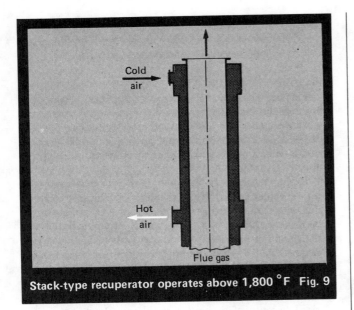

Cold air

Hot air

Flue gas

Stack-type recuperator operates above 1,800 °F   Fig. 9

by an other than gray gas to its enclosure. The following formula will suffice [1]:

$$Q_{gs} = \sigma A_s \left[\frac{T_g^4}{(1/\varepsilon_s) + (1/\varepsilon_g) - 1} - \frac{T_s^4}{(1/\varepsilon_s) + (1/\alpha_g) - 1} \right] \quad (3)$$

Eq. (3) predicts a radiative heat flux of 3,400 Btu/(h) (ft²) for the same situation ($\varepsilon_s = 0.8$, $T_s = 1,600°R$) in which Eq. (2) predicts 3,830 Btu/(h) (ft²). The linearized radiation coefficient predicted by Eq. (3) is 5.67 Btu/(h) (ft²) (°R). The gray gas model apparently caused the 13% overestimation because α_g was assumed equal to ε_g. Since real gases are not gray, Eq. (3) is preferable, when it can be used, to Eq. (2).

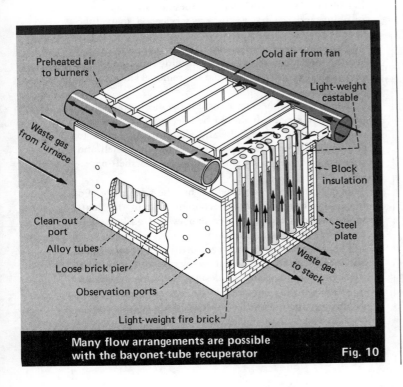

Preheated air to burners

Cold air from fan

Light-weight castable

Waste gas from furnace

Block insulation

Clean-out port

Alloy tubes

Steel plate

Loose brick pier

Waste gas to stack

Observation ports

Light-weight fire brick

Many flow arrangements are possible with the bayonet-tube recuperator   Fig. 10

A close examination of Eq. (3) shows that when T_s and T_g are equal, α_g and ε_g (i.e., the absorptivity and emissivity of the gas, respectively) must also be equal, in order that Q_{gs} be zero. ($Q_{gs} \neq 0$ when $T_g = T_s$ would be a clear violation of the second law of thermodynamics.) Fortunately, the procedure used to determine α_g does indeed predict that $\alpha_g = \varepsilon_g$ when $T_g = T_s$. Therefore, it may be concluded that the gray gas model will not cause large errors when T_g is close to T_s.

In the example just examined, an isothermal gas volume is surrounded by a single isothermal surface. Although no heat exchanger in reality could ever be this simple, a radiative stack recuperator could be divided up into several slices in the axial direction, so that each slice could be approximated by an isothermal gas surrounded by a single isothermal surface. Although the gas and wall temperatures would vary from slice to slice (i.e., in the axial direction), a separate calculation could be performed for each slice to determine the heat radiated from the gas to the wall at the slice.

In a radiative stack recuperator, part of the heat is transferred from the gas by convection, but the convection coefficient is usually about 2 Btu/(h) (ft²) (°F), which is roughly a third as large as the linearized radiation coefficient. Therefore, on the gas side of the exchanger, about three-fourths of the heat transfer is attributable to gas radiation. On the air side, however, there are no absorbing gases (trace amounts of CO_2 and H_2O in the air can be neglected), so that no radiation occurs, and the standard convective correlations may be used to determine the rate of heat transfer.

Not discussed are many other aspects to gas radiation, such as transmittance from one surface to another through an absorbing gas, and many other types of gas-to-gas exchangers in which gas radiation plays a significant role. In virtually all of these exchangers, however, a calculation of the amount of heat transferred by gas radiation is made extremely complex by the geometry (e.g., gas flow over a bank of tubes) and by the presence of refractory walls. Temperature calculations are also complicated by the fact that radiation introduces nonlinear terms (T^4) into the heat-balance equations, whereas convection and conduction are linear in temperature.

Fortunately, it is not always necessary to perform a detailed calculation of the radiant flux at every surface in a heat exchanger. Sometimes, of sole interest is the effect of radiation on the temperatures of those surfaces expected to be close to their temperature limit when only convection and conduction are considered. In that event, the radiant flux at the surface of interest could be calculated as in the previous example.

When the radiant flux is added to the convective flux, a new surface temperature can be obtained from a heat balance. Those surfaces that would rise the most in temperature because of radiation would, in general, be the ones exposed to a large volume of high-temperature gas containing large quantities of CO_2, H_2O or other absorbing gases.

For example, consider a tubular crossflow exchanger with gas flowing on the outside of the tubes and air on the inside. An exchanger of this type will usually have a large gas inlet-plenum directly upstream from the tubes, so that the first (and maybe the second) row of tubes is exposed to a large volume of high-temperature gas.

Furthermore, if it is assumed that the exchanger is operating under "turndown" conditions (i.e., gas and air flowrates are both reduced to some fraction of nominal values, but the gas temperature remains high), the temperatures of the first and second tube-rows could exceed their permissible limits, because the radiant flux would still be high but the reduced air flow inside the tubes would not provide the normal amount of cooling.

Thus, a simple calculation of the radiant flux onto the first tube-row from the gas and the walls of the inlet plenum would help to predict whether or not these tubes would fail due to excessively high temperatures.

High-temperature applications

When high gas temperatures (above 1,500°F) exist in combination with a significant amount of triatomic gases, radiation transfer becomes an important factor. In the furnace industries, for example, iron, steel and aluminum melting and forming result in products of combustion from natural gas, oil, coke-oven gas, blast-furnace gas, or mixed gas (all of which are radiating), and in waste-gas temperatures of 1,500–2,600°F. Additionally, these applications are typically natural-draft processes. The cost of adding an induced-draft fan and controls usually is not justified purely on the basis of achieving a higher allowable pressure-loss through the exchanger. The exchangers are, therefore, designed with low gas-side pressure drops (typically, 0.5 in. H_2O, or less), and hence low gas-side velocities and convective heat-transfer coefficients.

With the exception of ceramic checkerwork-type regenerators (used primarily with blast furnaces), the two types of recuperators for high temperatures are the radiation-, or stack-type, and the tubular. The tubular is subdivided into the single tube, fixed tubesheet and the bayonet tube.

The stack-type recuperator consists of an inner cylinder, usually of a high-grade alloy, through which the hot combustion gases flow at low velocity (Fig. 9). Flows can be parallel or countercurrent, or a mixture of both. Radiation heat transfer is the primary mechanism on the gas side, with the ratio of radiation to convection usually on the order of 3 to 1, or higher. For this reason, the stack type is found in services above 1,800°F, and with continuously high temperatures. Typical applications involve preheating incoming air. Because the stack-type recuperator depends mainly on radiation transfer, it produces little preheat in cyclic applications at lower gas temperatures, and hence does not give as good fuel savings over the total cycle as does a tubular exchanger.

The important differences between high-temperature and low-temperature tubular exchangers are that the tube diameters and spacings are larger for the first, so as to (1) take advantage of the radiation transfer that can be achieved with larger beam lengths and (2) keep the pressure loss very low. The tubes are typically austenitic or high-alloy ferritic steels, with substantial wall thickness (12 gage to $\frac{1}{4}$ in.), and without extended surfaces on the gas side. The combination of low gas velocities,

substantially dirty flue gases, and negligible gain from radiant heat transfer, make the use of fins on the gas side undesirable.

The bayonet recuperator is an example of a high-temperature bayonet tubular exchanger used primarily in the steel and aluminum industries (Fig. 10). It serves as a crossflow exchanger with the air- or tube-side fluid unmixed, as a multipass crossflow and counterflow exchanger (again, air unmixed within a pass), as a multipass crossflow and parallel flow (air unmixed), or as a simple counter- or parallel-flow exchanger. All of these flow arrangements are possible in the bayonet tube exchanger.

As shown in Fig. 10, the bayonet tube is a tube-within-a-tube. The cold air enters a plenum, then turns and flows down the annular space formed by the outer and inner tubes. After turning 180 deg. at the bottom of the sealed outer tube, the air flows up the inner tube into a hot-air collector plenum. The advantages of this design over a single tube having tubesheets at each end of the tube are: (1) freedom of expansion, (2) high air-side convection coefficient, and (3) an air flow that can be selectively varied within any tube in the bank of tubes.

Being fixed only at their flanged end, both the inner and outer tubes are free to expand. At tube metal temperatures averaging 1,400°F, 300-series stainless tubes in lengths of 10 ft. will extend 2 in. A tube-to-tube variation of as little as 50°F in average temperature will result in a differential growth of $\frac{1}{16}$ in. Tubesheets must be designed to accommodate the 2-in. growth and the individual tube-to-tube variations caused by uneven heat fluxes or maldistributed flow, otherwise severe stressing will eventually cause failure. At very high temperatures, the tubesheet has proved to be a weak link. Thermal distortion followed by weld failure when tubes are attached at two tubesheets is a common occurrence. The bayonet design solves this problem.

Selective sizing of the diameter of the inner tube permits very high velocities, hence high convective coefficients, in the annular space. Convective coefficients there are typically on the order of 12 to 18 Btu/(h) (ft²) (°F).

To achieve a high effectiveness, the tube area must be enlarged. When design limitations on tube length, as well space restrictions, limit expanding tube area by going to longer tubes, the only recourse is to increase the number of tubes. In a two-tubesheet, single-pass tubular exchanger, a point is reached, as tubes are added, at which the air velocity inside the tubes is decreased (lowering the air-side convective coefficient) at a rate that results in a negligible increase in heat recovered. This can be avoided by the multipass configuration. Although this arrangement boosts heat recovery by raising the air-side convective coefficient, it increases ducting problems, and the additional exchanger-housing cost may not be warranted.

However, as the tube surface area of the bayonet tube exchanger is increased by adding tubes, the air-side convective coefficient can be kept at high levels by enlarging the diameter of the inner tube (hence, decreasing the annular gap). Additionally, in a tubular single-pass unit, the air flow through all the tube rows is

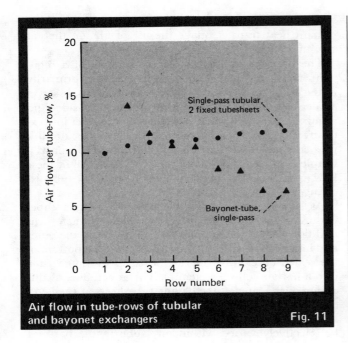

Air flow in tube-rows of tubular and bayonet exchangers   Fig. 11

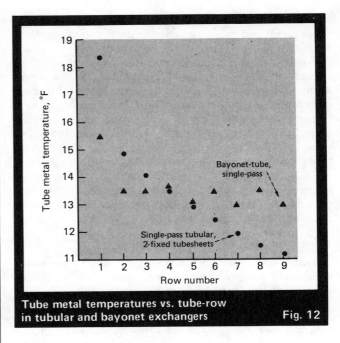

Tube metal temperatures vs. tube-row in tubular and bayonet exchangers   Fig. 12

essentially equal (with the same flow area, the only flow deviation is caused by density changes due to tube pressure loss), hence the row of tubes at the hot end of the exchanger will run much hotter than the row at the cold end. Fig. 11 shows air flow per tube-row and Fig. 12 tube metal temperatures at tube-rows for both the tubular single-pass and bayonet single-pass exchangers.

In the fixed-tubesheet design, the percentage of flow per tube-row is nearly constant, with slightly lower flows in the front tubes. The higher gas temperatures at the inlet, plus the high heat flux to rows 1 and 2, result in hotter air preheating, hence lower average density in these rows. Because tube-rows form a series connection between the cold and hot tubesheets, the lower density (higher velocity at a given mass flow) lessens the flow through the hotter tubes. The net result of (1) the density change on the flow per tube-row, (2) the temperature distribution through the exchanger and (3) the higher radiant heat fluxes to the first rows is that the tube metal temperatures are very much hotter in the first rows than elsewhere in the exchanger (Fig. 12).

Via selective sizing of annulus widths in the bayonet tube exchanger, the air flow in each row can be manipulated. The annulus width in the front rows is made

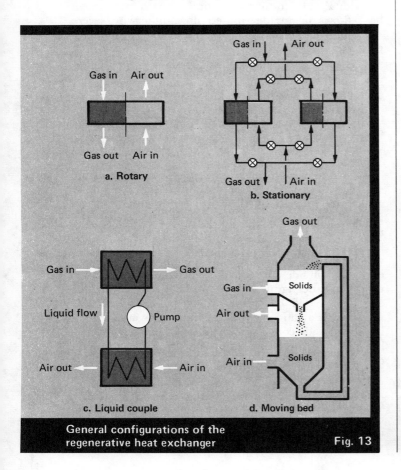

General configurations of the regenerative heat exchanger   Fig. 13

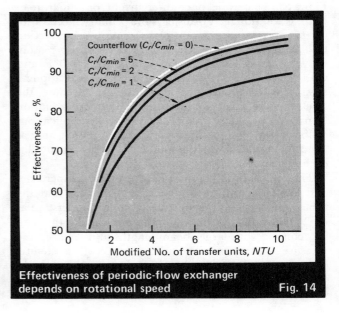

Effectiveness of periodic-flow exchanger depends on rotational speed   Fig. 14

much larger than that of the back rows, resulting in an air flow distribution of high flow at the hot end (Fig. 11). This produces controlled metal temperatures with a better uniformity than the two-tubesheet tubular exchanger. Additionally, placing the high air flow in the high-heat-flux, high-gas-temperature front rows results in a slightly greater heat-transfer effectiveness than do uniform flows.

It has been mentioned that heat fluxes are higher at the first row and, to a lesser extent, in the second or third rows than at rows further back. This is because the leading rows are exposed to the large volume of radiating gas, and to refractory walls that are sources of secondary radiation (gas radiates to a wall, which then radiates energy to the tube surfaces).

Fig. 12 shows the much higher tube-metal tempera-

tures that result from this radiation pattern. The leading tube-rows must be designed for the higher fluxes, either via higher alloys or, in the case of the bayonet tube, higher air flows.

Advantages of the regenerative exchanger

In the regenerative exchanger, an intermediate material transfers energy between two fluids of different temperature levels. The material is usually a solid (metal or ceramic), although it can be a liquid, as in the liquid couple.

Unlike the recuperative exchanger, heat need not pass through tube or plate walls. It is simply absorbed and released from the same surfaces. This direct heat transfer results in uniformly higher metal temperatures than with recuperative preheaters. Thus, lower exit-gas

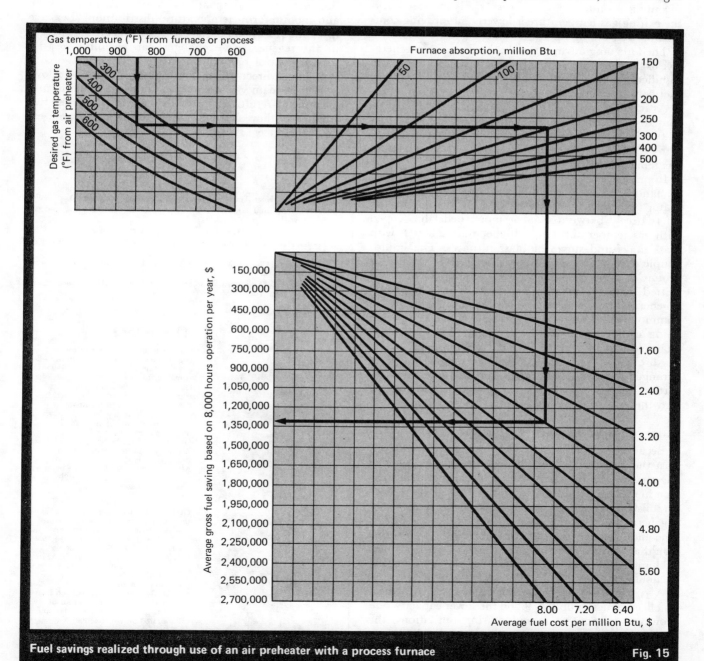

Fuel savings realized through use of an air preheater with a process furnace   Fig. 15

temperatures and a measure of high heat recovery are possible with minimized danger of dewpoint corrosion. In addition, because of the compact heating surface arrangement, the regenerative preheater is less than half the size and weight of a conventional tubular recuperative air preheater performing at the same conditions.

Magnitude of fuel savings possible

Fig. 13 shows general configurations of the regenerative exchanger. The rotary type has been widely used as an air preheater with boilers, as well as with a variety of process furnaces and ovens. As the rotor revolves, the heating surface alternately passes through the gas and air stream. Heat absorbed from the hot gas stream is released to the air stream as the heating surface traverses the respective streams. The temperature of the incoming combustion, or process, air is increased and the exiting gas temperature is lowered, improving combustion and hiking operating efficiencies.

The foregoing advantages obtained via the regenerative exchanger applied to a process furnace can result in significant savings from higher throughput and reduced fuel consumption.

As indicated in Fig. 14, a fuel saving of $1,270,000 (based on 8,000 h/yr of operation, at a fuel cost of $4.00/million Btu) can be realized for a furnace absorption of 200 million Btu/h. This assumes that the outlet gas temperature of 850°F is reduced to 400°F.

In practice, gas temperatures as high as 1,200°F can be handled by metal exchangers. The stationary regenerator (which has valves for switching the gas stream alternately between two, or more, beds) has coped with gas temperatures even higher than 2,000°F with glass tanks and stoves of blast furnaces. The liquid-couple system (generally an organic fluid) is suitable for remote location but is constrained to temperatures permitted by the heat-transfer liquids. The moving-bed regenerator finds limited application, although its use is common with fluid catalytic crackers, the heating effect being coupled with catalyst activity. An application example is shale heating in the Tosco process.

All of these types of regenerative exchangers can be arranged for countercurrent flow of the gases, thus affording the advantage of efficient surface heat-transfer and the potential for high thermal performance.

Regenerative exchanger design

Because the hot and cold streams flow alternately past the same exchanger surfaces, the temperatures of the surfaces and the gases vary with time and their location in the regenerator. After steady operation has been established, the temperature at any one location in the regenerator will be the same as that a full cycle earlier. The end-of-cycle conditions, together with the assumption that the flows are well distributed through the heat exchanger, serve as the basis for the performance predictions.

A regenerative exchanger theoretically can achieve an effectiveness of up to 0.99 on the lower capacity side. However, because of manufacturing limitations, this maximum value cannot normally be realized.

The rotary regenerator's effectiveness varies with the rotational speed of the solid mass. Fig. 15 shows how effectiveness declines with lower C_r/C_{min} ratios, with C_r = (revolutions/unit time) (solid mass) (C_{solid}), and $C_{min} = \dot{m}$ (air) C_p (air).

In the design or sizing of a regenerator, values of C_r/C_{min} greater than or equal to 5 are used because these yield an effectiveness very near the theoretical limit. Lower values of C_r/C_{min} lead to an effectiveness significantly lower than the maximum obtainable.

For predicting metal temperatures, it is accepted practice to approximate the temperature of the cold-end metal (where corrosion problems are possible) by the arithmetic average of the air-inlet and gas-outlet temperatures, except when cold-end bypass or hot-air recirculation are used. The flow redistribution inherent in these schemes requires additional consideration.

The rotary regenerative exchanger has two additional characteristics that have to be evaluated: leakage and nonuniform temperature distribution. Leakage from one gas stream to the other, which can be expected in any regenerative exchanger, can be minimized by careful control but not eliminated. The leakage is of two types: direct—from the high-pressure to the low-pressure stream via any opening between the two; and entrained—resulting from the mixing of the two streams as they alternately flow through the same exchanger space.

References

1. Jakob, M., "Heat Transfer," John Wiley & Sons, Inc., New York, 1957.
2. Kays, W. M., and Londin, A. L., "Compact Heat Exchangers," 2nd ed., McGraw-Hill, New York, 1964.
3. McAdams, W. H., "Heat Transmission," McGraw-Hill, New York, 1954.
4. Hottel, H. C., and Sarafin, A. F. "Radiative Transfer," McGraw-Hill, New York, 1967.
5. Hottel, H. C., Chap. 4 of W. H. McAdams, "Heat Transmission," McGraw-Hill, New York, 1954.
6. Wiebelt, J. A., "Engineering Radiation Heat Transfer," Holt, Rinehart, and Winston, New York, 1966.

The authors

Carl Scaccia is manager of the engineering technologies group at C-E Air Preheater, Wellsville, NY 14895. He is responsible for the development of new technologies in fluid mechanics, heat transfer, and applied mechanics. Mr. Scaccia has a B.S. in aerospace engineering from the State University of New York at Buffalo, an M.S. in mechanical engineering from the University of Rochester, and a Ph.D. in engineering science from S.U.N.Y. at Buffalo. He is a member of ASME, AIChE, and the Professional Engineers Soc. He is the author of over 20 technical articles and several patents.

Greg Theoclitus is chief staff engineer in the engineering technologies group of C-E Air Preheater, Wellsville, N.Y., and has been with the company since 1950. He has a B.Ch.E. from Rensselaer Polytechnic Institute, and an M.A. in physics from Alfred University. He is a member of ASME, and holds a professional engineer license in N.Y. and Pa. He also has seven patents on heat-exchanger designs, is the author of seven technical papers on heat transfer, and has collaborated on two reference volumes dealing with nuclear reactor data.

Specifying and selecting

Thermal duty, mechanical design and materials of construction are key elements for choosing a heat exchanger to fit process conditions. The guidelines developed here will prove useful in specifying the best unit.

Abe Devore, George J. Vago and **G. J. Picozzi**, *C-E Lummus*

□ Process layout of a new plant, with accompanying mass and energy balances, is the responsibility of either the process-design engineering group of a central engineering department or of an engineering contractor.

Decisions as to the mode of energy transport (e.g., fired heaters, air-cooled exchangers, shell-and-tube exchangers, etc.) are also normally the province of the process-design engineer.

Whether performed in-house or by a contractor, job specifications for heat-transfer equipment must be prepared and a technical evaluation of competitive bids made. These two factors are our principal concern here. For reasons to be detailed later, emphasis will be on shell-and-tube exchangers.

Job specifications for shell-and-tube exchangers may be subdivided into three broad categories:
- Thermal (process) data.
- Mechanical-design considerations.
- Materials of construction.

Thermal (process) data

The most important consideration is whether the mode of operation is steady-state or unsteady-state (i.e., a batch process). Quite often, a steady-state process for a "hypothesized worst case" of an actual unsteady-state mode of operation is specified. Under no circumstances should such a fictionalized steady state be used. Cycling, transients, etc., are crucial to the mechanical design of an exchanger [1,2]. Further, subtle thermal factors must be considered that do not normally enter into the design of steady-state exchangers [3].

For both unsteady- and steady-state modes of operation, it is necessary to evolve a complete specification. What does this constitute? Even for systems undergoing no phase change, the currently available specification formats, e.g., TEMA [4], are not always complete. For example, no provision is made for unsteady-state storage capacity, elapsed time, variable transfer rates, etc.

A complete specification of process requirements must define:

1. Flowrates of both streams. For the unsteady-state mode, storage volume, circulating rates and elapsed times must be indicated.

2. Composition (including phase definitions) of both streams. It is important to indicate here whether composition is expressed in terms of mol-fraction or weight-fraction.

3. Terminal temperatures.

4. Duty. This must be consistent with the information

Nomenclature

A	Surface area, ft^2, (m^2)
C_F*	The relevant force coefficients or product of these coefficients
D_1	Inside shell dia. in., (mm)
D_3	Outer tube limit, in., (mm)
d_2	Outside tube dia., in., (mm)
f_n	Fundamental (first mode) natural frequency of a tube run
g	Acceleration of gravity, ft/s^2, (m/s^2)
g_c	Conversion factor in Newton's second law of motion, (lb$_{mass}$) (ft)/(lb$_{force}$) (s^2), or (kg$_{mass}$) (m)/(kg$_{force}$) (s^2)
H	Enthalpy, Btu/lb, (kcal/kg)
m	Mass/unit length of a tubespan, lb/ft, or
n	Degrees of freedom of a vibrating system, no.
P	Tube pitch, in., (mm)
$(\Delta P_s)_{NN}$	Shellside fluid pressure drop from nozzle to nozzle, psi, (kg/cm^2)
$(\Delta P_t)_{NN}$	Tubeside fluid pressure drop from nozzle to nozzle, psi, (kg/cm^2)
q	Duty, Btu/h, (kcal/h)
q_s	Shellside duty, Btu/h, (kcal/h)
q_t	Tubeside duty, Btu/h, (kcal/h)
T	Hot-fluid temperature, °F, (°C)
t	Cold-fluid temperature, °F, (°C)
$(\Delta t)_{avg}$	Average value of $(T-t)$, °F, (°C)
$(\Delta t_m)_{corr.}$	Corrected mean temperature difference, °F, (°C)
U	Overall coefficient of heat transfer, Btu/(h) (ft^2) (°F), or kcal/(h) (m^2) (°C)
V	Velocity, ft/s, (m/s)
V_c	Critical crossflow velocity above which an undesirable amplitude response is likely for a specific excitation mechanism, ft/s, (m/s)
W	Mass flowrate of hot stream, lb/h, (kg/h)
w	Mass flowrate of cold stream, lb/h, (kg/h)
y_0	Cumulative (first mode) amplitude response to all forms of flow-induced excitation, in., (m)
γ	A constant that is unique for each mechanism of flow-induced excitation, dimensionless
δ_0	Log-decrement factor of the coupled fluid-structural system, dimensionless
$\widehat{\rho}$	Fluid density on shellside, lb$_{mass}$/ft^3, (kg/m^3)
Z	Height above datum plane in gravitational field, ft, (m)

furnished under Items 1, 2 and 3 above. This most elementary consideration of the first law of thermodynamics is violated with unbelievable frequency.

We strongly urge that the process-design engineer check the thermal balance prior to issuing the specification. The following basic formula must be satisfied:

$$q = W(\Delta H)_{hot} = w(\Delta H)_{cold} \qquad (1)$$

Errors in this simple heat balance usually lead to confusion and loss of time for the thermal-design engineer (e.g., computer crunch when an attempt is made to input the data to a design program). Then, the thermal designer is compelled to find out from the process engineer which of the variables (W, w, T_1, T_2, t_1, t_2, etc.) were incorrectly specified.

Eq. (1) is valid for most industrial flow processes. It is a simplification of the more general form (written for the unit mass of each stream):

$$\Delta H + \frac{\Delta(\bar{V}^2)}{2g_c} + \Delta\left(\frac{gZ}{g_c}\right) = q - w_a \qquad (1a)$$

where w_a = work term.

Eq. (1) neglects the work term and the kinetic and potential energy terms. With rare exceptions, this is a valid procedure.

The process-design engineer is inclined to design an exchanger for the anticipated "worst" operating conditions—e.g., a cooler with maximum heat rejection, maximum fouling resistance, and maximum inlet-cooling-water temperature.

Such a decision should take into account the controllability of the process. If the range of operation is extreme, this information should be conveyed to the designer, along with the mode of startup and shutdown, upset conditions, variations in feedstock composition, etc.

The usual design solution for extreme variations involves multiple units rather than a single unit. On the other hand, small variations in any steady-state process are always present. The usual way of providing for these small variations is to specify "percentage overdesign" factors.

In many instances, an overdesign multiplier is specified in order to anticipate a future increase in plant capacity. It is not uncommon to see a specification with up to 50% overdesign. This makes the task of the thermal-design engineer unnecessarily difficult. To make things worse, the overdesign is sometimes defined as "on surface only" or "on duty only." A rational specification of overdesign should always mean "on duty and flows" with due regard to the corresponding pressure drops.

Extreme overdesign factors lead to difficulties in process control, especially when the lower limit of flows and duties prevails. In cases where phase changes occur, this situation may be aggravated.

We suggest that any overdesign multiplier greater than 10% be carefully scrutinized. These multipliers should never be specified to take care of uncertainties in heat-transfer coefficients, because such a decision is the sole responsibility of the thermal-design engineer.

5. Transport properties of both streams (for both gas and liquid phases). For common fluids such as steam, cooling water and air, this is unnecessary. Abundant data exist in the open literature.

6. Operating pressures.

7. Allowable pressure drop. After the process engineer completes the energy balance for each stream, the available pressure head has to be distributed over the heat exchangers, piping and the associated equipment. The portion of the available head that is allotted to the heat exchanger becomes the allowable pressure drop. This is an important factor in heat-exchanger design, and in most instances may indeed be the controlling factor. It represents the potential energy that can be expended to transport the fluid through the exchanger, starting at

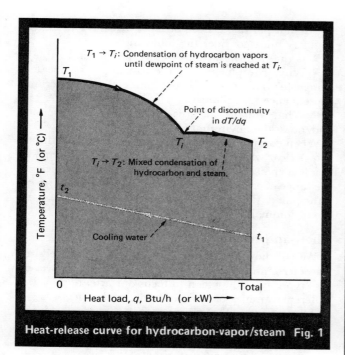

Heat-release curve for hydrocarbon-vapor/steam   Fig. 1

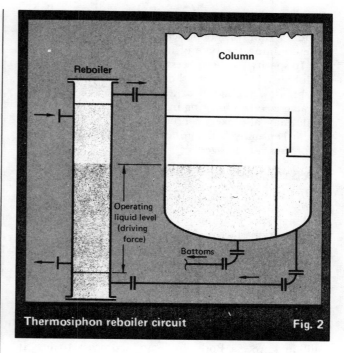

Thermosiphon reboiler circuit   Fig. 2

the inlet nozzle and terminating at the outlet nozzle.

Some portion of the pressure loss is used up in contraction, expansion and change of direction associated with the nozzles, headers, bundle entry and bundle exit. This type of loss does not contribute to heat transfer. Accordingly, nozzles that connect to exchangers should be specified to allow for this fact.

As a rule of thumb, 10 to 20% of the overall pressure drop is associated with nozzle, bundle entry, etc., losses in a good design.

It should be clear that the allowable pressure drop is a constraining factor in the optimization process. Unfortunately, pressure drop is invariably a stronger function of velocity than is heat transport.

8. *Fouling resistance.* For an understanding of what fouling resistance is, consult Ref. 9 and 10.

Usually, the best guide is data feedback from equipment in operation. However, where velocities are low, specifying high fouling resistance can be a self-fulfilling prophecy. Generally, it is a good idea to make a more generous allowance for pressure drop in order to avoid such cases. In high heat-transfer-rate situations, dragging down the clean coefficient with high fouling resistance may result in complete loss of controllability.

In the absence of data, TEMA [4] is the best available guide for evaluating appropriate values of the fouling resistances for various services.

Supplementary data for other equipment

The following information should supplement the previous listing whenever nonisothermal condensers (i.e., multicomponent condensing systems) are specified:

■ A heat-release curve (i.e., a plot of temperature vs. duty, as shown in Fig. 1).

■ A cumulative plot of mass flow of condensate vs. temperature.

■ A plot of condensate composition vs. temperature for a binary system.

■ A table of condensate composition vs. temperature for a condensable having three or more components. These data should correspond to at least three points of a smooth heat-release curve. It is particularly important to define these points if there are discontinuities in the curve, i.e., for dT/dq, where T = temperature and q = duty. (The data for the discontinuities are crucial to ascertaining heat-transfer mechanisms.)

If available, the following data are helpful (and may be most important in some cases) to the designer: (a) surface tension—especially relevant to reflux condensers, and (b) compressibility factor—important in high-pressure applications.

For proper design of vaporizers and reboilers, the supplementary data (as listed for condensers) are also necessary. Simply substitute "boiling fluid" for "condensate" in the previous discussions for the necessary data for condensers.

For natural-circulation (thermosiphon) reboilers, it is also necessary to specify:

■ Available static head for the process feed.

■ Variation of vapor pressure with temperature for a single-component boiling fluid. For a multicomponent system, a dewpoint/bubblepoint vs. temperature and composition table is necessary. For a binary system, a graphical plot is feasible.

For very low operating pressures (e.g., vacuum services), these data are extremely important.

■ The equivalent length of inlet and exit process lines are necessary in order to evaluate the thermal and hydraulic loop properly. Even better, a sketch of the tower with elevations, liquid levels, etc., clearly indicated, as shown by Fig. 2.

It is usually better practice to omit the desired recirculating rate. If the rate must be fixed (e.g., for tower equilibrium), then a forced-circulation reboiler should be considered. Fair [5] has given an excellent in-depth discussion for the variables that must be taken into consideration in thermosiphon reboiler design. His article includes a table of reboiler characteristics that is useful in determining the type of unit to be selected.

■ In specifying the process-outlet nozzle size of thermosiphon reboilers, care should be exercised to avoid the slug-flow regime, if at all possible. This regime is characterized by chugging instabilities [6]. As the outlet nozzle may also exert a crucial choking effect on the flow, it is usually better to allow the designer to select the process-nozzle sizes.

Mechanical-design considerations

After completing the thermal (process) specification, the process-design engineer must prepare the mechanical-design specifications. In sequence, this involves: (a) determination of a suitable type of exchanger, (b) determination of applicable codes, standards and regulations, and (c) quantitative formulation of mechanical-design factors (design pressures and temperatures).

Determining type of exchanger

In general, we are dealing with two-fluid heat exchangers that can be described by the function they fulfill in a process. Both fluids can be process streams. If only one process stream is involved, the other fluid is usually steam for heating, and water or air for cooling. Where air is used as the cooling medium, we speak of air-cooled exchangers. The decision to use air-cooled heat exchangers has to be made at an earlier stage of process design, and depends on the utility cost evaluation and optimization studies for the particular plant-site.

Selection of a suitable exchanger is occasionally dictated by one dominant factor. For example, hygienic requirements in the biochemical and foodstuff industries make frequent disassembly and cleaning mandatory. Because leaks to the environment are benign here, the plate-and-frame exchanger is usually an attractive choice. As another example, let us consider a low volumetric flowrate of a very viscous fluid with an associated wide temperature range (i.e., high duty). If this is coupled with a need for counterflow, a spiral-plate exchanger may be the indicated choice [7].

The three steps for mechanical design are interdependent. For example, a design pressure equal to or greater than 150 psig normally excludes spiral-plate and plate-and-frame types, and mandates the use of shell-and-tube exchangers. For surface areas equal to or less than 500 ft², double-pipes or coils may also be suitable. Here, we can generalize: The shell-and-tube exchanger is the optimum configuration from the standpoint of mechanical integrity, range of allowable design pressures and temperatures, and versatility in type of service.

Shell-and-tube exchangers may be designed to safely handle pressures ranging from full vacuum to approximately 6,000 psig (420 kg/cm²), and for temperatures in the cryogenic range up to approximately 2,000°F (1,100°C).

In implementing the steps for mechanical design, the process engineer must keep in mind the question: What are the consequences of failure?

Heinze [1] has clearly defined the problem of specifying design parameters for pressure vessels. The design engineer must consider the following:

■ Is the fluid hazardous (toxic and/or flammable)?
■ What equipment will the exchanger be located closely to? This may well determine the most serious consequences of failure?
■ What are the hazards of leakages to personnel and/or to the environment?

When considering responses to these questions, it becomes clear why chemical engineers have usually made as a first choice the shell-and-tube exchanger. In view of the range of environmentally hazardous fluids, it can be anticipated that such exchangers will become even more dominant.

Shell-and-tube exchangers

We will now consider in more detail the procedures for determining a suitable type of exchanger. However, by "type," we now mean to focus our attention on the different varieties of shell-and-tube exchangers.

The first factor that must be taken into account is the size of the exchanger. (We will be discussing an estimate, not a rigorous evaluation.) How does the process-design engineer approximate the size of an exchanger? We begin by writing the basic transfer-rate equation:

$$A = q/U(\Delta t)_{avg} \qquad (2)$$

If the distinct zones exist (e.g., desuperheating, condensing, subcooling, etc.) in the heat-release curve, we separately evaluate the areas in each zone.

What do we do about approximating the overall coefficient, U, in Eq. (2)? The "Chemical Engineers' Handbook" [8] has tabulated a variety of U-values for a wide range of fluids. With this tabulation, it should be possible to approximate almost any conceivable combination of service U-values.

Since q is fixed by process demands, the only remaining problem is to calculate $(\Delta t)_{avg}$. For this purpose, we suggest a straightforward log-mean temperature difference, $LMTD$, whenever q is a linear function of temperature, as indicated in Fig. 3. A counterflow setup is assumed. For nonlinear duties, a counterflow orientation is again assumed, retaining the minimum terminal Δt, but averaging the heat-release curve (Fig. 4).

Eq. (2) is now used to obtain an approximate surface area. If the resulting surface $A < 20$ ft² (≈ 2 m²), then a coil-type unit is probably an optimum choice. If the surface area falls in the range of 20 ft² $\leqslant A \leqslant 500$ ft² (≈ 2 m² $\leqslant A \leqslant 50$ m²), then double-pipe units should be considered as an alternative to shell-and-tube designs. For values of $A > 500$ ft² (≈ 50 m²), a shell-and-tube design should be specified.

The procedure outlined above dictates flexibility on the part of the design engineer when specifying borderline cases. For example, if 400 ft² is computed as the approximate surface, then shell-and-tube exchangers should be specified as an alternative to double-pipe units.

Thermal expansion and fouling

The logic flowchart of Fig. 5 indicates two factors of overriding importance: thermal-expansion stresses and

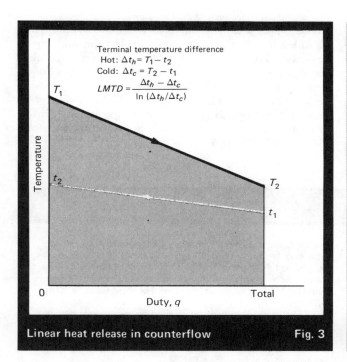

Terminal temperature difference
Hot: $\Delta t_h = T_1 - t_2$
Cold: $\Delta t_c = T_2 - t_1$

$$LMTD = \frac{\Delta t_h - \Delta t_c}{\ln(\Delta t_h / \Delta t_c)}$$

Linear heat release in counterflow   Fig. 3

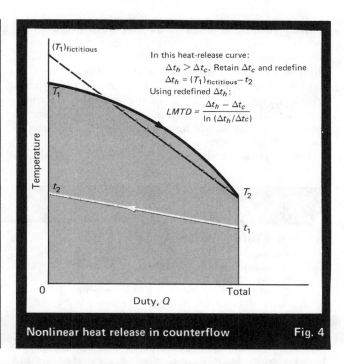

In this heat-release curve:
$\Delta t_h > \Delta t_c$. Retain Δt_c and redefine
$\Delta t_h = (T_1)_{\text{fictitious}} - t_2$
Using redefined Δt_h:

$$LMTD = \frac{\Delta t_h - \Delta t_c}{\ln(\Delta t_h / \Delta t_c)}$$

Nonlinear heat release in counterflow   Fig. 4

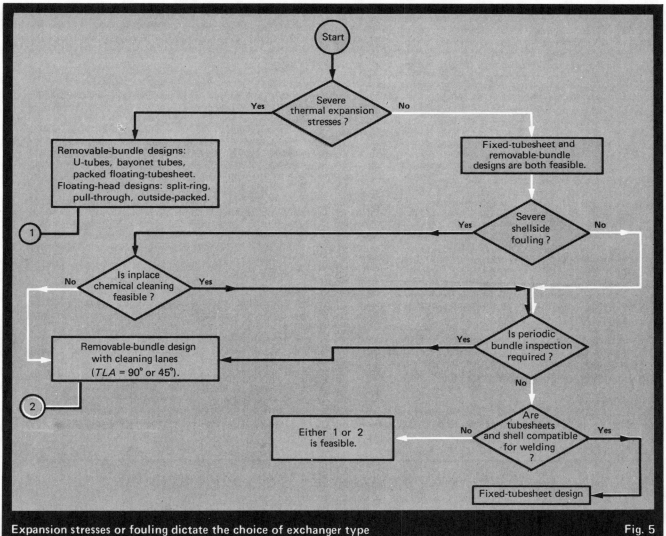

Expansion stresses or fouling dictate the choice of exchanger type   Fig. 5

shellside fouling. If the process is subject to severe excursions in flowrates and/or temperatures, the design engineer will have no difficulty recognizing the potential for thermal-expansion stresses. If plant experience in allocating a process stream exists (i.e., routing the process stream through the tubeside or the shellside), the question of shellside fouling can be resolved. In the general case, both decisions will have to be made on the basis of analytical procedures.

Let us consider the problem of thermal-expansion

stresses. This is almost nonexistent in two types of single-tubesheet heat exchangers: the U-tube and bayonet-tube. Bayonet exchangers actually have two tube-sheets, but only one is a pressure member. For a description with outline drawings of the types of shell-and-tube exchangers, Kern [9] and TEMA [4] are recommended. We will illustrate some specific assemblies of shell-and-tube exchangers in Fig. 6 along with our discussion of Table I.

(text continues on p. 25)

Advantages and limitations of shell-and-tube exchangers   Table I

Fixed-Tubesheet Design

Advantages

1. Lowest first cost, two-tubesheet (i.e., straight tube) shell-and-tube exchanger.

2. All shellside gasketed joints are eliminated.

3. No test rings needed to perform shellside hydrostatic testing. (Test rings are used to observe leaking tube-to-tubesheet joints.)

4. Individual tubes are replaceable and cleanable.

Fig. 6a illustrates a TEMA Type NEN unit, normally the most economical design. It is well suited for tubeside fouling fluids. The bolted-on channel-covers are removable, providing access for mechanically cleaning the tubes. Note that the channel nozzles (T1 and T2) need not be disconnected for this operation.

5. The term $(D_1 - D_3)$ is the diametral clearance between the inside diameter of the shell and the diameter of the tangent circle to the bundle (referred to as the OTL = outer tube limit). This has two important consequences: (a) for a given shell diameter and tube-field geometry, the tube-count is maximized, i.e., for a given surface area, the required shell diameter is minimized, and (b) the bypass stream is minimized in baffled flow—thus improving bundle penetration.

6. Depending on shell size, any number of tube passes (usually limited to a maximum of 16) are possible, including single pass (countercurrent or cocurrent).

TEMA Type AEL (Fig. 6b) is easier to fabricate than the other types for multipass units, because the (simple rectangular shaped) channel pass-partitions may be welded from both ends. This consideration is more important in smaller (pipe size) units where labor costs dominate.

7. Inspection of tubesheets is readily done by removing the channel cover(s). Types AEL and NEN are best in this regard.

8. When interleakage between shellside and tubeside fluids cannot be tolerated (this may occur at a tube-to-tubesheet joint), double-tubesheet construction is used. The fixed-tubesheet design is readily adapted to this. (See Fig. 6c).

9. Two-pass shells may be fabricated by welding a longitudinal baffle in place. No fluid leakage between shell passes is possible (TEMA Type BFM).

10. TEMA Type BEM (Fig. 6d) is advantageous for higher tubeside pressures and/or clean tubeside fluids. Fig. 6d shows a typical stacking arrangement for two shells in series. This is a much better arrangement than a TEMA Type BFM from a mechanical and thermal standpoint, although more costly.

Limitations

1. The unit is sensitive to temperature excursions and cycling. (Differential expansion between the shell and the tube

bundle is the problem.) A shell-expansion joint of the bellows type may be costly. Fig. 6b shows a detail of a less costly flanged-and-flued expansion joint. However, this type of joint cannot accommodate as much movement as the bellows type, being limited to a maximum movement of approximately $\frac{3}{16}$ in. (about $4\frac{3}{4}$ mm).

2. The shellside cannot be cleaned by mechanical means. Flushing with a solvent must be resorted to.

3. Vertical units must be vented through the top tubesheet, and usually require an external loop seal in order to avoid a vapor-liquid interface. (See Ref. 2.)

4. Because $(D_1 - D_3)$ is a minimum, bundle entry poses a problem in pressure losses, and erosion and vibration hazards when relatively high-volume flows are involved. Annular distributors (belts) or divided flow orientations (TEMA Type AJM) are usually used for this situation. Both add a good deal to the first cost.

5. Baffle assembly is more difficult than with the removable bundle types (except, possibly, the U-tube types).

6. With a large number of tube passes, channel assembly may be difficult. If permissible, switching of streams should be investigated.

7. Type B channels (usually referred to as bonnets) necessitate breaking connections in order to have direct access to the tubesheet. (This is true for all types of bundles, not just fixed tubesheets.) Unless a large number of tube passes is involved, Types BEM and NEN are less costly than the Type A channels.

8. Very large temperature ranges on the tubeside require careful design for multipass units. As a rule of thumb, a maximum temperature differential of 50°F (about 28°C) across passes should not be exceeded [12]. A ribbon-type (all partitions are parallel) tube field should be used in such cases. (This is also true for all types of bundles, not just fixed tubesheets.)

U-Tube Design

Advantages

1. Embodying only one header and one tubesheet, it is the lowest-first-cost exchanger. Fig. 6e illustrates the typical TEMA Type BEU unit.

2. As compared with the various straight-tube (floating head, outside packed, etc.) designs, the number of gasketed joints is minimized.

3. This design is advantageous for high tubeside pressure. Fig. 6f depicts a typical TEMA Type CEU design. A "false" head is used to get around the problem of short-circuiting between tube passes (because of channel-cover deflection).

(table continues on p. 22)

In Fig. 6, T1 represents tubeside inlet; T2, tubeside outlet; S1, shellside inlet; S2, shellside outlet.

The Type designation for TEMA exchangers is a three-letter combination that describes the stationary head, the shell, and the rear head in that order. Representative examples for such designations are shown in Fig. 6. For additional details, see the TEMA [4] standards.

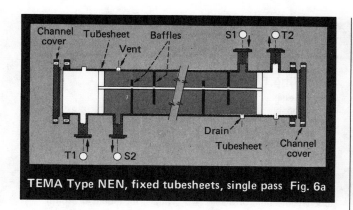

TEMA Type NEN, fixed tubesheets, single pass   Fig. 6a

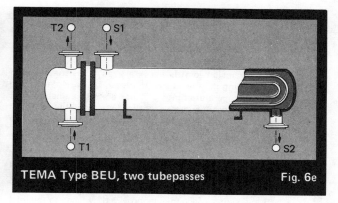

TEMA Type BEU, two tubepasses   Fig. 6e

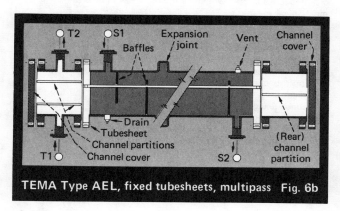

TEMA Type AEL, fixed tubesheets, multipass   Fig. 6b

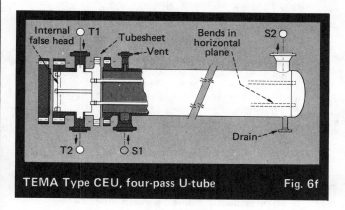

TEMA Type CEU, four-pass U-tube   Fig. 6f

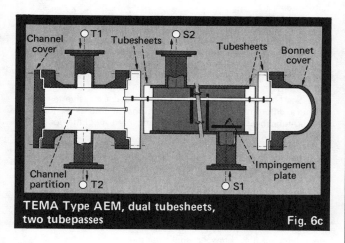

TEMA Type AEM, dual tubesheets, two tubepasses   Fig. 6c

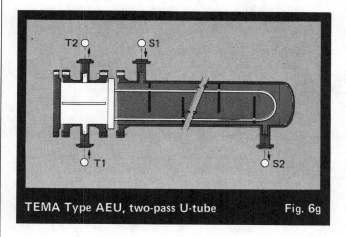

TEMA Type AEU, two-pass U-tube   Fig. 6g

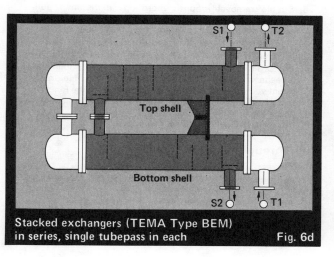

Stacked exchangers (TEMA Type BEM) in series, single tubepass in each   Fig. 6d

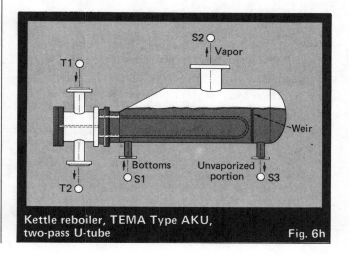

Kettle reboiler, TEMA Type AKU, two-pass U-tube   Fig. 6h

4. No expansion joint is necessary, as tubes may expand without creating thermal stress. However, be wary of too large a tubeside temperature range because a steep gradient across the pass partition may create localized stress problems in the tubesheet, and unequal expansion of the straight legs may also overstress the bends.

5. See Advantage 5 for fixed-tubesheet design.

6. TEMA Type AEU (Fig. 6g) is easier to fabricate than the Type BEU (see Advantage 6 for fixed-tubesheet design), and also provides access to the tubesheet (by removing the channel cover) without having to break tubeside connections.

7. Bundle entry space may be provided by locating the inlet nozzle beyond the bends. However, this should be carefully scrutinized for larger bundle diameters. (See Limitation 2 for U-tubes.)

8. The bundle is removable. Hence, a square-pitch tube field may be mechanically cleaned. Since only one tubesheet is involved (and no rear header), it is the lightest-weight removable bundle. Fig. 6h depicts a TEMA Type AKU, a kettle reboiler, that takes advantage of these characteristics.

Limitations

1. Usually restricted to clean tubeside service because of difficulty in cleaning.

2. In very large shell diameters, support of the bends is difficult. (The U-tube bundle becomes susceptible to vibration hazards.)

3. Baffle assembly is somewhat more difficult than for straight-tube removable bundles.

4. Single pass is impossible. When large tubeside volumetric flows are involved, this is a distinct disadvantage.

5. With the exception of the outermost tubes, individual tubes cannot be replaced.

Bayonet-Tube Design

Advantages

1. This is probably the most underused type of exchanger design. It represents the only perfect solution to the differential-expansion problem. Individual tubes are free to expand or contract at will.

2. Every tube is replaceable.

3. Unlike the U-tube design, there is no problem in supporting bends.

4. Bundle entry is facilitated by locating the inlet nozzle beyond the bundle. Further, entry into the tube nest is in an axial direction. This minimizes flow-induced vibrations that are so common in other types of exchangers. Kern [9] gives an example for an application of this type for a vacuum condenser.

5. Removable bundle facilitates cleaning.

6. For vaporizing at low temperatures of the boiling fluid with heating steam, this unit has a unique feature. Transfer of heat from the steam in the conveyor tube to the steam/condensate mix in the annular space prevents condensate freezeup. Fig. 6i shows a typical application.

7. Baffle assembly is extremely simple.

8. See Advantage 5 for fixed-tubesheet design.

Limitations

1. More costly than U-tubes because of the conveyor tubes. However, the latter are usually thin-walled because they are not pressure members.

2. An extra tubesheet is required for the conveyor tubes. But this is not a pressure member either. (Note the thin tubesheet in Fig. 6i.)

3. Rather difficult to adapt to multipass on the tubeside. Generally used for isothermal condensation or boiling on tubeside. However, it is possible to evaluate a correction for the mean temperature difference in nonisothermal applications. (See Kern's discussion [9] of Hurd's analysis.)

4. Sealing of the tube ends is an expensive operation.

Split-Ring Floating-Head Design

Advantages

1. Probably the most widely used design. The movable (floating) tubesheet accommodates differential tube-bundle expansion. Fig. 6j is a Type AES. Type A channels for all designs have the advantages discussed under Item 4 for fixed tubesheets.

2. Because the bundle is removable and consists of straight tubes, this design is widely used for fluids having fouling tendencies on both shellside and tubeside. We note here that a tube-layout angle of 90° or 45° with $\frac{1}{4}$ in. (about 6.3 mm) cleaning lanes is used whenever the bundle is to be cleaned by mechanical means.

3. See Advantage 6 under fixed tubesheets. The remarks on the advantages of an A channel also apply.

4. The floating-head gasketed joint is designed with either a metallic or metal-jacketed gasket. With proper design, interleakage of tubeside and shellside fluids is rare. Leakage at this floating tubesheet to the environment is impossible.

5. The term $(D_1 - D_3)$, although larger than for the fixed-tubesheet design, is a good deal smaller than for the pull-through floating-head design, to be discussed later.

6. The TEMA Type BES that has a bonnet (usually a bumped-head closure) in place of the A channel is usually less costly than Type AES, and finds application for higher tubeside design-pressures and/or clean fluids.

7. Usually less costly than the pull-through floating-head.

8. Baffle assembly is simpler than for fixed-tubesheet design. This is true, in general, for any removable bundle.

Limitations

1. The largest end-zone (i.e., space between an extreme baffle and its adjacent tubesheet) of any of the designs is inherent in the split-ring floating-head orientation. This has possibly two serious consequences: (a) low velocities reduce the effective shellside heat transfer (and may increase fouling tendencies), and (b) the long span may be susceptible to flow-induced vibration hazards.

2. See Limitation 8 under fixed tubesheets. Excessive temperature differentials may induce torsional moments at the floating-tubesheet end and/or destroy the integrity of the floating-head joint.

3. Since $(D_1 - D_3) \gg (P - d_2)$, the bundle bypass stream must be diverted into the bundle by means of sealing strips. Small shell sizes are particularly vulnerable to this effect.

4. Single-pass designs require an expansion joint and special flanges through the shell cover. Fig. 6k illustrates one possible way to design this but adds considerably to the cost.

5. The floating-head gasket is a "hidden" gasket. As long as the shell-cover joint is sound, interleakage cannot be immediately detected (i.e., between shellside and tubeside streams).

6. Bundle removal requires more operations than pull-through floating-head design. In addition to its somewhat high first cost, its maintenance is high.

7. Test rings are required to perform shellside hydrostatic testing (in order to observe leaking tube-to-tubesheet joints).

8. Two-pass shells are more difficult to fabricate than the fixed-tubesheet design. They are subject to possible interleakage between shell passes. It is possible to weld the longitudinal baffle to the shell, but this requires a costly split floating-head assembly.

9. Double-tubesheet design is not feasible.

Pull-Through Floating-Head Design

Advantages

1. Advantages 2, 3 and 4 for the split-ring design apply here as well.

2. The bundle is easier to remove than in the split-ring design (see Fig. 6l). If desired, the rear shell-cover does not have to be removed for bundle removal. Instead, the bundle can be

Advantages and limitations of shell-and-tube exchangers   Table I (cont.)

removed together with the floating head by first breaking the channel connections.

3. See Advantage 4 under split-ring design. If anything, the gasketed floating-head joint is even more reliable than the split-ring. Maintenance, in general, is less costly than for the split-ring.

4. See Advantage 6 for the split-ring design. TEMA Type B has the same advantage over the Type A unit.

5. See Advantage 8 for the split-ring design.

6. An inspection of Fig. 6l makes it evident that the clearance between the bundle and the shell is very large. In fact, $(D_1 - D_3)$ is larger than for any other type of exchanger.

When large volumes of shellside fluid are involved, this large clearance is an advantage. The vibration-excitation forces due to fluid flow are reduced because bundle entry and exit areas are increased, and the shell size for a given tube field is maximized. In turn, this requires tighter baffle-spacing for a given velocity, resulting in a more favorable range of bundle frequency-response.

7. This design does not have Limitation 1 of the split-ring design.

8. It is admirably adapted to kettle-reboiler design (see Fig. 6m), where all the advantages of a removable straight-tube bundle apply.

(table continues on p. 24)

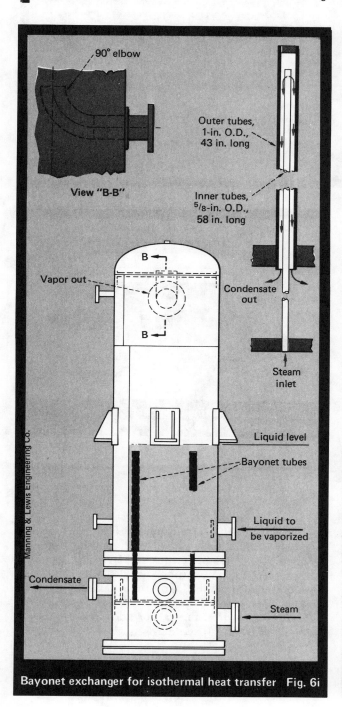

Bayonet exchanger for isothermal heat transfer   Fig. 6i

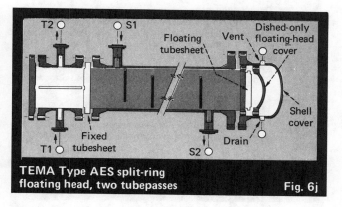

TEMA Type AES split-ring floating head, two tubepasses   Fig. 6j

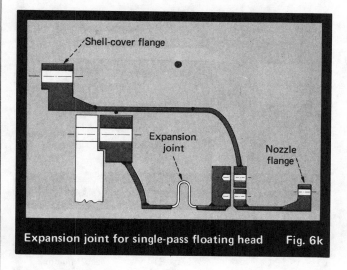

Expansion joint for single-pass floating head   Fig. 6k

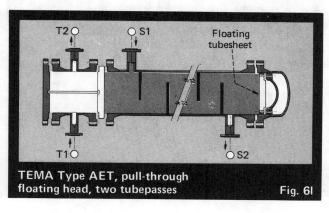

TEMA Type AET, pull-through floating head, two tubepasses   Fig. 6l

Advantages and limitations of shell-and-tube exchangers   Table I (cont.)

Limitations

1. Limitations 2, 4, 5, 7 and 8 for split-ring design apply with equal force here.
2. Since the inequality, $(D_1 - D_3) \geqslant (P - d_2)$, is even worse than for the split-ring, Limitation 3 for the latter applies with even greater force.
3. The design is usually more costly than that for a corresponding split-ring design.

Outside-Packed Floating-Head Design

Advantages

1. Advantages 2, 3, 5, 7 and 8 for the split-ring design apply equally here.
2. Double-tubesheet design is feasible.
3. Any number of tube passes are possible.
4. Leakage at the floating-head joint is to the environment and quickly detectable. (See, however, Limitation 1.)
5. Bundle removal is easier than split-ring design. Fig. 6n depicts a Type BEP—one possible variant for this design.
5. Bundle removal is easier than for the split-ring design.

Limitations

1. This type of packed joint should not be used when lethal, toxic or flammable fluids are to be contained. Leakage through a stuffing box is more likely than is failure of a gasketed joint.

2. This unit is very limited in its design pressure and temperature ranges. Design pressures $\leqslant 600$ psig, and design temperatures $\leqslant$ approximately 600°F.
3. The $(D_1 - D_3)$ term is approximately the same as in the split-ring design. (See Limitation 3 for the split-ring unit.)
4. This exchanger is more costly than a packed floating-tubesheet unit, and may even be more costly than the split-ring in small pipe sizes.

Packed Floating-Tubesheet Design

Advantages

1. Lowest first-cost straight-tube removable bundle. Widely used for lube-oil coolers and air-compressor intercoolers in small pipe sizes.
2. Individual tubes are easily replaceable and cleanable.
3. The term $(D_1 - D_3)$ is only slightly larger than the fixed-tubesheet design, so that sealing strips are not normally used.
4. Bundle removal and maintenance are simple. Fig. 6o shows a typical rear-head detail for this construction.

Limitations

1. Constrained to a maximum of two tube passes.
2. Same as Limitations 1 and 2 for outside-packed floating-head design. In these respects, this design is even more limited.

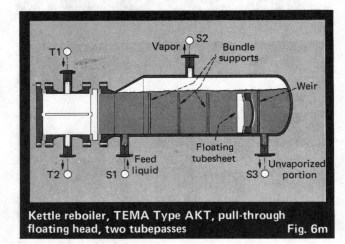

Kettle reboiler, TEMA Type AKT, pull-through floating head, two tubepasses   Fig. 6m

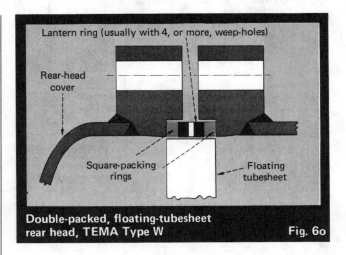

Double-packed, floating-tubesheet rear head, TEMA Type W   Fig. 6o

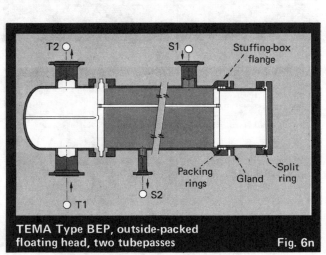

TEMA Type BEP, outside-packed floating head, two tubepasses   Fig. 6n

In Fig. 6, T1 represents tubeside inlet; T2, tubeside outlet; S1, shellside inlet; S2, shellside outlet.

The Type designation for TEMA exchangers is a three-letter combination that describes the stationary head, the shell, and the rear head in that order. Representative examples for such designations are shown in Fig. 6. For additional details, see the TEMA [4] standards.

The two-tubesheet exchanger design that is known as fixed tubesheets may have thermal-expansion stress problems. At this point, the logical question may arise: Why bother with fixed-tubesheet design? The answer is simple: It is the most economical of the two-tubesheet designs (i.e., of all straight-tube designs).

Evaluation of thermal-expansion stresses is a complicated affair, and may be arrived at from the following:

1. Thermal stresses arise from differences between shell-metal and tube-metal temperatures.

2. Determining these metal temperatures under normal operating conditions depends on establishing the heat-transfer rates at the specified steady-state process conditions.

3. Evaluating excursions or transients (e.g., thermal shock associated with startup), requires a more sophisticated analysis. For example, see Appendix of Sect. III of the ASME Code.

4. Thermal stresses must be superposed onto hydrostatic loading stresses.

Any stress analysis for two-tubesheet exchangers must be based on the fact that the tubesheets transmit their loading onto the bundle, which acts as an elastic foundation. See the discussion in TEMA [4].

Ultimately, the thermal-expansion problem is best handled by engineers specializing in stress analysis. The information transmitted to the stress analyst must include metal temperatures that are determined by the heat-transfer rates. Normally, most manufacturers of shell-and-tube exchangers will perform the entire set of computations.

Computations to determine thermal-expansion stresses are much too involved for the selection of a type of heat exchanger. In the absence of severe thermal cycling in the process itself, we suggest that the design engineer assume that the fixed-tubesheet unit may be a viable one—provided it meets the subsequent tests that are shown in Fig. 5. The job specification must then clearly indicate to the vendor the responsibility for a complete analysis of the thermal-expansion stresses.

Implicit in this discussion is the decision on routing the hot stream through the shell, and the cold stream through the tubes, or vice versa. Kern [9] is useful in this regard. A recent paper by Taborek [11] is another excellent general reference. If the allocation of the fluids may be switched, it is important to let the vendor know.

Shellside fouling

The next decision, "Severe shellside fouling?", is straightforward, as indicated in Fig. 5. The other decisions (feasibility of in-place chemical cleaning, and periodic inspection of the bundle) tied to this are related to the chemical nature of the fouling and to the corrosive action on the bundle. Consultation with in-house corrosion experts and metallurgists is necessary at this juncture. The latter are also indispensable to the final decision leading to a fixed-tubesheet design: Are tubesheets and shell compatible for welding? An excellent guide, in tabular form, was developed by Sieder and Elliott [13]. This guide covers most of the important metals and alloys used in fabricating exchangers. When fixed tubesheets are ruled out, Fig. 5 indicates that a removable-bundle design can be used.

Table I summarizes the advantages and limitations of the various types of shell-and-tube exchangers listed in Fig. 5.

High-pressure closures

We have not attempted to cover the entire spectrum of shell-and-tube exchangers in Table I. For example, omitted are the TEMA Type D channel that involves high-pressure closures; the TEMA Type X shell that finds wide application in vacuum condensers; the TEMA Type G and H shells used in reboilers. However, the basic guidelines in relation to functional application are covered in Table I.

Fluids under high operating pressures (over 2,000 psi) will require a high-pressure closure that is almost always chosen to be on the tubeside as the most economical way to design the exchanger.

The most common design will be a U-tube bundle of TEMA Type D. Situations when both shellside and tubeside fluids are under high pressure seldom occur. An important exception takes place in the ammonia-synthesis loop. In this and similar closed process-loops, the tubesheets are usually designed for differential pressure, i.e., the pressure drop across the loop. Generally, high-pressure closures require special attention to gasket design, and sophisticated stress analyses.

Materials of construction

We will emphasize some materials-of-construction factors important to the specification and design of shell-and-tube heat exchangers.*

The selection of materials of construction for heat exchangers is ideally the result of an interchange of expertise between process-design engineers, corrosion experts or metallurgists, and pressure-vessel design engineers. Let us briefly examine the principal factors involved in materials selection:

1. Physicochemical properties of the fluids—Since heat-exchanger tubing is relatively thin-walled, resistance to corrosive attack is a prime consideration. Unlike other pressure members of a shell-and-tube exchanger, corrosion allowance is never specified for tubing. There is also a strong interdependence between the selection of tubing and tubesheet materials because the tube-to-tubesheet joint is the most crucial element in the mechanical integrity of the exchanger.

Uhlig [18] is a useful general reference. For example, problems of galvanic corrosion that may arise from the use of dissimilar metals in exchanger construction are discussed.

In large exchangers, fluid-induced vibration may be an important factor. Since structural damping plays a strong role in absorbing this excitation energy, the consequences of a failure in the tube-to-tubesheet joint must be taken into account. This influences not only the choice of material but also the mode of tube-to-tubesheet attachment. In some cases, it may be necessary to select tube/tubesheet combinations that are compatible for a welded joint.

2. Choice of alloys—A conflict may occur in the choice of alloys between shellside and tubeside fluids. The tube-

*This topic is discussed in depth by B. N. Greene in another article, starting on p. 31.

sheet problem may be resolved by resorting to cladding. However, bimetallic (or duplex) tubes usually pose problems in securing the tube-to-tubesheet joint. The usual stratagem is to call for an extra-heavy tube wall of some compromise material, with an exchanger design that facilitates tube replacement.

As an example, let us consider the selection of a naval-brass tubesheet and a carbon-steel shell as optimum materials. Where a fixed tubesheet may have been acceptable by other criteria, welding incompatibility eliminates this design. One of the heat-exchanger designs having removable bundles must then be used.

3. Operating temperatures and pressures—The tube bundle (tubes plus tubesheets) is subject to the combined action of both streams for all types of shell-and-tube exchangers. Hence, it is customary to set the design temperature of the tubesheet at the more extreme value of the shellside and tubeside temperatures. This is supposedly justified by assuming that the controlling temperature is that for one stream flowing alone while the other stream has zero flow. However, a good operating procedure should prevent this condition from arising.

A more rational procedure would consist of evaluating heat-transfer rates in and around the tubesheets at the worst anticipated operating conditions. This would lead to a more meaningful assessment of extreme-temperature conditions that are relevant to the selection of a suitable alloy.

4. Code acceptability—It is not a good idea to specify a chemically suitable alloy that is not recognized by Sect. VIII of the ASME Code, unless there is simply no alternative. A time-consuming "Special Case" ruling must then be obtained from the ASME Code Committee.

At this point, all the information necessary for preparing the specifications should be complete. It remains to fill out the necessary data sheets for transmittal to heat-exchanger manufacturers. On receiving this information (plus all relevant additional in-house specifications), the design engineers of the manufacturer will normally go through the following sequence:

a. Check the heat and mass balances.

b. Check the correctness (if possible) of the specified physical properties and phase equilibria.

c. Examine the specified geometrical constraints for completeness.

Any errors or inconsistencies have to be resolved with the process-design engineer before the heat-exchanger design engineer commences the actual thermal and mechanical design of the exchanger.

Codes, standards and regulations

There is a voluminous literature on determining the applicable codes, standards and regulations. A symposium on these was organized as recently as December 1979 by Rubin [14]. In the U.S., the most frequently used Codes and Standards for shell-and-tube exchangers are:

■ ASME Code, Sect. VIII, Div. 1. Pressure Vessels.

■ Standards of Tubular Exchanger Manufacturers Assn. (TEMA).

■ Heat Exchangers for General Refinery Service, American Petroleum Institute Standard, API 660.

■ Standards of the Heat Exchange Institute (HEI).

Of these, only the ASME Code can be said to have legal status. Until recently, this Code was concerned with an exchanger solely as a pressure vessel, i.e., it addressed itself to the integrity of the pressure envelope, with little or no regard for the internals. This situation is changing. There is now a special group working on shell-and-tube exchangers. Additionally, there now exist ASME standards on tube-to-tubesheet joints and on tubesheet design, although the latter differ from the TEMA approach.

Yokell [15] has described how the TEMA standards cover the mechanical design factors that are beyond the scope of the ASME Code. Some of the most important factors:

1. Nomenclature is clearly defined for all exchanger parts, and for size and type designations.

2. "Out of roundness" permitted by the ASME Code (1% of the diameter for a rolled plate) is too large for a good baffle fit.

3. Internal welds must be ground flush (for bundle insertion and removal).

4. Tolerances for the location and angularity of nozzles are specified.

5. The ASME Code flat-plate formula is usually inadequate (due to deflection) for the channel covers of multipass units. TEMA gives an equation for this.

6. TEMA addresses itself to the problems associated with tube rolling, ligaments in the tubesheets and, in general, tubesheet distortion.

7. Gasketed joint details are clearly presented.

8. An important "Recommended Good Practice" section is included in TEMA.

In general, there is a wealth of quality-assurance detail in the TEMA Standards that is specific to shell-and-tube exchangers.

Rubin [16] has observed that a specification of "TEMA" by itself is insufficient. There are three categories: TEMA R, B and C. It is important to distinguish which of these categories is applicable because they affect many internal mechanical details and, therefore, the quality and cost of the exchanger.

The API 660 standard was originally developed in order to improve the (most demanding) quality assurance of TEMA-R exchangers. Some examples:

■ Packed floating-head construction is expressly prohibited.

■ Bundle entrance area must be equal to, or greater than, nozzle cross-section area.

■ Minimum bend radius must be at least 1.5 tube diameters for the U-tube design.

■ Preferred minimum tubewall thicknesses of TEMA are made mandatory.

■ Hub (welding neck) flanges are mandatory for girth flanges. (Exceptions are permitted for pressure/temperature ratings less than 150 psig.)

Power-plant exchangers (feedwater heaters and surface condensers) are normally covered by the HEI standards. Quality control on feedwater heaters is particularly demanding. The HEI standards differ from those previously discussed in one very important aspect: They concern themselves with thermal and hydraulic factors.

Federal and local codes may also have to be taken

into account. Vervalin [17] has compiled a useful list of environmental organizations that may be helpful.

Design pressures and temperatures

The quantitative formulations of mechanical-design factors involving pressure and temperature have been thoroughly treated by Heinze [1]. A careful study of this article is worthwhile. We summarize here some of his recommendations relevant to heat-exchanger design:

■ An analysis of the process should be carefully made in order to determine the maximum (minimum, if vacuum service is involved) fluid pressures that can be anticipated. Factors such as pump shutdown and surging must be taken into account. Anticipated life expectancy and future service should also be considered.

■ Full-vacuum design should be specified for vacuum service. (If the maximum pressure is above atmospheric, this will be in addition to the design pressure evaluated from the next item.)

■ For internal design-pressures (D.P.) above atmospheric, whichever of the following relations that yields the greater value is recommended:

D.P. = 1.1 (maximum operating pressure)
D.P. = (25 psig + maximum operating pressure)
D.P. = ($\approx$ 1.75 kg/cm^2 + maximum operating pressure)

Our in-house shell-and-tube exchanger specification calls for basically the same safety margin (30 psig).

■ Design pressures evaluated from the preceding may not be sufficient to satisfy Code requirements for relief-valve settings, and should be adjusted accordingly. In addition, extra margins of safety are urged for hazardous materials.

■ Full understanding of the reaction kinetics that may be involved in the system is extremely important. From both safety and environmental standpoints, it may make sense to specify an exchanger design-pressure that is sufficiently high for total containment.

■ It is important to determine both minimum and maximum anticipated operating temperatures in order to obtain the design temperature. In combination with the specified design pressure, this will affect the flange rating. The deterioration of allowable stress with increasing temperature plays an important role in selecting a suitable alloy. Future services may also be relevant here. And possible process upsets should be taken into consideration.

For the widely used carbon steel in the temperature range from $-20°F$ to $650°F$ ($\approx -29°C$ to $343°C$), it is suggested that the design temperature (D.T.) be evaluated from:

D.T. = (50°F + maximum anticipated temperature)
D.T. = ($\approx$ 28°C + maximum anticipated temperature)

For operating temperatures less than $-20°F$ ($\approx -29°C$), a lower-limit design temperature should be specified in addition to the upper limit.

Evaluation of competitive bids

It is necessary to state clearly that the best procedure for a technical analysis of bids requires that a heat-exchanger design engineer be consulted. Engineering contractors normally follow such a procedure. Our discussion here is intended to give the process-design engineer some insight into the analytical process involved.

An essential ingredient for an intelligent evaluation of a heat-exchanger design is an understanding of how the problem is likely to be formulated by the vendor's design engineer.

Taborek [11] has clearly expressed this in a simple group of inequalities that involve the interdependency of heat-transfer rates and fluid pressure drop. To this, we would add only a qualifier for mechanical integrity against flow-induced vibration excitations. This means that a triad of interdependent modes of energy transport is to be constrained within specified limits. Thus, we write an extension of Taborek's formulation as:

$$q_t = q_s = q \geqslant q_{process\ requirement} \tag{3a}$$
$$(\Delta P_t)_{NN} \leqslant (\Delta P_t)_{NN,\ specified\ maximum} \tag{3b}$$
$$(\Delta P_s)_{NN} \leqslant (\Delta P_s)_{NN,\ specified\ maximum} \tag{3c}$$
$$V < V_c \tag{3d}$$
$$V_c/(f_n d_2) = \gamma[(m\delta_o)/(\hat{\rho}d_2{}^2)]^{1/2}/(C_F{}^*)^{1/2n} \tag{3e}$$
$$\dot{y}_o/d_2 < 0.05 \tag{3f}$$

Eq. (3d) through (3f) are due to Brothman [19,20], and express the requirement that the hazards of interspan collision and fluid-elastic swirling be ruled out. Other formulations (of a proprietary nature) are possible for the vibration-risk problem [21,22].

Eq. (3a), (3b) and (3c) express the requirement for achieving the needed process heat transfer within the constraint of the allowable pressure drops for both streams.

For small shell sizes (i.e., pipe shells), vibration-risk analysis is rarely necessary. For large shell sizes with accompanying large-volume flows, it is almost always a problem. This is clear when one thinks of the bundle assembly as a series of continuous beams, subject to fluid-induced excitations.

This fact should give the process-design engineer a clue as to why so many proposals of exchanger designs in the large plate sizes specify multisegmental baffle assemblies. The idea is to reduce the crossflow component of the vector velocity field, the V term in Eq. (3d), and at the same time reduce the supporting span-length of the "continuous beam" (i.e., the bundle baffle-spacing is reduced) for achieving a given velocity.

In the extreme case, it may be necessary to reduce the crossflow component to zero and resort to purely axial flow. An ingenious variation of this idea is embodied in the rod-baffle assembly [24].

The dimensionless term on the lefthand side of Eq. (3e), known as the Strouhal number, is strongly dependent on tubefield geometry. This accounts for occasional designs having square or rotated-square tubefields in fixed-tubesheet exchangers. Situations arise where the Strouhal numbers for the latter geometries are more favorable.

A length-constraint term is conspicuous by its absence from the Eq. (3) set. Normally, the process-design engineer will specify an overall length limitation along with the equivalent of Eq. (3a), (3b) and (3c). In a general sense, the problem is then overspecified, i.e., the additional boundary condition may not be consistent with any of the solutions to the Eq. (3) set. In actual practice, the design engineer will input the desired length into the design program and attempt to satisfy all of the constraints of Eq. (3). What this all signifies is: There is no assurance that a feasible solution will al-ways exist when the problem is constrained beyond Eq. (3). This means keeping one's mind open to alternative proposals from vendors who must cope with such a problem.

Specification of a given tube diameter, pitch or tube layout angle may also conflict with the limits imposed by Eq. (3). Still another frequently encountered example involves the specification of a minimum tubeside velocity that may be incompatible with Eq. (3b). In this case, it is a good idea to allow flexibility in the upper limit of Eq. (3b) if it is essential to maintain the tube-side velocity above a certain minimum value.

While Eq. (3a) through (3c) may be processed through a thermal-design program, the correct imple-mentation of Eq. (3d) through (3f) depends on output from a mechanical-design program. For example, the frequency term, f_n, is strongly influenced by the state of stress of the tube bundle.

Having a qualitative understanding of the previous discussion, the process-design engineer is better pre-pared (with the help of a heat-exchanger design engi-neer) to evaluate bids.

Systematize bid evaluation

Baker [23] prepared a comprehensive exchanger-comparison sheet (see p. 143 of his paper). There are several items that we suggest as additions—most impor-tantly, the type of baffle, i.e., is it segmental, double-segmental or triple-segmental? We strongly recommend that a listing be made in order to systematize the evalu-ation process (see Table II), and to ascertain that each vendor has included all of the necessary details.

Most vendors today make use of the data banks and correlations of the Heat Transfer Research Institute (HTRI), the British Harwell group (HTFS), or both. It must be borne in mind, however, that these programs are tools, not "magic black boxes" that generate unique (optimum) solutions. Accordingly, it is not surprising when disparities in bids appear. In particular, it is possi-ble for several bids, all from reputable vendors, to be "in-line" with, say, one lone bidder apparently differing to some degree. However, the latter may have the only correct solution.

To check a situation like this, carefully scrutinize all the parameters in Table II that may influence transfer rates, pressure drop (and, possibly, vibration hazard). It is important that the vendor's data sheet be complete. Pay careful attention to notes under the "Remarks" sec-tion. If a study of the deviant proposal fails to yield any clue as to its justification, it is only fair to contact the vendor and ask for clarification.

From the preceding discussion, it is evident that bid evaluation is an involved and time-demanding process. Frequently, more than one design turns out to be ac-ceptable. Such alternative designs must then be individ-ually check-rated. It is not good practice to judge ac-ceptability by a statistical comparison of the proffered designs.

This raises the question: Should heat exchangers be designed by in-house experts or by a contractor, and should vendors then be required to bid on a so-called "hardware" basis? For larger projects, it may be advan-tageous to take this route in terms of time and econom-

Checklist for comparing competitive bids for shell-and-tube heat exchangers	Table II

TEMA size
TEMA type
Surface/unit, A
Shells/unit, number
Arrangement, (number in parallel by number in series)
Shell-passes/shell, number
Tube-passes/shell, number
Crossflow shellside velocity, nominal
Window shellside velocity, nominal
Tubeside velocity
$(\Delta P_s)_{NN}$
$(\Delta P_t)_{NN}$
Shellside fouling resistance
Tubeside fouling resistance
Duty, q
Corrected mean-temperature difference, $(\Delta t_m)_{corr.}$
Service coefficient, U_s
Clean coefficient, U_c
Total number of tubes, N_{TT}
Tube O.D., d_2, (bare/finned), describe type
Tube thickness (minimum-wall/average-wall)
Tube length
Tube layout angle, TLA
Pitch, P
Shell material
Shell I.D. (or O.D.)
Shell-cover material
Channel (or bonnet) material
Channel-cover material
Stationary-tubesheet material
Floating-tubesheet material
Baffle material
Baffle type (segmental/double, segmental/other)
Inlet-baffle spacing
Central-baffle spacing
Baffle cut (diameter/area), %
Longitudinal/baffle/method of sealing
Tube-bundle support material
Impingement-plate material
Annular-distributor material
Slot area of distributor
Shell nozzle-inlet rating
Shell nozzle-outlet rating
Channel nozzle-inlet rating
Channel nozzle-outlet rating
Corrosion allowance (shell/tube)
Code
TEMA class
Job specifications
Weight
Delivery time
Price
Freight
Spares, extras, etc.
Total cost
Escalation formula

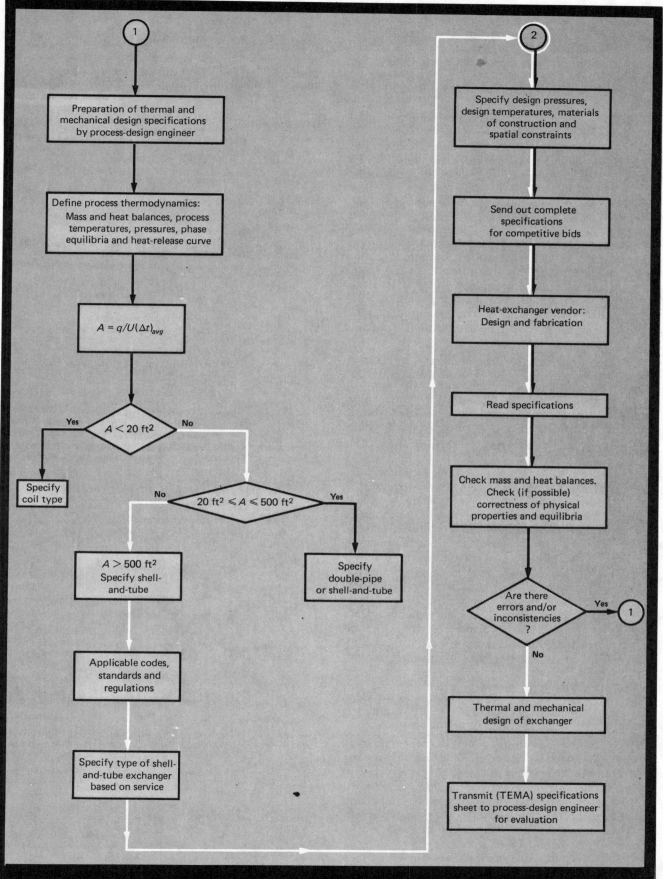

Procedure for technical specification and technical evaluation of exchangers   Fig. 7

ics. But such bids should not be solicited solely on a "hardware" basis. The vendor should be given the freedom to submit alternative designs. This assures a more active contribution from the vendor in the competitive bidding process.

Allow the vendor a reasonable time for a competitive thermal design. As a rule of thumb, allow one man-day per rating for no phase-change heat transfer, two man-days otherwise.

It is good practice to encourage alternative bids, but make sure to delineate the design factors that must be adhered to. If the alternatives are not acceptable, the vendor should be so informed. The "most square feet for the money" should not be an acceptable criterion, particularly when unacceptable exceptions are included in the bid.

Whenever possible, "package" bids should be considered, because they seem to result in the most reliable service. In the long run, this may turn out to be more economical than singling out individual items. Packaging usually is advantageous when like-materials, like-pressures, like-sizes, etc. are grouped. Grouping can be done at a later stage in order to give vendors a second opportunity.

It is good practice to avoid requesting a bid from a vendor who is either unequipped or noncompetitive for a particular requirement. If uncertain on this point, it makes good sense to simply ask the vendor rather than receive a useless uncompetitive bid. A vendor who is competitive should not be excluded for want of minor details, but should be afforded the opportunity of completing the offering.

Escalation and delivery are growing in importance. Care in evaluating the escalation formula should be taken—since firm quotations have virtually disappeared. Longer delivery times aggravate the problem.

The proper evaluation of escalation requires that the initial cost be split into material and labor components, and that the "validity date" of the bid be considered. If necessary (and if possible), set the validity date that must be adhered to and/or that is firmed for some preset period in which to place the order.

It is wise to anticipate revisions, and to obtain from the vendor the unit-pricing of nozzles, rates for engineering, etc. To anticipate the possibility of cancellations, set the cancellation formula for drawings, materials, labor, etc.

The point of delivery should be carefully noted. Is it f.o.b. cars, f.o.b. jobsite, f.a.s. dockside, or f.o.b. shop with freight allowed? This will guard against additional charges and/or unforeseen delays.

Great care should go into compiling an "acceptable bidder" list. Shop inspection, credit rating, and reputation are all considerations. If this is not done, delays in delivery time and great expense can be incurred later.

When vendors request information on bid status, it is only fair to let them know whether they are out of line. This may help them determine their future competitive chances on particular types of exchangers.

Quite often, bids will be requested from vendors for estimating purposes only, i.e., for order-of-magnitude estimates. Such requests should not be on a rush basis, unless all of the basic information (surface area, size,

materials of construction, etc.) has been worked up in-house.

If the vendor is required to do the thermal design, then a reasonable time to accomplish this should be allotted. The details of the procedures that we have outlined are shown in Fig. 7 in the form of a flowchart.

References

1. Heinze, A. J., *Hydrocarbon Process.*, May 1979, pp. 181–191 (see also errata on p. 71 of Aug. 1979 issue).
2. Lord, R. C., and others, *Chem. Eng.*, June 1, 1970, pp. 153–160.
3. Clasen, L. J., Practical Aspects of Heat Transfer, in *Chem. Eng. Prog. Technical Manual*, pp. 26–41, AIChE, New York, 1978.
4. "Standards of Tubular Exchanger Manufacturers Assn." 6th ed., Tubular Exchanger Manufacturers Assn., Cleveland, 1978.
5. Fair, J. R., *Pet. Refiner*, Vol. 39, pp. 105–123 (1960).
6. Tong, L. S., "Boiling Heat Transfer and Two-Phase Flows," p. 48, Wiley, New York, 1965.
7. Auth, W. J., and Loiacono, J., Practical Aspects of Heat Transfer, in *Chem. Eng. Prog. Technical Manual*, pp. 108–138, AIChE, New York, 1978.
8. Perry, R. H., and Chilton, C. H., eds., "Chemical Engineers' Handbook," 5th ed., p. 10–39, McGraw-Hill, New York, 1973.
9. Kern, D. Q., "Process Heat Transfer, McGraw-Hill, New York, 1950.
10. Taborek, J., Aoki, T., Ritter, R. B., and Palen, J. W., *Chem. Eng. Prog.*, Feb. 1972, pp. 59–67.
11. Taborek, J., *Heat Transfer Eng.*, Jan. 1979, pp. 15–29.
12. Afgan, N., and Schlunder, E. U., eds., "Heat Exchangers: Design and Theory Sourcebook," Hemisphere Pub. Co., Washington D.C., 1974.
13. Sieder, E. N., and Elliott, G. H., *Pet. Refiner*, Vol. 39, pp. 223–227 (1960).
14. Rubin, F. L., ed., The Interrelationships Between Codes, Standards and Customer Specifications for Process Heat Transfer Equipment, Winter Annual Meeting, ASME, Dec. 1979.
15. Yokell, S., Practical Aspects of Heat Transfer, in *Chem. Eng. Prog. Technical Manual*, pp. 139–167, AIChE, New York, 1978.
16. Rubin, F. L., *Chem. Eng.*, Apr. 8, 1968, pp. 130–136.
17. Vervalin, C. H., *Hydrocarbon Process.*, Oct. 1979, pp. 89–94.
18. Uhlig, H. H., "Corrosion Handbook," Wiley, New York, 1948.
19. Brothman, A., others, *AIChE Symp. Ser.*, Vol. 70, 190–204 (1974).
20. Brothman, A., others, Practical Aspects of Heat Transfer, in *Chem. Eng. Prog. Technical Manual*, pp. 55–85, AIChE, New York, 1978.
21. Chenoweth, J. M., and Kissel, J. H., Flow-Induced Tube Vibration in Shell-and-Tube Heat Exchangers, *AIChE Today Ser.*, 1977.
22. Whittle, R. H., and Soper, B. M. H., Tube Vibration in Shell-and-Tube, Cross Flow, Multi-Baffled Heat Exchangers, Heat Transfer and Fluid Flow Service, Report DR48, Harwell, U.K., Mar. 1979.
23. Baker, W. J., *Hydrocarbon Process.*, June 1966, pp. 141–143.
24. Small, W. M., and Young, R. K., *Heat Transfer Eng.*, Feb. 1979, pp. 21–27.

The authors

A. Devore   **G. J. Vago**   **G. J. Picozzi**

Abe Devore is a senior thermal-design engineer in the Heat Transfer Div. of C-E Lummus, 1515 Broad St., Bloomfield, NJ 07003. He joined the company in 1976, and has a background of 30 years experience as chief rating engineer with Industrial Process Engineers and Davis Engineering Co. Mr. Devore has published many articles in the chemical engineering literature. He has a B.Ch.E. from Cooper Union and an M.Ch.E. from Brooklyn Polytechnic Institute. He is a member of AIChE and its Energy Conversion Division.

George J. Vago is a senior design engineer in the Heat Transfer Div. of C-E Lummus, Bloomfield, N.J. His work includes the design of heat-transfer equipment, air-cooled and shell-and-tube heat exchangers. Mr. Vago has an M.S. in mechanical engineering from Czech Technical University (Prague), and is a member of ASME and the ASME-Heat Transfer Div. Committee for heat-transfer equipment.

G. J. Picozzi is a senior project estimator with C-E Lummus, Bloomfield, N.J. He is the equipment specialist supervisor for the Bloomfield Division, which he joined in 1972. Mr. Picozzi has a B.S. in mechanical engineering from Georgia Institute of Technology.

Materials of construction

Choosing construction materials is always difficult.
It involves weighing many factors to achieve an optimum
balance between short-term costs and long-term profits.
Heat exchangers present a doubly difficult problem.

B. N. Greene, C-E Lummus

☐ In the design of pressure vessels, tanks, piping and pumps, the engineer usually need only consider one set of operating conditions. Materials selection then normally proceeds routinely.

With a heat exchanger, however, two sets of operating conditions must be taken into account. Further, one set may influence the second, creating, in effect, a third. An example of this is an exchanger cooling with water the residuum from a vacuum distillation unit. If the water is from a cooling tower, it could lower the tube metal temperature enough so that the residuum would start to congeal, changing it into a liquid-solid mass. The residuum might also raise the tube metal temperature sufficiently to cause the solids in the cooling-tower water to deposit on, and corrode, the inside of the tubes.

The most frequent cure for this problem is to resort to warm or tempered water, low in solids content, to prevent congealing. The point is that the character of one process stream can modify that of the other. In this case, the process had to be altered. In others, the material of construction or a mechanical design feature may have to be changed.

Complexity of exchanger design

Although selecting materials for equipment other than heat exchangers is difficult, the factors involved can be evaluated separately. Corrosion resistance at all possible operating temperatures, suitability of structural strength (including against vacuum), welding ease, availability, conformance to all applicable laws, codes and insurance requirements—all must be considered in every instance. Numerous examples can be cited to show that when any one of these factors was not considered, unacceptable materials were selected.

When choosing materials for shell-and-tube exchangers, the engineer follows the same procedures, but must do it again for the second process fluid. This is the case even for air-cooled exchangers, even though virtually every material might be considered suitable for air.

Yet, even this is not enough, for the engineer must also consider whether the material selected for the shellside will be compatible with that chosen for the tubeside.

If, for example, carbon steel were chosen for brackish cooling water on the tubeside, and the shellside condensing phase contained acid, the tube-bundle alloy would have to be reevaluated. In this instance, the shellside contents could be handled by stainless steel, a high-nickel alloy, or titanium. Stainless steel would be the most economical of the three, all of which are resistant to the acid phase.

However, if carbon steel were accepted for the tubeside, and stainless steel for the condensing acid side, a problem of incompatibility would arise. Tubes of carbon steel would not resist the acid corrosion. And tubes of stainless steel might not be acceptable as sufficiently resistant to pitting corrosion or stress-corrosion cracking resulting from the brackish cooling water. Thus, the tubes and tubesheets might have to be fabricated from a more corrosion-resistant alloy than stainless or carbon steel.

Ultimately, this at first apparently simple exchanger might end up having a carbon steel channel, stainless steel shell, and nickel alloy or titanium tubes. The tubesheets might be of the same alloy as the tubes, or might be of carbon steel clad with stainless steel on the shellside.

And the selection procedure is not finished yet. Other factors must be considered. Titanium tubing, because of its excellent corrosion resistance, is often less expensive than some nickel-alloy tubing on a cost per linear foot basis. Titanium's excellent corrosion resistance allows tube walls to be thinner than do many nickel-base alloys. However, titanium's elastic modulus is lower than that of nickel alloys. Therefore, a vibration analysis would have to be performed, and its results might indicate that baffles should be added or that existing baffles should be thickened. Because the baffles would be of an alloy material, the additional cost would be significant.

Additional complicating factors

The materials selection procedure does not end with the evaluation of the two sets of operating conditions, or with the evaluation of the compatibility of the materials for these differing conditions. The joint between the tube and tubesheet may accumulate corrosion deposits and may be depleted of dissolved oxygen. Both factors could lead to crevice corrosion, pitting, or stress corrosion cracking.

Additionally, the stress at the tube roll, and the difference in temperature between the tube and the tubesheet at their joint, as well as the development of unfavorable stress in the remainder of the tube could all undermine the corrosion resistance of the exchanger.

The concentration of corrodents at the tubesheet-tube joint could alter operating conditions. An example of the worst case could be a reboiler returning vapor to a tower from a liquid stream containing as little as 1 ppm chloride. If the liquid level frequently drops below the upper rows of tubes, the fluid will boil, leaving chloride and other potentially corrosive ions on the tubes, and in the tube-tubesheet joints. After the level is restored and the tubes again covered with liquid, the chloride ions will be removed from the tubes. However, they will remain in the tube-tubesheet joint. As this is repeated, the chloride content at the tube rolls will concentrate, reaching several tenths of a percent, and will become extremely corrosive.

Velocity fluctuations also alter conditions inside exchangers. Tubeside velocities are usually uniform and predictable. However, flow on the shellside varies widely from one part of the exchanger to another. Around baffles, it can be highly turbulent, but virtually stagnant at tubesheet and shell junctions. Both stagnant and turbulent streams may cause higher rates of metal loss than a continuous stream flowing at a modest velocity, such as 5 ft/s.

"Shock" or film condensation will also alter operating conditions. When a warm vapor contacts relatively cold tubes (below the dewpoint of the warmer stream), a condensate film, which may be corrosive, will form on the tubes.

This can occur in the so-called "dry condenser" for condensing naphtha while preheating the crude feed to an atmospheric-distillation column. Only hydrocarbon condensate is normally present in this stream. However, if the temperature of the tubes is overlooked, or underestimated, an aqueous condensate may form on the outside of the tubes. This condensate, being acidic, will rapidly corrode carbon steel tubes.

Another such example is the unexpected desublimation of ammonium chloride or ammonium bisulfide. If excessive ammonia is injected into an atmospheric overhead system, the partial pressure of ammonia will increase while that of HCl and H_2S will remain essentially unchanged. This will raise the temperature at which solid NH_4Cl or NH_4HS can exist in stable equilibrium with the vapor stream. In these instances, the temperature of the tubes may be low enough that solid NH_4Cl or NH_4HS may deposit on the tubes. Fouling, even plugging, can follow, resulting in higher metal temperatures, promoting corrosion.

This represents another instance of new conditions being created by the interaction of two other sets of conditions.

Economic influences on materials selection

Of course, heat exchangers, as well as any other process equipment, must be designed with full consideration of cost. However, some additional factors and different economics enter into the design of heat exchangers.

The first economic question that arises with exchangers is whether to select seamless or welded tubing. Both grades are available in carbon steel, nickel steels, 12 Cr steels, stainless steel, many high-nickel alloys, titanium, and some zirconium alloys. In Cr-Mo steels, copper alloys, and some nickel alloys, only seamless tubing is available.

For certain exchanger sizes, the welded grades may be one-third the price of the seamless grades. Usually, this ratio runs about 2 to 1. It, therefore, may be more economical to substitute welded 12 Cr tubing for seamless 5 Cr–1/2 Mo tubing, and obtain greater corrosion resistance and longer life. This approach introduces concern over the integrity of welded tubing. However, welded tubing has proven reliable in service.

In addition to the differences in sizing, tubing differs from piping in availability and cost. Tubing is readily available in thin wall sizes and small diameters, piping in large diameters ($\geq$ 2 in.) and thicker walls (10 Birmingham Wire Gage, 0.134 in., or Schedule 20). Tubing is normally more economical for the heat-transfer surface, and is generally selected.

Tubing, available in many weights and sizes, affords almost total flexibility in size selection. Its cost is roughly proportional to its thickness. Only rarely will tubing thinner than 18 BWG (0.049 in.) be chosen, because of concern about vibration, and difficulty in handling and forming sound tube-to-tubesheet joints. Most often, the 16-BWG (0.065 in.) tube is the minimum wall thickness selected. Additional thickness above that required for internal and external pressure is often necessary as a corrosion allowance. This added thickness is, however, limited by the effect on pressure drop and heat-transfer efficiency.

Design changes caused by materials

The materials engineer usually selects exchanger materials after the process engineer has determined the best process configuration.

Nevertheless, the former can suggest modifications of the exchanger design or of the process, which will improve performance or reduce costs.

One example of this is when alloy selection could affect pressure drop and equipment size. In the case of an exchanger that preheats crude and simultaneously cools a distillation column side-stream (such as atmospheric gas oil), the corrosivity of the streams could be such that carbon steel may be indicated for the shell, channel and associated piping. However, carbon steel tubing may have to be 12 BWG (0.109 in. wall), whereas only 16 BWG (0.065 in. wall) would be required for 5 Cr – 1/2 Mo steel.

Changing to the thinner 5 Cr – 1/2 Mo tubes would, therefore, reduce the weight of the bundle by about 40% and drastically reduce the pressure drop. For turbulent-flow conditions, which normally exist, the pressure drop is essentially proportional to the fifth power of the inside diameter. Therefore, if the pressure drop through a 3/4-in., 12-BWG tube were 10 psi, the pressure drop through a 3/4-in., 16-BWG tube would be only 4.65 psi.

If pressure drop were limiting the design, the change to thinner-walled tubing would allow a reduction in the number of tubes in the bundle. This would, of course, result in a smaller shell and a lighter bundle.

In one instance, a more resistant alloy reduced the initial cost of the exchanger by 30%, as well as enhancing its reliability.

Another common design change that results from materials considerations is the reversal of the roles of the shellside and tubeside. In a water-cooled exchanger, cooling-tower water typically flows on the tubeside, and the process stream on the shellside. This is normally preferred. If, however, the shellside would require an expensive alloy, it may be advisable to alter the design. Shells, being longer than channels, require more material. Therefore, it would be less costly to purchase alloy channels and carbon steel shells.

Of course, the materials engineer would have to confirm that the lower velocity on the shellside would not cause fouling or pitting, and that the higher velocities on the tubeside would not cause erosion-corrosion. With this assurance, the cost of an exchanger could be significantly reduced without affecting its performance.

Velocity limits can also affect materials selection and heat-exchanger design. Often, an overhead condenser will handle a two-phase-flow, containing a corrosive process condensate, at velocities up to 60 ft/s. Many of these condensates are not corrosive to carbon steel at low velocities, or at moderate velocities if the stream is inhibited.

The materials engineer may elect the more costly material in order to retain the exchanger's efficiency and minimize its size. The size reduction would reduce the overall cost of the installed exchanger and could even lower the initial cost of the exchanger itself. This would be particularly true for the airfin condenser, because of the large investment in structural steel that would be required.

In some cases, vibration can influence the selection of materials. Tube walls of corrosion-resistant alloys can be thinner than those of carbon steel tubes. However, thinner walls are more subject to mechanical problems, especially from vibration. The choice will be among decreasing the baffle spacing, increasing the tube thickness, or reversing the operating roles of the shellside and tubeside. Normally, it is easiest to increase the number and thickness of baffles.

Another instance in which materials-related factors may cause a change in equipment design is the desublimation of solids from a vapor. In some overhead condensers, an acid-gas removal system or a sour-water stripper, the partial pressures of ammonia, hydrogen chloride and hydrogen sulfide may be high enough to cause deposition of ammonium chloride or ammonium hydrogen sulfide. In the presence of water, NH_4Cl and NH_4HS deposits are highly corrosive and could plug an airfin condenser. When the air and metal temperatures are low (as in winter), these solids may deposit in the tubes. To prevent this, the engineer may have to specify a recirculating air system that will allow only warm air to cool the tubes. This will reduce the winter efficiency of the exchanger, but not below its summer efficiency. The exchanger, therefore, will still meet all of the design criteria.

The author

Barry N. Greene is a senior materials specialist in the Lummus Technical Center, C-E Lummus, 1515 Broad St., Bloomfield, NJ 07003. For the past five years, he has been involved in all aspects of materials engineering including design, materials selection, fabrication, and failure analysis. He has a B.A. from Adelphi University, and a B.S. and M.S. in metallurgy from Massachusetts Institute of Technology. He is a member of the National Assn. of Corrosion Engineers and the American Soc. for Metals.

Section II
Shell-and-Tube Equipment

Latest TEMA standards for shell-and-tube exchangers

Recent revisions of the TEMA rules will make an important impact on the design and fabrication of this important class of heat exchangers. In many instances, the changes will facilitate their specification.*

Frank L. Rubin, DM International, Inc. (A Davy Company), and
Nathan R. Gainsboro, Southwestern Engineering Co.

☐ TEMA (Tubular Exchanger Manufacturers Assn.) is a trade association comprised of twenty-nine companies that manufacture shell-and-tube heat transfer equipment, primarily for the process industries. The TEMA standards were developed in 1941 and are revised regularly by the Technical Committee.

The standards are not intended to be a "code" in the sense that the ASME standards are, but instead represent the combined experience and judgment of the members in producing functionally and structurally reliable equipment at reasonable costs.

TEMA Standards are often included in bid requirements and purchase specifications for shell-and-tube heat exchangers. They provide for three classes of construction, each with its own level of quality.

The sixth edition (issued in 1978) defines the three classes as: Class R, "for the generally severe requirements of petroleum and related processing applications;" Class C, "for the generally moderate requirements of commercial and general process applications;" and Class B, "for general process service." The standards assure a high quality level in shell-and-tube exchanger fabrication.

By contrast, there are no industry standards for fuel-oil heaters, hot-water heaters and air-cooled heat exchangers. The quality levels vary enormously for these products because of the lack of a published standard that has widespread acceptance.

It is not common practice to apply TEMA standards to this group of equipment because only moderate services are required for hot-water and fuel-oil heating. Exchangers built below TEMA standards will usually be quite satisfactory.

When TEMA standards are referenced by a user's specification, compliance is required. Any exceptions or reinterpretations must be spelled out and agreed upon by the parties involved.

The new edition

The sixth edition contains new material as well as some clarification, re-editing and revision of previously covered topics. Included in the standards are comprehensive requirements for the mechanical design, fabrication and testing of shell-and-tube heat exchangers. In these sections, the experience of fabricators and equipment users has been combined with the ASME Boiler & Pressure Vessel code.

In addition, background information on thermal design, nomenclature and guarantees also appears. Each section has an identifying alphabetical letter that appears with each paragraph. This code letter and a paragraph number enable quick location of a pertinent section of the standard (Table I).

Upon examining the new standards, one notes that the *mechanical* design requirements for all three classes of construction are essentially identical (see box). The principal differences among the classes are summarized in sections R, C and B.

*Portions of this article have been reproduced from "Standards of the Tubular Exchanger Manufacturers Assn.," 6th ed., New York (1978), with permission of the publisher.

Quick guide to important sections of the TEMA standards		Table I
Section	**Number**	**Letter**
Nomenclature	1	N
Fabrication tolerances	2	F
General fabrication and performance information	3	G
Installation, operation and maintenance	4	E
Mechanical standards, Class R	5	R
Mechanical standards, Class C	6	C
Mechanical standards, Class B	7	B
Materials specifications	8	M
Thermal standards	9	T
Physical properties of fluids	10	P
General information	11	D
Recommended good practice*	12	RGP

*Section 12 includes several subparagraphs that address subjects not covered elsewhere

Highlights of the mechanical standards

1. *Scope of standards.* Definition of class, construction codes, testing, corrosion allowances, service limitations.

2. *Tubes.* Lengths, diameters, gages, U-tubes, tube pitch, layout.

3. *Shells and shell covers.* Diameters, tolerances, minimum thicknesses.

4. *Baffles and support plates.* Types, tube-holes, clearances, thicknesses, spacing, U-tube rear support, tube bundle vibration, impingement baffles and erosion protection, tie rods and spacers, sealing devices.

5. *Floating heads.* Internal floating heads, outside-packed floating heads, and externally sealed floating tubesheet.

6. *Gaskets.* Types, materials, peripheral gaskets, pass-partition gaskets, gasket joint details.

7. *Tubesheets.* Thickness, shell and tube longitudinal stresses in fixed-tubesheet exchangers, special cases, tube-holes in tubesheets, expanded tube joints, welded tube joints, pass partition-grooves, pulling eyes, clad and faced tubesheets.

8. *Channels, covers and bonnets.* Thickness, depth, effective cover thickness, pass-partition grooves.

9. *Nozzles.* Construction, installation, pipe-tap connections, stacked units, split-flange design, nozzle loadings.

10. *End flanges and bolting.* Minimum bolt size, bolt circle layout, wrench and nut clearances, bolt type.

Of importance to most chemical engineers, however, are the revisions to the TEMA rules. These revisions generally fall into the following categories. (TEMA paragraph designations are included in parentheses for easy reference.)

Nomenclature (Sec. 1)

This is doubtless one of the most useful and widely accepted features in the standards. It enables all suppliers and users to speak a common language in describing the general configuration of equipment. It also provides a concise selection of the various configurations that are commonly used to meet the needs of particular conditions of service.

In the new standard, exchanger configuration is still defined by a three-letter designation describing the stationary head, the shell and the rear head. The definition for a Type "X" shell has been added (Fig. 2). Type "C" now relates only to removable-bundle exchanger constructions.

Fabrication tolerances (Sec. 2)

Compliance with the recommended tolerances assures the purchaser that the delivered equipment will fit in place with respect to piping connections and foundation placement. These tolerances represent good fabrication practice within the limits of reasonable care and cost. Shop inspection is facilitated by consistent application of these rules.

The revisions include a rotational tolerance on nozzle faces of ±1 deg. This minimizes misalignment with prefabricated piping. Also, for nozzles over 36 in. there is a maximum misalignment factor for the flange surface. The other revision of importance is a clarification of "confined gasket" requirements.

Supplemental nameplate data (G-3.12)

Special design/operating limitations are often built into an exchanger, and these data should be included for the guidance of operating personnel. The two most important items of this nature are (1) differential design and test pressure, and (2) differential temperature limitations for fixed-tubesheet exchangers.

Although these items are covered in the new specifications under supplemental nameplate data, there are broader implications. Item (2) affects thermal stresses in the tubes and shell, as well as tube-to-tubesheet joint loads. In the case of differential-pressure design cases, it is vital that field personnel be aware of hydro-test precautions.

Guarantees (G-5)

Thermal performance tests are unlikely to prove meaningful because actual fouling resistances are unknown. Damage from flow-induced vibration is not covered under the warranty because of the lack of a definitive prediction method. However, most fabricators will respond to a problem if it occurs. Under the new standard, by agreement, specific guarantee terms may replace the published ones.

The guarantee section has very few changes. These include new criteria covering the supports required for vertical units. Seismic design appears for the first time; the purchaser is required to specify in the inquiry any seismic requirements that are to be considered in the design.

TEMA has chosen to retain the statement that "the manufacturer assumes no responsibility for deterioriation of any part of the equipment . . . . due to flow-induced vibration." This first appeared in the 5th edition. The organization is conducting research in this area in the hope of solving this problem.

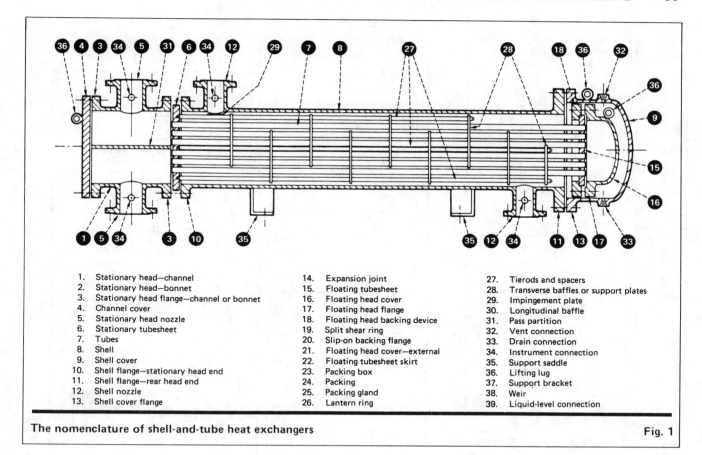

1. Stationary head—channel	14. Expansion joint	27. Tierods and spacers
2. Stationary head—bonnet	15. Floating tubesheet	28. Transverse baffles or support plates
3. Stationary head flange—channel or bonnet	16. Floating head cover	29. Impingement plate
4. Channel cover	17. Floating head flange	30. Longitudinal baffle
5. Stationary head nozzle	18. Floating head backing device	31. Pass partition
6. Stationary tubesheet	19. Split shear ring	32. Vent connection
7. Tubes	20. Slip-on backing flange	33. Drain connection
8. Shell	21. Floating head cover—external	34. Instrument connection
9. Shell cover	22. Floating tubesheet skirt	35. Support saddle
10. Shell flange—stationary head end	23. Packing box	36. Lifting lug
11. Shell flange—rear head end	24. Packing	37. Support bracket
12. Shell nozzle	25. Packing gland	38. Weir
13. Shell cover flange	26. Lantern ring	39. Liquid-level connection

The nomenclature of shell-and-tube heat exchangers   Fig. 1

Mechanical standards (Sections 5, 6, 7)

Scope (R, C, B-1.11). There is a new definition (none previously existed) that limits the standards to services where considerable fabrication and operating experience exists. All tables have been limited to units having a 60-in. shell diameter.

Some designers have been attempting to apply TEMA standards to very-large-diameter and/or high-pressure units. These are beyond the scope of the standards and the extrapolations could lead to problems. The new definition provides an explicit scope (Fig. 3):

"The TEMA Mechanical Standards are applicable to shell-and tube heat exchangers with inside diameters not exceeding 60 inches, a maximum product of nominal diameter (inches) and design pressure (psi) of 60,000, or a maximum design pressure of 3,000 psi. The intent of these parameters is to limit shell wall thickness to approximately 2 inches and stud diameters to approximately 3 inches. Criteria contained in these Standards can be applied to units constructed with larger diameters. For units outside this scope, refer to Section 12, Recommended Good Practice. Subsections and paragraphs with additional coverage in Section 12 are identified by an asterisk."

Tubesheets (R, C, B-7)

This section contains rules for the design of a pressure part not covered by the ASME Code, Sec. VIII, Div. 1. It also defines shell-and-tube axial stresses in fixed-tubesheet exchangers. Although U-tube and floating-head units are fairly straightforward, fixed-tubesheet units involve additional complications:

■ The tubes act as an elastic foundation for the tubesheets, which furnish tube support.

■ Differential thermal growth between the shell and the tubes may increase or decrease the load on the tubesheets.

■ If the tubesheet is extended to bolt to an adjacent flange, the bolting moment adds another load to the tubesheet.

All of the above factors are considered in the calculation of an "effective design pressure" for the shellside and the tubeside. Use of an expansion joint often is required to minimize thermal expansion loads and stresses.

Fixed tubesheets (R, C, B-7.19)

Calculation of effective differential expansion and design pressures now incorporates a value of the J factor for expansion joints that have spring constants greater than zero (i.e., the expansion-joint stiffness is now taken into account). J can now vary between zero and one, whereas previous methods set J equal to zero *or* one. (J is defined as the ratio of the spring rate of the expansion joint to that of the shell and expansion joint together.)

Shell-and-tube longitudinal stress formulas are now located in this section and have been revised to add a "C" factor that cuts calculated stress in half when tension exists. Also, a new procedure is presented for calculating allowable tube compressive stress.

Since the methods presented for design of tubesheets are based on certain simplifying assumptions, a paragraph has been added to call attention to various spe-

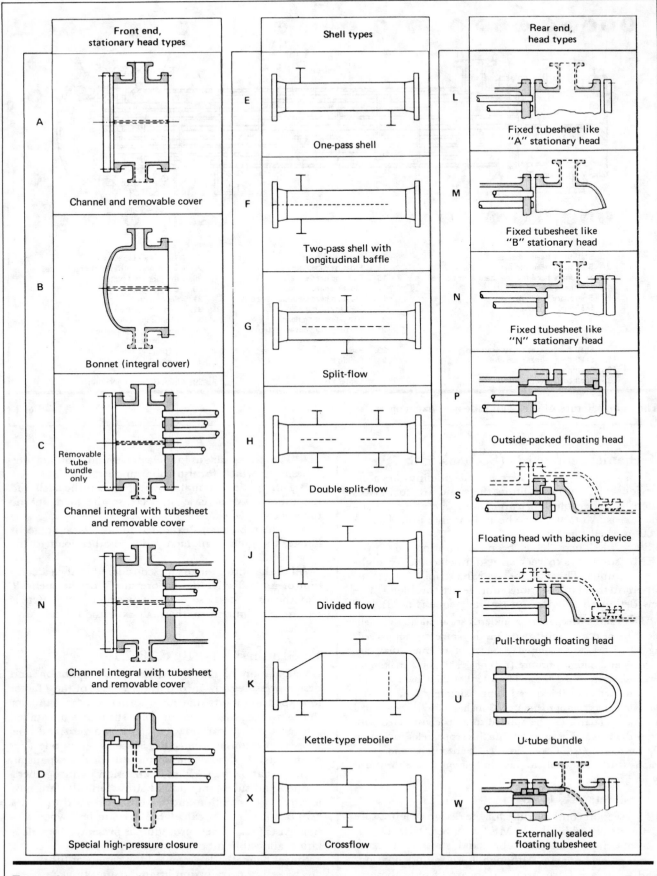

Types of heads and shells commonly used in heat exchangers   Fig. 2

cial cases that may require more rigorous or additional analysis.

Packed floating head (R, C, B-7.161)

For Type P designs, the formula for calculating effective design pressure has been replaced by two new formulas for bending and shear. These result from definitive analyses of this outside-packed configuration. They assure proper selection of tubesheet thickness.

Tube-hole grooves (R, C, B-7.44)

Although used almost universally, tube-hole grooves in tubesheets serve only to increase the holding effect of the tube-to-tubesheet joint. They do not increase sealing capability, and in fact may actually worsen it, as compared to a smooth tube-hole.

The tube-projection allowance (from the tubesheets) has been liberalized to allow a maximum of one-half the tube diameter. The old requirement was $\frac{1}{8} \pm \frac{1}{16}$ in.

Tube-holes in baffles (R, C, B-4.2)

Tube holes in widely spaced baffles or tube supports are still drilled $\frac{1}{64}$ in. over the outside diameter of the tubes. The new standard specifies $\frac{1}{32}$ in. at all times for tubes larger than 1.25 in. outside diameter (paragraph R, C, B-4.2). This makes fabrication easier than in the past, when only $\frac{1}{64}$ in. was required for such spacing.

All drilled holes are larger in diameter than the drill bit. The new standard adds a maximum over-tolerance of 0.010 in. on holes drilled in baffles and tube supports.

Bare tubes (R, C, B-2.21)

The standard now allows the use of "average wall" tubes, one BWG (Birmingham Wire Gauge) number heavier than the recommended "minimum-wall" tubing. The previous edition required the average-wall tube to be at least as thick as the recommended minimum wall.

This change represents a considerable simplification for purchasers since it eliminates the problem of average-wall tube thickness. In the past, there was difficulty in relating an average-wall tube to a minimum-wall tube requirement.

Baffles; erosion protection (R, C, B-4.6)

This important section can be confusing. Entrance and exit areas for both shell and bundle are separately and explicitly defined. (Note that potential tube vibration is not considered in the numerical values of velocity head.) The design details of the entrance and exit configurations must be considered carefully to comply with these requirements and to assure the structural integrity of the bundle under all known service conditions.

Transverse baffles. The multi-segmental type of baffle or tube-support plate, as well as the single-segmental baffle, are now defined as standard. Before this edition, only single-segmental baffles were standard.

Impingement baffles. Entrance and exit areas for both the shell and the bundle are now redefined to reflect the intent of the standards. Previous editions were not too clear on these points, which are of great importance in assuring the performance and integrity of the tube bundle.

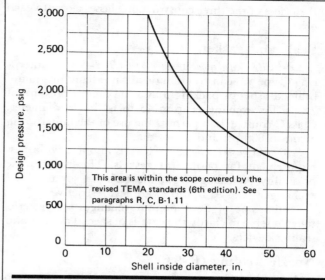

The range of heat exchangers now covered by TEMA standards   Fig. 3

In addition, both double-segmental and triple-segmental baffles are now included as standard constructions. Diagrams of these are provided. (Previously, only single-segmental baffles were standard.)

Clearances (R, C, B-4.3)

The clearances specified are intended to limit fluid leakage around transverse baffles while maintaining practical fabrication capability. Note that these clearances may be doubled for cases where the shellside coefficient or the MTD is not affected. Such cases include essentially isothermal boiling or condensing.

Note also that the clearance is based on the "design" I.D. of the shell—not the actual I.D., which may be up to $\frac{1}{8}$ in. larger as per paragraphs R, C, B-3.122. No tolerance has been prescribed for the specified baffle and support clearances.

In the case of pipe shells, there could theoretically be a condition where the I.D. of the pipe is smaller than the O.D. of the baffles. However, pipe usually runs thinner than nominal, and such a mis-match would be rare. In any case, it is up to the manufacturer to provide a correct fit.

Maximum spacing of supports (R, C, B-4.52). The allowable spans are based on limiting the deflection of the tubes to a fraction of the ligament when they are filled with water. There is no correlation of these values with potential tube vibration effects, which must be considered separately.

Postweld heat treatment (R-5.12, R-8.14)

A significant change that results in cost savings is the deletion of the mandatory heat-treatment requirement. Heat treatment of welded carbon-steel channels and floating heads in Class R service was required in the past. Updated welding methods, primarily automatic, do not create residual stresses of a magnitude comparable to those induced by older techniques. Classes B and C never required heat treatment. This change brings the prices of the three construction classes closer to each other.

Seals and gaskets

Externally sealed floating tubesheet (R, B-5.3). Lantern-ring (Type W) construction details now call for square-braided asbestos packing and a maximum pressure of 75 psi for exchangers from 43 to 60 in. in diameter. Class C construction permits the use of the lantern ring at up to 600 psi and 60 in. dia.

Peripheral gaskets (R-6.3). There is also a new flatness requirement for peripheral gaskets. This reduces the possibility of leakage from the gasketed joints of the exchanger. Classes B and C are less restrictive.

While some of the remaining changes are esoteric, they do serve to clarify the details of special designs and sealing devices.

Nozzle loading (R-9.6)

A mostly self-explanatory but important paragraph is the one on nozzle loadings. Heat exchangers are not intended to serve as anchor points for piping; therefore, the purchaser is responsible for minimizing all piping loads to the heat exchanger.

TEMA is taking steps to protect internal components as well.

Area requirements for the shell or bundle entrance are given in terms of the product of density and velocity squared. Because normal process-line velocities are greater than the allowable nozzle velocities, a piping transition from high velocity to a lower velocity at the shell inlet often is required. The new standard states that "a properly designed diffuser may be used to reduce line velocities at the shell entrance."

Material specifications (Sec. 8)

In some cases, new materials are being used in the manufacture of TEMA exchangers. Technical information on these materials is now listed in the standards. Included are data on titanium, zirconium, Alloy 20, welded carbon steel, welded alloy steel, welded copper, and copper alloys. The standard also gives additional data on thermal conductivity and modulus of elasticity.

Thermal standards (Sec. 9)

With the increased sophistication of computerized thermal-rating methods, TEMA has gradually phased out that portion of the standards dealing with calculation of film coefficients and pressure losses. Only the basic relationships are included. However, these are more rigorous and include a list of conditions necessary to establish the validity of the log-mean temperature difference and associated correction factors. Attention is called to the temperature efficiency charts that enable rapid estimation of exchanger performance under various operating loads or conditions.

The section on fouling resistances, developed long ago, remains the only guide of its type and has proven quite useful over the years. There has been no significant change in these resistances even though fouling is now known to be affected by many factors.

In general, thermal standards have been modified to be more rigorous.

Tube wall resistances are given for both bare and integral-fin tubes. Although fouling resistances are little changed, the table of resistances for water has been modified to delete specific river waters and to add closed-cycle condensate.

The section on fluid-temperature relations has a number of changes. The text has been rewritten to reflect current knowledge and to better define the limitations of the LMTD (log-mean temperature difference) correction-factor charts. A number of selected references are listed for cases not covered by the charts.

Installation, operation, maintenance (Sec. 4)

Industry terminology is changing and this is reflected in the standards. For example, the term "vent valves" replaces "vent cocks," and "pulsation and vibration" replaces "surge drums."

This section also includes warnings about hazardous and toxic materials. There are new details on shutdown operations and instructions for drainage of exchangers. Tube bundle removal and handling are dealt with, as well as how to protect gaskets and packing-contact surfaces.

Other changes alter the standards for safe loading of eye-bolts and lifting and pulling mechanisms. These significant revisions make the standard an excellent basic instruction manual for operating personnel.

General information (Sec. 11)

A number of improvements and additions have been made to this section. The principal changes are:

1. Bolting data (Table D-5) have been extended to four-inch bolt size.

2. The external working-pressure table for tubing has been deleted because of Code changes that require a more rigorous calculation.

3. Material-property tables have been updated and expanded.

A highlight of this section is a re-done table of conversion factors relating English and SI/metric units in a very clear manner. The table will undoubtedly be expanded as additional questions arise from users of the standards. A gradual conversion to the SI/Metric system is contemplated, and toward this end, the table provides an outstanding, compact reference.

Recommended good practice (RGP)

This completely new section extends some of the tables to larger shell diameters (100 in.) and furnishes additional guidance on topics not specifically covered in the main sections of the standards. It is, in a sense, a nonmandatory appendix to the code.

The largest single portion of the RGP Section is devoted to the subject of tube vibration. It supplies basic information, especially regarding tubes' natural frequencies. No attempt is made to define the probability or likelihood of flow-induced vibration, because a definitive procedure is not yet a reality. Included is an excellent list of references that deal with vibration problems and solutions. Also, there is a brief discussion of various mechanisms that cause heat-exchanger tubes to vibrate.

Shell-and-tube heat exchangers

Many factors affect the design of this equipment. Here are the construction details, and guidelines for placing streams on the shellside or tubeside, based on their particular physical properties.

Davinder K. Mehra, Brown & Root, Inc.

☐ Shell-and-tube heat exchangers are the most common type of heat-transfer equipment in the chemical process industries (CPI). Their mechanical design is governed by Tubular Exchanger Manufacturers Assn. [1], American Soc. of Mechanical Engineers [10] and American Petroleum Institute [11] standards. These usually complement one another. TEMA standards are widely accepted worldwide for the thermal and mechanical integrity of shell-and-tube exchangers.

TEMA standards have a convenient system for designating shell-and-tube exchangers. For example: a fixed-tubesheet exchanger with removable channel and cover, bonnet-type rear head, two-pass shell, 33-1/8 in. I.D., with tubes 8 ft long, will be designated as:

Size 33-96 Type AFM

Here, the number 33 is the nominal inside diameter in inches, rounded off to the nearest integer, and the number 96 is the nominal length of the tubes in inches. For straight tubes, the tube length is the actual overall length, and for U-tubes, the tube length is the length from the end of the tube to the bend tangent. The three letters, AFM, represent the front end (stationary head type), the shell type, and the rear end (head type), respectively.

For this example, A represents a channel and removable cover, F represents a two-pass shell with longitudinal baffle, and M represents a fixed-tubesheet rear end. Other arrangements are covered in the TEMA standards [1]. These designations are well established as a means of communication between users, engineering contractors, and manufacturers of the equipment.

Removable tube bundles

There are two basic types of construction, which are related to service conditions: removable and nonremovable bundles. The salient features of each type and their applications will be discussed in this article in rela-

tion to maintenance, performance and economy [2,3,4].

There are two types of removable bundles: one made with U-tubes, and the other of straight tubes, in several variations, for handling the floating head.

U-tube exchangers

U-tube exchangers provide the simplest construction for removable bundles (Fig. 1a) and, hence, are the most economical. The exchanger consists of straight tubes bent in the form of a U, and attached to the tubesheet. The U-bend section of the bundle is free to expand in the shell, and the need for an expansion joint, when thermal relief is required, is eliminated.

Advantages

1. Can be used in services requiring extremely high pressures on one side.

2. Inlet nozzle on shell can be located beyond U-bends. This eliminates the need for an impingement plate that would otherwise be necessary.

3. Least expensive of all types of shell-and-tube exchangers.

Disadvantages

1. Suitable for clean services only, because U-bends are difficult to clean by mechanical methods.

2. Individual tubes are generally difficult to replace except for those in the outer rows.

3. Tube passes are restricted to even numbers only. (This is not a serious disadvantage because most exchangers have an even number of passes.)

Floating-head exchangers

Floating-head exchangers are the most sophisticated type of shell-and-tube heat exchangers used in those CPI plants where regular maintenance is done. These exchangers have straight tubes, secured at both ends to tubesheets. One tubesheet is free to move, providing for differential thermal expansion between the tube bundle and shell.

Originally published July 25, 1983

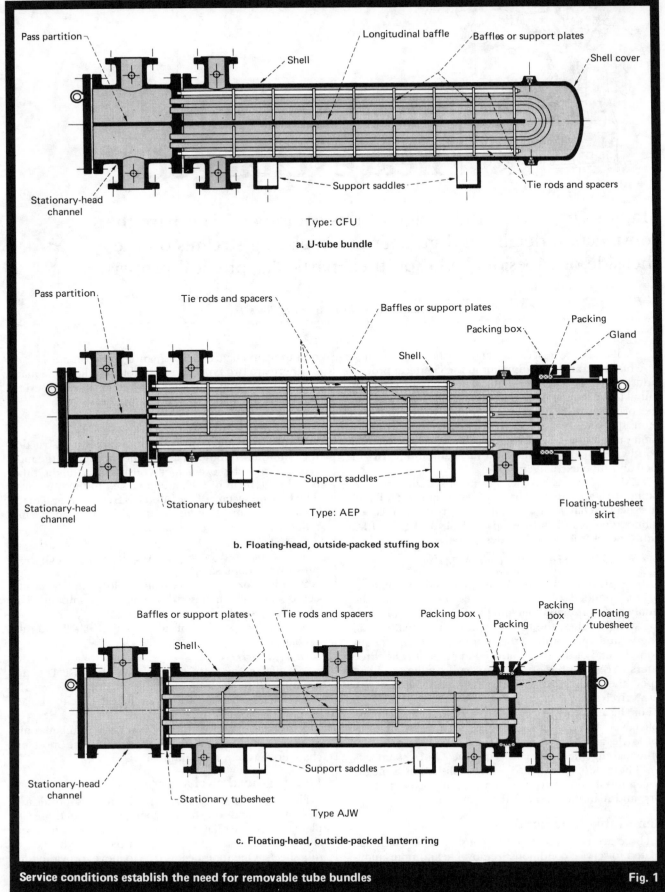

Type: CFU

a. U-tube bundle

Type: AEP

b. Floating-head, outside-packed stuffing box

Type AJW

c. Floating-head, outside-packed lantern ring

Service conditions establish the need for removable tube bundles

Fig. 1

TEMA [1]

Tube bundles may be removed for inspection, replacement, and external cleaning of the tubes. Likewise, tubeside headers, channel covers, gaskets, etc., are accessible for maintenance and replacement, and tubes may be cleaned internally. The two basic types differ in the method of sealing the shellside and tubeside fluids—one method relies on an outside-packed stuffing box; the other, on an outside-packed lantern ring.

Outside-packed stuffing box—The shellside fluid in this exchanger (Fig. 1b) is sealed by rings of packing compressed within a stuffing box by a packing-follower ring. The packing allows the floating tubesheet to move back and forth. Since the stuffing box contacts only the shellside fluid, there is no mixing of shellside and tubeside fluids if leakage occurs through the packing. This arrangement is limited to (a) shellside services up to 600 psig and 600°F, and (b) nonhazardous and nontoxic services—because leakage of shellside fluid to the environment is possible.

Outside-packed lantern ring—The shellside and tubeside fluids are each sealed by separate rings of packing, or O-rings, (Fig. 1c) and separated by a lantern ring provided with weep holes. Hence, leakage through either packing will be to the outside. The width of the tubesheet must be sufficient to accommodate the length of the two packings and the lantern ring, plus whatever length is needed for differential thermal expansion between the exchanger's shell and tube bundle.

Sometimes, a small skirt is attached to the floating tubesheet to provide a bearing surface for the packings and lantern ring. This design is limited to (a) 150 psig and 500°F and (b) one or two tube passes. It is not applicable in services where leakage of either tubeside or shellside fluid is unacceptable, or where possible mixing of tubeside and shellside fluids cannot be tolerated.

Pull-through bundle

A separate bundle is bolted directly to the floating tubesheet, as shown in Fig. 2a. Both the assembled tubesheet and head are small enough to slide through the shell. The tube bundle can be removed without breaking any joints at the floating end. Clearance requirements (the largest for any type of shell-and-tube exchanger) between the outermost tubes and the inside of the shell must provide for the gasket bolting at the floating tubesheet.

For an odd number of passes, a nozzle must extend from the floating-head cover through the shell cover. Provision for both differential thermal expansion and tube-bundle removal requires, for example, packed joints or internal bellows.

This design is very expensive, and its application restricted to services where failure of the internal gasket (between the floating tubesheet and its head) is acceptable—i.e., where mixing of shellside and tubeside fluids does not present a hazardous situation.

Inside split backing-ring

The floating cover is secured against the floating tubesheet by bolting it to a well-secured split backing-ring (Fig. 2b). This closure, located beyond the end of the shell, is enclosed by a shell cover of larger diameter. The shell cover, split backing-ring and floating-head cover must be removed for the tube bundle to slide through the shell.

Clearances between the outermost tubes and the inside of the shell (about the same as those for outside-packed stuffing boxes) approach the inside diameter of the gasket at the floating tubesheet. This type of construction is ideal for high shellside pressures and temperatures where service conditions require removable tube bundles. For an odd number of tube passes, a nozzle must extend from the floating-head cover through the shell cover.

With the exception of special designs, this is the most expensive type of shell-and-tube heat exchanger.

Nonremovable tube bundles

The most popular heat exchanger in the CPI is the fixed-tubesheet unit (Fig. 3). This design has straight tubes secured at both ends to tubesheets welded to the shell. The exchanger can be designed with removable channel covers (TEMA Type AEL), bonnet-type channels (TEMA Type BEM), or integral tubesheets (TEMA Type NEN).

Advantages

1. Maximum protection against leakage of shellside fluid to the environment.

2. Minimum shell diameter of all shell-and-tube heat exchangers for a given heat-transfer surface, with the same diameter, length and number of tubes, and tube passes.

3. Relatively inexpensive.

Disadvantages

1. Shellside not accessible for mechanical cleaning, hence, application limited to clean service on shellside.

2. Expansion joint required in shell to relieve stresses due to differential thermal expansion.

Shell designs

Various shell designs are available for heat-transfer applications. These can be broadly classified as: single-pass, two-pass, split-flow and divided-flow shells.

Single-pass (TEMA Type E) is the most common design. However, use is restricted to service conditions where there is no temperature cross.

Two-pass (TEMA Type F) is used for services where temperature cross is unavoidable due to process considerations, and where space limitations exclude the use of two, or more, shells in series. This shell is available with removable and nonremovable longitudinal baffles.

A major problem with removable baffles is fluid leakage through the clearance between the longitudinal baffle and shell. In extreme situations, this clearance can make the unit ineffective in its intended service. The usual practice for this type of unit is to provide longitudinal seal strips (bars) to prevent fluid leakage, or weld the baffle to the shell.

Split-flow and divided-flow (TEMA Types G, H, J and X) are used in services where shellside heat transfer is not controlling, and low shellside pressure drop is desired.

Mechanical design

The mechanical design of most shell-and-tube heat exchangers is governed by standards *[1,11,12]*. Surface condensers for steam must meet the standards of the

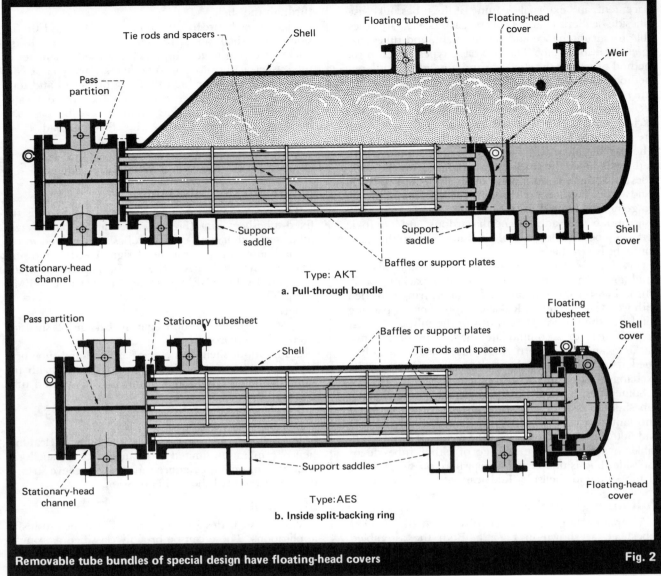

Type: AKT

a. Pull-through bundle

Type: AES

b. Inside split-backing ring

Removable tube bundles of special design have floating-head covers   **Fig. 2**

TEMA [1]

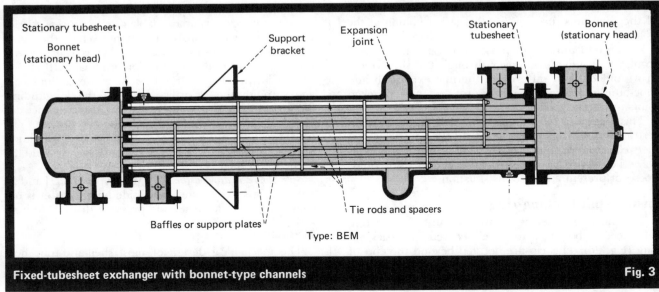

Type: BEM

Fixed-tubesheet exchanger with bonnet-type channels   **Fig. 3**

TEMA [1]

Heat Exchange Institute (HEI) [12]. Some proprietary designs may not be covered by these standards.

The TEMA mechanical standards are applicable to shell-and-tube heat exchangers having: inside diameters not exceeding 60 in.; a maximum product for the nominal diameter (in.) multiplied by the design pressure (psi) of 60,000; or a maximum design pressure of 3,000 psi. The intent of these parameters is to limit shell thickness to approximately 2 in., and stud diameters to approximately 3 in. The criteria contained in the standards can also be applied to units constructed with larger diameters.

The TEMA Standards define three classes (R, C and B) of heat exchangers for processsing applications in the CPI. Each class is specified in terms of design, fabrication and materials of construction for unfired shell-and-tube heat exchangers, based on severity of service. In brief:

TEMA Class R Standards satisfy the requirement of petroleum-processing and related applications.

TEMA Class C Standards satisfy the moderate requirements of commercial and general process applications.

TEMA Class B Standards satisfy the requirements for chemical process services.

Rubin [5] presents a table showing the differences among the three classes of exchangers.

Materials of construction for shell-and-tube heat exchangers are selected on the basis of compatibility with process fluids, anticorrosion properties, economics, and client preference.

Tubeside materials

Tubes—ASME specifications are used for designating tube material. Tube sizes used in the CPI are 3/4, 1, 1 1/2, and 2 in. Tubewall thicknesses range from 18 to 10 Birmingham Wire Gage (BWG). Preferred tube lengths are 8, 10, 12, 16 and 20 ft. Other lengths may be used if delivery and handling are not a major factor. In such cases, even tube lengths are preferred.

The minimum preferred tube pitch is 1.25 times the outside diameter of the tube. However, for tubes in a square pattern, a minimum lane of 1/4 in. is required for mechanical cleaning.

Tubesheets—These elements perform the important function of separating the shellside and tubeside fluids, and provide the anchor point for tube ends. Tubesheets are usually machined from material similar to that specified for tubes. However, for stainless steels and other alloys for the tubes, the tubesheet may be fabricated from carbon steel, and cladded with the alloy material.

Channel and floating-head covers, and channels—The metallurgy of these components must be compatible with the tubeside fluid. Cladding, lining, or weld-overlay on base metal may be used to reduce costs for stainless steels or other alloys.

Shellside designs

Shells, shell cover and vapor belt—These components are usually made from carbon steel, unless process conditions and mechanical-design considerations mandate exotic metallurgy. For shell sizes up to 24 in., standard pipe is preferred for economic reasons. Above 24 in., the components are fabricated from metal plates. Cladding or lining can reduce the cost of alloy exchangers.

Baffles and support plates—The purposes of baffles are to: (1) deflect the shellside fluid over the tubes in its passage through the exchanger, and (2) act as tube supports to maintain tube pitch, prevent sagging of tubes, and prevent flow-induced vibrations. Materials of construction for baffles are usually the same as for the shell. Table I lists some of the common baffle types.

Other shellside components

Tie rods and spacers are used to retain all transverse baffles and tube-support plates securely in position. Their materials of construction are similar to those for baffles. The TEMA Standards for Class R, C and B set out the tie-rod diameter and minimum number to be used.

Sealing strip and dummy tubes are used to prevent excessive bypassing of the fluid around or through the tube bundle. This ensures that the shellside fluid passes effectively over the tube bundles in crossflow. Sealing strips are not usually required in fixed-tubesheet exchangers, or where bundle-to-shell clearance is less than 1 in. Materials of construction are similar to those for baffles.

Impingement plates are located in the shellside inlet-flow area to prevent suspended solids or two-phase mixtures from impinging on the tubes to cause damage by ero-

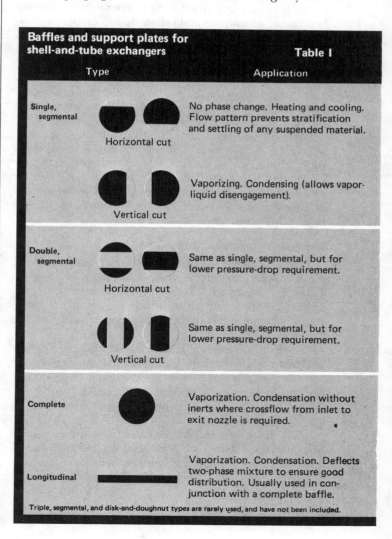

Baffles and support plates for shell-and-tube exchangers — Table I

Type	Application
Single, segmental — Horizontal cut	No phase change. Heating and cooling. Flow pattern prevents stratification and settling of any suspended material.
Single, segmental — Vertical cut	Vaporizing. Condensing (allows vapor-liquid disengagement).
Double, segmental — Horizontal cut	Same as single, segmental, but for lower pressure-drop requirement.
Double, segmental — Vertical cut	Same as single, segmental, but for lower pressure-drop requirement.
Complete	Vaporization. Condensation without inerts where crossflow from inlet to exit nozzle is required.
Longitudinal	Vaporization. Condensation. Deflects two-phase mixture to ensure good distribution. Usually used in conjunction with a complete baffle.

Triple, segmental, and disk-and-doughnut types are rarely used, and have not been included.

sion. Guidelines are detailed in the relevant section of the TEMA Standards [1].

Thermal design and rating

The thermal design of shell-and-tube heat exchangers is a complex function of various parameters that sometimes work against each other. These parameters can be broadly classified as related to physical properties of the process streams undergoing heat transfer, and to the geometry of the exchanger. For a given set of conditions, there are many alternatives for selecting heat-transfer equipment for a specified service. The following discussion covers the several parameters and their influence upon the design and rating of a shell-and-tube exchanger for a specified service.

Process data

Process data for the thermal design and rating of shell-and-tube heat exchangers must be listed on a specification sheet, such as that in the TEMA Standards [1]. All data on the fluid streams must be given, because incomplete data lead to delays in designing the heat exchanger, and to assumptions by a designer that may be totally off the mark. Heating and cooling curves are critical in the design of condensers and reboilers.

More exchangers have been known to fail due to errors in computing an effective temperature difference because of incomplete data than due to errors in computing heat-transfer film coefficients. For multiple duties in an exchanger, a separate data sheet must be prepared for each service, so that the unit can be checked for all known operating conditions.

Density, viscosity, specific heat and thermal conductivity of the process streams undergoing heat transfer have the most profound effect on the size and type of heat exchanger, because such data enter directly into the calculations for heat-transfer film coefficients and pressure drop. Hence, it is imperative that physical-property data be as accurate as possible. Najjar and others [6] give an excellent treatise on the impact of physical-property data on heat-transfer film coefficients and, hence, on the design of heat-transfer equipment.

Geometry effects: tubeside

Tubeside diameter—Thermal efficiency and economic considerations require as small a diameter as possible. However, the practical limit is the requirement for mechanical cleaning of tubes, which necessitates a minimum diameter of 3/4 in. for straight tubes, and 1 in. for U-tubes.

For clean streams, or if chemical cleaning is feasible and tubeside pressure drop is not critical, smaller-diameter tubes may be used—with the potential of significant improvement in heat-transfer performance.

The smallest tube size normally considered for a process heat exchanger is 5/8 in., although there are applications where 1/2, 3/8, or even 1/4-in. tubes are the best selection. Tubes of 1 in. dia. are normally used when fouling is expected because smaller tubes are impractical to clean mechanically. Falling-film exchangers and vaporizers generally are supplied with 1 1/2 and 2-in. tubes.

Tube length—Generally, the longer the tubes, the lower the cost of a given surface, because a smaller number of tubes is required and the size of tubesheets and flanges is reduced. (The maximum tube length that can be handled by most manufacturers is about 40 ft.) However, this does not imply that the longest possible tube always produces an optimum exchanger, as other length-dependent factors will influence the efficiency of the overall design.

Tube lengths of 20 ft are often considered a maximum where tubeside mechanical cleaning is required. This limit is imposed by the practical length of cleaning drills. On the other hand, short tube length will be inherently expensive, and may result in an unnecessarily large number of tube passes, small baffle spacings, and high maintenance-labor costs.

Tube arrangement—Tubes are arranged in triangular, square or rotated-square pitch. Triangular tube-layouts result in better shellside coefficients and provide more surface area in a given shell diameter, whereas square-pitch or rotated-square-pitch layouts are used when mechanical cleaning of the outside of the tubes is required. Sometimes, widely spaced triangular patterns facilitate cleaning. Both types of square pitches offer lower pressure drops, but lower coefficients, than triangular pitch.

Geometry effects: shellside

The shellside heat-transfer coefficient and pressure drop depend on the flow pattern of the process stream (i.e., laminar, transition, turbulent) and on the flow distribution across the tube bundle.

The tolerances and clearances necessary for the mechanical design of the exchanger—such as tube-to-baffle hole, baffle-to-shell, bundle-to-shell, and the orientation of pass partitions—lead to fluid bypassing the tube bundle. In extreme cases, excessive tolerances and clearances can render the unit ineffective.

To achieve good shellside heat transfer, bypassing of the fluid on the shellside must be reduced or eliminated by using: proper baffle cutouts and baffle spacing, sealing strips, and proper arrangement of pass-partition lanes along with dummy tubes.

In many cases, and especially for shellside laminar flow, it may be necessary to specify the tightest mechanical tolerances in order to minimize fluid bypassing, which tends to be large in the laminar-flow regime. In turbulent flows, large momentum-change effects cause less leakage and bypassing relative to crossflow, and better mixing reduces the harmful effect of such streams on the mean temperature difference—allowing somewhat looser tolerances. For condensing vapors, the effects of fluid bypassing are relatively unimportant.

Baffles establish the flow path of the shellside fluid, which depends on the type and arrangement of the baffles (Table I). The minimum baffle spacing recommended by TEMA [1] is 0.2 of the shell diameter, or 2 in., whichever is greater. The maximum baffle spacing is such that the unsupported tube length does not exceed the value indicated for TEMA Class R, C and B exchangers for the tube material used [1]. The baffle cut is specified as a percentage of the shell diameter. Size of the baffle cut is determined by a combination of the effects of pressure drop vs. heat transfer.

Sealing strips prevent bypass around a bundle by blocking the clearance area between the tubes and the inside of the shell. Common types include:

- Tie rod and spacers that hold the baffles in place but that can be located on the periphery of the baffle to prevent bypassing.
- Longitudinal sealing strips extending from baffle to baffle.
- Tie roads with "winged" spacers. The wings are extended longitudinal strips that are attached to the spacers.

Dummy tubes are tubes that do not pass through the tubesheet. Generally, closed at one end, they are used to prevent bypassing through lanes parallel to the direction of fluid flow inside the bundle. Strips or tie rods and spacers can also be used to prevent bypassing through a bundle.

Tube-pass arrangement represents the number and orientation of pass-partition channels that directly affect the heat-transfer efficiency and pressure drop on the shellside. Pass-partitions parallel to the flowpath produce channels for short-circuiting flow between baffle tips—resulting in inefficient heat transfer. Therefore, the number of pass-partitions parallel to the flow must be minimized.

Fluid allocation

Many factors must be taken into consideration in order to determine which fluid should be on the shellside and which on the tubeside of an exchanger. These are:

1. Viscosity—Higher heat-transfer rates are usually obtained by placing a viscous fluid on the shellside.

2. Toxic and lethal fluids—Generally, the toxic fluid should be placed on the tubeside, using a double tubesheet to minimize the possibility of leakage. The ASME Code requirements *[10]* for lethal service must be followed.

3. Flowrate—Placing the fluid having the lower flowrate on the shellside usually results in a more economical design. Turbulence exists on the shellside at much lower Reynolds numbers than on the tubeside.

4. Corrosion—Fewer costly alloy or clad components are needed if the corrosive fluid is placed inside the tubes.

5. Fouling—Placing the fouling fluid inside the tubes minimizes fouling by permitting better fluid-velocity control. Increased velocities tend to reduce fouling.

Straight tubes can be physically cleaned without removing the tube bundle. Chemical cleaning can usually be done better on the tubeside. Finned tubes on square pitch are sometimes easier to clean physically. Chemical cleaning is usually not as effective on the shellside because of bypassing.

6. Temperature—For high-temperature services requiring expensive alloy materials, fewer alloy components are needed when the hot fluid is placed on the tubeside.

7. Pressure—Placing a high-pressure stream in the tubes will require fewer (though more-costly) high-pressure components.

8. Pressure drop—For the same pressure drop, higher heat-transfer coefficients are obtained on the tubeside. A fluid having a low allowable pressure drop should be placed there as well.

Flow arrangement

In an exchanger having one shell pass and one tube pass, the two fluids transfer heat in either cocurrent or countercurrent flow. This affects the value of the log mean temperature difference (LMTD). There is a distinct thermal advantage to counterflow, except when one fluid is isothermal.

In cocurrent flow, the hot fluid cannot be cooled below the cold-fluid outlet temperature. Thus, the ability of cocurrent flow to recover heat is limited. Nevertheless, there are instances when cocurrent flow works better than counterflow, such as when cooling viscous fluids, because a higher heat-transfer coefficient may be obtained. Cocurrent flow may also be preferred when the temperature of the warmer fluid may reach its freezing point.

Fouling

Fouling is the deposition of undesirable materials on the heat-exchanger surface, which increases resistance to heat transmission. Fouling is a complex phenomenon and may be due to sedimentation, crystallization, chemical reaction, polymerization, coking, growth of organic material such as algae, and corrosion. These fouling mechanisms may operate independently of each other or in parallel. The rate of fouling is controlled by physical and chemical relationships that, in turn, are affected by the operating conditions. The operating variables that have important effects on fouling processes are:

1. Flow velocity—Very strong to moderate effect on a majority of fouling processes.

2. Surface temperature—Affects most fouling processes, and in particular crystallization and chemical reaction.

3. Bulk-fluid temperature—Affects rate of reaction and crystallization.

4. Materials of construction—Possible catalytic action and corrosion.

5. Surface—Roughness, size and density of cavities will affect crystalline nucleation, sedimentation, and the adherence tendency of deposits.

The materials of construction and the surface nature have the greatest effect in initiating fouling rather than in continuing and sustaining it.

As already noted, deposits on the heat-transfer surface due to fouling increase the overall thermal resistance; in addition, they lower the overall heat-transfer coefficient of the exchanger. For heat exchangers to maintain satisfactory performance in normal operation, and a reasonable service time between cleanings, it is important during design to provide a sufficient surface via a fouling allowance appropriate to the expected operating and maintenance conditions.

Appropriate values for fouling resistances involve physical and economic considerations, which vary from user to user—even for identical services. The user should specify design fouling resistances on the basis of past experience and current, or projected, costs. In the absence of such information, the user may be guided by the TEMA Standards *[1]*.

Flow-induced vibration

Shellside flow may produce forces that result in destructive tube vibrations. In most exchangers, the intensity of vibration becomes a problem when it:

- Causes some part of an exchanger to fail.
- Upsets the process conditions.
- Creates a condition that endangers those who must work nearby.

Vibration becomes evident in an exchanger when there is: (a) mechanical failure due to metal fatigue, collision, baffle damage, and/or tube-joint failure at the tubesheet; (b) excessive noise exceeding safety limits; and (c) excessive shellside pressure-drop.

Several methods have been proposed for predicting tube vibration. These are based on techniques involving vortex shedding, turbulent buffeting, and fluid whirling. However, existing correlations are inadequate to ensure that any given design of exchanger will be free of vibration problems. All that is known is that vulnerability of an exchanger to flow-induced vibration depends on the flowrate, tube and baffle materials, unsupported tube spans, tube-field layout, shell-diameter and inlet/outlet configurations.

A number of changes can be tried at the design stage to reduce or alleviate vibration problems: (a) reduce velocity of shellside fluid entering the bundle as well as the velocity inside the bundle, (b) change tube-field layout and baffle spacing, (c) change baffle type and/or add detuning baffles, and (d) use special designs such as rod/baffle heat exchangers.

Condensers: shellside

Condensation takes place in many CPI applications—often being the condensation of overhead vapors in distillation columns. While practically all types of heat exchangers have been used for condensation, the discussion here will be limited to shell-and-tube condensers. Basically, these can be classified as shellside and tubeside condensers, depending upon the allocation of the process vapors to be condensed.

Shellside condensers are the most common type in the CPI. Shell orientation may be horizontal or vertical:

Horizontal configuration is used when the coolant flows on the tubeside for reasons of high pressure or fouling, and when design may be limited by pressure drop, in which case TEMA Type J or X shells can be used.

Advantages

1. Type E (one-pass) shell is the common configuration that has simple construction and low cost.

2. Tubes are easier to clean because coolant (usually water) is in the tubes. The shellside vapor is usually clean, and a nonremovable bundle can be used.

3. Baffles can be adjusted to produce a high velocity if sufficient pressure drop is available, greatly increasing the condensation heat-transfer coefficient—especially in the presence of noncondensables.

Disadvantages

1. Condensate flooding due to improper design of baffles or small liquid-drain lines—causing accumulation of condensate that will blanket the lower tubes on the shellside.

2. Accumulation of inerts in the unit, due to low shellside velocities and absence of appropriate vent nozzles. Hence, can lead to poor performance.

3. Not ideal for condensate subcooling.

Vertical configuration is preferred whenever the coolant is a boiling fluid. Here, the process-vapor flow is usually downward for better drainage. This configuration of single-pass, upflow, tubeside boiling against downflow, shellside condensation eliminates many problems inherent in handling a boiling coolant in other configurations.

Advantages

1. Higher heat-transfer coefficients compared with vertical tubeside condensation, because shellside baffles interrupt buildup of condensate film.

2. Subcooling is more easily accomplished.

3. Ideal arrangement wherein a wide range of capacity can be obtained by controlled flooding of the tube surface.

Disadvantages

1. Mechanical cleaning is awkward and may be impossible for long tubes.

2. Support structure may be more expensive than horizontal orientation.

Condensers: tubeside

Tubeside condensers are applicable when condensation of the process vapors requires special metallurgy. The difficult problem of shellside cleaning of cooling-water deposits is tolerated to eliminate the need for special metallurgy of the shell. Orientation of the unit may be horizontal or vertical.

Horizontal configuration for tubeside condensation is probably the least effective arrangement in terms of thermal performance. Usually, a single tubepass, or a two-pass U-tube, design is preferred. In multitube-pass arrangements, there is always the uncertainty of liquid dropping out in the pass turnarounds. This may not affect the heat-transfer coefficient, but it can have a disastrous effect on the mean temperature difference for wide-condensing-range mixtures.

Advantages

1. Higher heat-transfer coefficients can be expected than with vertical orientation (tubeside condensers).

2. Easier maintenance.

Disadvantages

1. Condenser operation may be unstable, due to the blanketing effect of condensate accumulation in the tubes.

Vertical configuration for tubeside condensation is the best in relation to heat transfer and flow dynamics. The process-vapor flow is usually down the tubes—occasionally, vapor flow may be up the tubes (e.g., reflux or knockback condensers).

In the downflow arrangement, vapors enter at the top, and the condensate plus noncondensables or inerts are removed from the bottom channel. A baffling arrangement is sometimes used to isolate the noncondensables from the condensate. Vapor vent-nozzles must be provided even for pure-component total condensers to assure venting of trace amounts of inerts that otherwise tend to accumulate.

Advantages

1. Effective and stable operation.

2. Inert accumulation is not a problem.

3. Condensate subcooling can be easily handled and accurately predicted.

4. True countercurrent flow enables use with a temperature cross.

Disadvantages

1. Support structures may be expensive, and maintenance difficult.

Reflux-type condensers are used because this arrangement minimizes piping and support structure. In distillation columns, the condenser can sit directly on top of the column. Hence, both piping and structural components are minimized, and reflux pumps eliminated. However, this arrangement gives poor heat transfer, and entrainment can be a serious problem.

Reboilers

Boiling-heat-transfer services occur in evaporation and distillation processes. Here, the discussion will deal only with reboilers that are: (1) pool-boiling units such as kettle and internal reboilers, and (2) high-velocity recirculation units such as thermosiphon and pump-through reboilers.

Thermal performance and fouling tendencies, as well as construction characteristics and unit costs, vary widely for different types. Therefore, careful selection of the type most consistent with process requirements is important [7,8,9]. The operation of each reboiler type depends on the design for the external piping, which must be carefully reviewed.

Kettle and internal reboilers

The kettle reboiler is a pool-boiling unit, and has a tube bundle that is usually two-pass tubes and/or U-tubes. The tube bundle is inserted into an enlarged shell that serves as a reservoir for column bottoms, and as a disengaging space for vapor (Fig. 4a). An overflow weir maintains the liquid level above the top of the tubes.

Advantages

1. Reboiler operation is insensitive to large changes in process operating conditions, because there is no two-phase fluid recirculating back to the column.

2. High heat fluxes, up to 20,000 Btu/(h)(ft²) are possible.

Disadvantages

1. Fouls more readily than other types, as heavy residual materials accumulate continuously in the vaporization section.

2. Expensive design.

The internal reboiler has a tube bundle that is inserted directly into the distillation column. Operating characteristics are almost the same as for kettle reboilers.

Advantages

1. Since the unit requires neither a shell nor process piping, it is the least expensive reboiler type—allowing even for the expensive flange required to support the end of the bundle.

2. Lower tendency to foul than kettle reboilers.

Disadvantages

1. Bundle length limited by column diameter, which often means a high diameter-to-length ratio. This produces a lower maximum-allowable heat flux.

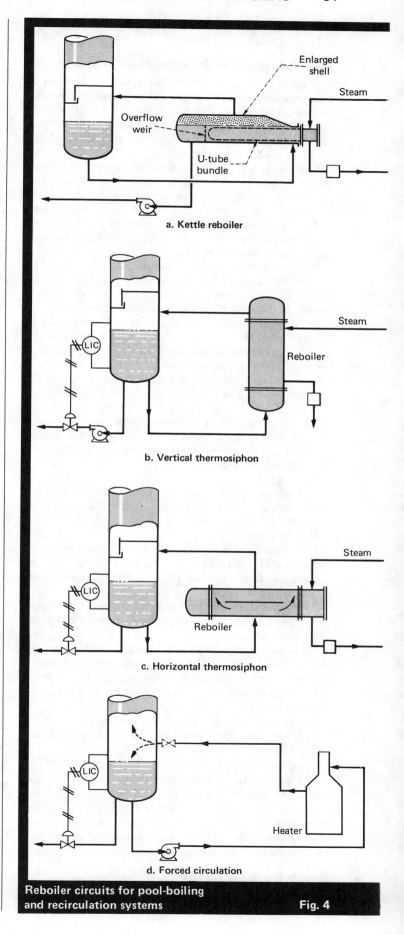

a. Kettle reboiler

b. Vertical thermosiphon

c. Horizontal thermosiphon

d. Forced circulation

Reboiler circuits for pool-boiling and recirculation systems   Fig. 4

2. Maintenance is a problem because the column has to be shut down and drained.

Thermosiphon and forced-flow reboilers

The operation of thermosiphon reboilers depends on a balance between two-phase friction and acceleration losses in the reboiler flow loops, and the net static head developed by the liquid in the distillation column. Vapor-to-liquid ratio is critical to the operation of the reboiler (usually 5 to 35% vapor).

Therefore, the unit is sensitive to changes in process operating parameters such as composition or temperature. These changes can lead to violent oscillations in the flow loop, and contribute to poor overall column control and low heat-transfer efficiency. Operating characteristics and relative merits of two types (vertical and horizontal) of thermosiphon reboilers:

Vertical thermosiphons are single-pass shell-and-tube units with standard 8-, 10- or 16-ft tube length, with vaporization occurring in the tubes. A typical circuit is shown in Fig. 4b. The tube length depends on the amount of recirculation and the available static head. Shorter tubes are used when the available static head is too small to maintain sufficient vapor-liquid circulation for satisfactory operation. Longer tubes are desirable to keep exchanger cost low, but the hydrodynamics of the flow circuit must be carefully checked for mist flow.

Tube diameters of 3/4 to 2 in. are typical—3/4-in. tubes are preferred in moderate-pressure systems and clean service to keep exchanger cost low. Larger-diameter tubes are usually necessary for vacuum services and/or fluids having high viscosity.

Advantages

1. Less expensive than any other type of reboiler, except internal ones.

2. Less fouling tendency.

Disadvantages

1. Unsuitable for highly viscous materials, and wide boiling-range mixtures.

2. Unsuitable for large turndown requirements (greater than 2 : 1).

3. Maintenance difficult due to vertical orientation.

4. Size of the unit limited because constructional considerations are a factor.

Horizontal thermosiphons are characterized by process flow on the shellside. Multiple piping inlets and outlets, and split horizontal baffles, are often used in order to provide good distribution and higher velocities (Fig. 4c).

Advantages

1. Less-sensitive operation than vertical type.

2. More area can be placed in a single shell than with vertical unit.

Disadvantages

1. Piping must be carefully laid out to equalize pressure in all parallel branches.

2. More expensive than vertical type because fixed-tubesheet construction can rarely be used since most fouling is on the shellside, and because of the complicated nature of the piping and supporting structure.

Pump-through reboilers

A pump-through reboiler is a special case, and has application mostly in highly viscous and/or highly foul-ing services. These units require careful economic analysis because a balance must be made between initial pump capital and operating costs, and the cost of the exchanger and its maintenance. (See Fig. 4d for a typical circuit.)

Advantages

1. Can be designed for extremely high heat fluxes.

2. Operation is insensitive to changes in process conditions.

Disadvantages

1. Vapor fraction has to be kept very low (less than 5%) and, sometimes, even eliminated by placing a backpressure valve in the exit piping. Therefore, circulation rates are usually very high.

Special designs

Included in the category of special designs are falling-film evaporators, gas chillers, bayonet-type exchangers, scraped-surface exchangers, and proprietary designs in special materials such as graphite, glass, ceramics and polytetrafluoroethylene.

The underlying principles for the thermal rating of such exchangers are the same as those for shell-and-tube heat exchangers. Due to their special features, however, the mechanical design may not be covered by the prevailing standards. Most of these designs are proprietary, and have been developed for specific applications in the CPI.

References

1. "Standards of Tubular Exchanger Manufacturers Assn.," 6th ed., Tubular Exchangers Manufacturers Assn., Tarrytown, N. Y., 1978.
2. Lord, R. C., Minton, P. E., and Slusser, R. P., Design of Heat Exchangers, *Chem. Eng.*, Jan. 26, 1970, pp. 96–118.
3. Lord, R. C., Minton, P. E., and Slusser, R. P., Guide to Trouble-Free Heat Exchangers, *Chem. Eng.*, June 1, 1970, pp. 153–160.
4. Kern, D. Q., "Process Heat Transfer," McGraw-Hill, New York, 1950.
5. Rubin, F. L., What's the Difference Between TEMA Exchanger Classes?, *Hydrocarbon Process.*, June 1980, p. 92.
6. Najjar, M.S., Bell, K. J., and Maddox, R. N., Influence of Improved Physical Property Data on Calculated Heat Transfer Coefficients, *Heat Transfer Eng.*, Vol. 2, No. 3 and 4, Jan.-June 1981.
7. Mathur, J., Performance of Steam Heated Exchangers, *Chem. Eng.*, Sept. 3, 1973, pp. 101–106.
8. Orrell, W. H., Physical Considerations in Designing Vertical Thermosiphon Reboilers, *Chem. Eng.*, Sept. 17, 1973, pp. 120–122.
9. Shah, G. C., Trouble Shooting Reboiler Systems, *Chem. Eng. Prog.*, July 1979, p. 53.
10. "ASME Boiler and Pressure Vessel Code," Section VIII, Div. 1, American Soc. of Mechanical Engineers, New York, 1980.
11. "Heat Exchangers for General Refinery Service," API Standard 660, American Petroleum Institute, Washington, D.C.
12. 'Standards for Steam Surface Condensers," Heat Exchange Institute, Cleveland, Ohio, 1978.
13. Devore, A., Vago, G. J., and Picozzi, G. J., Specifying and selecting heat exchangers, *Chem. Eng.*, Oct. 6, 1980, pp. 133–148.

The author

Davinder K. Mehra is the engineering manager for heat-transfer equipment at Brown & Root, Inc., Post Office Box Three, Houston TX 77001. Mr. Mehra has had more than 17 years of experience in process engineering in various capacities. He has a B.S. in chemistry from the University of Bombay (India) and a B.S. in chemical engineering from the University of London (U.K.). He is a member of AIChE, a representative to Technical Advisory Committee of Heat Transfer Research Inc. for Brown & Root, and a licensed professional engineer in Texas.

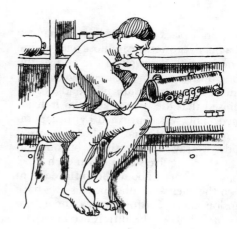

Troubleshooting shell-and-tube heat exchangers

Many things can go wrong with heat exchangers, but most problems can be corrected once you find the cause. Here is an explanation of the things that can go wrong, and the ways to find and fix them.

Stanley Yokell, Energy and Resource Consultants

☐ Trouble in a heat exchanger seems to show up as either a process or mechanical problem. However, thermal performance depends upon the construction of the exchanger, and mechanical functioning depends upon how the thermal regime affects the structure. Thus, you cannot treat problems of mechanical failure and inadequate performance separately.

The most frequent complaints are: (1) the unit does not perform as required; and (2) it leaks. Performance deficiency is obvious when duty or pressure-drop requirements are not met, but there are more-obscure process problems. For example, a refrigerated condenser may adequately remove latent heat without exceeding the permissible pressure drop. Yet, if the tube-wall temperature is too low, condensate may freeze on the tubes. This can lead to progressive loss of capacity, increase in pressure drop, or uneven operation. The frozen condensate may cause mechanical damage to the tubes or tube-to-tubesheet joints. Another example is an exchanger that transfers heat at both the design rate and pressure-drop but, because of high tubewall temperature, causes unacceptable stream degradation.

Similarly, a leak to the atmosphere is obvious, but there are other kinds of mechanical problems. You may not discover a leak between the shellside and tubeside until you find gross mixing of the streams, or a drastic decline in performance. Small leaks between the sides may lead you to conclude, erroneously, that the tubes are progressively fouling.

Performance deterioration usually comes from deposits on the tube surfaces. However, an exchanger's capacity to transfer heat may decline because of internal leakage between the passes. Such a leak slowly increases as the gasket surface erodes. The resulting capacity-loss may also appear to be due to fouling. Often, the problem will remain hidden until you remove the channel- and

return-cover in preparation for cleaning the tubes. You may then discover erosion of the pass partitions and the tubesheet, as illustrated in Fig. 1. Because erosion and corrosion are interactive, you may not be able to determine which caused the damage.

Performance is affected by increases in clearance between the shell and the cross-flow baffles, but this structural ailment is not obvious. Baffle corrosion, or damage during reassembly of a unit, may be the cause. Increases in the clearance between the baffle- or support-holes and the tubes result from corrosion. They affect the performance but may also contribute to causing vibration damage to the tubes. Other kinds of mechanical problems are excessive distortions and damage due to vibration-forcing by the fluid regime.

The trouble may be the result of faulty thermal or mechanical design, poor construction or misuse of the equipment [1, 2]. Defects in thermal design show up quickly. However, the effects of mechanical misdesign and shoddy work usually appear after the manufacturer's guarantee has expired.

Determining that an exchanger has been misused may make you unpopular. Nevertheless, you have to diagnose the problem dispassionately. You may have to get information from the specifier, the designer, the manufacturer, the installer, the user and the maintainer. Each has a legitimate interest in the outcome of your work. Of course, the information you receive may be colored by its source.

Assembling the facts

Before you start your analysis, collect as much of the following information as possible:
- The process flowsheet.
- The original exchanger specification, and all subsequent revisions.
- The manufacturer's original specification sheet and all subsequent revisions.

This article is adapted from a chapter in the author's forthcoming book, "A Working Guide to Shell-and-Tube Heat Exchangers," to be published by McGraw-Hill, probably toward the end of 1984.

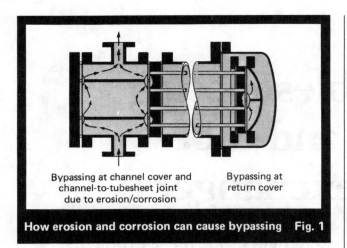

Bypassing at channel cover and
channel-to-tubesheet joint
due to erosion/corrosion

Bypassing at
return cover

How erosion and corrosion can cause bypassing   Fig. 1

■ All correspondence about the thermal and mechanical design.

■ Thermal-design calculations.

■ Mechanical-design calculations.

■ Setting plan, and assembly and detail drawings.

■ Manufacturer's data-report forms and nameplate stampings.

■ Manufacturer's installation, operating and maintenance instructions.

■ Shop and field inspection reports.

■ Radiographs, nondestructive-examination reports, and records of hydrostatic and leak tests.

■ Startup and shutdown procedures, and procedures for dealing with upsets.

■ Operating logs, temperature- and pressure-recorder charts.

■ Maintenance records.

It is often helpful to visit the installation and to interview the operating and maintenance personnel. Frequently, these people make changes that improve the way a heat exchanger functions or that reduce maintenance costs and downtime. But, because they rarely have time, and may not be trained, to consider the side-effects, their actions may cause problems.

Incident in a plant

An apparently innocent modification to a procedure may cause serious damage. The following illustrates what can happen. A pharmaceutical plant had several long-tube, fixed-tubesheet evaporators, built with bellows-type shell expansion joints. After each run, the specified procedure was to brush the inside of the tubes to remove organic matter, then sterilize. Sterilization was accomplished by filling the tubes with water and supplying steam to the shell to boil the water, which was then blown down with dry steam.

After a period of uneventful operation, the evaporators began to exhibit buckled tubes and leaks in the tube-to-tubesheet joints. The leaks and tube failures were randomly distributed in the tube field. Examining the available information led nowhere. It was obvious that differences in the tubemetal temperatures within the bundle were causing the failures, but not why there were differences.

Interviewing the shift operators and maintainers shed

no light. However, on a plant visit, the heat-exchanger specialist noticed a neatly stowed steam hose with a pipe inserted into its outlet end. Screwed to the end of the pipe was a reducing bushing that appeared to have been turned down to fit into the evaporator tube-ends. It was immediately surmised that the tubes had been individually steam-cleaned. This was confirmed by the operator's reply when he was asked what the hose was used for. "Oh," he said, "we save the brushing, boilout and blowdown steps by shooting live steam down each tube. And we don't have to worry about the difference in expansion between the shell and tubes because we have these great expansion-joints." Unfortunately, the operator also had not worried about the difference in expansion between steamed tubes and adjacent cold ones.

By examining the flowsheet and specifications, you may learn why there are differences between measured and design-point terminal conditions. Subtle modifications may have been made to the stated requirements in order to reduce exchanger size and cost. By reading the technical correspondence, you may be able to spot a design compromise that has led to the present problem.

Computer-designed exchangers

Most exchanger calculations are done by computers using thoroughly debugged and tested programs. Well-written programs warn the user of undesirable or unworkable aspects as illustrated in Fig. 2*. However, the user usually has the option of ignoring warnings and allowing program execution to continue. You may find that unheeded warnings are the source of the trouble. On the other hand, the program designer may have assumed that the construction would follow the physical model upon which the program was based. Warnings that certain construction features are essential may, therefore, have been omitted. The program user may have had a different mental image and not provided the necessary features. For example, a program designer might assume that the user would employ seal strips or dummy tubes but the user omitted them because they were not shown on the computer output.

Most programs repeat input data in an orderly arrangement. Here, you may find conceptual errors by examining the data fed to the machine. In addition, far more output information is provided than is available on heat-exchanger specification sheets. Finding out what is wrong may be simplified by having access to the calculated film-rates for each zone, the amount of excess surface provided and the calculated interpass and metal-wall temperature. Rubin has pointed out the importance of tabulating design data by zones in multizonal condensers [3]. It is equally valuable to examine individual zones in troubleshooting condensers.

Reviewing mechanical-design calculations may be as important for what was not calculated as for what was. For example, the thickness calculation for the tubesheet may show that it is adequate to withstand the applied loads without exceeding the allowed stress levels. However, at the allowed stresses, the tubesheet may deflect so much that there is pass-partition bypassing, as shown in Fig. 3. When the deflection is severe, it may cause flanged joints to leak.

*Courtesy B-Jac Computer Services, Midlothian, Va.

Flow arrows on the setting plan may alert you to incorrectly connected piping. By comparing assembly drawings with the installed exchanger, you may find that it has been put together incorrectly. When you look at the construction details, you may find a mechanical source of a performance failure. An example is a horizontal condenser in which the vapor condenses on the shell side, noncondensables are present, and a shell vent is provided at one end, but there are no ears (see Fig. 4) on the baffle to prevent vapor bypassing. Such an omission may be very serious if the vapor inlet is large and if tubes have been eliminated from the layout to provide for impingement protection.

Use the manufacturer's data reports and nameplate stamping to confirm information shown on the drawings. Read the installation, operating and maintenance instructions for clues to what has gone wrong. Examine inspection reports for notes of deviations from specifications, and both the drawings that were accepted and those that required correction. Either may have led to the malfunction. By studying radiographs and records of tests and nondestructive examinations, you gain insight into what took place during fabrication.

Scrutinize the procedures for startup, shutdown and dealing with upsets, to search out possible errors and omissions. Compare operating logs and charts with nominal values, to pinpoint the onset of the disfunction. Search the maintenance logs for alterations or substitutions made in hope of improving the equipment, or getting it back on line quickly.

Diagnosing exchanger ailments

When a problem occurs soon after startup of a new, or recently overhauled, exchanger, ask these questions:
- What is the nature of the problem?
- Did anything happen during transporting, handling, storing, installing, testing or starting the unit that could have caused the trouble?
- Were the manufacturer's installation and operating instructions followed?
- Are any external loads being applied that were not considered in the design?
- Were specified startup and operating procedures followed?
- Is the piping hooked up correctly?
- Is the unit assembled correctly?
- Are the vents and drains connected and working?
- Is there blockage or fouling resulting from inadequate cleaning after installation?
- Was the trouble found at the specified design-point conditions?
- Are all needed controls and safety devices installed and operating?
- Is the problem with the exchanger, or is it really elsewhere in the system?

When equipment that has been in satisfactory stable operation exhibits problems, the questions to ask are:
- What is the nature of the problem?
- Were any changes made to operating procedures?
- Were any changes made to stream compositions, flowrates or operating conditions?
- Are the vents and drains working?
- Are all controls and safety devices operating?

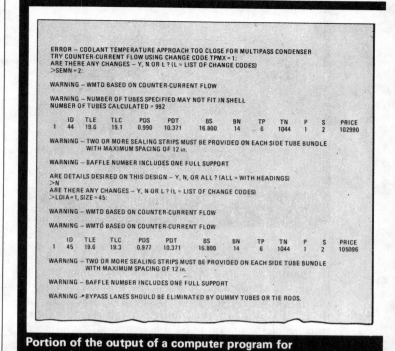

Portion of the output of a computer program for condensing heat-exchanger design, showing warnings Fig. 2

- Did an upset precede the onset of the problem?
- Is the problem with the exchanger, or is it really elsewhere in the system?

Diagnostic techniques/instruments/tools

The techniques and instruments for diagnosis of heat-exchanger ailments include looking, feeling, listening and measuring. Most of the instruments used extend your capacity to see, feel and hear. You can learn most of what you need to know about the health of a heat exchanger by taking its temperatures and pressures, and by metering flowrates and measuring dimensional changes.

Testing exchanger performance

Procedures for testing performance have been published by AIChE [4]. The American Soc. of Mechanical Engineers has published procedures for measuring temperature, pressure and flow [5]. Full-fledged performance testing is costly and may not be practical. But it might be the only way to ascertain that the problem is in the exchanger and not elsewhere in the system.

The need to measure terminal conditions is self-evident, but it may be equally important to measure intermediate temperatures. For example, you may be able to discover why a condenser is not working as desired, by comparing a temperature profile with the assumed condensing curve. You can find out if there is vapor and noncondensable maldistribution in horizontal units with condensing in the shell, by measuring the temperature from the top to bottom (at intervals along the shell length). By comparing the liquid temperature in a thermosiphon reboiler with the temperature of the two-phase flow to the column, you can estimate the extent of boiling-point elevation in the reboiler.

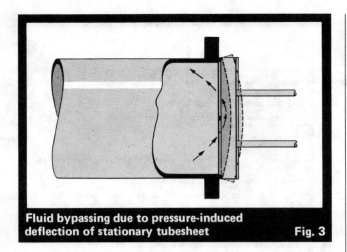

Fluid bypassing due to pressure-induced
deflection of stationary tubesheet   Fig. 3

When condensing-steam, or other vapor, is used in a heater that is acting erratically, the fault may lie with ineffective condensate drainage. By installing a recording flowmeter in the condensate drain line, you can detect wide swings in the condensate flowrate, indicating trap problems or the need for a condensate pump or barometric leg.

Measuring pH

Unexpectedly high fouling- and corrosion-rates may be caused by acidity or alkalinity in the streams. The pH meter is the instrument to use for monitoring acidity. When a gas mixture contains a large percentage of hydrogen, a change in composition can affect the duty and rate of heat transfer. And when hydrogen is used for heating or cooling, dilution with another gas may impair performance. This is because the thermal conductivity and specific heat of hydrogen are about ten times those of most other gases.

The gas chromatograph may help you to unravel such problems.

Detecting leaks

Leaks of hydrocarbon vapors are seldom seen, but are usually smelled. However, the olfactory sense tires quickly; therefore, operators may not become aware of a serious leak. Use a hydrocarbon leak detector. Upon shutdown, you may locate leaks by hydrostatic, halogen or helium testing.

Examining distortions and deformations

Many problems are associated with distortions or deformations. Look for the following: (1) bolts that are permanently stretched; (2) channel covers that are bowed; (3) extensions of tubesheets used as flanges, that are cupped; (4) flanges that are bent, cupped or distorted between bolt holes, giving a scalloped appearance; (5) ligaments between tubeholes that are distorted or extruded; (6) nozzles and supports that are bent; (7) pass partitions that are bent or crippled; (8) tubes that are bowed; (9) a tubesheet face at the stationary or inlet end that is out of parallel with the tubesheet at the floating or return end; (10) tubesheet rotation about the longitudinal axis; (11) tubesheet waviness; and (12) a shell expansion joint that is deflected angularly or squirmed.

Extending examination capability

To extend your capacity to see the condition of various parts, use the following: (1) high-intensity lights; (2) inspection mirrors; (3) magnifying lenses; (4) borescopes; (5) fluid-penetrant examination; (6) eddy-current examination; (7) magnetic-particle examination; (8) radiography; and (9) ultrasonic scanning.

You may measure the extent of corrosion, without disassembling, by ultrasonic scanning or by trepanning and plugging. To find out what lies under fouling- or corrosion-deposits, scrape the surface with an inspection scraper.

You can hear some of what goes on inside an exchanger by placing your ear against the shell. But interpreting what the sounds mean is difficult. Flashing liquids, gases and vapors flowing at sonic velocities, and sand and silt flowing through a unit, make characteristic sounds. You can sometimes estimate the degree of fouling on the shell side by tapping the shell lightly with a small hammer.

You may expect to encounter the following kinds of troubles: (1) overdesign; (2) underdesign; (3) maldistribution; (4) externally caused problems; and (5) mechanical ailments. Overdesign and underdesign are discussed here for units in sensible heat transfer, and process condensers and reboilers. Maldistribution, external factors and mechanical ailments are considered as they apply generally.

Overdesign

Many problems originate with the assumption that an overdesigned exchanger will work well at the design point. New and freshly cleaned exchangers are somewhat oversurfaced, relative to design-point conditions, because the designer has provided excess surface to allow for fouling. The resulting deviation from design-point outlet conditions is seldom objectionable because you can adjust flowrates and inlet conditions somewhat. The fouling allowance is intended to permit you to operate at design conditions up to the point where the unit is normally shut down for cleaning.

Excess fouling allowance

However, the excess surface may be so great that you cannot compensate enough to control the system. This may come about as follows: (1) You do not know enough about the fouling characteristics of the fluids. To be safe, you specify excessive resistances; (2) The maintenance group specifies a very long interval between shutdowns for cleaning, and your extrapolation from records of fouling rates results in your providing a large fouling allowance; and (3) A review committee decides that more is better.

Excess capacity

Even more troublesome is the situation that comes about because excess capacity has been arbitrarily specified. Here, you first calculate the clean surface required at design-point conditions. You add additional surface to allow for dirt or scale buildup. Then you increase the surface as a percentage of the subtotal.

The best way to add surface is to increase tube length (within the allowable pressure-drop limits) because it lets the fluid regime remain unchanged. However, if the reason for the overdesign is to provide for future increases in flowrates, you must increase the tube count to carry the increased volume and still stay within the pressure-drop limits. When you do this, the fluid regime at the design point is not optimal. Hence, the rate of fouling may exceed your expectations.

Overdesign in sensible heat transfer

When you oversize a unit designed for sensible heat transfer on both sides, the cold-fluid outlet becomes hotter than designed. Tubemetal temperature is also hotter at the cold-fluid outlet, and colder at the hot-fluid outlet, than anticipated. Interpass fluid and metal temperatures are similarly displaced.

As a result, the incoming cold fluid may be charred, or the warm fluid frozen, at their outlets. If the warm-fluid viscosity increases rapidly as temperature decreases, the pressure drop will be more than calculated. Tubemetal temperature displacement is significant because it changes the assumptions that govern the design of fixed-tubesheet equipment. When the design is based on restraining the difference in expansion between the shell and the tubes, a substantial deviation can overstress the tubes, tube-to-tubesheet joints, and shell. If an expansion joint is used to accommodate the difference, the deviation may cause more deflection than planned. This may reduce joint life.

You can check for overdesign by measuring the flowrates, terminal temperatures and pressures after startup. If a unit that is designed for cocurrent operation appears to be overdesigned, it likely was piped for countercurrent operation, permitting a temperature cross. Conversely, if too much surface in a countercurrent unit is causing charring or freezing, the cure may be to repipe for cocurrent flow.

When more tubes have been provided than are needed to carry the flow at optimum velocity, it may be possible to blank off some of them. Alternatively, you may be able to modify channel and return construction to bypass a pair of tube passes. Before deciding to do either, investigate what will happen to the tubemetal temperatures as well as the effects on the structure of the exchanger.

You may reduce the effective surface by placing long ferrules or inserts of inert material in the inlet ends of each pass. You can thus control the amount of effective surface without materially changing the flow regime. (The ferrules also protect the tube-ends from erosion.) To minimize the effect of turbulence at the exit from the ferrule bevel, insist on a long taper. (An angle of 15 deg is desirable, but may not be practical to attain.) When you consider using inserts, make sure that the tube material is not sensitive to crevice corrosion in the fluid being handled.

If you cannot turn down flow, or blank off tubes or tube surface, you may reduce effective surface by relocating shell connections to create dead zones near the tubesheets. This is an extreme measure that will almost certainly reduce tube and tubesheet life because of corrosion in the dead zones. Leave the original shell connections in place, valved off, and drain the dead zones frequently.

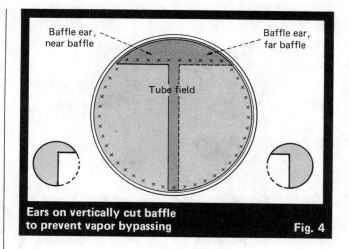

Ears on vertically cut baffle to prevent vapor bypassing   Fig. 4

Overdesign in process condensers

In a process condenser, the symptoms of overdesign are difficulty in balancing the column, and reflux subcooling. To visualize what happens, consider a correctly sized condenser of area A, condensing a pure vapor of latent heat λ, at rate W. Condensation takes place at temperature T and system pressure P. Cooling water, flowing at rate w, having specific heat c, enters at temperature t_1 and leaves at temperature t_2. The condensing duty Q is established by the vapor flowrate and latent heat, and is equal to the heat removed by the cooling water. This is shown in Eq. (1) and (2).

$$Q = W\lambda \tag{1}$$
$$Q = wc(t_2 - t_1) \tag{2}$$

The basic heat-transfer relationship is:

$$Q = UA\,\Delta T_m \tag{3}$$

The logarithmic mean temperature difference, ΔT_m is:

$$\Delta T_m = \frac{t_2 - t_1}{\ln\dfrac{(T - t_1)}{(T - t_2)}} \tag{4}$$

For simplicity, assume that the ideal gas law holds true in the vapor supply, and λ, c and U remain constant. Now, examine what will happen if you increase condenser surface A. The feedrate to the column, the reboiler surface and the heat supply to the reboiler, fix W. Therefore, by Eq. (1), Q is fixed. Because Q is fixed and U is constant, by Eq. 3, ΔT_m must be reduced. You can reduce ΔT_m by raising t_1 or t_2 or both. Operators are most likely to choke back on the water supply to reduce t_2. This reduces water velocity, which promotes organic fouling and inorganic scale deposition. If t_2 rises much above 120°F(49°C), dissolved inorganic salts may precipitate rapidly and form a hard scale.

On the other hand, if you maintain w, t_1 and t_2 unchanged, the only way to reduce ΔT_m is to reduce T. Because the vaporization rate fixes the vapor volume, P must be reduced to satisfy the ideal gas law. This may affect column and reboiler operation.

You may choose to use the excess condenser surface to subcool the condensate, thereby increasing the condenser duty. The subcooled reflux will cool the column,

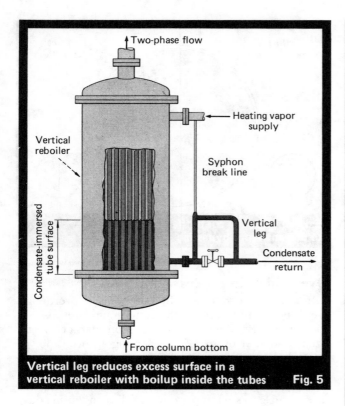

Two-phase flow

Heating vapor
supply

Vertical
reboiler

Syphon
break line

Condensate-immersed
tube surface

Vertical
leg

Condensate
return

From column bottom

**Vertical leg reduces excess surface in a
vertical reboiler with boilup inside the tubes**   **Fig. 5**

causing the controls to demand more heat from the reboiler. Not only is it wasteful to deliver excess heat to the reboiler and reject it in the condenser, but forcing the reboiler may lead to its operating at a higher heat-flux than desirable. High reboiler heat-flux may cause vapor binding and uneven operation, accompanied by rapid fouling. A better remedy is to limit t_2 to 120°F (49°C) and temper the cooling water to raise t_1.

The following horror story illustrates how costly it can be to overdesign a process condenser. A fixed-tubesheet process condenser for corrosive vapor was designed for condensing inside 1-in.-O.D.×0.029-in.-thick titanium Grade 2 tubes. The required shell size was 47 in. I.D. The required tube length, including an allowance for fouling, was 11.6 ft. The designer selected 12-ft-long tubes, but management decided arbitrarily to increase tube length to 16 ft to provide for future increased capacity. No consideration was given to how the 39% oversize would affect column operation. The operators balanced the column by choking the cooling-water supply.

Some years later, column throughput was increased 30% but the condenser could not handle the increase. Upon investigation, the shell was found to be filled with so much scale that cleaning was not practical. Shortly thereafter, the condenser was scrapped.

If a process condenser is designed for horizontal operation, you may be able to correct for overdesign by rearranging the unit for vertical flow. Kern compared a vertical unit condensing n-propanol in the shell with a horizontal condenser of the same tube count and arrangement [6]. His calculations show a vertical clean, overall coefficient that is 63% of the clean, overall coefficient of the horizontal arrangement. For a specific problem, the difference may be more or less than in

Kern's example. Therefore, it is necessary to calculate the vertical overall coefficient when you consider such a change.

A way sometimes used to control overdesigned surface condensers is to reduce the overall coefficient by deliberately introducing air into the steam. The film of air—through which the steam must diffuse before it makes contact with the cold surface—may be considered an additional resistance in series with the original film and fouling resistances. Standiford reworked data of Meisenburg, Boarts and Badger to correlate their data in the form of an air-film resistance vs. % air in the steam [7, 8]. These correlations are repeated here.

(SI Units)
$$F = 4 \times 10^{-5} \, C \tag{5}$$
$$F = 6.5 \times 10^{-5} \, C' \tag{6}$$
$$F' = 0.00023 \, C \tag{7}$$
$$F' = 0.00037 \, C' \tag{8}$$

In these correlations F is fouling resistance in SI units, F' is the resistance in English units, C is weight % air and C' is mol %.

The air concentration increases as vapor traverses the condenser because vapor is condensed, and condensing takes place over a range, reducing ΔT_m. The fouling resistance of the air, therefore, increases from inlet to outlet, and the effective driving force decreases.

In addition to the symptoms described, you will observe low suction- and head-pressures in overdesigned refrigerated equipment. The refrigerant vaporization temperature should be chosen to be above the process-stream freezing point. If it is not, process fluid may freeze on the surfaces during startup. If the unit is sized correctly, the deposit may melt as design flow is reached and tubewall temperature rises above freezing. However, in overdesign, the coating will continue to build up. Because pressure drop increases with the inverse of the square of the flow area, an early symptom of overdesign in refrigerated units may be an increasing process-side pressure drop. In condensers, the pressure-drop rise raises the system pressure and boiling temperature in the vaporizer. Reboiler output decreases, further increasing relative condenser overdesign, which increases both the length of the condenser tubes coated with frozen condensate and the thickness of the deposit. This further increases pressure drop. The ratchet effect eventually leads to shutdown.

You may remedy overdesign by using a higher-boiling refrigerant, or operating the refrigerant system at a higher pressure. In vertical refrigerated units, you can control the refrigerant level to reduce effective surface. In horizontal coolers, with refrigerant in the shell, you may reduce the level to expose tubes (if there is no adverse differential expansion between submerged and exposed tubes). But in horizontal condensers, the exposed tubes would bypass uncondensed vapors to the vent. You would then have an overdesigned condenser that does not condense all of the vapor! You may correct these units by: (1) blanking tubes, if pressure-drop and differential expansion permit; or (2) inserting sleeves of inert, low-thermal-conductivity material in the tube inlets. Sleeve length is set by how much excess surface is furnished. (You may also coat the ends of the tubes.)

When vapor enters the tubes of a vertical downflow condenser, inert gas is swept through and separated in the bottom head by means of a funnel or baffle plate. This provides the option of controlling system pressure by controlling vent-gas pressure. If there is not enough inert gas in the system, you can bleed some into the vapor inlet stream. For operation below atmospheric, you can adjust pressure by bleeding air into the jet.

If the excess surface is causing the condensate to be subcooled too much, repipe the unit to bring coolant in at the top. This cocurrent flow arrangement ensures that there will not be a temperature cross. Another alternative is to loop-seal, to flood the bottom of the tubes. This is not a good choice, because: it may be difficult to achieve a desired amount of subcooling; you cannot get rid of the noncondensables; system pressure control is indirect; and there is a possibility of hydraulic-hammer damage.

Overdesign in reboilers

The principal symptom of overdesign in horizontal thermosiphon reboilers is the phenomenon called breathing. It comes about as follows. Liquid enters the shell of the reboiler, where it meets the hot tubes, and flashes. More liquid enters to replace the flashed material and cools the tubes. The fresh liquid remains briefly until it too flashes. Instead of even, continuous two-phase flow, puffs of vapor and entrained liquid leave the reboiler.

Ordinarily, there is a modest amount of excess surface (the difference between required clean and fouled surface, plus surface added by rounding calculations upward) and you can control the boiling rate by adjusting the temperature of the heating medium to reduce the driving temperature difference. But with gross overdesign, you may not be able to turn down the heat input enough to prevent more feed from boiling up than enters. This depletes the holdup volume, reduces the boiling point, and leads to breathing.

You can mitigate the problem by installing a restriction in the vapor outlet. This increases the pressure in the reboiler, elevating the boiling point. In doing so, you reduce the driving temperature differential, at the same time causing some energy to be transferred by sensible heat at a lower rate than by nucleate boiling. The disadvantage of this remedy is the flashing that takes place across the restriction.

Always place the restriction directly at the reboiler outlet or provide a flash pot to ensure that the column is not used as a flash chamber.

When the configuration is a vertical thermosiphon with vaporizing in the tubes, excess surface can lead to drying out if the tubes are long. The low heat-flux can also create insufficient pumping, making reboiler and column operation unstable.

Reducing the temperature of the heating medium would seem to be the way to control the effects of the excess surface, but doing so further reduces pumping because the smaller temperature difference further reduces the heat flux.

Fig. 5 shows a way often used to handle overdesign in this kind of reboiler. The vertical pipe leading from the condensate-outlet causes tube surface to be submerged in condensate to the height of the leg. Heat is transferred from the condensate by natural convection at a much lower rate than by vapor condensation. Therefore, the excess surface is rendered ineffective. If you use this scheme, provide for blowdown at the inner surface of the bottom tubesheet (to avoid corrosion in the region of the stagnant condensate pool). The height of the leg should be adjustable to accommodate fouling buildup or operating variations. A level control may also be used.

The response time of this procedure is slow. A faster way to reduce the boiling rate is to introduce inerts into the heating medium. Although blanketing the surface with inerts is effective, it presents the problem of disposing of the inerts downstream.

If the boiling range is wide, and heat is supplied in the sensible mode, you can reduce effective temperature difference by piping for cocurrent flow. If the range is small, piping cocurrently improves circulation because it generates more vapor at the bottom. Therefore, you can turn down the inlet temperature, to reduce temperature difference, with less decline in pumping than if heat were supplied by a condensing vapor.

Oversized kettle and internal reboilers unbalance the column. The overhead condenser may appear to be too small because of the volume of vapor generated. Here, too, column operation becomes unstable. Another problem that may be overlooked is that oversized kettles foul rapidly. For this reason, excessive allowance for fouling is self-defeating.

Because kettles operate well at both low and high temperature-differences, you can most easily control overdesigned units by reducing the heating-medium temperature. When the heating mode is sensible, bring the hot fluid in at the bottom to reduce the effective temperature-difference somewhat. For condensing vapors, you can install a loop seal in the condensate to flood tubes in the outlet pass. You can also introduce inert gas in the tubes to blanket some surface, but you will also have to dispose of it later.

Forced-circulation reboilers essentially operate in the sensible-heat-transfer mode in the tubes. When the unit is overdesigned, the symptoms at the reboiler itself are the same as for overdesign in a liquid heater. However, excessive vapor is flashed in the flash chamber, which both makes the column operation unstable and overloads the condenser. There are no ill effects on the reboiler itself, and the rate of fouling does not increase when it is overdesigned.

The best way to control an overdesigned forced-circulation reboiler is to reduce the driving temperature differential. You can bypass and remix a portion of the feed before the flash drum, but this may lead to more-rapid fouling and some vaporization in the tubes. If the heating mode is a condensing vapor, you can flood some of the heating surface by loop sealing, provided that the differential temperature between the tubes is not excessive. You can also introduce inerts to blanket the surface, again being faced with the problem of downstream removal.

Underdesign

Underdesign—defined as inadequate surface to produce design outlet conditions when the fluids enter at

Solvent outlet temperature		Corrected LMTD		Actual LMTD as a percent of design LMTD	Percent of required surface provided
°F	(°C)	°F	(°C)		
80	(26.7)	17.31	(9.6)	100.0	100.0
81	(27.2)	19.09	(10.6)	110.3	90.7
82	(27.8)	20.00	(11.1)	115.4	86.5

Illustration of a small outlet-temperature deviation associated with substantial underdesign in a 1-2 solvent cooler   Table I

design flowrates and conditions—is probably the heat-exchanger problem that occurs least frequently. In the author's 25 years of experience with two shops, not one of more than 6,000 exchangers built had to be modified or replaced because of failure to perform.

Heater and cooler underdesign

Heater and cooler underdesign is signaled by failure of the stream outlet temperatures to reach design values. This may be accompanied by higher pressure-drop than allowed. Another kind of underdesign is excessive pressure loss when operating at design flows and temperatures. Underdesign may not be immediately apparent—unless the exchanger is tested for its clean-performance rate—because the tubes are new and clean. But the unit appears to foul more rapidly than anticipated because the underdesign eliminates the fouling allowance.

Sometimes, what seems to be a small deviation from design in the outlet temperature is symptomatic of substantial undersurface. The following shows how deviations of 1 or 2°F (0.56 and 1.1°C) can be associated with 10 to 15% less surface than required. A 1-2 cooler is used to cool a volatile solvent from 120° to 80°F (48.9° to 26.7°C), with cooling water entering at 70° and leaving at 80°F (21.1° and 26.7°C). Table I shows that, with a constant overall coefficient and no change in cooling-water temperatures, 90.7 and 86.5% of the required surface has been provided.

In a liquid heater, when there is a large viscosity decrease with increasing temperature, failure of the outlet temperature to reach the design point may cause excessive pressure drop. Another situation where you may observe pressure drop beyond design values is when the fluid circulated consists of a liquid and a gas phase. If the design basis was stratified flow, but the actual regime is slug flow, there will be a substantial increase in pressure drop over the design value.

When gases are circulated, a decrease in system pressure can also cause excess pressure drop. The pressure loss varies inversely with gas density, which is a direct function of pressure. A possible remedy for excess pressure drop of gases that are flowing in the shell is to rearrange the piping and nozzles to an approximation of a divided-flow configuration. Here, a new inlet is relocated opposite the cutoff segment of a baffle in the middle of the bundle. The previous inlet becomes an outlet, manifolded to the original outlet. The scheme is shown in Fig. 6 for an even number of baffles, but it works equally well for an odd number.

Dividing the flow reduces shellside pressure-drop to about one-eighth of the loss in a shell having the inlet at one end and the outlet at the other. You have to provide impingement protection under the new inlet and do some surgery to the piping. Before you proceed with this change, analyze the thermal performance of the revised configuration.

Underdesign in process condensers

When a condenser seems to be too small, take into account that—in addition to the condenser's configuration—the system pressure, flowrate and nature of the entering vapors determine its performance. The vapors may be generated from a pure liquid, or from two pure immiscible liquids. For these conditions, condensing is isothermal. For all other combinations of phases and liquid composition (e.g., vapors of miscible substances, pure vapors and noncondensables, vapors of miscible liquids with noncondensables, vapors of miscible and immiscible liquids, and vapors of miscible and immiscible liquids in the presence of noncondensables), vapor temperature decreases from vapor inlet to outlet.

A small change in system pressure, inward leakage of noncondensables, or a difference in the entering-vapor composition can make a correctly sized condenser underperform. An unsuitable arrangement of vapor vent and condensate nozzles, poorly arranged internals and inadequate venting can render an otherwise appropriate amount of surface area ineffective. And if you install a condenser designed for horizontal operation in the vertical position, it will probably not be large enough. Therefore, to troubleshoot a malperforming condenser, investigate the vapor-stream composition, system pressure, venting, possibility of vapor bypassing, noncondensable inleakage, and how the unit was installed relative to how it was designed.

Some indications that a condenser is underdesigned: (1) loss of volatile vapor out of the vent; (2) high vent-gas temperature; (3) an unusual temperature profile between vapor inlet and condensate outlet; and (4) failure to reach design condensate-subcooling-temperature in a condenser-subcooler. However, before you conclude that the unit is too small, eliminate the possibility of a system pressure that deviates from design, excessive air inleakage, maldistributed vapor- and dam-baffle bypassing, or improperly arranged loop-seal piping.

Shellside condensing pressure-drop is affected by total system pressure in the same way that gas flow pressure-drop is affected. You can, therefore, also reduce excessive shellside pressure-drop in horizontal units by rearranging the connections to approximate divided flow. However, the vapor inlet is larger than the condensate outlet, and the change requires more modification. In addition, because gaseous inerts cannot be removed at the top of the shell, vent gases must exit from the sides of the condenser below the bundle centerline.

The principal reasons that divided-flow condensers underperform are maldistribution and inert-gas buildup.

If the vapor is led directly to the bundle, and the tube length is long, noncondensables and more-volatile materials may accumulate near the tube-ends, reducing effective surface and increasing pressure drop.

You can rectify maldistribution by replacing the single inlet with a vapor belt—thereby bringing the vapors into the bundle through several openings in the shell on either side of the divider baffle. You can rectify accumulation of inerts by adding additional vents in the regions of the tube-ends.

Adapted from surface condensers, process condensers designed for pure crossflow may be provided for services where the absolute pressure is below 0.5 psi (3,447 N/m^2), available pressure-drop is below 0.15 psi (1,034 N/m^2) and the vapor condenses over a range of 20°F (11°C) or less.

Ideally, the path for the vapor to follow as it flows into the bundle is a gradual transition, as shown in Fig. 7. But the construction is expensive and exchangers are sometimes built with the nozzle entering a vapor dome. Such units underperform because of maldistribution. You can alleviate the problem by providing additional vapor inlets near the tube-ends, thereby splitting the vapor into three streams. You have to make sure that nearly equal quantities of vapor enter the inlets.

In vertical reflux and knockback condensers condensing in the tubes, there is seldom inadequate surface to handle the duty. The overload is, instead, one of excess vapor-velocity and consequent flooding. You see this when condensate spews material out the vent. This is described inelegantly as "puking." You can reduce flooding considerably by tilting the condenser 30 deg off the vertical.

Underdesign in reboilers

The way a reboiler works when you start up is the clue to underdesign. Vapor output increases as the system heats up, but before the design rate is reached, the output levels off and may decrease. This results from excessive heat-flux and consequent film boiling and vapor blanketing of the tubes, instead of nucleate boiling. If you consider reboiler overload in terms of heat flux, you can visualize how failing to control an overdesigned condenser can make an adequate reboiler underperform. Heat removed in the condenser must be supplied by the reboiler. Excess removal overloads the reboiler.

A thermosiphon reboiler may appear to be underdesigned when its static head is higher than planned for in design. Under this condition, the boiling point in the reboiler is elevated, narrowing the temperature difference between the reboiler contents and the heating medium. The excessive static head increases the circulation rate, driving the heat-transfer mode toward sensible heat transfer. Vaporization then occurs by flashing in the liquid, rather than by nucleate boiling on the tubes. The rate in sensible heat transfer is lower than in nucleate boiling. The combination of reduced driving force and poorer heat-transfer rate lowers the effective capacity of the unit.

Excessive static head develops when the column liquid level is higher than planned for in design. A less frequent cause is higher bottoms-liquid density than estimated. Low piping-friction losses result from larger-than-planned pipe diameters and shorter-than-estimated pipe lengths in the circulating loop.

Therefore, when your thermosiphon reboiler appears

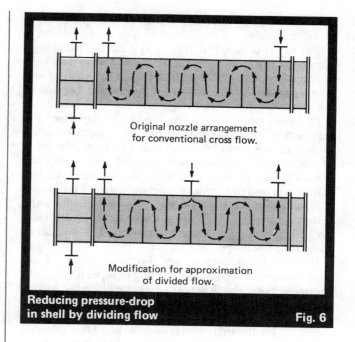

Original nozzle arrangement for conventional cross flow.

Modification for approximation of divided flow.

Reducing pressure-drop in shell by dividing flow   **Fig. 6**

to be undersized, check for excessive circulation by installing a variable restriction between the column and the reboiler inlet. As you choke-down flow, performance may reach the design level. If variations in column feed-rate and composition are small, you may install an orifice plate instead of a variable restriction.

Large horizontal thermosiphon reboilers underperform when the feed is not distributed uniformly along the length of the tubes. Before you conclude that this kind of unit is too small, examine the arrangements for dividing the feed. If the boiling range is large, and the design is not a split or double split arrangement, there may be a kind of internally generated maldistribution. What happens is that, when the inlet stream strikes the hot tubes, the low-boiling liquids flash and travel to the outlet. This concentrates the high-boilers in the ends, reducing the effective driving temperature-difference. If the bundle is a removable multipass one, you can convert either to split or double split flow in order to correct the situation.

Vertical thermosiphon units are usually unsuitable for

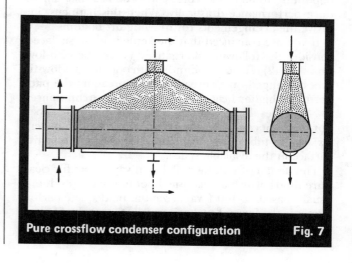

Pure crossflow condenser configuration   **Fig. 7**

operation under vacuum and at very high pressure. If you cannot balance a vacuum or high-pressure column in which vapor is supplied from a vertical thermosiphon, consider converting to forced circulation. You may have to increase the size of the line between the column and pump and adjust the position of the reboiler because of net-positive-suction-head requirements.

In forced circulation, if the fluid being vaporized has several components with a wide spread of normal boiling points, an excessive circulation ratio may cause too much of the volatile components to boil off. This raises the overall boiling point, reducing the driving temperature-difference, and thus reducing performance. Thus, when a forced-circulation reboiler fails to perform, check the pump output volume. You can correct excessive circulation by installing an orifice plate or other restriction in the pump discharge.

Insufficient circulation may also be the cause of low performance (as well as contributing to rapid fouling). In thermosiphons, the flow may be too small for the following reasons: (1) insufficient static head; (2) excess friction in the piping; (3) in tubeside flow, outlet-nozzle flow-area smaller than tube flow-area; and (4) in tubeside flow, inlet-nozzle flow-area less than half of tube flow-area. In these units, rapid fouling takes place if the rate of vaporization is more than 25% of the circulation rate.

These reboilers foul rapidly when the weight rate of vaporization exceeds the weight rate of circulation. If the pump characteristics have not been matched to the actual backpressure on the pump discharge, you may have to install a restriction in the discharge to make it operate at the design rate. For example, a pump designed to deliver 2,500 gpm (9,463 L/h) at a head of 20 psi (137 kPa) may not be able to deliver the design volume at a head of 1 psi (6.89 kPa), and the reboiler will underperform. You can check pump performance by shutting down and bypassing the pump. If there is no further decrease in performance, the unit has been operating in natural circulation.

Kettle reboilers are rarely underdesigned. In fact if the fluid has a wide boiling-range, and the available driving temperature-difference is low, a reboiler will probably be oversized. This is because effective temperature difference will probably have been based on the vapor temperature. However, less-volatile material tends to collect, depositing fouling and corrosion products that can reduce performance drastically. When this happens, improve the arrangements for drawing off bottoms.

If you are convinced that the kettle is undersurfaced, consider the following alternatives: (1) retube with low-fin tubes; (2) replace the bundle with one that has low-fin tubes arranged on a spread pitch. (Spreading the pitch permits a higher heat flux.) Under most circumstances, low-fin tubes do not foul more rapidly than bare tubes. For the right combination of heating and boiling materials, low-fin tubes provide a marked increase in effective surface in the available space.

The design of natural-circulation vaporizers for clean pure fluids, which are assumed not to leave a residue, is usually based on 100% vaporization. Inadequate performance in the form of surges is usually the result of failure to maintain a liquid level that will ensure submergence of all of the tubes.

Corrective measures are: (1) installation of a weir in the vaporizer shell; (2) installation of an internal or external standpipe; or (3) installation of a liquid-level control. However, if the tube-ends of a vaporizer designed for full vaporization become fouled, you know that the feed is not pure. To correct the fouling, rearrange piping to provide a natural circulation rate of three times the vaporization rate. When working with forced-circulation equipment, double the pumping rate and recirculate half the flow.

Flow maldistribution

A major cause of underperformance in all kinds of exchangers is maldistributed flow. Typical is a falling-film evaporator that does not work as predicted if either there is uneven distribution of feed to the tubes or the liquid descending each tube does not wet the tube surfaces uniformly.

The causes of maldistribution to falling-film evaporator tubes are: (1) an installation that is out of plumb; (2) inadequate calming of the feed to the top tubesheet; (3) variations in weir height from tube to tube; and (4) a combination of these factors. If the full circumference of each tube is not wet, the unit is probably out of plumb. To reset the evaporator properly, insert a level in the tubes at quarter and eighth points to verify their perpendicularity to the horizontal.

Maldistribution in heat-exchanger tubes probably occurs under the following circumstances: (1) axial nozzle-entry velocity exceeds tube velocity; (2) radial nozzle-entry velocity is twice the tube velocity. Indications of tubeside maldistribution are uneven tube-end erosion at the inlet and tubesheet, and underperformance.

You can correct maldistribution, when the fluid enters a bonnet axially, by installing a distributor baffle (target plate). Alternatively, you may replace the bonnet with a conical enlarger. To be effective, the included angle must be quite small (somewhat less than 10 deg) and the construction may be impractical.

Some bonnets with axial inlets are made from truncated right cones having included angles as great as 60 deg. With this construction, you may anticipate some reduction in the calculated performance. You can improve distribution somewhat by replacing the right cones with eccentric cones, as in the arrangement shown in Fig. 8. To estimate whether the cost is justified, compare the calculated overall coefficient with the measured coefficient of the clean unit. A quick way to assess the situation follows.

Assume that metal resistance of the clean unit is negligible compared with the sum of shell- and tubeside film resistances. Then U_c, the clean overall coefficient, in terms of the outside film coefficient related to tube O.D., h_{io}, is given by Eq. 9.

$$U_c = \frac{h_o h_{io}}{h_o + h_{io}} \qquad (9)$$

Define K as the ratio h_o/h_{io} and substitute in Eq. 9 to derive Eq. 10. Eq. 10 relates the calculated clean overall coefficient to the calculated inside film coefficient.

$$h_{io} = \frac{K + 1}{K} U_c \qquad (10)$$

By using the calculated values of h_o and h_{io} to evaluate K, and the measured value of U_c in Eq. 10, you can estimate how much improvement of inside film coefficient is required.

There is less likelihood of maldistribution when the fluid enters a channel radially. However, if there is maldistribution, you can correct it by using a distributor. There is better distribution in vertical units if fluid enters the bottom channel. For either axial or radial entry to the channel, the distribution is better in vertical bottom-fed units than in horizontal equipment.

Maldistribution of shellside fluids may also cause exchangers to underperform. Condensing vapors may bypass available surface, traveling directly to the vent. Gases and liquids may flow parallel to the tubes, in open channels created by the spreading of tube rows to accommodate U-tube bending and pass-partition location. Especially prone to shell fluid maldistribution is the pull-through floating-head configuration.

Maldistribution in horizontal two-phase flow can cause poorer heat transfer and higher pressure-drop than anticipated. Instead of uniformly mixing, the liquid and gas phases stratify. You may overcome the problem by switching the position of the exchanger to vertical, with the mixed stream entering at the bottom. You may improve gas-liquid mixing by feeding gas and liquid coaxially, as shown in Fig. 9.

You are justified in suspecting shellside maldistribution as a cause of poor performance when: (1) tubes have been dropped out of a full layout to accommodate impingement protection; (2) construction drawings show no seal strips or dummy tubes, or very few distribution devices; and (3) temperature probes reveal temperature profiles different from those that were assumed.

The effects of shellside maldistribution may be dramatic. If tubes grouped at the periphery of a segment of a fixed-tubesheet unit are substantially hotter than the rest, and the shell is fitted with a bellows expansion joint, the unit may assume the shape of a banana. You may see the bending, or notice cracked insulation and leaks in the connections to the piping. In floating-head equipment, the banana shape of the bundle is not visible, but shows up as leaks in the tube-to-tubesheet connections, broken tubes and bent baffles.

The distortion may be symptomatic of an excessive temperature range between the tubeside inlet and outlet of a multipass exchanger. However, if you have a horizontal unit that is cooling gas in the shell over a wide range—say 1,000 to 400°F (538 to 200°C)—and the gas is stratified into upper hot and lower cold layers, the bundle will bend.

The most likely cause of stratification is location of both shellside connections on the top, combined with a baffle system that does not ensure complete gas mixing. Examine the baffle system for one or more of the following: (1) vertical cuts instead of horizontal ones; (2) location of a baffle where gas can bypass through the expansion joint (when a sleeve is not provided); (3) too few baffles; (4) excessive clearance between baffles and shell.

To correct the situation may require the following surgery: (1) rotating the unit to change baffle-cut position to horizontal, providing an outlet 180 deg from the

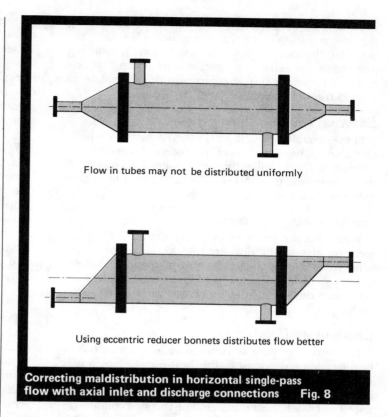

Flow in tubes may not be distributed uniformly

Using eccentric reducer bonnets distributes flow better

Correcting maldistribution in horizontal single-pass flow with axial inlet and discharge connections   Fig. 8

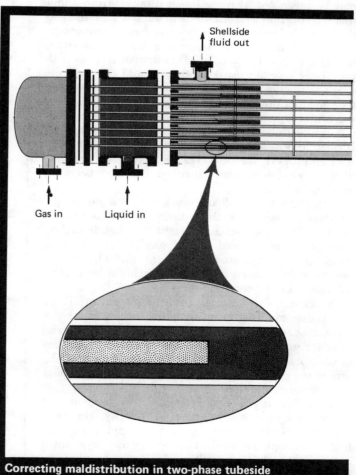

Shellside fluid out

Gas in   Liquid in

Correcting maldistribution in two-phase tubeside flow by installing coaxial gas distributor   Fig. 9

inlet, and blanking off the previous outlet; (2) rotating the unit and installing a distributor belt; (3) sealing the baffles to the shell; (4) installing seal strips; (5) installing a sleeve in the expansion joint; and (6) adding baffles. Some of these measures may require you to replace or rebuild the tube bundle.

External causes of problems

Some troubles caused by factors external to heat-exchanger design have been discussed. An example is maldistribution in falling-film-evaporator feed caused by not leveling the unit. A brief discussion of a few other external causes of difficulty follows:

Inerts in heating steam

When you use steam as a heating medium, the problems on the steam side are the same as for steam condensers. Noncondensables may blanket the surface, thereby reducing capacity. In addition, unvented CO_2 or NH_3 may dissolve in the condensate and corrode the exchanger.

Condensate flooding

If you do not remove condensate as it forms (when heating vapor condenses), the heat-transfer surface is flooded with liquid. The lower rate of sensible heat transfer at the flooded surfaces reduces capacity—compared with latent heat transfer. The retained condensate may be subcooled. If the vapor is supplied from a unit that is recovering heat elsewhere in the system, returning the subcooled condensate may unbalance the vaporizer.

Flooding results if the condensate discharge nozzle is too small or poorly placed, or if condensate traps or seal-pot systems do not work as they should. To find the cause of flooding, check the condensate line to be sure that it slopes downward. Recalculate the nozzle size for the actual flow and pressure conditions. If these check out satisfactorily, look for dirt in the condensate trap. If it is clean, the trap is probably too small or has not been selected properly for the service. Before you come to one of these conclusions, make sure that it has not been put in backward.

Nonreturn traps will not work if the vapor supply pressure is lower than the pressure in the condensate receiver. When the receiver pressure is slightly higher, buildup of liquid head in the condensate pool may supply enough pressure to operate the trap. This uneven operation can make vaporizers and reboiler surge, unbalancing the downstream operation.

On startup, there may be insufficient pressure if the vapor demand is greater than the supply valve can accommodate. The remedy is to install a bypass and valve for manual startup, or a different kind or capacity of control valve. If you throttle the supply to bring the unit to temperature gradually, you may create a similar low-pressure condition.

Here, flooding is not significant, because you are seeking a temporary reduction in capacity. However, when you are at temperature—and the supply pressure is high enough to operate the trap—there may be so much condensate backed up that the trap is temporarily overloaded. For this reason, there should be a manual condensate trap-bypass.

Inverted-bucket traps may not work if you insulate them, because they have to be primed with condensate. If steam is not condensed in a trap, it may take a long time for the trap to become primed. Because most hot lines are insulated for safety and heat conservation, it is a common error to insulate inverted-bucket traps as well. Under these conditions, if such a trap is not performing well, strip the insulation and install a personnel guard.

Hydraulic hammer

Hydraulic hammer can damage condensate discharge lines and knock tubes loose from the joints to the tubesheets. If a steam trap is oversized and the pressure drop between the trap and condensate receiver is large, you will probably hear the noise and see the accompanying shaking produced by a water hammer. The hammer is an unsteady-state phenomenon that you can relieve with appropriate protective devices, but it is usually more desirable to install a correctly sized trap.

Instrument and control problems

Problems arising from instruments and controls are vexing because these devices are so reliable that they are the last to be suspected. Before a startup, check the calibration, zero point and damping of the instruments used to measure and control level, flow, temperature and pressure. When problems arise, look for condensate in sensing lines, plugging of inlets, or mechanical damage to sensing or transmitting elements. If an element is placed close to an inlet, or unprotected from fluid forces, you can anticipate a malfunction. You can correct condensation problems and some causes of plugging by insulating or steam-tracing lines. You may be able to avoid repeated mechanical damage by shielding or relocating elements. If the instruments and controls are not damaged and test out properly, but operation is erratic, the sensing elements are probably located where a stream is not completely mixed or a constant flow-regime has not been established.

Piping problems

You may find a startling variety of foreign materials in piping and equipment on startup. Typical are nuts, bolts, marking crayons, soapstone sticks, weld-rod stubs and slag, crushed stone, sand and gravel, chunks of cinder block, pieces of construction blueprints, newspapers, twine, birds' nests, lunch bags, small tools, work gloves, articles of clothing and personal hygiene, small dead animals, etc. If you draw cooling water from ponds, rivers or bays—salt, sand and marine life may enter. When you use cooling-tower water, look for silt deposits in the heat-exchanger's return channel.

It is a good practice to install a dual strainer in the cooling-water feed. This permits you to keep foreign matter out and clean the strainer without interrupting production. However, single strainers with bypasses are frequently installed and it is usual to provide a bypass around dual strainers. In the course of operation the strainer may be plugged, starving the exchanger for cooling water. Instead of cleaning the strainer, the operator opens the bypass. After some time, the exchanger fails to perform, necessitating an unplanned shutdown.

A piping problem peculiar to kettle-type vaporizers is

the manometer effect of connecting the column bottom to the kettle bottom. If the pipe leading the two-phase material from the reboiler to the column is too small or too long, the backpressure may depress the kettle's liquid level below the column's liquid level. Kettle designs assume that liquid surfaces are at the same height, and kettles are positioned accordingly. Because of the manometer effect, top tubes may be exposed when the kettle level is depressed. This is insidious because you may not become aware of it until there is product degradation or mechanical difficulty with the bundle.

Mechanical ailments

The symptoms of leakage between the shellside and tubeside vary with the process parameters. Generally, the fluid from the higher-pressure side appears downstream in the low-pressure stream. But a leak in a distillation column reboiler may cause steam distillation of some high-boilers, and thus alter the overhead product composition; a leak in a condenser may be indicated by water in the product and reduced condenser performance; and rapid corrosion may be symptomatic of acid-cooler leaks.

Most leaks occur in the region of the tubesheets. The list of possible causes is long and includes the following: (1) erosion/corrosion at the tube inlets; (2) cavitation damage at the tube outlets; (3) tubesheet erosion; (4) ligament cracking in one or both tubesheets; (5) tube-end fatigue; (6) tube-end stress-corrosion cracking; (7) tube-end crevice corrosion; (8) corrosion of the tubes in the regions between the tubesheets and adjacent baffles; (9) tube pullout or pushout; (10) relaxation of residual stress required for tightness in expanded joints; (11) root cracking in welded joints, as illustrated in Fig. 10; (12) improper tubehole finish, e.g., axial scores in hole surfaces; (13) tubehole damage and ligament distortion occurring during retubing; (14) erosion damage at the shell inlet; (15) cavitation at the shell outlet; (16) inadequate bundle support at the floating end of a floating-head unit; and (17) vibration damage.

You may also find leaks throughout the length of the tubes. The causes of such leaks are: (1) general corrosion; (2) improper tube fabrication; and (3) vibration damage.

Erosion/corrosion at tube inlets

The evidence of erosion/corrosion at the tube inlets is thinning of the tube-ends, usually on one side, and longitudinal rounded grooves. The cause may be maldistribution or excessive turbulence. If the inlet pass is free of erosion/corrosion, but subsequent passes are not, the crossover area is too small. You can rectify maldistribution as previously discussed. You may protect the tube-ends with sleeves or with perforated plates drilled on the tubesheet pitch-pattern to the tube I.D. When crossover area is inadequate, replace the channel and return cover. Partial retubing is justifiable for this kind of failure, provided that you take corrective steps in order to prevent recurrence.

Impingement attack

This is a form of erosion characterized by sharply defined pits with smooth walls. The pits may be elon-

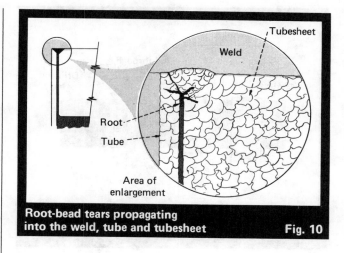

Root-bead tears propagating into the weld, tube and tubesheet   Fig. 10

gated in the direction of flow. They are usually concentrated in the inlet region but may extend throughout the tubes.

Cavitation damage at tube outlets

Cavitation damage at tube outlets appears as erosion, and damage to the tube-ends and tubesheet. You can suppress cavitation by installing a leg or standpipe in the outlet piping to make sure that the outlet chamber is full of liquid under hydrostatic head. Alternatively, you can install a wasting plate drilled on the tubesheet pitch-pattern to the tube I.D. and fitted with the tubesheet. Partial retubing may also be justified for cavitation damage if the tubesheet can be repaired and you take prophylactic action.

Tubesheet erosion

Tubesheet erosion may result from maldistribution, erosive particles in the tubeside inlet stream, leakage past pass-partitions, and from the wire drawing that occurs when tube-to-tubesheet joints leak. If you cannot remove particles and they will not settle out at low flow velocities, increase the inlet nozzle size. Otherwise, install an impact plate in the way of the inlet nozzle.

Leakage past pass-partitions

The causes of leakage past pass-partitions are: (1) bending of the pass-partitions because of differential pressure; (2) a permanently bowed tubesheet; (3) tubesheet deflection under pressure; (4) bowing due to the temperature difference between the channel face and shell face; and (5) insufficient bolting to seat the gasket at the pass-partition ribs.

Bent pass-partitions can be straightened and reinforced. You can repair erosion, and mill and machine the gasket surfaces of dished tubesheets, if they are thick enough. However, if the tubesheets bend so much (under pressure or for thermal reasons) that you cannot stiffen them, the problem will recur. You may reduce the leakage by installing a different kind of gasket if the service conditions, bolting, and flange design permit.

Wire drawing damage

Wire-drawing damage to tubeholes, when the joints leak, is also called "wormholing" (to describe its appear-

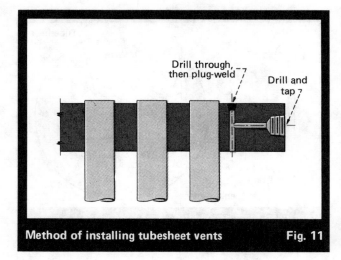

Method of installing tubesheet vents   **Fig. 11**

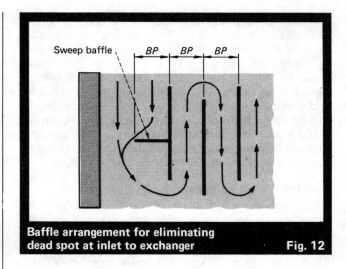

Baffle arrangement for eliminating dead spot at inlet to exchanger   **Fig. 12**

ance). It is most prevalent when the joints are front-face-welded only and there is a high pressure-difference between the shell and the tubes. Thus, this kind of damage is often found in feedwater heaters. You may be able to repair the damage, but it is essential to find out why the joint leaked.

Ligament cracking

The main causes of ligament cracks are excessive heat-flux through the tubesheet, thermal shock (an extreme case of excessive heat-flux), and corrosion from the shell side. The heat flux through the tubesheet thickness is excessive when thermally induced forces in the ligaments combine with pressure forces to create a condition of overstress [10, 11].

Therefore, if the tubesheet is thick, the temperatures of the fluids in contact with the tubesheet faces are far apart, and you find cracking in the ligaments, the cracking has probably resulted from high heat-flux. If the inlet tube-pass temperature is very far from the outlet pass temperature, and ligament cracks are confined to the tube field at the inner edges of the passes, the heat flux between passes is probably too high. Other than redesigning the exchanger or changing the operating parameters, there is little you can do to correct this situation.

In vertical units, ligaments in the top tubesheet may crack because the underside of the tubesheet is not thoroughly washed by the shellside fluids. A layer of noncondensable gases, which may be corrosive, blankets the tubesheet. As a result, the tubesheet metal temperatures may exceed the design basis temperature. After repairs, you can rectify inadequate tubesheet venting by installing vents, as shown in Fig. 11.

Ligament failures in bottom tubesheets are due mostly to corrosion initiated by failure to blow down sediment. Such a tubesheet failure is usually preceded by tube failures near the bottom. However, if you find cracks in the bottom tubesheet of a unit used to generate vapors—a waste-heat boiler, for example—the probable cause is thermal shock. Typically, this results from not providing means to ensure that the bottom tubesheet will always be immersed in shell fluid. Fresh feed falling on the bottom surface cools it so rapidly that the tubesheet is shocked. The effect is equivalent to impact loading.

Another cause of heat shock is fluctuation, on the bottom tubesheet surface, between vapor binding and nucleate boiling. Therefore, to investigate such thermally caused problems, calculate the tubesheet metal temperatures, boiling coefficient on the tubesheet surface, and heat flux in the metal and boiling liquid.

Tube-end fatigue

You may find tube-end fatigue in the rolled regions of joints of tubes that work-harden rapidly. When vibration causes tubes to fatigue, the breaks usually appear where the tubes emerge from the tubesheet. If the joint is front-face-welded only, vibration may cause the welds to crack. Weld cracks usually start in the root and propagate by tearing the tube and tubesheet, as shown in Fig. 10.

Little can be done to repair fatigue failures. You can avoid fatigue in replacement bundles by using a different expanding method or eliminating the cause of vibration. Similar to fatigue failure is tearing of the outer edges of the tube-ends. It occurs when the rolling-tool collar burnishes and hardens the edge. As the tube is expanded, the edge tears or is sensitized to tearing.

Crevice corrosion

Crevice corrosion causes tube failures within the tubesheet. You can minimize the problem by expanding the tube-end so that it makes full contact with the depth of the tubesheet.

Stress-corrosion cracking

Stress-corrosion cracking takes place in the transition between the expanded and unexpanded sections of the tubes, where the tubemetal is in tension. To reduce stress-corrosion in replacement tubes, use equipment and techniques that will produce a tight, strong joint with the least possible wall reduction.

Tube-end corrosion

When your search reveals that the cause of stream mixing is tube leaks between the tubesheets and adjacent baffles, it is likely that the tubes were corroded because there was little or no circulation in that region. You may also find inner-face tubesheet corrosion behind the inlet pass, which indicates that there is acceleration of the corrosion rate due to higher temperature in that part of the tubesheet.

Smith [12] has described the principal causes of dead spots: (1) assembling the bundles to the shells upside down; and (2) construction in which the distance between the tubesheets and adjacent baffles is greater than the baffle pitch in the body of the unit. (This is done to accommodate inlet and outlet nozzles larger in diameter than the desired baffle pitch.) The need to blow down condensate from the shells of vertical units in which a liquid level is maintained has already been discussed.

In addition to creating dead spots (which reduce performance) upside-down assembly eliminates impingement protection, leading to tube erosion at the inlet, and cavitation damage at the outlet. Therefore, when your examination finds reduced performance, tube erosion at the tubes in the way of the inlet nozzle, and tube corrosion at the end-regions, will reveal whether the unit was assembled correctly.

There is little you can do to correct dead spots due to baffle arrangement. But when you order replacement bundles, you should consider some means of ensuring that flow sweeps the end-compartments, such as the arrangement shown in Fig. 12.

Tube pullout or pushout

Leaks between the exchanger sides may occur if the tubes pull out or push out of the tubeholes. High tube-loads are as apt to stretch or collapse the tubes as they are to pull them or push them out of the tubesheets. Therefore, when you discover pullout or pushout, look for accompanying tube damage.

Applying design pressure will not cause tubes to pull or push out of tubesheets if the exchanger meets TEMA or HEI standards and satisfies the ASME Code requirements [13]. To investigate, map the positions of the affected tubes, indicating which pulled loose and which pushed out.

Unidirectional failures, located roughly in the center region, indicate that excessive pressure may have been applied. This may be due to a pressure-relief-valve malfunction, storing a shutdown unit—filled with liquid and sealed—in the sun, or failing to drain water from a shutdown unit when the temperature is below freezing. Overpressure may bend the tubesheets permanently. Also, overpressure resulting from fluid expansion or freezing, may split tubes and damage flanges and bolting. Therefore, examine these parts of the unit, too.

Unidirectional failures located mostly in the periphery of the tube field of a fixed-tubesheet exchanger, built without a shell expansion joint, indicate differential expansion that the tubejoint is not strong enough to restrain. When you see failures in this region, reexamine the calculations of tube- and shell-metal temperatures, and the tubesheet, tubejoint and tube-strength calculations.

When tubes in the hot-pass area of a multipass straight-tube unit push out, and those in the cold-pass field pull out, the cause is too great a temperature range. Other indications are the bundle's assuming a banana shape in floating units, and tube buckling in the hot pass.

Randomly distributed pulled-out tubes in straight-tube flue-gas/air exchangers indicate that some tubes are becoming fouled more rapidly than others. Flue gas ordinarily flows inside the tubes; therefore, the fouled-tubemetal temperatures are closer to the cool-air temperature than the temperatures of the clean tubes, and pull loose.

The duration of residual stress in an expanded joint is a function of the tube and tubesheet creep-behavior during service. When residual stress relaxes, after a substantial period of successful operation, the tubes usually leak—well before the joint strength is seriously reduced. If the tubes and tubesheet appear to be in good condition, you may restore tightness by re-expanding.

Root cracking in welded joints [16]

When you discover leaks in welded tube-to-tubesheet joints, the temptation is to make a quick fix by chipping out the crack, cleaning the area and rewelding. But unless you determine the mechanism and cause of the failure, the repairs will solve the problem only temporarily and there will be a continuing series of failures. Repeated fixes will eventually make the tubejoints incapable of being repaired.

Failures found soon after the exchanger goes into service are usually due to improper fabrication, but may be due to faulty joint-design. The kinds of failures to expect are subsurface porosity and wormholing (used here to describe spiral leak-paths in the weld), burn-throughs in the tubes, and weld-root cracks. Failures after a period of successful service can usually be traced to weld-root cracking and propagation in the following locations, listed in order of frequency: (1) through the weld; (2) through the weld and tubewall; and (3) through the weld, tubewall and tubesheet, as shown in Fig. 10.

Whether you inspect visually, or employ radiography or other nondestructive examinations, you will probably not find out how or why root-bead tear originated. For a complete analysis, you should subject samples of the failed joints to chemical and metallurgical study. In light of the knowledge gained, you may be able to prescribe appropriate repairs, and measures to prevent recurrence.

Improper tube-hole finish

You can hardly ever expand tubes tightly into holes when the contact surfaces are axially scored. If you do get a joint tight, it will probably leak soon after it goes into service. It is frustrating to find a few joints in the tube field that cannot be sealed by moderate increases in hydraulic expanding pressure or rolling torque. Because the tube- or hole-damage is hidden, you cannot know that the problem stems from a scratched hole or tube. A frustrated worker may be tempted to increase pressure or torque enough to deform the ligaments to the point of loosening adjacent tubejoints.

During construction, axial damage incidental to drilling may escape detection. The tool used for machining annular grooves into the holes may upset the groove-edge metal. If the upset material is not removed, the tube will be scratched during insertion. In addition to the joint-sealing problem, tube life may be reduced by premature corrosion along the scratches.

The clue to improper finish is randomly located leaks that persist after expanding and reexpanding. The rem-

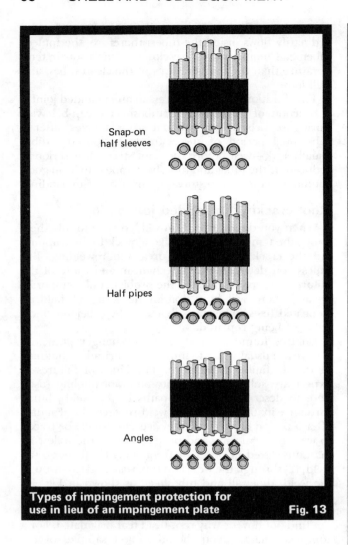

Snap-on
half sleeves

Half pipes

Angles

**Types of impingement protection for
use in lieu of an impingement plate**   **Fig. 13**

edy is to extract the tubes from the leaking holes, refinish the holes and install new tubes.

Tubehole damage

During retubing, you may score the holes when you remove the old tubes. The edges of annular grooves may be upset as you pull the tubes, subsequently scratching the new tubes as you insert them. The conditions under which field retubing is done are frequently difficult, and axial scoring and upset metal may go unnoticed. Detect and treat this problem as above.

Ligament distortion during retubing

The tendency during retubing is to assume that you must apply higher rolling torque or expanding pressure than for building a new unit, in order to get the joints tight. If the hole finish has not been damaged or if damage has been repaired, the assumption is not justifiable. Excessive rolling can misshape the holes along the lines of the pitch pattern. You can detect this kind of damage to the ligaments by rotating a tube micrometer inside the tube at the tubesheet. There is little you can do to correct the situation. Depending upon the service conditions, it may be satisfactory to ream the distorted holes oversize and expand both the tube and an interposed ferrule into the hole.

Erosion damage at the shell inlet

The symptoms of erosion damage at the shell inlet are pits, similar to tubeside impingement attack, and leaks in the tubes located at the inlet region of the shell. Some causes are: (1) lacking, or inadequate, impingement protection; (2) unforeseen flow conditions; (3) insufficient escape area around the impingement plate; and (4) insufficient calming length when there is an increase from the supply pipe diameter to the inlet pipe size.

Impingement protection is required whenever liquid can be entrained in an entering vapor; when solid particles are suspended in the inlet stream; under most two-phase flow conditions; and when the product of density ρ and square of velocity V exceeds the value of the design standard. In the TEMA Standards ρV^2 may not exceed 1,500 lb/(ft)(sec^2) [22.32 kg/m-s^2] for noncorrosive, nonabrasive single-phase fluids and 500 lb/(ft)(s) [7.44 kg/m-s] for all other liquids, including liquids at their boiling points. For all other gases and vapors, including nominally saturated vapors, and for liquid-vapor mixtures, the standards require impingement protection. You may use a properly designed diffuser to reduce line velocities under the TEMA Standards.

Under appropriate conditions, it is proper to omit an impingement plate. When you find erosion damage and no impingement protection, compare the actual ρV^2 with the ρV^2 assumed in design. If an impingement plate has been omitted, but is now required, there will probably not be enough room to install one that allows adequate fluid escape area (equal to inlet cross-sectional area). Alternatives are: (1) snap commercially available tube protectors over the first two or three tube rows in the way of the inlet; (2) shield the tubes with formed half-cylinders or angles—see Fig. 13; and (3) install a dome, diffuser-reducer or distributor belt.

When you find erosion despite the use of an impingement plate, either the impingement plate is too small or there is not enough escape area. The edges of the plate should extend about 1 in. (25 mm) beyond the nozzle I.D. You may have to install a larger plate or make the changes described above.

It is a reasonable design option to omit an impingement plate by using a large inlet nozzle to hold ρV^2 to an acceptable level. This may be done when the available pressure-drop is very small and the turbulence created by an impingement plate cannot be tolerated. But, if the supply pipe to the shell is smaller than the inlet, you can anticipate erosion unless the enlargement reducer (diffuser) is placed approximately 20 inlet-nozzle diameters before the inlet. Alternatively, a long-radius elbow, connecting the supply to the inlet, provides protection.

Cavitation at the shell outlet

Cavitation may occur when the shell-discharge flushes down a vertical pipe without restriction, and the resulting reduced pressure induces bubbles of vapor to form in the fluid. In addition to cavitation, there is usually vortexing. After you repair the damage, install a vortex breaker in the outlet and a trap in the outlet piping.

Inadequate bundle support

When a floating tubesheet overhangs, the tubes may sag. This permits the floating assembly to drag along the shell during bundle installation and pulling. If the unit is a packed type, the sealing surface (which requires a 63 rms finish to seal properly) is marred. No amount of extra packing or gland-follower pressure will cure the leak. In all types of floating equipment, failure to provide a support at the floating end can lead to tube damage in the regions of the last baffle and floating tubesheet.

It is not practical to install a drilled support-plate to correct rear-end bundle support deficiency. If geometry permits, you may be able to fasten a shoe to the inner face, or bottom, of the tubesheet to provide some support.

Vibration damage

Symptoms of vibration damage are: (1) tube leaks at baffle penetrations (this may also indicate crevice corrosion); (2) a pattern of leaks midway between baffles or between the baffles and adjacent tubesheets; (3) leaks at U-bends; and (4) cracks in the tubes where they emerge from the tubesheets. If there are indications of vibration damage, it will probably be necessary to perform a complete vibration analysis to determine what to do about it.

Some remedies are: (1) installing deresonating baffles (when the vibration-inducing mechanism is acoustical coupling); (2) lacing the tubes with wire between the baffles; (3) sliding flat bars between the tubes at locations between the baffles; (4) changing the flow pattern to divided-, split- or cross-flow when thermally possible; (5) altering the character of the shellside stream; (6) retubing with tubes that have a natural frequency in a safe range; and (7) replacing the bundle with one designed to be free of vibration damage.

Improper tube fabrication

It is hard to determine that a tube has failed because of improper manufacture, after an exchanger has been in service. The starting point of the investigation is tube cleaning, followed by borescopic examination. But, to confirm your suspicion, you must extract the tube without damaging it. In U-tube bundles, you can pull only tubes in the outer row. Even when you can extract tubes, the prospect is unpalatable because it shuts down production. The common practice is to plug the ends when a few tubes leak and return the unit to service until a scheduled maintenance-shutdown period.

(Remember that if you operate the exchanger subsequent to installing the plug, you may remove evidence of faulty manufacture.)

When you examine tube failures in welded tubes, look for incomplete fusion, inclusions, and burn-throughs. In seamless and welded-and-drawn tubes, look for die and mandrel scratches. These are longitudinal gouges in the tubewall. Tubes may also fail because of the mill straightening operation or improper heat treatment. The innermost bends of U-tube bundles are the most severely stretched and compressed of all the tubes in the assembly. Improper bend-fabrication in the inner rows appears as tube flattening and crimping.

In addition to the troubles symptomized by leakage between the sides, you may have to deal with the following conditions.

Leaking gasketed joints

If the force required to seal the gaskets at the pass-partition ribs was neglected in the design, the bolting may be insufficient, and the flanges (or rims of tubesheets extended as flanges) may bend excessively. This condition is evidenced by stretched bolts, cupped flanges and uncompressed gasket ribs. Short of modifying the flanges and bolting, all you can do is to substitute a more easily compressed gasket after truing the parts.

Weld shrinkage tends to cup ring-flanges and fixed tubesheets at their joints to the shell. Expanding tubes into tubesheets tends to bow tubesheets. Because of these tendencies, the ring of gasket that seals the mating parts may be pinched at its outer or inner edge. A small relaxation of bolt stress or gasket compression permits leakage. You may stop the leak by retightening, but it will probably recur. If you torque the bolts too much, you may deform the flanges beyond repair. The cure is to machine the gasket surfaces flat. If you cannot shut down to find the cause of the leak, insert a spacer between the flanges at the edge beyond the bolt holes, to avoid bending the flanges when you retighten.

If you disassemble a unit for cleaning and you cannot get it tight upon reassembly, the obvious suspicion is that the gasket surface is scratched or dirty. However, the most frequent cause is flange-warping caused by an improper unbolting procedure. If you unbolt consecutively around the bolt circle, you are very likely to bend the flanges. To prevent the bending, unbolt in the criss-cross pattern recommended in the TEMA and HEI Standards for bolting up. Loosen the bolts in several passes through the pattern, never completely unbolting a bolt on the first or second pass.

Unexplained stream mixing when the tubes and tube-to-tube-sheet connection are tight may be due to leakage through the flanged joint at the floating end. In split-ring units, see if the split ring has twisted or if the bolts are bent. These conditions reduce the force available to hold the return cover tightly to the tubesheet. The gasket may then relax and the seal will be broken. To restore tightness, remachine or replace the split ring. You can prevent recurrence of the bending by machining ·the tubesheet and split ring to the configuration shown in Fig. 14.

When flanged joints leak in high-temperature service, the bolts have probably relaxed. If you find the leaks early enough, the gasket may not be damaged and you can usually seal the leak by carefully retorquing all of the bolts in the original bolting sequence. To prevent flange damage during retorquing, insert the previously described split spacer between the flange edges. You may increase the interval between retorquing bolts that creep, by installing spring washers under the nuts. But remember that spring washers are intended to take up bolt thermal expansion and will not eliminate the need to retighten. The least bolt relaxation will occur if all of the bolts are loaded identically. Therefore, consider using a hydraulic bolt tensioner and ultrasonic bolt-stress measurement.

Failures in shell expansion joints

Expansion joints fail because of corrosion, fatigue, overpressure, larger-than-design axial deflections, unan-

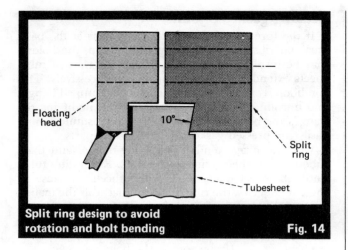

Split ring design to avoid rotation and bolt bending   Fig. 14

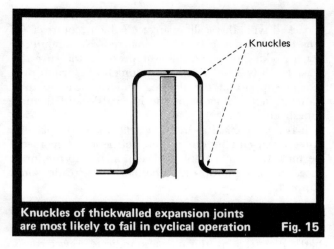

Knuckles of thickwalled expansion joints are most likely to fail in cyclical operation   Fig. 15

ticipated angular deflections and offsets, or some combinations of these causes. The elements of bellows expansion joints are thin—0.035 to 0.093 in. (0.89 to 2.36 mm). Furthermore, most configurations cannot be fitted with vents and drains. Therefore, liquids flowing through horizontal shells are entrapped in the bottom of the joint, and noncondensables in the top. The directions of pressure and deflection stresses, and entrapment of process fluids, create optimum conditions for stress-corrosion cracking of austenitic steels (from which bellows are ordinarily fabricated). Fluids high in chloride-ion content cause stress-corrosion cracking. Therefore, in running down the source of failure, investigate the fluid composition and the bellows material.

Thick expansion joints, made from flanged-only heads, flanged-only heads with flue holes, or similar construction, are ⅛ in. (3.18 mm) or thicker. There is usually room in the straight flanged section for vents and drains without impairing joint life. Furthermore, thick joint elements are often made of carbon steel, which is less sensitive. Therefore, stress-corrosion cracking is seldom seen in this kind of element. Corrosion in thick joints usually results from not piping up the vents and drains.

Testing expansion joints

The Standards of the Expansion Joint Manufacturers Association (EJMA), which apply to bellows expansion joints, require fatigue analysis and testing [17]. For this reason, fatigue failures in bellows joints are improbable if the directions and amplitudes of deflections do not exceed design specifications, and cycle life has not expired. However, a small discontinuity, such as a dent or weld drag, can drastically reduce cycle life.

With some restrictions, expansion joints meeting the EJMA Standards may be stamped for Code construction. The joint manufacturer is not required to hydrostatically test the joint; the responsibility falls on the heat-exchanger manufacturer who installs it. A leak in a bellows joint can be disastrous, because there may be no acceptable way to repair it. Most bellows expansion-joint manufacturers will hydrostatically test on demand.

Thick-walled joints are not yet subject to specific Code rules for design analysis. The manufacturer is responsible for details of design and construction as safe as those provided by Code rules. Therefore, when a thick-walled joint fails, part of the effort to discover the cause is a review of the manufacturer's design analysis, with emphasis on how cycle life was established. Most thick-walled-joint fatigue failures occur in the knuckles, as shown in Fig. 15. Joints in which the knuckle radii are less than three metal thicknesses are candidates for low cycle life.

The probability of fatigue failure is increased when deflections are imposed that the design did not anticipate. Examples are: (1) shop precompression or extension during fit-up; (2) offsets in installation; (3) anchoring both supports, which prevents free movement and causes angular deflection; and (4) twists imposed by nozzle loads.

When bellows joints are overpressured, the side wall may be permanently bulged and the joint may squirm. These troubles are visible—in sidewall bulge there is a permanent reduction in the space between sidewalls of 7%, and in inplane squirm, the change of the pitch between adjacent convolutions is 15% at any measurement location. Within these limits there is not a catastrophic failure, but the joint life is severely reduced. Beyond this point the root of the joint bulges and there is general and continued deformation of the bellows with small increases in pressure, until the joint bursts. After an episode of overpressure, it is essential to inspect the bellows for damage.

The most highly stressed areas of thick-walled joints are at the outer knuckle or edge, and at the connection to the shell. Overstress may relieve itself by yielding at these localities; nevertheless that is where to look for signs of overpressure damage.

When you investigate why a joint fatigued, bear in mind that not only the frequency of deflection but also the amplitude and direction affect joint life. You may not be able to measure the actual deflection to compare it with the assumed; therefore, review the calculated deflection based on design conditions and compare it with deflection calculated for actual operations. Look for the possibility that the free end is restrained. This can bend the unit at the expansion joint, thereby reducing joint life.

What can you do about a broken shell expansion joint? It is futile to try to repair bellows joints, and nearly so to repair thick joints. You may extend the useful life of the exchanger by replacing thick joints in sections; dead-heading the expansion joint and operating without it; installing a larger bellows joint over the cut-out broken bellows; or replacing the joint in a retubing operation, using existing tubes shortened somewhat. Aside from the first option, it is necessary to examine the thermal and mechanical effects of the repair.

Packed floating-head problems

Packed joints are *supposed* to leak slightly during operation. You create problems when you try to eliminate the last vestige of a leak. By squeezing the packing too tightly against the floating tubesheet barrel, you restrain the bundle from moving relative to the shell. The restraint of movement may show up as either tubejoint leaks or bent tubes.

If a packed joint leaks excessively (more than five to ten drops a minute), tighten the gland-follower bolts uniformly. If a moderate amount of tightening does not correct the problem, replace the packing. If you simply shove an additional ring of packing into the packing box, you may jam the inner, hardened rings into the barrel and score it. Once the barrel is scratched or scraped along its length, the joint will always leak (unless you refinish the barrel).

Packing problems often stem from cocking the gland follower as you tighten it against the packing. If you have cocked the follower, back it off, loosen the packing, and retighten uniformly.

Summary

In troubleshooting exchanger problems, consider the interactions of thermal and mechanical design. Gather as much design, construction and operating information as you can. Consider the nature of the problem and the possibility that the trouble lies elsewhere in the system.

Match design-assumptions with actual operating conditions. Search for upsets and flow regimes that deviate from those anticipated. Look into how clean conditions and excess surface affect the system and structure. When overcapacity has been provided, find out how the excess surface was furnished and determine its effect on normal operation.

Before you conclude that equipment is underdesigned, determine if the piping and mechanical arrangements are suitable. Examine equilibrium relationships, heat-release curves, effects of noncondensables and venting, and boiling-point elevation. Search also for incorrect channel or bundle assembly, fluid bypassing around channel pass-partitions, short-circuiting on the shell side, leakage between baffles and shells and tubes and baffle holes, and inadequate seal strips or dummy-tube placement.

Extend your senses with instruments and nondestructive testing techniques to detect the causes of symptoms. To troubleshoot leaks between the shell and tube sides, scrutinize the regions near the tubesheets for corrosion, erosion, fatigue and vibration damage. Locate the positions of tube perforations relative to baffle positions, for vibration indications. Examine the tubes for possible manufacturing flaws and look into floating-head closure integrity.

When flanged joints leak, examine flange calculations. Determine if pass-partition gasket compression was considered. Look also for deformations resulting from weld shrinkage and improper bolting and unbolting procedures.

When you troubleshoot expansion-joint failures, consider how drainage and venting affect the corrosion resistance of the material of construction. Examine the amplitude, direction and frequency of deflections. Inspect joints after excursions of excess pressure.

Packed-expansion-joint problems arise from over-tightening gland followers, scoring floating-head barrels and cocking gland followers. Replace hardened packing when joints leak, in preference to adding rings.

References

1. Gilmour, C. H., Troubleshooting Heat-Exchanger Design, *Chem. Eng.*, June 19, 1967, p. 22.
2. Discrepancies Between Design and Operation of Heat Transfer Equipment, *Chem. Eng. Prog.*, Jan. 1961, p. 71.
3. Rubin, F. L., Multistage Condensers: Tabulate Design Data to Prevent Errors, *Hydrocarbon Process.*, June 1982, pp. 133-135.
4. "Equipment Testing Procedure, Heat Exchanger—Section 1 Vaporization, Condensation, and Sensible Heat Transfer in Shell-and-Tube and Double-Pipe Heat Exchangers," 2nd ed., American Institute of Chemical Engineers, New York, N.Y., 1980.
5. "ASME Performance Test Code, PTC 19.2. Pressure Measurement (1964), PTC 19.3, Temperature Measurement (1974), and PTC 19.5, Flow Measurement (1972), American Soc. of Mechanical Engineers, New York, N.Y.
6. Kern, D. Q., "Process Heat Transfer," McGraw-Hill, New York, N.Y., 1950.
7. Standiford, F., Effect of Non-condensibles on Condenser Design and Heat Transfer, *Chem. Eng. Prog.*, July 1979, pp. 59–62.
8. Meisinberg, S. J., others, *Trans. AIChE*, Vol. 31, pp. 622–38 (1935), and Vol. 32, pp. 100–104, 449–450 (1950).
9. Shah, G. C., Troubleshooting Reboiler Systems, *Chem. Eng. Prog.*, July 1979, pp. 53–58. (Presents reboiler problems from a user standpoint.)
10. Gardener, K. A., Heat Exchanger Tubesheet Temperature, *The Refiner and Natural Gasoline Manufacturer*, Mar. 1942.
11. Gupta, J. P., others, An Approximate Method for Estimating the Temperature Field in Tubesheet Ligaments of Tubular Heat Exhangers Under Steady State Conditions, Private Paper, Joseph Oat Co., Camden, N.J.
12. Smith, H. E., The Interrelationships Between Codes, Standards, and Customer Specifications for Process Heat Exchangers—A User's Standpoint. Presented at the Winter Annual Meeting of the ASME, New York, N.Y., Dec. 2–7, 1979.
13. "Standards of Tubular Exchanger Manufacturers Association," 6th ed., The Tubular Exchanger Manufacturers Assn., Tarrytown, N.Y., 1978.
14. "Standards for Power Plant Heat Exchangers," 1st ed., The Heat Exchanger Institute, Cleveland, Ohio, 1980.
15. "ASME Boiler and Pressure Vessel Code," American Soc. of Mechanical Engineers, New York, N.Y., 1983.
16. Norris, E. B., others, Considerations in Design of Tube-to-Tubesheet Joints in High Temperature Heat Exchangers Equipment, ASME Paper 68-PUP-11. Presented at the Joint Conference of the Pressure Vessel and Piping Div. Dallas, Tex., Sept. 22-25, 1968.
17. "Standards of the Expansion Joint Manufacturers Association," 5th ed., The Expansion Joint Manufacturers Assn., Inc., Tarrytown, N.Y., 1980.

The author

Stanley Yokell is Senior Consultant for Energy and Resource Consultants, P.O. Drawer O, Boulder, CO 80306 (phone: 303-449-5515) and Suite 5e, The Livery, 209 Cooper Ave., Upper Montclair, NJ 07043 (phone: 201-744-5545), where he is engaged in management and engineering consulting. He holds a B.Ch.E. from New York University, and has done postgraduate work at Newark College of Engineering. He is a member of AIChE, a licensed professional engineer in New Jersey and Colorado, and a member of the Special Working Group on Heat Transfer Equipment, Sect. VIII, ASME Pressure Vessel Code Committee.

Determining tube counts for shell-and-tube exchangers

To design a heat exchanger, the engineer must know the total number of tubes that can fit into a shell of a given diameter—that is, the tube count. Here is a mathematical method and a set of tables for determining tube counts, or for solving other problems that involve laying out small circles within a larger circle.

P. S. Phadke, Consultant

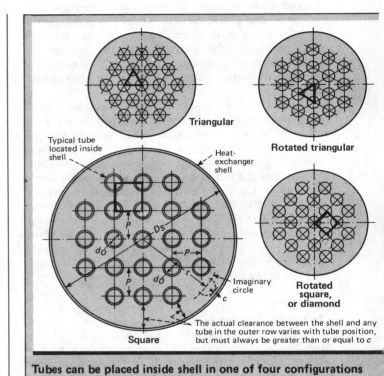

Tubes can be placed inside shell in one of four configurations

☐ The number of tubes that can be accommodated in a shell of a given inside diameter is known as the tube count. This parameter varies considerably, depending on: (1) the outside diameter of the tubes, (2) the tube pitch, (3) the layout configuration—square or triangular, or rotated square or rotated triangular, as shown in the figure, (4) the minimum clearance between the outermost tubes and the inside of the shell, and (5) the number of exchanger passes.

The conventional method of obtaining the tube count—plotting the layout and counting the tubes (thus, the term "tube count")—is cumbersome, time-consuming and prone to errors, and tables often cover only certain standard combinations of the parameters listed above. This mathematical approach and set of tables eliminates the disadvantages of the plot/count method, and can accommodate any configuration. Because the equations are presented in dimensionless form, the procedure is completely independent of units, so any consistent system (English, metric or SI) can be used.

Basis

This technique is based on number theory, and the detailed analysis is too complicated to go into here. How-

ever, some interesting generalizations can be made: Tubes are located only at certain radii. Triangular layouts are governed by the prime number 3 and all primes of the form $6m + 1$, and square layouts are governed by the prime number 2 and primes of the form $4m + 1$ (where m is any integer ≥ 1). For single-pass arrangements, the tube counts themselves are of the form $6m + 1$ for triangular configurations and $4m + 1$ for square ones.

The tube counts obtained for single-pass exchangers are exactly as calculated by the theory. However, for 2-, 4-, 6- and 8-pass arrangements, a certain number of tubes will have to be removed from the theoretical single-pass shell to accommodate the pass partition plates. For these cases, the following assumptions have been made:

■ The partition plate is located where a row of tubes would have been, and only one row, or at the most two, will be affected.

■ The thickness of the partition plate is less than 70% of the tube outside diameter.

■ The distance between the centerline of the partition plate and the centerline of the nearest row of tubes is equal to the pitch.

Originally published September 3, 1984

Procedure

For single-pass situations, the calculations are simple and straightforward. These are discussed first. For other arrangements, the initial steps are the same, but some further calculations are required.

Triangular-layout tube counts — Table I

N_S	C_1	N_S	C_1	N_S	C_1	N_S	C_1	N_S	C_1
0	1	169	613	364	1333	577	2095	787	2869
1	7	171	625	367	1345	579	2107	793	2893
3	13	172	637	372	1357	588	2125	796	2905
4	19	175	649	373	1369	589	2149	804	2917
7	31	181	661	379	1381	592	2161	811	2929
9	37	183	673	381	1393	597	2173	813	2941
12	43	189	685	387	1405	601	2185	817	2965
13	55	192	691	388	1417	603	2197	819	2989
16	61	193	703	397	1429	604	2209	823	3001
19	73	196	721	399	1453	607	2221	829	3013
21	85	199	733	400	1459	613	2233	831	3025
25	91	201	745	403	1483	619	2245	832	3037
27	97	208	757	409	1495	624	2257	837	3049
28	109	211	769	412	1507	625	2263	841	3055
31	121	217	793	417	1519	628	2275	844	3067
36	127	219	805	421	1531	631	2287	847	3079
37	139	223	817	427	1555	633	2299	849	3091
39	151	225	823	432	1561	637	2335	853	3103
43	163	228	835	433	1573	643	2347	859	3115
48	169	229	847	436	1585	651	2371	867	3121
49	187	237	859	439	1597	652	2383	868	3145
52	199	241	871	441	1615	657	2395	871	3169
57	211	243	877	444	1627	661	2407	873	3181
61	223	244	889	448	1639	669	2419	876	3193
63	235	247	913	453	1651	673	2431	877	3205
64	241	252	925	457	1663	675	2437	883	3217
67	253	256	931	463	1675	676	2455	889	3241
73	265	259	955	468	1687	679	2479	892	3253
75	271	268	967	469	1711	684	2491	900	3259
76	283	271	979	471	1723	687	2503	903	3283
79	295	273	1003	475	1735	688	2515	907	3295
81	301	277	1015	481	1759	691	2527	912	3307
84	313	279	1027	484	1765	700	2539	916	3319
91	337	283	1039	487	1777	703	2563	919	3331
93	349	289	1045	489	1789	709	2575	921	3343
97	361	291	1057	496	1801	711	2587	925	3355
100	367	292	1069	499	1813	721	2611	927	3367
103	379	300	1075	507	1831	723	2623	931	3403
108	385	301	1099	508	1843	724	2635	937	3415
109	397	304	1111	511	1867	727	2647	939	3427
111	409	307	1123	513	1879	729	2653	948	3439
112	421	309	1135	516	1891	732	2665	949	3463
117	433	313	1147	523	1903	733	2677	961	3481
121	439	316	1159	525	1915	739	2689	964	3493
124	451	324	1165	529	1921	741	2713	967	3505
127	463	325	1177	532	1945	751	2725	972	3511
129	475	327	1189	541	1957	756	2737	973	3535
133	499	331	1201	543	1969	757	2749	975	3547
139	511	333	1213	547	1981	763	2773	976	3559
144	517	336	1225	549	1993	768	2779	981	3571
147	535	337	1237	553	2017	769	2791	988	3595
148	547	343	1261	556	2029	772	2803	991	3607
151	559	349	1273	559	2053	775	2815	993	3619
156	571	351	1285	567	2065	777	2839	997	3631
157	583	361	1303	571	2077	784	2857	999	3643
163	595	363	1309	576	2083				

Nomenclature

c — Minimum clearance between the outermost tube and the inside of the shell, in. or mm

C_1 — Tube count for single-pass exchanger

C_2 — Tube count for 2-pass exchanger

C_4 — Tube count for 4-pass exchanger

C_6 — Tube count for 6-pass exchanger

C_8 — Tube count for 8-pass exchanger

C_x — Number of tubes along horizontal axis

C_y — Number of tubes along vertical axis

d_o — Outside tube dia., in. or mm

D_s — Inside dia. of shell, in. or mm

e — Dimensionless constant; 0.265 for 6-pass arrangement and 0.404 for 8-pass arrangement

N_r — Largest integer equal to or less than r

N_s — Largest integer equal to or less than s

P — Tube pitch, in. or mm

r — Dimensionless radial span, or the distance at which the center of the farthest tube may be located from the center of the shell, in order to maintain the minimum clearance, c

s — r^2 (dimensionless)

u, u_1, u_2, v, w, w_1, w_2, z, z^*, z_1 and z_2 are dimensionless dummy variables.

The relationship between v and N_v, w and N_w, z and N_z, z_1 and N_{z_1}, and z_2 and N_{z_2} is the same as the relationship between r and N_r and s and N_s; i.e., N_x is the largest integer equal to or less than x.

Table I applies to triangular and rotated triangular layouts; Table II, to square and rotated square (or diamond) ones.

Single-pass arrangements

First, calculate the two basic dimensionless quantities:

$$r = \frac{0.5(D_s - d_o) - c}{P} \qquad (1)$$

$$s = r^2 \qquad (2)$$

Then determine N_r and N_s, the largest integers equal to or less than r and s, respectively.

Locate, in the N_s column of the appropriate table, the calculated value of N_s, or use the next smaller value listed. The corresponding value in the C_1 column is the tube count for a single-pass arrangement.

Triangular layouts: 2- or 4-pass

First calculate:

$$w = \frac{2r}{\sqrt{3}} \qquad (3)$$

Then calculate the corresponding integer N_w; as well as C_x and C_y by the following equations:

$$C_x = 2N_r + 1 \qquad (4)$$

$$C_y = 3N_w \qquad \text{if } N_w \text{ is even} \qquad (5a)$$

$$C_y = 3N_w + 1 \qquad \text{if } N_w \text{ is odd} \qquad (5b)$$

For *triangular* 2-pass layouts:

$$C_2 = C_1 - C_x \qquad (6)$$

and for *rotated triangular* 2-pass layouts:

$$C_2 = C_1 - C_y - 1 \qquad (7)$$

For either *triangular* or *rotated triangular* 4-pass configurations:

$$C_4 = C_1 - C_x - C_y \qquad (8)$$

Triangular layouts: 6- or 8-pass

For *triangular* configurations, calculate:

$$v = \frac{2er}{\sqrt{3}} + 0.5 \qquad (9)$$

and the corresponding value of N_v. Then calculate:

$$u = \frac{\sqrt{3}N_v}{2} \qquad (10)$$

$$z = \sqrt{s - u^2} \qquad \text{if } N_v \text{ is even} \qquad (11a)$$

$$z = \sqrt{s - u^2} - 0.5 \qquad \text{if } N_v \text{ is odd} \qquad (11b)$$

and the corresponding integer, N_z.
For a 6-pass layout:

$$C_6 = C_1 - C_y - 4N_z - 1 \qquad (12)$$

For an 8-pass arrangement:

$$C_8 = C_4 - 4N_z \qquad (13)$$

For *rotated triangular* layouts, let:

$$v = 2er \qquad (14)$$

Calculate N_v, and then:

$$u_1 = 0.5N_v \qquad (15)$$

$$z = \sqrt{s - u_1^2} \qquad (16)$$

$$w_1 = \frac{2z}{\sqrt{2}} \qquad (17)$$

$$u_2 = 0.5(N_v + 1) \qquad (18)$$

$$z^* = \sqrt{s - u_2^2} \qquad (19)$$

$$w_2 = \frac{2z^*}{\sqrt{3}} \qquad (20)$$

$$z_1 = 0.5(w_1 + 1) \qquad \text{if } N_v \text{ is odd} \qquad (21a)$$

$$z_1 = 0.5w_1 \qquad \text{if } N_v \text{ is even} \qquad (21b)$$

$$z_2 = 0.5w_2 \qquad \text{if } N_v \text{ is odd} \qquad (22a)$$

$$z_2 = 0.5(w_2 + 1) \qquad \text{if } N_v \text{ is even} \qquad (22b)$$

and the integers N_{z_1} and N_{z_2}, in the usual manner.
For a 6-pass layout:

$$C_6 = C_1 - C_x - 4(N_{z_1} + N_{z_2}) \qquad (23)$$

For an 8-pass arrangement:

$$C_8 = C_4 - 4(N_{z_1} + N_{z_2}) \qquad (24)$$

Square layouts: 2- or 4-pass

For *square* configurations, calculate C_x, using Eq. (4), and then find:

$$C_y = C_x - 1 \qquad (25)$$

Use Eq. (6) or (8) to determine C_2 or C_4, respectively.

Square-layout tube counts — Table II

N_S	C_1	N_S	C_1	N_S	C_1	N_S	C_1	N_S	C_1
0	1	164	517	365	1153	578	1813	794	2501
1	5	169	529	369	1161	580	1829	797	2509
2	9	170	545	370	1177	584	1837	800	2521
4	13	173	553	373	1185	585	1853	801	2529
5	21	178	561	377	1201	586	1861	802	2537
8	25	180	569	386	1209	592	1869	808	2545
9	29	181	577	388	1217	593	1877	809	2553
10	37	185	593	389	1225	596	1885	810	2561
13	45	193	601	392	1229	601	1893	818	2569
16	49	194	609	394	1237	605	1901	820	2585
17	57	196	613	397	1245	610	1917	821	2593
18	61	197	621	400	1257	612	1925	829	2601
20	69	200	633	401	1265	613	1933	832	2609
25	81	202	641	404	1273	617	1941	833	2617
26	89	205	657	405	1281	625	1961	841	2629
29	97	208	665	409	1289	626	1969	842	2637
32	101	212	673	410	1305	628	1977	845	2661
34	109	218	681	416	1313	629	1993	848	2669
36	113	221	697	421	1321	634	2001	850	2693
37	121	225	709	424	1329	637	2009	853	2701
40	129	226	717	425	1353	640	2017	857	2709
41	137	229	725	433	1361	641	2025	865	2725
45	145	232	733	436	1369	648	2029	866	2733
49	149	233	741	441	1373	650	2053	872	2741
50	161	234	749	442	1389	653	2061	873	2749
52	169	241	757	445	1405	656	2069	877	2757
53	177	242	761	449	1413	657	2077	881	2765
58	185	244	769	450	1425	661	2085	882	2769
61	193	245	777	452	1433	666	2093	884	2785
64	197	250	793	457	1441	673	2101	890	2801
65	213	256	797	458	1449	674	2109	898	2809
68	221	257	805	461	1457	676	2121	900	2812
72	225	260	821	464	1465	677	2129	901	2837
73	233	261	829	466	1473	680	2145	904	2845
74	241	265	845	468	1481	685	2161	905	2861
80	249	269	853	477	1489	689	2177	909	2869
81	253	272	861	481	1505	692	2185	914	2877
82	261	274	869	482	1513	697	2201	916	2885
85	277	277	877	484	1517	698	2209	922	2893
89	285	281	885	485	1533	701	2217	925	2917
90	293	288	889	488	1541	706	2225	928	2925
97	301	289	901	490	1549	709	2233	929	2933
98	305	290	917	493	1565	712	2241	932	2941
100	317	292	925	500	1581	720	2249	936	2949
101	325	293	933	505	1597	722	2253	937	2957
104	333	296	941	509	1605	724	2261	941	2965
106	341	298	949	512	1609	725	2285	949	2981
109	349	305	965	514	1617	729	2289	953	2989
113	357	306	973	520	1633	730	2305	954	2997
116	365	313	981	521	1641	733	2313	961	3001
117	373	314	989	522	1649	738	2321	962	3017
121	377	317	997	529	1653	740	2337	964	3025
122	385	320	1005	530	1669	745	2353	965	3041
125	401	324	1009	533	1685	746	2361	968	3045
128	405	325	1033	538	1693	754	2377	970	3061
130	421	328	1041	541	1701	757	2385	976	3069
136	429	333	1049	544	1709	761	2393	977	3077
137	437	337	1057	545	1725	765	2409	980	3085
144	441	338	1069	548	1733	769	2417	981	3093
145	457	340	1085	549	1741	772	2425	985	3109
146	465	346	1093	554	1749	773	2433	986	3125
148	473	349	1101	557	1757	776	2441	997	3133
149	481	353	1109	562	1765	778	2449	1000	3149
153	489	356	1117	565	1781	784	2453		
157	497	360	1125	569	1789	785	2469		
160	505	361	1129	576	1793	788	2477		
162	509	362	1137	577	1801	793	2493		

For *rotated square*, or *diamond*, layouts, let:

$$w = \frac{r}{\sqrt{2}} \qquad (26)$$

and find the integer N_w as usual. Then calculate:

$$C_x = 2N_w + 1 \qquad (27)$$

and C_y from Eq. (25).

Use Eq. (6) or (8) to determine C_2 or C_4, respectively.

Square layouts: 6- or 8-pass

For *square* layouts, calculate:

$$v = er + 0.5 \qquad (28)$$

and N_v, as well as:

$$z = \sqrt{s - N_v^2} \qquad (29)$$

and N_z.

Use Eq. (12) or (13) to find C_6 or C_8, respectively.

For *diamond* configurations, calculate:

$$v = \sqrt{2}er \qquad (30)$$

and N_v, in the standard manner. Find:

$$u_1 = \frac{N_v}{\sqrt{2}} \qquad (31)$$

and z, by Eq. (16). Then calculate:

$$w_1 = \sqrt{2}z \qquad (32)$$

$$u_2 = \frac{(N_v + 1)}{\sqrt{2}} \qquad (33)$$

and z^*, by Eq. (19), and:

$$w_2 = \sqrt{2}z^* \qquad (34)$$

If N_v is odd, use Eq. (21a) and (22a) to calculate z_1 and z_2; if N_v is even, use Eq. (21b) and (22b). Then find the integers N_{z_1} and N_{z_2}, in the usual manner.

To determine the tube count C_6 or C_8, use Eq. (23) or (24), respectively.

Adjustments

The tube counts calculated using this method represent the maximum number of tubes that could fit inside the shell. In practice, some tubes would have to be removed to accommodate tie-rods. Thus, the actual tube count is the calculated value less the number of tie-rods. In this manner, the tie-rods automatically get placed at tube locations, making fabrication easier.

Example

A heat exchanger has a shell inside diameter (D_s) of 1,200 mm and uses 28-mm-O.D. (d_o) tubes that are laid out with a pitch of 36 mm (P) in a rotated square (diamond) configuration.

Determine the tube counts for 1-, 2-, 4-, 6- and 8-pass arrangements. How many extra tubes can be accommodated if the pitch is reduced to 35 mm?

Before the problem can be solved, one more piece of information is needed, and that is the minimum clear-

ance (c). For a shell of 1,200-mm dia., the Tubular Exchanger Manufacturers Assn. recommends a value of 5/16 in., or 8 mm.*

Solution

Single-pass. From Eq. (1) and (2), and the definition of the N_r and N_s, $r = [0.5(1,200 - 28) - 8]/36 = 16.0556$; $N_r = 16$; $s = 257.78$; and $N_s = 257$. From Table II, for $N_s = 257$, $C_1 = 805$.

2- and 4-pass. From Eq. (26), $w = 16.0556/\sqrt{2} = 11.353$; thus, $N_w = 11$. Eq. (27) and (25) yield $C_x = 2(11) + 1 = 23$ and $C_y = 23 - 1 = 22$. Using, Eq. (6) and (8), the 2- and 4-pass-exchanger tube counts are, respectively, $C_2 = 805 - 23 = 782$ and $C_4 = 805 - 23 - 22 = 760$.

6-pass. By definition, $e = 0.265$ for a 6-pass layout. Using the procedure described above, the following parameters are obtained: $v = \sqrt{2}(0.265)(16.0556) = 6.017$ [Eq. (30)]; $N_v = 6$; $u_1 = 6/\sqrt{2} = 4.243$ [Eq. (31)]; $z = \sqrt{257.78 - (4.243)^2} = 15.485$ [Eq. (16)]; $w_1 = \sqrt{2}(15.485) = 21.899$ [Eq. (32)]; $u_2 = 7/\sqrt{2} = 4.950$ [Eq. (33)]; $z^* = \sqrt{257.78 - (4.95)^2} = 15.274$ [Eq. (19)]; and $w_2 = \sqrt{2}(15.274) = 21.600$ [Eq. (34)]. Since N_v is even ($N_v = 6$), $z_1 = 0.5(21.899) = 10.949$ [Eq. (21b)] and $z_2 = 0.5(21.6 + 1) = 11.3$ [Eq. (22b)]; $N_{z_1} = 10$ and $N_{z_2} = 11$. Finally, from Eq. (23), $C_6 = 805 - 23 - 4(10 + 11) = 698$.

8-pass. For an 8-pass arrangement, $e = 0.404$. The calculated parameters are as follows: $v = \sqrt{2}(0.404)(16.0556) = 9.173$ [Eq. (30)]; $N_v = 9$; $u_1 = 9/\sqrt{2} = 6.364$ [Eq. (28)]; $z = \sqrt{(257.78 - [6.364]^2)} = 14.740$ [Eq. (16)]; $w_1 = 2(14.74) = 20.846$ [Eq. (32)]; $u_2 = 10/\sqrt{2} = 7.071$ [Eq. (33)]; $z^* = \sqrt{(257.78 - [7.071]^2)} = 14.415$ [Eq. (19)]; and $w_2 = \sqrt{2}(14.415) = 20.385$ [Eq. (34)]. This time, since $N_v = 9$ is odd, $z_1 = 0.5(20.846 + 1) = 10.923$ and $z_2 = 0.5(20.385) = 10.193$ [Eq. (21a) and (22a)]; $N_{z_1} = 10$ and $N_{z_2} = 10$. From Eq. (24), the tube count $C_8 = 760 - 4(10 + 10) = 680$.

35-mm pitch. For $P = 35$ mm, the procedure is exactly the same, and the tube counts are as follows: $C_1 = 861$; $C_2 = 838$; $C_4 = 816$; $C_6 = 750$; and $C_8 = 732$.

An important point to note is that here, by reducing the pitch by just 1 mm, the tube counts increase by 56! An error of 0.2 mm on the pitch could easily cause the tube count to be off by 8 to 16 tubes. This clearly demonstrates why the plotting/counting method can produce inaccurate results.

The author

P. S. Phadke, Satyajit Society 30/1, Erandavana, Pune—411 004, India, is an independent consultant to the chemical process industries, specializing in design and troubleshooting work. Previously, he was a Process Engineer with R. L. Dalal and Co. (Bombay) from 1969–1973, and head of the Chemical Engineering Div. of Kirloskar Consultants, Ltd. (Pune) from 1973–1977. He received a B. Tech. in chemical engineering from the Indian Institute of Technology in 1967, and an M. Tech. in 1969. He is a registered engineer in India and an Associate Member of the Indian Institute of Chemical Engineers.

*"Standards of Tubular Exchanger Manufacturers Assn.," 6th ed., Tubular Exchanger Manufacturers Assn., Tarrytown, N.Y., 1978.

Heat-exchanger tube-to-tubesheet connections

Most heat-exchanger failures occur at the point where the tube is secured in the tubesheet. If you want tight, long-lasting joints, you must specify the proper manufacturing and quality control procedures.

Stanley Yokell, Energy and Resource Consultants, Inc.

☐ The connection of the tubes to the tubesheets is the most critical element of a shell-and-tube heat exchanger because its reliability depends upon the integrity of the many parallel tube-to-tubesheet joints. Consequently each of the many joints must be virtually free of defects.

The part of the tube held to the tubesheet is stressed more severely than the main body of the tube. The configuration of joints allows only limited nondestructive examination. For these reasons, the tube-to-tubesheet joint is the site of most failures.

To understand the degree of reliability needed, consider the functions of the joint and the consequences of failure.

Joint functions and requirements

The main function of tube-to-tubesheet joints is to seal the tubes tightly to the tubesheets. And for most equipment, an additional major function is to support the tubesheet against pressure-induced loads.

Leaking joints may cause:

- Erosion of tube ends and tubehole walls.
- Corrosion of the lower-alloy side.
- Poisoning or fouling of the atmosphere.
- Fire or explosion.
- Tube-wall fouling.
- Catalyst poisoning.
- Product adulteration and degradation.
- Yield reduction.
- Power-generation-capacity reduction.
- Plant shutdown and power outage.

When joint leakage is intolerable, consider using double tubesheets. You can justify the extra cost: (1) when the hazard caused by a leak is great; (2) when joint failures are more probable than failures in tube bodies.

The degree of tightness you need depends upon the service conditions. Minor leaks in commercial low-pressure water heaters may be tolerable. Here, if you see a drop of water at the tube joint after a half-hour on hydrostatic test, it is hardly significant. To try to reduce the leakage rate below watertightness would hardly be worthwhile.

On the other hand, consider that the permissible chloride ion concentration in a surface condenser is 0.1 ppm. A typical brackish cooling-water supply might have a chloride ion concentration of 5×10^{-3} ppm [1]. A surface condenser producing 5,000 gpm (3.15 m^3/s) of condensate could therefore tolerate a total leak of approximately 0.1 gpm (6.3×10^{-6} m^3/s) of brackish water. The number of tubes in a two-pass condenser that could handle the steam load would be approximately 50,000, making the average permissible leak of brackish water through each joint 10^{-6} gpm (6.3×10^{-9} m^3/s).

Measuring joint tightness

One way you can assess the quality of joints is to measure how tight they are. It is reasonable to assume that if you set the acceptable measured leak-rate below what you can tolerate in service, you will assure that high-quality joints have been made.

Depending upon the service conditions, you may use the following to gauge tightness:

- Visual observation during liquid pressure-testing.
- Bubble-formation testing.
- Halogen-leak testing.
- Helium-leak testing.

Liquid pressure testing—For most equipment, the *Pressure Vessel Code* [2]* hydrostatic test is adequate. When you remove the channels or bonnets, the tube ends are visible. You can then see any leaks of pressurized shell-side water. You may be satisfied that if no leaks appear after a specified time (one-half hour is the minimum acceptable period), the joints will be watertight.

However, when the tube-side design pressure is higher than that of the shell side, the direction of leakage will probably be from the channel or bonnet into the shell. When you hydrostatically test the tube side, small leaks behind the tubesheet are hard to see in U-tube and floating-head units; they are impossible to see in fixed-tubesheet exchangers.

For these conditions, test for joint leakage indirectly by measuring loss of pressure over an extended period

*Hereafter, this ASME publication will be referred to simply as the *Code*.

Originally published February 8, 1982

The Kynex Corp.

(usually 24 h). If the pressure falls, assume there is a leak. However, you must remember that: (1) ambient temperature changes may cause the pressure to vary; and (2) small leaks in fittings and connections may falsely indicate joint leaks.

By using a test fluid more searching than water, you may test for a higher level of tightness than with the extended-time hydrostatic test.

Bubble-formation testing—After you test the shell side hydrostatically, you may further test tube-joint tightness by performing the leak tests described in Article 10 of Section V, "Non-destructive Examination," of the *Code*.

To perform a gas-and-bubble-formation test, fill the shell with inert gas or air at the design pressure. After 15 min., flow bubble solution over the joints and tube ends. The joints are acceptable if there is no continuous bubbling.

The test solution must not break away from the test surfaces or break down rapidly because of air drying or low surface tension. Household detergents and soap solutions are not suitable. You may obtain proper test solutions from suppliers of materials for nondestructive examination.

Halogen leak-testing—For exchangers handling lethal or noxious fluids, halogen diode-detector testing may be suitable. For *Code* purposes, the method is not considered quantitative. However, with a 10% (by volume) tracer-gas concentration, the largest actual measured leak-rate allowed in *Code* tests is 1×10^{-12} std m^3/s.

The detector or "sniffer" sucks air through a tubular probe and then passes it over a heated platinum element (the anode). This element ionizes the halogen vapor. The ions flow to a collector plate (the cathode). Current, proportional to ion-formation rate, is indicated on a meter.

The sniffer must first be calibrated against a capillary halogen standard that has a leakage rate of 0 to 1×10^{-14} std m^3/s of refrigerant gas, and again at intervals of not more than 2 h during use.

Specific test procedures vary; a typical procedure follows: Clean and dry the shell. Fill it with a mixture of 10% (by volume) tracer gas and clean, dry air or inert gas, usually at 30 to 50 psi (207 to 345 kPa). Allow at least 30 min for tracer dispersion. Traverse each weld and insert the probe into each tube end. Be careful to keep the probe tip within $\frac{1}{8}$ in. (3.2 mm) of the test surface.

Determine the scanning rate by passing the probe over the leak-standard orifice at a rate that detects leakage of 1×10^{-13} std m^3/s.

Alternatively, encapsulate each tube with a funnel connected to the probe. Determine the response time by encapsulating the standard and measuring its response time.

The preferred tracer-gas is Refrigerant 12, but you may substitute Refrigerant 11, 21, 22, 114 or methylene chloride.

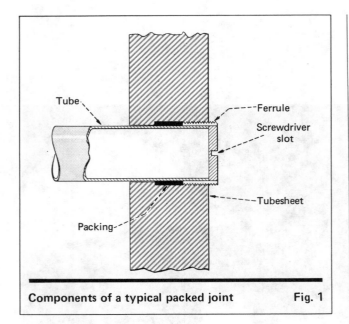

Tube

Ferrule

Screwdriver slot

Packing

Tubesheet

Components of a typical packed joint   **Fig. 1**

Helium leak-testing—When the hazard presented by a leak is unacceptable, or when the effect of a leak can justify a higher testing cost, consider helium mass-spectrometer leak-testing.

Helium-leak testing is done with portable mass spectrometers sensitive to minute traces of helium. Two methods are used. The cheaper one is the "sniffer" method. The more-costly hood method can detect smaller leaks than the sniffer.

The sniffer method is not considered quantitative for *Code* purposes. However, with a 10% (by volume) helium tracer-gas concentration, the largest actual measured leak-rate allowed in *Code* tests is 1×10^{-13} std m^3/s.

The hood method is quantitative. The *Code* acceptance standard, total allowable, integrated leak-rate is 1×10^{-14} std m^3/s, unless otherwise specified. However, the method is difficult to apply to individual tubes.

Testing joint strength [3]

The pull-out or push-out strength of a tube-to-tubesheet joint cannot be conveniently measured directly in a heat exchanger. Therefore, to find the typical strength of a given joint, you must test joints in a model. The parameters of the model—tubesheet and tube materials, tube diameter and gage, tubesheet thickness, hole-drilling tolerance and surface finish, hole grooving, weld procedure, etc.—must match those of the production exchanger.

A shear-load test [4] is used. Both the size of the model and the number of tubes tested affect the significance of the results. Results also vary because of the effects of the range of tolerances in hole drilling, tube manufacturing, material specifications, and joint-producing techniques.

Heat-exchanger manufacturers should have test results available to support their tube-to-tubesheet-joint procedures. When you specify or buy equipment, scrutinize the joint-fabrication procedures and supporting

test results at least as carefully as you scrutinize mill test reports of the chemical and physical properties of the metals and the welding procedures.

When you do a shear-load test at room temperature, you do not account for the effects of operating temperature [5]. At operating temperature, the different expansion rates in the tubes may combine with reduction of tube and tubesheet strength to produce an unsatisfactory joint. So for critical service, consider performing tests at operating temperature.

In cyclical operation, joints may be periodically loaded and unloaded, or the direction of the load may be reversed. Hence, failure may occur well below the point of failure in a static test. When the operation of an exchanger is to be cyclical, calculate the joints' required load-bearing capacity [6]. Test specimens by alternately loading and unloading to this value, or by reversing loads until they fail. Determining the number of cycles to failure in this way is expensive, but it may avoid serious problems.

When you test tube-to-tubesheet specimens, the effects of tubesheet deflection, interpass temperature differences, tubesheet temperature gradients, and vibration are excluded. You must consider these effects when you design an exchanger. If you do not consider them during design, you may have to consider them when the exchanger fails.

How tube-to-tubesheet joints are made

Tube-to-tubesheet joints are made by:
- Stuffing the space between the tube and the hole with packing.
- Sealing the tube to the hole with an interference fit.
- Welding or brazing the tubes to the tubesheets.
- Using a combination of methods.

Packed joints

Fig. 1 is a sketch of a typical packed joint. At the inner side of the tubesheet, the clearance between the tube and hole is just enough to let the tube slide through. The counterbored recess at the outer side of the hole is threaded for approximately half its depth. A slotted, threaded ferrule is used to squeeze packing rings into the chamber. The friction of the compressed packing against the tube and hole determines the strength and tightness of the joint.

The advantage of packed joints is that they are easy to assemble and retube. The tradeoff is that either the tubesheet ligaments are thin or the tube pitch is spread. The first alternative requires you to use thick tubesheets. The second reduces the number of tubes you can install in the shell and also may reduce the shell-side coefficient.

It is not always as easy to replace tubes as you may have assumed when you decided to use packed joints. The threaded connections may become frozen. Consequently, you may destroy the ferrules when you try to remove them. Also, the packing may dry out, becoming stiff and hard to remove.

More economical and positive methods have largely replaced packed joints. Aside from some small auxiliary exchangers, their principal current use is in vertical, low-pressure heat recuperators. In this equipment a hot gas, which may be laden with pollutants, flows down through vertical tubes. Cold combustion air flows upward through the shell. Baffles in the shell direct the air back and forth across the tubes.

Pollutants may settle on the walls nonuniformly, resulting in different rates of fouling that may cause the average temperature of adjacent tubes to be markedly different. If the tubes are firmly fastened to both tubesheets, the forces created by differential expansion may rupture the joints, damage the tubesheets and buckle the tubes.

With loosely packed joints at the lower tubesheet and welded joints at the upper, tubes may expand individually. You can pack the joints loosely because heat recuperators usually run at low pressure, and some hot and cold stream mixing is generally acceptable.

The tubes of many heat recuperators are large enough for the ferrules to be designed as stuffing-box gland followers with wrench flats.

Alternatives to packed-end recuperator joints are: (1) expanding the lower tube end to sliding contact with the tube hole; and (2) making each lower tube-to-tubesheet connection through a bellows expansion joint.

Interference-fit joints

If you were to shrink the holes in a tubesheet onto the tubes, interference between the tubes and holes would create interfacial pressure. The hydraulic tightness and strength of the joints would depend upon:
- Interfacial fit pressure.
- Surface finish of the tube and hole.
- Static coefficient of friction.
- Length of tube embedded in the hole.
- Hole diameter.
- Poisson's constant.
- Other properties of the pair of metals.

An equation that relates fit pressure of shrink-fit joints to the interference, when the tubes are thin-walled (ratio of diameter to thickness more than 10), is [7]:

$$P_o = \frac{IE_t}{2\phi_t(1 + E_t\phi_p/E_p\phi_t)} \quad (1)$$

where: P_o = interfacial fit pressure, psi (Pa); I = interference, in. (m); b = outside tube radius, in. (m); E_t = tube elastic modulus, psi (Pa); E_p = tubesheet elastic modulus, psi (Pa); $\phi_p = b(1 + v_p)$; $\phi_t = a^2(1 - v_t)/2(b - a)$; v_p = Poisson's constant for the tubesheet; v_t = Poisson's constant for the tubes; and a = inside tube radius, in. (m).

The pullout force is related to the interfacial fit pressure by:

$$F = 2\pi P_o b(1 - e^{-\alpha f l})/\alpha \quad (2)$$

where: F = pullout force, lb (N); $\alpha = v(c^2 - b^2)/b(c^2 - a^2)$; c = radius of an imaginary ring of tubesheet surrounding the tube, equivalent to a shrunk-fit ring, in. (m) (c may be chosen to be b + the ligament); f = static coefficient of friction; and l = thickness of tubesheet, in. (m). Other symbols are as previously defined.

In the derivation of this equation it was assumed that Poisson's constant was the same for both tubes and tubesheet. The effects of the assumption are inconsequential if you use the average value.

The tensile force that will cause a tube to yield is:

$$F_{yt} = \pi(a + b)(b - a)S_{yt} \quad (3)$$

where: F_{yt} = force to cause tube-yield, lb (N); and S_{yt} = yield stress of the tube, psi (Pa)

If you set the pullout force calculated by Eq. (2) equal to the force that will cause tube-yield, calculated by Eq. (3), you can determine the interfacial fit pressure associated with the strongest joint. This is:

$$P_o' = \alpha(a + b)(b - a)S_{yt}/2b(1 - e^{-\alpha f l}) \quad (4)$$

where: P_o' = fit pressure for strongest joint, psi (Pa).

By rearranging Eq. (1) and substituting P_o' for P_o you can estimate the interference needed to produce the strongest joint, based on an assumed coefficient of friction.

$$I' = \frac{2P_o'(1 + E_t\phi_p/E_p\phi_t)}{E_t} \quad (5)$$

In this equation: I' = interference to make the strongest joint, in. (m).

It would appear that this could be obtained by shrink fitting. However, the sum of the hole-drilling and tube-diameter tolerances is greater than the interference that gives the strongest joint. For example: Based on a friction coefficient of 0.4, 1-in. (25.4-mm) O.D. tubes and 1-in. (25.4 mm.) thick tubesheets, the strongest joint interference for various tube gages and metal pairs varies in the range 0.0005 in. (0.013 mm) to 0.0015 in. (0.038 mm). The permissible tube undersize is 0.006 in. (0.1524 mm). The permissible hole oversize is 0.002 in. (0.0508 mm) for 96% of the holes and 0.01 in. (0.254 mm) for 4%. The combined tolerance is 0.008 in. (0.2032 mm) to 0.016 in. (0.4064 mm). Therefore, in spite of attempts to size holes and tubes, shrink fitting has not been a commercial success.

The practical way to achieve an interference fit is to expand the tube into the hole. To create interference, the tube must be permanently enlarged to a diameter greater than the hole. You do this by applying radial force inside the tube.

When you first load the tube into position there is clearance between the tube and hole. As you apply pressure, the tube bulges out. If the tube contacts the hole before the expanding pressure enlarges it beyond its elastic limit (yield stress/$\sqrt{3}$) and you then release the pressure, the tube will spring back to its original dimensions.

However, if the pressure needed to cause contact exceeds its elastic limit, the tube will be stretched permanently. Now, if you release the pressure, the tube will recover somewhat, but not to its original size. The amount of recovery depends upon the tube properties and tube and hole dimensions.

The recovery is elastic. If you reestablish the pressure and again release it, the tube will return to its previously relaxed size.

Measurements of the tube before and after you

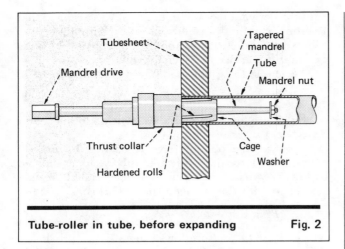

Tube-roller in tube, before expanding   **Fig. 2**

stretch it permanently will show that the increase in diameter is accompanied by a length reduction and by barely perceptible wall thinning. The relative amounts of these dimensional changes are determined by Poisson's constant, the length of tube-end expanded, tube diameter and wall thickness.

After contact, as you raise the pressure, the stress distribution in the tube changes. If you ignore the effects of adjacent holes and differences between tube and tubesheet properties, you can imagine the assembly as a large plate with a hole in its center, with pressure acting in the hole.

With increasing pressure, enlargement continues and the plastic zone spreads outward. Beyond the plastic zone, the stress is elastic. The stress in the elastic zone corresponds with the stress at the boundary of the plastic zone, not the pressure in the tube. When the plastic zone reaches the tube exterior, the pressure on the hole is equal to the tube elastic-limit stress.

In the model of a hole in a large plate, the plastic zone radius increases beyond the outer tube-wall radius as you continue to raise the pressure in the tube. When the plastic zone radius reaches 1.75 times the hole inside radius ($1.75a$), you cannot obtain further enlargement because the tube interior begins to extrude. The corresponding expansion pressure is 2(tube yield stress)/$\sqrt{3}$, or $1.155 \, S_{yt}$.

After you release the pressure, there is some recovery, which is very nearly elastic. The inside of the tube is now larger than when you started. The assembly is permanently strained and there is a zone of residual stress beyond the interior of the tube. Residual stress increases from zero at the inside of the tube to a maximum, then declines with increasing radial distance.

The residual stress at the outside tube radius is the interfacial fit pressure, analogous to the pressure you would get by shrink fitting.

Expanding pressure is probably seldom applied to the point of fully developing yield in the tubesheet, because the resulting tubesheet distortion may be unacceptable. Furthermore, the permanently enlarged holes could make it difficult to retube a unit.

When the tube and tubesheet materials are different, differences in modulus of elasticity, yield stress and Poisson's constant affect the residual pressure. The fit

pressure is also affected by the ratio of outside to inside tube radius (b/a).

At one extreme, the plastic limit may be reached in the interior of a thick tube, while there is not enough stress at the outer wall to deform the hole. At the other, a thin springy, strong tube may enlarge elastically enough to permanently deform a surrounding hole in a low-yield-strength, high-elastic-modulus tubesheet. For such a metal pair, when you release the pressure, the tube recovers its original size, but the hole does not; therefore clearance is *increased*. At neither extreme can you produce an interference fit.

When you use Appendix A, Sect. VIII, Div. 1 of the *Code* to establish allowable loads in expanded tube joints, it is prudent to establish reliability factors for conditions near these extremes by test.

Expanding is the most frequently used way to join tubes to tubesheets. It is the standard method used for exchangers built to TEMA Standards [9].

Tube expanding methods

You may use the following ways to obtain an interference fit:
- Expanding the tube by rolling.
- Exploding charges in the tube ends.
- Compressing an elastomer axially in the tube ends, to create radial pressure.
- Applying hydraulic pressure directly to the tube end.

Roller expanding

The tube roller shown in Fig. 2 consists of a cylindrical cage with equally spaced longitudinal slots. Three to seven rollers, nested in the slots, are made of hardened steel. The tapered mandrel fits between the rolls. The drive end of the cage is threaded to receive a thrust collar and locking nut. You adjust the position of the rolls in the tube by adjusting the position of the thrust collar.

To expand the tube, you push the mandrel forward, driving the rolls outward to press on the tube, and then rotate the mandrel. Friction between the mandrel and rolls causes the rolls to turn.

Before the advent of power-driven rolling, hammer blows were used to drive the mandrel forward, and you turned the mandrel with a wrench. In power-driven tube rollers, torque is supplied to the mandrel. As the mandrel tightly presses the rolls to the tube, the surface under each roll is slightly depressed. The tube wall is squeezed as the rolls ride up the side of the depression.

You can supply the force to insert and retract the mandrel by a self-feeding arrangement. If you set the slots in the cage to make an angle with the longitudinal axis of the cage, the mandrel will self-feed. You withdraw it by reversing the direction of rotation.

However, setting the rolls at an angle changes the motion of the rolls to a combination of sliding and rolling. If the tubes are soft and the tubesheet hard, self-feeding may cause the tube to take an hourglass shape. The opposite condition may cause a barrel shape. Either reduces the amount of contact surface, making the joint less satisfactory.

With self-feeding, the tube is pulled in reaction to the

thrust of the rolls. You can use the thrust collar to hold the tubes in their axial position as you expand.

Torque is supplied by electric, air or hydraulic motors. Hydraulically driven equipment is shown in Fig. 3. Unlike self-feeding rolling tools, its roll slots are in line with the tube length. You insert and retract the mandrel hydraulically. There is no roller thrust or reaction collar, so you must hold the tube in its axial position some other way.

When you roll a tube, the effect of the high contact-pressure of the rolls on the tube is added to the effect of radial stretching. There is markedly more tube wall reduction than if you apply pressure uniformly. This is accompanied by axial extrusion of the tube end.

It is hard to measure the *actual* wall reduction. In production, you deduce the reduction by measuring the hole and tube before rolling, and the tube interior after rolling. However, the after-rolling measurement includes the stretching of the hole. Therefore, what you measure is more appropriately termed "apparent wall reduction."

Wall reduction and tube extrusion have each been used as indicators of joint strength. However, when you roll thin, springy, strong tubes into a tubesheet, the reduction may be too small to be a significant indicator of rolling degree.

To establish a repeatable procedure, you sense the torque being drawn by the roller, and set the control to stop rolling torque at a predetermined value. Many shops determine the torque cutoff point from measurements of wall reduction produced at a given torque level. The wall reduction is in turn correlated with strength and tightness tests. However, the current trend is to relate torque values directly to strength and tightness tests.

Roller expanding demands careful attention to:
■ Cleanliness of the roller, tube interior, and exterior tube hole.
■ Number of rolls.
■ Angle of rolls relative to tube axis.
■ Roller rotational speed.
■ Lubrication and cooling of the roller.
■ Condition of cage, rolls and mandrel.
■ Shape of rolls.
■ Measurements.
■ Maintenance of precise torque cutoff settings (using a torque analyzer).
■ Technique of rolling.
■ Effects of worker fatigue.

When you have a heat exchanger with thick tubesheets it is customary to limit the depth of hard rolling to the amount that provides a joint strength equal to tube strength. This is considered to be achieved at a depth of $1\frac{1}{2}$ to $2\frac{1}{2}$ in. [10].

The current practice in the U.S. is to expand the tubes for a length not less than 2 in. or tubesheet thickness minus $\frac{1}{8}$ in. for TEMA Class R and B exchangers, and the least of twice tube diameter, 2 in., or tubesheet thickness minus $\frac{1}{8}$ in. for TEMA Class C.

However, to avoid crevice corrosion you may have to bring the tube into contact with the hole for the full thickness of the tubesheet.

A less obvious reason for full-depth contact is that

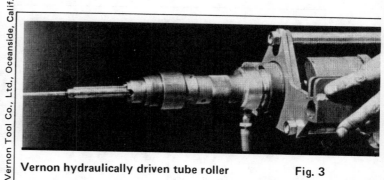

Vernon hydraulically driven tube roller   Fig. 3

Vernon Tool Co., Ltd., Oceanside, Calif.

when you consider the tubesheet to be a perforated plate, the effective size of the perforation is the inside diameter of the tube where it is in intimate contact with the hole. But, where there is clearance between the tube and hole, the effective perforation size is the diameter of the hole.

If you are concerned only with sealing the unexpanded space, the two alternatives are: (1) Roll the tube into the tubesheet at the shell-side face. Then roll the tube into the tubesheet at the tube-side face. This procedure leaves the intervening space unexpanded; and (2) Expand the full depth.

The maximum practical depth that can be rolled in one step is about 2 in., because of the torque required and tube extrusion. The first alternative (above) lets you seal the front and back of the tubesheet without excessive tube compression resulting from extrusion.

The second alternative requires special attention to the rolling technique. If you do the first step at the inner end of the joint and successive steps progressively outward, the tube end can move out of the hole without causing the tube to be compressed.

Extrusion caused by roller expanding will cause tubesheets to cock if an improper technique is used. You should establish the sequence of rolling with the fabricator as part of the procedure for tubing or retubing a unit. A typical recommended sequence is shown in Fig. 4.

When you roll tubes into double tubesheets, the effects of extrusion are intensified. The life of the exchanger will be shortened if the tubesheets are not set parallel with each other or if corresponding holes in adjacent tubesheets are not aligned accurately.

The sequence that will avoid these problems is:
(1) Fix the tubesheets at each end parallel with each other, with tube holes aligned.
(2) Set the pair of tubesheets at each end parallel with those at the other end, with holes aligned.
(3) Tube the bundle, following the specified cleaning procedure.
(4) Tack-expand the tubes in the front inner tubesheet in the order shown in Fig. 4.
(5) Fully expand the tubes in the front inner tubesheet, using progressive step rolling.
(6) Repeat Steps 4 and 5 at the rear inner tubesheet.
(7) Test the shell side.
(8) Repeat Steps 4 and 5 at the front outer tubesheet.
(9) Test the gap, if it is possible, or the channel side if the gap is exposed.

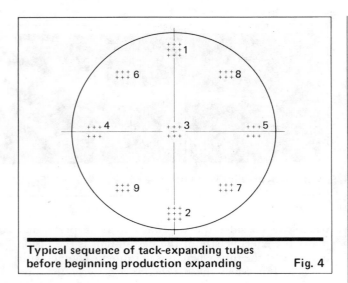

Typical sequence of tack-expanding tubes before beginning production expanding   **Fig. 4**

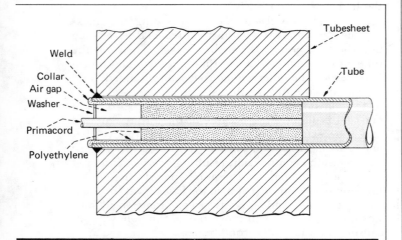

Explosive insert in tube-to-tubesheet assembly   Fig. 5

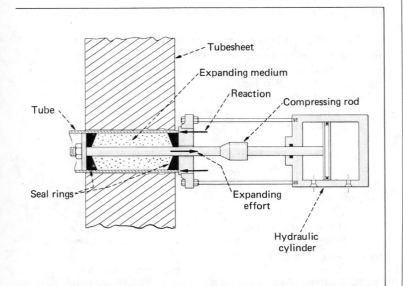

How Hitachi rubber expanding works   Fig. 6

(10) Repeat Steps 8 and 9 for the rear outer tubesheet.

When you fix the tubes at one end, then roll the tubes at the other, the friction of the rolls may cause the tube to twist. As a result, the bundle may have a noticeable twist after rolling is completed. You can avoid this problem by using a tapered drift pin to lock the tubes to the second tubesheet.

The other ways to expand tubes—exploding charges in the tube ends, compressing an elastomer in the tube ends, and directly applying hydraulic pressure—apply expanding pressure uniformly.

Exploding charges in tube ends

Setting off of a charge in the tube ends is called near-contact forming, kinetic expanding, near-explosive expanding or Detna-forming* [11].

In kinetic expanding, the tube is the workpiece. The tubesheet and the air gap in the clearance between the tube and hole make up a forming die. The explosive-charge package consists of Primacord contained in a cylinder of polyethylene. The polyethylene medium transmits the expanding force to the tube in a controlled way.

Polyethylene is used because it: (1) is cheap; (2) is available; (3) is easy to handle; (4) is flexible; (5) resists attack by water and most solvents; (6) has fairly high density; (7) has a high-enough melting point; (8) does not react with the tube metal; and (9) does not create a cleaning problem after expanding.

From the control standpoint, polyethylene is desirable because: (1) it is resilient, accommodating large elastic strains without cracking or bursting; and (2) stress waves are rapidly attenuated as they are propagated through it. The resiliency makes removing the insert easy. The rapid diminution of stress waves minimizes shocks to the tubes and tubesheets.

To get consistent expansions in the desired expansion region, you must fit the Primacord carefully into the polyethylene insert, leaving no air gap.

The pressures generated when you set off an explosion decrease very rapidly with distance, especially very close to the explosion. This makes it possible to control precisely the length of tube that you expand. You may contact-expand tubes into a tubesheet after you make a primary front-facejoint weld. Precise control of expanding distance lets you avoid deforming the weld.

Fig. 5 is a cross-section through a portion of a tubesheet into which a tube has been welded. It shows the arrangement of Primacord, polyethylene medium wrapper and air gap for controlled contact expanding of the tube without stressing the weld.

To establish the charge size, perform pull-out load tests. You may substantiate the charge size by measuring the strain in unrestrained tubes that have been expanded by setting off the selected charge.

The detonation in the tube produces less-severe surface distortion on the inside of the tube than does rolling. The tubesheet is not injured by the explosion in the tubes. The process is applied mostly to thick tubesheets.

The manufacturer must work out the expanding pro-

*A trademark of Foster-Wheeler Energy Corp., Livingston, N.J.

cedure. If you review the manufacturer's procedure, points of attention are: (1) how the charge is established; (2) what steps are taken to ensure that the connections to the charges are secure and reliable; (3) how misfires are to be prevented; (4) how misfires are to be corrected; (5) how the joint is to be tested; and (6) what is the procedure for cleanup after expansion.

One of the uses of near-explosion expanding is the contact expanding of previously welded-in tubes into thick tubesheets of high-pressure feedwater heaters. Another important use is to make expanded-only joints in low-pressure feedwater heaters.

A unique advantage in making explosive-expanded-only joints is that with one explosion you can apply full expansion pressure in the vicinity of annular grooves in the tube hole, and contact-only pressure in the balance of the tube end [12].

The other uniform-pressure expanding methods are more common.

Compessing an elastomer in the tube

Hitachi calls its procedure for compressing an elastomer in the tube ends to achieve radial expanding force, "rubber expanding." Fig. 6 is a schematic drawing, showing how the Hitachi Rubber Expanding Machine works. The expanding medium is a cylinder of elastomer. The pressing rod, connected to the hydraulic cylinder, passes through the medium. When you hydraulically retract the pressing rod, the medium is compressed. It bulges radially, exerting pressure uniformly on the tube interior. The radial force is uniform at any section perpendicular to the tube axis. However, it probably varies with axial distance in the medium.

The seal rings at the inner and outer faces of the elastomer cylinder prevent it from extruding during compression. The nut and washer on the inner end of the pressing rod transfer compressing force to the medium. Reaction to the compressive force is contained by the thrust bushing seated in the retainer held to the hydraulic cylinder by the tie rods.

When you rubber-expand, the degree of expansion is related to the hydraulic-cylinder retracting pressure. However, you cannot directly measure the expansion pressure in the tube end.

Applying hydraulic pressure directly

You can supply expanding force by applying hydraulic pressure directly in the tube ends [13,14]. Because there is no intervening medium, you can directly measure and precisely control the pressure. Furthermore, you can repeat the pressure within a very narrow range.

In direct hydraulic expanding, the working fluid is demineralized or distilled water. The basic system consists of a two-stage pump and reservoir assembly, operating gun, and mandrel, plus needed hydraulic piping.

Fig. 7 shows the HydroSwage* power unit with housing removed. The compressed-air-driven Haskel hydraulic water-pump feeds a fixed-ratio intensifier to produce hydraulic expanding pressure of approximately 40,000 psi (275 MPa). Adjusting the input to the

*A trademark of Haskel Inc. The worldwide distributor for HydroSwage equipment is Torque and Tension Equipment Inc., Campbell, Calif.

Haskel HydroSwage power unit, swivel fittings with hydraulic tubing, gun and mandrel   Fig. 7

intensifier lets you precisely control the expanding. To make the operation of the unit flexible, you may connect the low-pressure side to the intensifier by an umbilical hose.

High-pressure water is conveyed to the gun and mandrel through sections of hydraulic tubing joined by the four-axis swivel fittings shown in Fig. 7. To provide flexibility, you place the fittings on the power unit, between equal lengths of hydraulic tubing, and on the gun (or you can use flexible capillary tubing).

The gun accepts mandrels sized for all tube diameters, gages and expanding-lengths. It has three signal lights to indicate the stage of operation. An amber light on the bottom shows the start of expanding. A green light on the top signals completion of the cycle. A red light in the middle comes on only when there is a problem in reaching the preset expanding pressure. It lights up when the intensifier has completed its stroke without attaining the pressure setting. This may result from a defective tube, or a need for new mandrel seals.

A gun and mandrel are shown in Fig. 8. The finger that projects at the forward end of the gun operates an adjustable pressure switch. It must be depressed by contact with the tubesheet to activate the switch and turn the gun on.

The mandrel has front and rear O-rings with backup

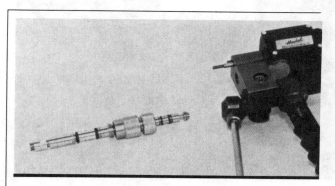

Haskel HydroSwage and mandrel   Fig. 8

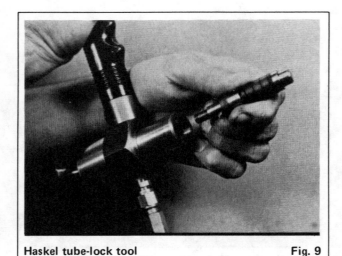

Haskel tube-lock tool   Fig. 9

rings. You can set the O-rings to seal on the tube interior exactly where you want expansion to stop. O-ring life is affected by the bead height if you use welded tubes. There is no limit to the mandrel length, therefore you can expand the full depth of the tubesheet in one operation.

The mandrel of the HydroSwage system has a self-centering feature that allows the inside tube diameter to increase as much as 0.040 in. (1.016 mm) during expansion, with one mandrel.

When you apply pressure to the tube, the changes in the condition of the tube metal do not take place instantaneously. Therefore, the gun is provided with an adjustable timer, which you set to an experimentally determined dwell time.

Use the following sequence to expand tubes with the HydroSwage: (1) thread the mandrel into the gun; (2) insert the mandrel into the tube; (3) depress the operating trigger button; (4) remove the mandrel when the green light comes on. If the red light comes on, investigate the cause. Correct it and repeat the procedure.

How the Haskel tube-test tool is used   Fig. 10

When you are swaging, the mandrel is too tightly locked to the tube to be moved manually. The axial forces are balanced, therefore you do not have to resist thrust.

You have to lock the tube in position before you make a uniform-pressure expanded joint, a joint made by a hydraulically driven roller with inline roll slots and a welded-first joint. For this purpose you may use a tapered drift pin, but if a power unit is available, it is more convenient, cleaner and faster to use the tube-lock tool illustrated in Fig. 9.

The tube-lock tool operates from the low-pressure side of the power unit. A hydraulic cylinder causes the polyurethane segments to expand when you apply pressure. After you release the pressure, the tube has been bulged out enough to hold it axially. However, the tube is not in contact for the full depth of the hole, nor is it tightly sealed.

You may use the tube-test tool to verify the pressure given in the expanding-procedure specification. The tool, pictured in Fig. 10, consists of a mandrel threaded at one end, and with a boss on the other. An O-ring is provided as a seal between the rod and tube near the threaded end. A second O-ring seals the back of the boss to the face of the tubesheet in the ligament space surrounding the tube.

The testing procedure is: (1) insert short stubs of tubing from the same production heat into the tubesheet; (2) insert the tool into tube end and tighten the nut against the thrust washer bearing against the tube stub; (3) introduce high-pressure test water through the mandrel; (4) inspect for leaks.

Although the manufacturer supplies the tool as part of a system, you may use it to verify the tightness of roller-expanded joints, welded- or brazed-only joints and welded- or brazed-first, expanded joints.

How temperature affects expanded joints

The effects of metal temperature on expanded tube-to-tubesheet joints have not been adequately investigated. Still, the following elementary discussion may be useful as a guide.

You expand tubes into tubesheets at room temperature, but usually operate heat exchangers at other temperatures. The temperature change affects the joint strength and tightness. Of concern are:
- Differences in expansion between tube and hole.
- Changes in metal properties.
- Creep.
- Changes in coefficient of friction.

As the temperature changes, if the tube expands more than the tubesheet, the joint becomes tighter until there is inelastic deformation. In the opposite case, the joint becomes looser. You can calculate the change in interference by using the following equation:

$$\Delta I = 2b\,(T_o - T_r)\,(\psi_p - \psi_t) \qquad (6)$$

where: ΔI = change in interference; T_o = operating temperature, °F (°C); T_r = room temperature, °F (°C); ψ_p = tubesheet mean coefficient of thermal expansion, in./(in.)(°F) or m/(m)(°C); ψ_t = tube mean coefficient of thermal expansion, in./(in.)(°F) or m/(m)(°C).

You can then appraise its effect by using Eq. (1) and (2).

The tube and tubesheet may also expand at different rates along the axis of the tube. Imagine a stainless-steel tube stub in a hole in a steel tubesheet, with the ends of the stub flush with the tubesheet faces. If you heated the assembly, the tube stub would expand more than the tubesheet and protrude from each face. The force to restrain this movement is supplied by friction.

You can appraise this thermal effect by estimating the interfacial pressure and interference needed. If you make the simplifying assumption that the tube is held in a ring of tubesheet having a radius equal to half the pitch, you can develop an equation for the unit equilibrium force or contact pressure. You can substitute this value in rearranged Eq. (1) to calculate the associated increase in interference needed.

The equations developed this way are:

$$P_o'' = \frac{(\psi_p - \psi_t)(T_o - T_r)(r^2 - b^2)(b^2 - a^2)E_p E_t}{fb[(r^2 - b^2)E_p + (b^2 - a^2)E_t]} \quad (7)$$

In this equation: P_o'' = interfacial pressure required to prevent axial movement of tube relative to tubesheet, psi (Pa); and r = one-half tube hole pitch, in. (m).

The interference that will develop this pressure is:

$$I'' = \frac{2P_o''(1 + E_t\phi_p/E_p\phi_t)}{E_t} \quad (8)$$

where: I'' = interference to create pressure psi (Pa), in. (m).

You can include the effects of temperature on metal properties by choosing the values of yield stress, modulus of elasticity and mean coefficient of thermal expansion at the anticipated operating temperature.

The main impediment to using expanded tube joints at high temperatures is the tendency of the strain to relax with the passage of time. This tendency to creep is enhanced as temperature increases. You may test the creep behavior of the pair of metals by simulating operating conditions in a model.

There does not appear to be much information available on the variation of friction coefficients with temperature. Values of the coefficient could readily be determined at various temperatures for given surface conditions and pairs. However there may not be a direct relationship between these values and the actual conditions after you expand a tube into a hole.

Probably the best way to determine the effect of temperature on expanded joints is to perform pull-out or push-out tests at the proposed design temperature after a suitable period of heat-soaking.

Grooves, flaring and beading

A general rule of thumb for expanded joints is that rough tube-holes make strong joints, and smooth holes make tight ones. You can enhance the strength and tightness of the joints by providing annular grooves in the holes.

Tests of the effects of grooving on roller-expanded boiler-tube joints showed a 39% enhancement for one

Specimen of a tube expanded into a double-grooved hole by uniform pressure   Fig. 11

groove and 53% for two [15]. When you roll tubes into grooved holes, the tube metal extrudes into the grooves. The shape of the extruded metal is approximately rectangular. Overexpanding, which causes axial tube extrusion, tends to shear these keys, resulting in weaker, less-tight joints.

When you use one of the uniform-pressure expanding methods, the tube bulges into the groove as shown in Fig. 11. The metal-to-metal interference at the points where the nearly parabolic shape of the tube-bulge meets the groove edges is responsible for making the joint tight.

Many groove configurations have been used. The TEMA Standards for Class R exchangers state, "All tube holes for expanded joints shall be machined with at least two grooves, each approximately $\frac{1}{8}$" wide by $\frac{1}{64}$" deep. When integrally clad or applied tubesheet facings are used, all grooves shall be in the base material unless otherwise specified by the purchaser." The standard for Class B units is identical except that ". . . by the purchaser" is deleted, and the standard for Class C equipment requires grooving for tubes $\frac{5}{8}$ inch O.D. and larger for ". . . design pressures 300 psi and/or temperatures in excess of 350°F . . ."

Because uniform-pressure expanding bulges the tube into grooves in the hole (in contrast to the extrusion that rolling causes), you can use explosive, rubber or hydraulic means to expand tubes into the gap between integral double tubesheets [16].

In uniform-pressure expanding, the groove width, depth and position affect the strength and tightness that you achieve [17].

You obtain the optimum width when the product βW is in the range of 1.5 to 3.0. Here:

$$\beta = \sqrt[4]{3(1 - \nu)^2/R^2 t^2} \quad (9)$$

where: W = width of groove, in. (mm); R = mean tube radius, in. (mm); and t = tube wall thickness, in. (mm).

The joint strength varies almost linearly with groove depth. The minimum depth of groove that you should specify for uniform-pressure expanding is $\frac{1}{64}$ in. (0.4 mm).

For maximum pull-out strength, locate the grooves near the outer tubesheet face. For maximum push-out, position them near the inner face. For alternating service place one groove near each face.

If the direction of the forces resulting from pressure on the tubesheets is outward, you will enhance the joint strength and tightness by flaring or beading the tubes.

In flared-end tubes, you increase the strength by the force needed to draw the tube down to its original size. On beaded-end tubes, you raise the strength by the force that will shear the bead.

Flaring and beading make the tube interfere with the outer edge of the hole. This metal-to-metal interference is a further barrier to leakage.

Another use for beading-over the tube ends is to provide a weld-joint preparation. When you do this, the beaded and welded joint is tight and strong in both directions.

Tube ends are sometimes flared or beaded to ease tubeside entrance effects. You may accomplish this more effectively by expanding the tubes flush with the front face of the tubesheet, then machining a smooth taper or radius from the bore of each tube to a point just short of the middle of the ligament.

Welded and brazed joints

Use welding, brazing, or welding and brazing to join tubes to tubesheets when:

■ The operating metal temperature is high.
■ Operation is highly cyclical.
■ Helium-leak tightness is specified.
■ Quality assurance requires nondestructive examination of each joint.
■ The dimensions and properties of the joints are unsuitable for expanding.

When you choose to weld or braze the tubes to the tubesheet, consider the following questions:

■ Can the metals be joined by welding or brazing?
■ What is the most suitable process?
■ Has a qualified procedure been established and used successfully in the shop and field?
■ How will the dimensions of tubes and tubesheet and hole layout affect the welding or brazing?
■ Will the proximity of the shell or pass partitions to tube holes interfere with producing good joints?
■ Will it be necessary to preheat or to post-weld heat-treat the joints? If so how will the exchanger be affected?
■ What nondestructive examinations can be used?

These questions cannot be answered independently of each other. Whether or not metals can be joined by brazing or welding is partly a question of metallurgy, but it is also a question of process and dimensions. It is not possible, for example, to fusion-weld titanium tubes to steel tubesheets, but they can be successfully explosion-welded. However, if the tubesheets are thin and the ligaments small, explosion-welding is not feasible.

The size of the exchanger may affect the procedure. In brazing and fusion-welding it is desirable to have the tubesheets horizontal (i.e., with tubes vertical) when you do the work. The advantage of this arrangement is that the position in which you work on the tube end is constant. By contrast, when the tubes are horizontal, your position continually changes as you traverse the tube.

However, few shops have facilities for setting large exchangers vertically. Furthermore, heat exchangers are usually horizontal during hydrostatic testing. Therefore, it may be impractical or too costly to use the most desirable fusion-welding or brazing position.

Basic information on welding and brazing is given in the "Welding Handbook" [18]. In addition, the American Welding Soc. (AWS) publishes articles and discussions in its Welding Journal. The New York-based Welding Research Council (WRC) pursues new developments. The council issues progress reports and interpretive reports. When a conclusion has been reached on a new process or development, the Council publishes a final report in a WRC bulletin.

Brazing and welding processes

The processes used most often to bond tube ends to tubesheets are: (1) brazing; (2) fusion-welding processes; and (3) explosive welding.

In brazing, you produce coalescence by heating the assembly to a temperature above 800°F (427°C). This melts a nonferrous filler metal that flows into the space between the tube and hole by capillary action. The base metals have higher melting points than the filler and do not melt.

The operating temperature at which you can use brazed joints depends on the filler metal as well as on the base metals. If you qualify a procedure satisfactorily under Section IX of the Code, then under Sect. VIII, Div. 1, the filler metal is considered to be suitable for operating temperatures of 200°F (94°C) or less.

You may use certain classifications of brazing filler metals for service temperatures as high as 300°F (204°C) if you meet additional conditions listed in the Code.

You must use a suitable flux, atmosphere or flux-atmosphere combination to exclude atmospheric gases that can oxidize or embrittle the braze metal. You can braze either with a torch or in a furnace. Furnace brazing may be in an open-flame or a closed furnace.

The joint design and brazing technique must ensure flow of braze metal into the joint. You cannot easily see evidence of such penetration. Consequently, it is prudent and customary to reduce the joint efficiency factor. You can use a higher efficiency by preplacing braze-metal rings behind the tubesheet and having visible evidence that the braze metal penetrated to the front.

If you follow the latter procedure, you eliminate the crevice between the tube and hole. Therefore, it may be done to prevent crevice corrosion. When you make the primary joint by front-end welding, you may eliminate crevice corrosion by back-brazing. However, the shell-side process fluids must be compatible with the braze deposit.

Cleanliness of the parts is essential in successful brazing. If there is any oil, grease, oxide scale or foreign matter present, you will get porous joints. When you examine a brazing procedure, be sure that a cleaning specification is part of it.

The Code prohibits using brazed joints for lethal-material service and for unfired steam boilers.

Most metallically bonded joints are made by fusion welding. The technology is highly advanced. Technical institutions, welding equipment manufacturers, fabricators and technical societies continually work to improve the processes and techniques. Here is a brief summary of the fusion-welding processes and their application to tube-to-tubesheet joints.

When you fusion-weld, you bring the base metals to a molten state. The metals then fuse, forming complex solutions.

You may supply heat-of-fusion by: (1) burning a mixture of gas and oxygen at the joint; (2) applying a voltage gradient to the parts, causing current to flow against resistance; (3) inducing an electric current to flow in the metals; and (4) striking an arc across a gap subjected to a voltage difference.

In fusion-welding tubes to tubesheets, the electric arc processes are normally used, sometimes supplemented by gas welding.

Electrodes intended to be melted into joints are termed consumable. Electrodes meant to be used only as terminals are called nonconsumable; if they melt into the molten puddle, they make inferior welds.

When you use nonconsumable electrodes, you may fuse the base metals only, or you may provide filler metal. Filler may be fed continuously from spools of wire at rates controlled to match welding speed, or you may manually feed a rod of filler metal. In some joints it is advantageous to set rings of filler in place before striking the arc.

Arc welds can be made in the atmosphere, but they tend to become oxidized, embrittled and porous. Therefore, air is excluded from the molten metals by fluxes or shielding gases. You may also use fluxes to contribute metal elements to the weld.

The fusion-welding processes used to bond tubes to tubesheets are:

■ Shielded metal-arc welding or SMAW ("stick").

■ Gas tungsten-arc welding or GTAW (or TIG, tungsten inert gas).

■ Gas metal-arc welding or GMAW (or MIG, metal inert gas).

■ Oxyfuel gas welding or OFW.

When you use shielded metal arc (SMAW), you clamp a stick of flux-coated metal rod in the jaws of a welding handle. The process is manual, which puts a premium on welding skill. The welder must follow the specified procedure faithfully. Joint quality varies with the skill and the physical and emotional state of the welder.

You need an adequate ligament width to stick-weld tubes to tubesheets; the welds are fillet welds. Their best application is welding heavy-wall tubes to tubesheets.

The largest number of welded tube-to-tubesheet joints is made by using the gas tungsten-arc welding process. The nonconsumable electrode is thoriated tungsten. You shield the weld from the atmosphere by blanketing it with inert gas (helium, argon, CO_2 or mixtures). The gas flows at a controlled rate through an annular space between the electrode and surrounding nozzle. You may also blanket the back side of the joint with shielding gas.

You must take great care not to let the electrode touch the weld puddle, because tungsten inclusions embrittle the weld.

A projection of the tube may melt to act as filler metal in the joint. You may also feed filler metal from a wire spool continuously through the nozzle, or hand-feed straight lengths to the arc.

A variety of systems is available for automatically welding tubes to tubesheets by the GTAW process [19].

Kynex automatic welding gun for welding tubes to the back side of the tubesheet   Fig. 12

The Ciber-Tig* system consists of a gas-tungsten-arc welding machine, programmer, automatic tube-to-tubesheet welding head, and motor speed controller.

The system automatically controls the flow of pre-purge gas, shielding gas and post-purge gas, wire feed-rate, rotational speed and weld voltage buildup and decline. Various accessories are available to control the welding remotely, stabilize the arc and record arc voltages, currents and sequences. The latter accessory is extremely valuable for establishing welding procedure specifications and controlling their application.

You can join the tubes to the front or rear face of the tubesheet by using the GTAW process. If you forge tube-hole projections on the rear face, you can butt-weld the tubes to the projection. This permits you to radiograph the tube-to-tubesheet weld [20,21].

When you weld the tubes to the back side of the tubesheet, the perforation in the tubesheet is the same as the tube I.D. This gives you flexibility in making future repairs.

Kynex Corp. of Rome, N.Y., provides an onsite automatic tube-to-tubesheet joint-welding service. Fig. 12 shows the Kynex GTAW process gun for welding near the back side of a tubesheet. Fig. 13 shows the Kynex automatic face-welding gun for performing welds where the tubes are flush with the front face.

In the gas-metal-arc welding process the electrode is consumed. You feed it from a spool through a nozzle. Shielding gases flow around the electrode through the nozzle. You may use gas up behind the joint as in

*A trademark of Hobart Bros. Co., Troy, Ohio.

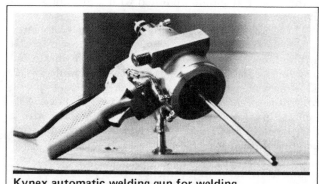

Kynex automatic welding gun for welding tubes to front tubesheet face   Fig. 13

GTAW. Or the electrode may be a small tube, filled with flux that contributes alloying elements to the weld. This is called flux-cored wire.

The most suitable use for GMAW is to make fillet welds that join large-diameter tubes to front tubesheet faces.

Reliable fusion welds must be free of porosity, non-metallic inclusions, and cracks. To achieve this quality, you must meet the following conditions: (1) the base metals must be compatible; (2) the tube ends and holes must be *completely* free of foreign matter; (3) the environment must be clean and dry; (4) the flux on coated rods must be dry; (5) shielding gas must be bone-dry; (6) the base-metal grain structure must be uniform; (7) gases generated by welding must be able to escape the weld puddle; (8) the tube-to-tubesheet temperature must be kept nearly constant after you begin welding; (9) you must exclude condensation moisture when you interrupt welding; (10) you must not let the voltage and current output of the welding source fluctuate; and (11) you must keep the welding area free of stray magnetic fields.

Metals to be fusion-welded must be able to form tough, crackfree solid solutions. When the tube/tubesheet pair does not meet this requirement, you may clad the tubesheet face with a metal compatible with the tubesheet and the tubes.

If the tube and tubesheet are not scrupulously clean before welding, you can count on porosity at best and complete joint failure at worst. Be sure that any weld procedure that you review requires mechanical cleaning of the tube ends and holes, followed by washing with a volatile chloride-ion-free solvent. To assure the quality of the joint, make this a quality-control hold point.

Ordinarily, clean, untreated atmospheric air is a suitable welding environment. But for some base metal pairs you must surround the work area with uniform-temperature, controlled-humidity, filtered air. You may do the work in a special room called "clean room" or in a housing that surrounds the tubesheet.

The consumables—weld rods and shielding gases—must be dry. When you evaluate a shop's capability of making sound tube-to-tubesheet fusion welds, make certain that it is standard shop practice to store opened containers of flux-coated rods in moisture-excluding ovens. On the working floor, opened packages of loose rods ought to be held in portable rod-warmers. Note also that, before a cylinder of shielding gas is connected to the welder, it should be tested with a moisture-sensitive paste.

The grain structures of tubesheets made from large rolled plates or forgings may vary across the surfaces. If a tube hole pierces the tubesheet where an unsatisfactory local condition exists, it can cause a faulty weld. This may show itself at a spot where the local carbon content is too high to make a tough weld. You can surmount this problem by cladding the tubesheet with a thin layer of weld deposit of acceptable composition before you drill it.

If the gases produced by fusion welding cannot freely escape from the weld puddle, the tube-joint welds may be porous and predisposed to cracking. When you make the joints at the front face of the tubesheet, using the geometries shown in Fig. 14, you must hold the tubes in place. You may do this by tack-welding the tubes, expanding the tube ends with a drift pin to make line contact, or using a device like the previously described Haskel tube-setting tool.

The practice of setting tubes by lightly rolling them before welding is likely to cause trouble by: (1) not permitting an adequate escape path for welding gases at the root of the weld; and (2) introducing foreign matter to the surfaces to be welded (lubricants and flakes from the roller and cage of the expander).

If you let the tubesheet temperature vary widely during welding, the size of the root opening will vary from joint to joint. As welding heat spreads throughout the tubesheet, the holes deviate from roundness. The amount they recover depends on the change in radial temperature gradient. As the welds solidify they shrink, which further distorts the holes near the joint being made. To reduce the amount of distortion, keep the tubesheet at a uniform temperature.

When welding is interrupted, it is advisable to warm the tubesheet before restarting. Re-start warming is separate from pre-heating and post-weld heat-treating, which serve different purposes. The re-start warming also helps to dispel condensation moisture.

If you interrupt the welding for a long time, condensation moisture may settle in the unwelded joints. It can make subsequent welds porous when the moisture vaporizes. Good practice is to put a cloth bag of desiccant in a plastic wrapping around the assembly.

Although you might not notice it, the line-voltage fluctuates. These variations may change current and arc-gap length, and cause unseen changes in the weld penetration and metal deposits. For best results use a voltage regulator on the supply to the welding machine.

Shops are full of stray magnetic fields (often generated by neatly coiled leads of welding cable). External magnetic fields add to the effect of arc blow, a deflection of welding current in direct-current welding caused by variations of magnetic flux in the work piece. Arc blow may cause skips and uneven tube-joint welds.

The reliability of fusion welds also depends upon the joint design. When you design a joint, major factors to consider are the leak path and the strength of the weld.

For tightness, the leak path through the weld throat should be at least as long as the tube wall thickness.

You can design the weld joint to be as strong as the tube in resisting axial loading, or to just meet the *Code* requirements. Tube welds probably fail more from localized stress than from axial loading [22]. Simply adding weld metal does not always help.

The joint design should allow the basic weld configuration to be replicated easily. Furthermore, there are so many joints that you have to consider it likely that some will fail on test or during the life of the exchanger. Therefore, joint geometry should permit repairing.

For joint designs for high-pressure exchangers or those in nuclear or other hazardous service, consider also [22]:

■ Joint weld-stress due to tubesheet flexure.

■ Interactions between the tubes and tubesheet resulting from differential thermal expansion.

■ Residual welding stress.

■ Effects of combining welding with expanding.

You may examine the fatigue strength of specific joint designs by using *Code* methods for these service requirements.

Explosive welding

When the metals are not compatible for fusion welding, but you need the tightness and strength of welded joints, consider explosive welding.

The AWS defines explosive welding (EXW) as [18] "a solid-state welding process wherein coalescence is effected by a high-velocity movement produced by a controlled detonation."

There are three requirements for explosive welding [23]. One, you must progressively bring together the two components, thereby producing a collision front that traverses the surfaces to be joined. Two, the velocity of the collision front must not exceed 120% of the

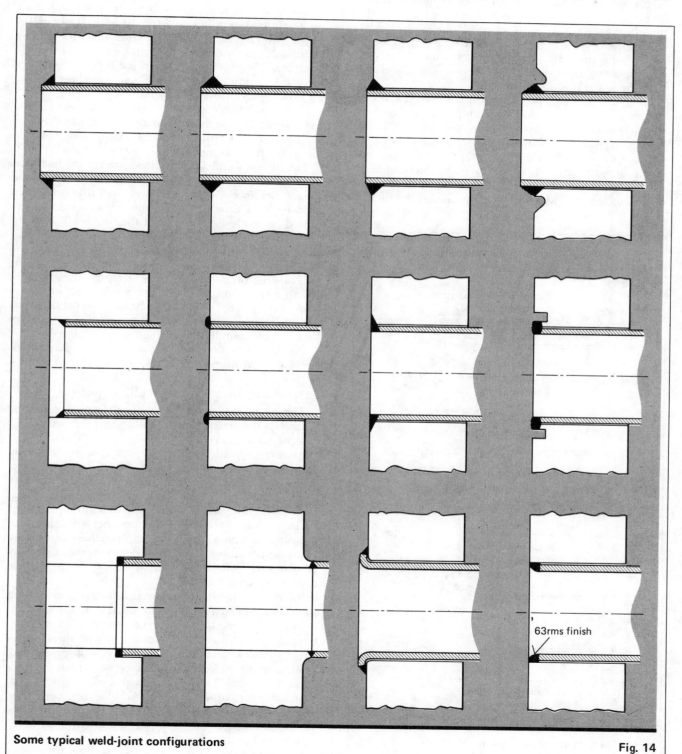

Some typical weld-joint configurations

Fig. 14

sonic velocity of the materials. And three, the pressure created at the interface must be several times the yield strength of the materials.

If you meet these conditions, the component surfaces become molten at the collision front. The molten metal, together with any surface contaminants, is propelled before the collision front in the form of a jet. The jet is ejected from the interface at the tube end. Upon passage of the jet, the cleaned surfaces, in contact under high pressure, diffuse into each other to form a metallurgical bond. Any tube projection is severed at the tubesheet face.

The bond interface has the shape of a sinusoidal wave. This prevents complete molten-metal ejection because some is entrapped by the vortices associated with the wave peaks and troughs.

Although the velocity front must not exceed 120% of the sonic velocity, explosives that produce a higher-velocity front are cheap and easy to handle. To use them you must make an angular joint preparation.

Fig. 15 is a schematic representation of the process. The tube hole, tapered outward toward the front face, makes an angle with the tube. When you detonate the explosion near the junction of the tube and the untap-

ered part of the hole, the blast makes the tube collide with the hole. The distance the tube must travel to the collision point is progressively greater along the tube because of the taper. This reduces the collision-point velocity below the detonation velocity and makes it possible to use high-velocity-front explosives.

The increasing gap between the tube and hole limits the surface area you can weld. However, this is acceptable because you can get enough welded surface to attain full weld strength.

The depth of the countersunk taper usually is $\frac{1}{2}$ to $\frac{5}{8}$ in. The angle used lies between 10 and 20 deg. High angles produce larger wave lengths than do smaller ones, and this causes less jet-metal entrapment at the interface. Therefore, you should use high angles when the combination of molten tube and tubesheet materials produces a brittle intermetallic.

However, when you must have small ligaments, you have to reduce the angle to machine less metal out of the ligament. You must make a compromise between the geometry and metal requirements.

The tube-wall thickness mainly governs the ligament thickness needed to avoid deforming the hole. Because thick tubes accelerate more slowly than thin ones, they need higher charges to reach the collision velocity required for explosive welding. If the tube wall-mass accelerated to this velocity distorts the ligament, the collision pressure will be reduced.

You may partially circumvent this problem by putting tapered plug supports in adjacent unwelded holes. Another expedient is to reduce the tube wall in the region of the joint. Tube-wall thicknesses versus ligament widths for unsupported and supported ligaments have been experimentally determined and tabulated [23].

Fig. 16 is a photograph of a cross section of an explosively welded tube-to-tubesheet joint.

Inspecting brazed and welded joints

The configuration of brazed and front-face fusion-welded joints restricts your ability to perform significant nondestructive tests.

An experienced welding inspector, using a strong light, can detect most surface flaws. Properly done, fluid-penetrant examination can reveal surface porosity and cracks that might not be found on the visual examination. Neither visual inspection nor fluid-penetrant examination will disclose the presence of subsurface porosity, inclusions and cracks. Nevertheless, they are important because: (1) surface defects reasonably may be assumed to indicate internal defects; and (2) these checks are simple to do.

Weld metal in which surface defects have been found should be removed until sound metal is found. Before a repair is attempted, take extreme measures to make sure the parts are cleaned meticulously. When visual examination under strong light discloses no visible evidence of foreign matter, clean the surfaces with at least three washes of distilled water. Follow this with three solvent washes and clean-air drying. Acetone has been used successfully. However, it is highly flammable and exposure to its vapors is undesirable; use it carefully.

You can examine radiographically joints made by welding the tubes to the rear tubesheet face [24]. Plac-

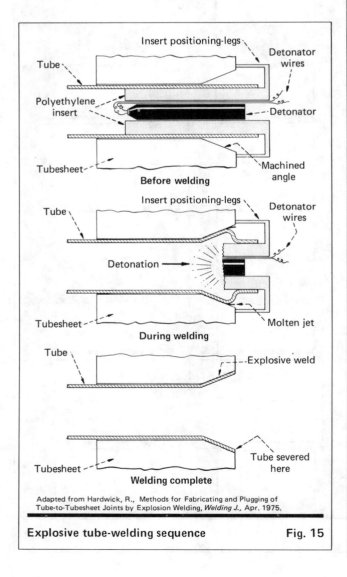

Before welding

During welding

Welding complete

Adapted from Hardwick, R., Methods for Fabricating and Plugging of Tube-to-Tubesheet Joints by Explosion Welding, *Welding J.*, Apr. 1975.

Explosive tube-welding sequence   **Fig. 15**

ing the film and source in the nest of tubes is difficult. There is also more complexity in interpreting the radiographs than in isolated butt-welded pipe joints.

Butt-welded joints made internally, socket-welded joints and explosively welded joints can be examined ultrasonically. Interpretation requires preparation of models with standard discontinuities for comparison.

You may visually examine inner surfaces of internal and rear-face joints with optical devices.

Combining joining methods

When you make the primary joint by welding at the front tubesheet face, you obtain the following benefits by also expanding the tubes to make light contact with the hole walls:

■ You close the crevice between the tube and hole. For the purpose of eliminating crevice corrosion, this is preferable to boring enlarged holes in the tubesheet behind the joint.

■ You reduce the effective perforation diameter, thereby increasing the tubesheet stiffness and its load-bearing capacity.

■ You isolate the weld vibration, thus increasing resistance to fatigue failure [25].

■ You reduce discontinuity flexural stress between tube and ligament.

In addition to these benefits, you make a tighter, stronger joint, and isolate the weld from axial loads when you strength-expand (expand enough to develop full expanded joint strength) the tubes after welding.

However, if the tube and tubesheet have different linear coefficients of thermal expansion and the operating temperature is high, the restraint of the strength-expanding will impose thermal stress on the weld.

If you make the primary tube-fastening by expanding, seal-welding provides a second barrier to leakage.

The sequence in which you combine welding with expanding may determine the results you get. There are several reasons to weld first. Foremost is the need to provide welding surfaces that are totally free of foreign substances. If you use hydraulic or rubber expanding you may obviate this problem.

There are other reasons to weld first. The previously discussed restraint of differential axial thermal expansion between tube and tubesheet may cause weld-root tears and cracks if you expand first. In addition, when the tube is tightly pressed against the hole wall, gases produced by welding must escape from the surface of the weld puddle. This increases the prospect of porosity.

If the tube-joint welding procedure requires you to post-weld heat-treat, there is no point in expanding first. The heat treatment will relax the tube.

It is inconvenient to weld first because weld metal may overlap the tube end, and weld shrinkage may reduce the tube I.D. so that it is difficult to insert expanders. Deal with these problems by reaming out excess metal at the tube mouth.

In non-U-tube exchangers, another way of combining joining methods is to weld the joints at one end and expand them at the other. You may choose this construction because one end operates at a high temperature and the other end is cool, or because the fluid environment makes it desirable.

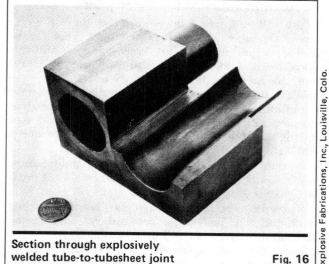

Section through explosively welded tube-to-tubesheet joint   **Fig. 16**

Explosive Fabrications, Inc., Louisville, Colo.

Joints can be made by combining expanding and packing (gasketing) [26]. Fig. 17 is a photograph of a section through a joint in an explosively clad tubesheet. To make this joint you prepare the hole by machining an annular groove into the base metal and another in the alloy cladding. The grooves are somewhat deeper than the TEMA standard depth, but are limited by the ligament thickness. A suitable rubber gasket (usually temperature-resistant silicone) is inserted into the grooves. The tube ends are cleaned, inserted, and the tube is expanded by one of the previously described methods.

The specific application of this combination is to deal with the case of high-strength, thin-walled, low-elastic-modulus tubes to be joined to low-strength, high-elastic-modulus tubesheets. When you design this kind of joint, you cannot assume that the tubes support the tubesheet.

A modification of this combination is to use O-ring grooves and O-rings. You do this at one end of a two-tubesheet unit, while fixing the other end. Each tube may then float individually to accommodate differential expansion between adjacent tubes.

Expanded and gasketed joint   **Fig. 17**

Explosive Fabrications, Inc., Louisville, Colo.

Failures in tube-to-tubesheet joints

You will find most failures in the region of the rear face of the tubesheet. That is where most tubes are anchored to the tubesheet and receive bending and torsional loads. Also, if the tubes are subjected to vibration, this area may fail.

When the tubes are expanded into the holes, there is a transition from the expanded diameter to the original diameter near the back face. This is also the location of the highest fluid shear on the tubewall and consequent erosion/corrosion in the tubes.

In thin-walled tubes, an error in expanding technique in which the tubes are expanded beyond the rear face may lead to cracking.

If you use a regressive (front to rear) roller-expanding technique, you will create compressive loads in the tubes and the reaction will be taken here.

Inside the tube, failures may result from work-hardening, stress-corrosion-cracking and fatigue. Outside the crevice, corrosion may attack.

When you fusion-weld austenitic stainless-steel tubes to tubesheets, there may be precipitation of complex carbides in a heat-affected zone behind the weld. An atmosphere that the stainless steel ordinarily resists may corrode this zone.

When a tube-joint welding procedure requires preheat or post-weld heat treatment, the emplacement of thermocouples to control temperature is critical to avoiding distortion and undesired metallurgical changes.

Fusion welds may fail because of: (1) hidden porosity or large gas pockets; (2) blowholes caused by burn-through of the tube wall; and (3) root bead tears.

Root bead tears may cause stress risers. Grain-boundary precipitation, underbead cracking, and thermal stress compound the effect.

High stresses in the joint may cause cracks to grow normal to the tube wall. Base metals sensitive to fissure cracking may sensitize the weld metal. Root tears may open the way to weld failures.

Summary

Pay careful attention to the design and production of tube-to-tubesheet joints, because their functions are essential and the consequences of failure are dire.

Make adequate tests of tightness and strength. Be sure to qualify testing and manufacturing procedures and personnel.

Relate the kind of joint, manufacturing processes, testing and nondestructive examinations to the service of the unit. Consider the size and shape of the exchanger when you select a joint design and joining method. Remember that the way the joints are made affects the overall structure of the equipment.

To produce acceptable joints, cleanliness and faithful adherence to qualified procedures are essential.

Take advantage of the best features of different kinds of joints by combining them.

Pay special attention to the part of the joint where the tube emerges from the rear face of the tubesheet, because it is the most likely place for failure to occur.

References

1. Sverdrup, H. U., others, "The Oceans," Prentice-Hall, Englewood Cliffs, N.J., 1942. [For the analysis of seawater.]
2. "ASME Boiler and Pressure Vessel Code," 1980 ed., American Soc. of Mechanical Engineers, New York, N.Y., 1980.
3. *Ibid.*, Sect VIII, Div. 1, Appendix A. [Describes testing to establish reliability factors to use in calculating the maximum allowable load in either direction on the tube-to-tubesheet joints of non-U-tube heat exchangers.]
4. "Tension Testing of Metallic Materials," American Soc. for Testing and Materials, Philadelphia, Pa., 1979. [Specification E-8 describes equipment and methods.]
5. "Standards for Power Plant Heat Exchangers," 1st ed., The Heat Exchanger Institute, Cleveland, Ohio, 1980. [On p. 10 are tabulated the maximum recommended metal temperatures for expanded joints in carbon steel tubesheets.]
6. "ASME Boiler and Pressure Vessel Code," *op. cit.*, Section VIII, Div. 1, Appendix A.
7. Yokell, S., "A Working Guide to the Shell and Tube Heat Exchanger," McGraw-Hill, N.Y., to be published. [This book will contain the derivation of this equation. The derivation is also obtainable from the author.]
8. Goodier, J. N., and Schuessow, G. J., The Holding Power and Hydraulic Tightness of Expanded Tube Joints: Analysis of the Stress and Deformation, *Trans. ASME*, July 1943. [A similar equation is developed for push-out.]
9. "Standards of Tubular Exchanger Manufacturers Association," 6th ed., The Tubular Exchanger Manufacturers Assn., New York, N.Y., 1978.
10. Goodier, J. N., and Schuessow, G. J., *op. cit.*
11. Berman, I., and Schroeder, J. W., Near Explosive Forming, "Explosive Welding, Forming, Plugging and Compaction," American Soc. of Mechanical Engineers, New York, N.Y., PVP-44, 1980.
12. Berman, I., and Schroeder, J. W., Detnaform at Power Plant Installations, paper delivered at Sixth International Conference on High Energy Rate Fabrication, Essen, West Germany, Sept. 12–16, 1977.
13. Krips, H., and Podhorsky, M., Hydraulic Expansion—A New Method for the Anchoring of Tubes, *VGB Kraftswerktechnik*, (West Germany), Vol. 56, No. 7.
14. Krips, H., and Podhorsky, M., Hydraulic Expansion of Tubes, *VGB Kraftswerktechnik*, (West Germany), No. 1, 1979. [Discussion of theory.]
15. Maxwell, C. A., Practical Aspects of Making Expanded Joints, *Trans. ASME*, July 1943.
16. Yokell, S., Double-Tubesheet Heat Exchanger Design Stops Shell-Tube Leakage, *Chem. Eng.*, May 14, 1973, p. 133. [Discusses double-tubesheet construction.]
17. Yoshitomi, Yuji, others, Tube-Hole Structure for Expanded Tube-to-Tubesheet Joint, U.S. Pat. No. 4,142,581, Mar. 6, 1979.
18. "Welding Handbook," 7th ed., American Welding Soc., Miami, Fla., Vol. 1, 2 and 3, 1980. [Sections 5 and 6 of the 6th ed. will be superseded by the 7th ed. Vol. 4 in 1982 and Vol. 5 in 1984.]
19. Automatic Process Streamlines Tube-to-Tubesheet Welding, *Welding J.*, July 1960.
20. Rowlands, E. W., Jr., and Cooksey, J. C., Internal Welding of Tubes to Tubesheets, *Welding J.*, July 1960.
21. Schwartzbart, H., In-Bore Gas Tungsten Arc Welding of Steam Generator Tube-to-Tubesheet Joints, *Welding J.*, Mar. 1981.
22. Lohmeier, A., and Reynolds, S. D., Jr., Carbon-Steel Feedwater-Heater Tube-to-Tubesheet Joints, ASME paper No. 65-WA/pwr-10, Nov. 1965.
23. Hardwick, R. C., Methods for Fabricating and Plugging of Tube-to-Tubesheet Joints by Explosive Welding, *Welding J.*, Apr. 1975.
24. Foster, B. E., and McClung, R. W., A Study of X-Ray and Isotopic Techniques for Boreside Radiography of Tube-to-Tubesheet Welds, *Mater. Eval.*, Vol. 35, No. 7, p. 43 (1977). [Description of a new technique.]
25. Sebald, J. F., and Hawthorne, L. H., Mechanical Effects Resulting From Welding Copper-Base Alloys to Tubesheets, presented at the American Soc. of Mechanical Engineers Annual Meeting, Dec. 1957.
26. Hardwick, R., Tube-to-Tubeplate Welding and Plugging by Explosives, paper No. 40, presented at the Symposium on Welding and Fabrication in the Nuclear Industry, of the British Nuclear Engineering Soc., London, 1979.

The author

Stanley Yokell is Senior Consultant for Energy and Resource Consultants, P.O. Drawer O, Boulder, CO 80306 (phone: 303-449-5515) and Suite 5e, The Livery, 209 Cooper Ave., Upper Montclair, NJ 07043 (phone: 201-744-5545), where he is engaged in management and engineering consulting. He holds a B.Ch.E. from New York University, and has done postgraduate work at Newark College of Engineering. He is a member of AIChE, a licensed professional engineer in New Jersey and Colorado, and a member of the Special Working Group on Heat Transfer Equipment, Sect. VIII, ASME Boiler Pressure Vessel Code Committee.

Pretreating mild-steel water-cooled heat exchangers

Here is a method for cleaning and passivating new mild-steel exchangers. This will help to eliminate waterside corrosion and fouling.

Raymond M. Pasteris, Mobil Oil Corp.

☐ Water-cooled heat exchangers made of mild steel are subject to corrosion and deposits of corrosion products on tube surfaces. However, by using proper pretreatment procedures, one can minimize such problems and guarantee long life and efficient heat transfer.

Mild-steel corrosion can be controlled effectively by the rapid and uniform formation of a corrosion-inhibitor film on a clean metal surface. However, new heat exchangers are not clean. Their surfaces contain lubricants and mill scale, which is the result of fabrication [2]. These contaminants must be removed before establishing a protective film. Pretreatment consists of two steps:

- Cleaning the metal surface.
- Passivating the surface with a chemical corrosion-inhibitor [1].

We will outline two simple procedures to effectively pretreat new mild-steel cooling-water heat exchangers. Before this, let us review the events leading up to the investigation and development of the two-step exchanger pretreatment program.

A common but ineffective pretreatment

Before placing a new exchanger in service, it is commonly flushed with water. Flushing removes extraneous materials such as dirt and loose rust, but does nothing to get rid of oils and tightly adhering mill scale.

After flushing, a corrosion inhibitor is added to the cooling water at a higher-than-normal concentration to establish corrosion protection. However, our experience has shown that even on clean surfaces such levels may be insufficient to provide good protection, and that on dirty surfaces the film cannot be applied uniformly. A nonuniform film results in localized pitting-attack [1].

This conventional method had been employed with success for years at one of our refineries. However, the tubes it was being used on were made of admiralty brass. A problem started to develop when mild steel began to replace the admiralty tubes, owing to process and cost considerations. Previously, there had been no corrosion or fouling, since admiralty brass has excellent resistance to corrosion caused by water. But now, an effective pretreatment program was needed, because the mild-steel units were corroding and fouling—despite good cooling-water treatment.

Fig. 1 shows the effects of corrosion and fouling on a shell-and-tube device, using the old pretreatment process. An analysis of the deposits appears in Table I.

Bench-scale study

A bench-scale study was done to develop an effective procedure. Various pretreatments were tried on mild-steel test-sample tubes, using actual refinery cooling water. Two sample tubes were used in the trials [2].

Test exchanger tube A was pretreated with a chromate-zinc corrosion-inhibitor solution at 1,000 ppm (as CrO_4^{-2}) at 130–140°F. The solution was mildly agitated for 1 h. This pretreatment simulated the conventional full-scale one-step procedure used on new mild-steel exchangers.

Fouling of mild-steel exchanger after two years' service using old pretreatment **Fig. 1**

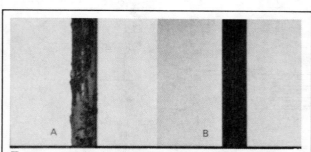

Two-step method reduces corrosion: Tube A had no initial cleaning; B did **Fig. 2**

Originally published November 16, 1981

Deposit analysis for Fig. 1		Table I
Metal - mild steel. Deposit appearance - brown		
	Percent	
Loss at 105°C	4	
Loss at 800°C	20	
Inorganic analysis		
	Percent	
Iron	71	
Chromium	13	
Phosphorus	5	
Silicon	4	
Sulfur	2	
Aluminum	1	
Zinc	1	

Typical analysis for cooling water used in test of two-step treatment		Table II
	ppm	
Calcium	520	
Magnesium	360	
Sodium	610	
Bicarbonate alkalinity	36	
Chloride	790	
Sulfate	640	
Silica	16	
Chromium (CrO_4^{-2})	13	
Zinc (Zn)	2.1	
pH (pH units)	6.9	
Conductivity (μmho/cm)	3,000	
Expressed as $CaCO_3$ except as noted		

Test exchanger tube B was pretreated in two steps. First, the tube was soaked in a 20,000-ppm (2.0%) (as PO_4^{-2}) solution of an alkaline-polyphosphate-based cleaner at 130–140°F, with mild agitation for 2 h. Both tubes were free of rust, but were coated with oil to simulate an actual new exchanger tube. After the cleaning procedure, tube B was passivated in a 1,000-ppm (0.1%) (as CrO_4^{-2}) chromate-zinc solution at 130–140°F. Mild agitation was again done for 1 h.

The two pretreated tubes were placed in refinery cooling-water service for 54 d under normal operating conditions. A typical cooling-water analysis is given in Table II. Fig. 2 illustrates the improved corrosion protection attained following this two-step pretreatment procedure. From these runs, full-scale treatments were developed.

We now use two methods for offstream pretreatment of cooling-water exchangers:

- Recirculation.
- Steam heating and compressed-air agitation.

Pretreatment by recirculation

Pretreatment by recirculation requires a portable, heated solution-tank and a circulating pump (see Fig. 3). As an alternative, a tank truck supplied by a chemical cleaning company may be used.

After installation of the new cooling-water exchanger, water-flush it for approximately 15 min. Cooling water or service water can be used.

Prepare the polyphosphate cleaning solution in the tank, and heat it to 130–170°F. Use the nomograph (see Fig. 4) to determine the amount of chemical required to attain a 10,000-ppm, or 1% (as PO_4^{-2}), cleaning solution. Adjust the solution pH to 6.0–7.0. In this range, the polyphosphate cleaner is most effective, and calcium phosphate scaling-potential is minimal.* Circulate the hot cleaning solution through the exchanger for a minimum of 4 h. A maximum recirculation flowrate along with proper exchanger venting ensures complete contact of the pretreatment solution with the metal.

After draining and flushing the cleaning solution from the exchanger and tank, prepare the chromate-zinc passivating solution and heat it to 130–170°F. Use the nomograph to determine the amount of chemical required to make a 1,000-ppm (as CrO_4^{-2}) passivating solution. Circulate this solution for a minimum of 2 h.

The exchanger is now ready for service. If it is not to be used for a long time, seal it with the passivating solution inside to ensure corrosion protection. When putting the unit in service, flush it and the solution tank, and add the effluent to the cooling-water system. This allows for partial recovery of the corrosion inhibitor, along with slower bleedoff to the facility's wastewater treatment unit. (Always follow proper safety and environmental procedures when mixing and draining any chemicals.)

Pretreatment by agitation

When a solution tank is not available, use steam heating and air agitation (see Fig. 5).

This procedure begins with a thorough 15-min flush of the exchanger, with cooling water or service water. Drain the exchanger. Next, prepare a slurry of the polyphosphate cleaner and pump it into the exchanger. A 55-gal drum is ideal for making up the slurry. Fill the exchanger using a pneumatic or manual drum-pump. The exchanger should be vented and filled with serv-

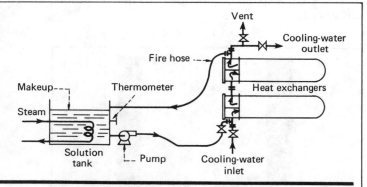

Pretreatment by recirculation requires a hot solution-tank and a pump Fig. 3

*Some pretreatment products are pH-buffered, thus eliminating acid handling and pH control during application. Consult a water-treatment-chemical supplier.

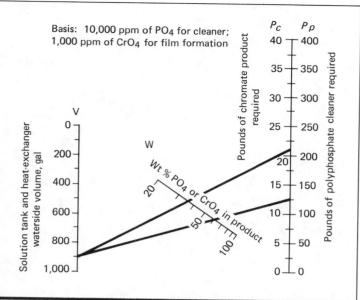

Basis: 10,000 ppm of PO₄ for cleaner; 1,000 ppm of CrO₄ for film formation

Polyphosphate-cleaner/chromate requirements for offstream pretreatment   **Fig. 4**

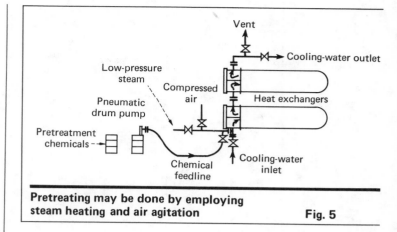

Pretreating may be done by employing steam heating and air agitation   **Fig. 5**

ice water rapidly to promote internal mixing of the cleaner.

Begin injection of low-pressure steam and compressed plant air to heat and agitate the cleaning solution. Maintain the solution temperature at 130–170°F by intermittent steam injection for a minimum of 4 h. Water-flush and drain the exchanger to prepare it for passivation.

Pump the liquid chromate-zinc corrosion inhibitor into the exchanger. Again, rapidly fill the exchanger to promote mixing of the corrosion inhibitor within the unit. Begin injection of the low-pressure steam and agitate with plant air. Maintain the corrosion-inhibitor solution at 130–170°F for a minimum of 2 h.

Nonchromate programs

While the outlined procedures employ a chromate-zinc corrosion inhibitor, the basic two-step procedure can be adapted to a nonchromate corrosion-control program. Consultation with your water-treatment-chemicals supplier, along with implementation of a bench-scale study, will aid in establishing an effective pretreatment procedure for a nonchromate program.

Acknowledgements

Special thanks to Nalco Chemical Co. and P. R. Puckorius for help in establishing this program.

References

1. Puckorius, P. R., Proper Startup Protects Cooling-Tower Systems, *Chem. Eng.*, Vol. 85, No. 1, Jan. 2, 1978, p. 101.
2. Nalco Reprint 98, Pretreatment—The Key to Effective Protection of Cooling Water Systems.

The author

Raymond M. Pasteris is a senior utilities engineer for Mobil Oil Corp., P.O. Box 874, Joliet, IL 60434. Tel: (815) 423-5571. His responsibilities include refinery engineering studies of boiler and cooling-water treatments, and steam-generation and cooling-tower operation for performance optimization. He holds a B.S. degree in chemical engineering from the University of Illinois, and is a registered professional engineer in Illinois. He is a member of the Joliet Junior College Faculty, the American Water Works Assn. and AIChE.

Nomograph

Example: A bank of two exchanger bundles is to be pretreated by the circulation method. The capacity of the solution tank is 700 gal. The total waterside capacity of the two exchangers is 200 gal. Estimate the pounds of polyphosphate cleaner and chromate product required for pretreatment. The polyphosphate cleaner is 60% by weight PO_4^{-2} and the chromate product is 35% by weight CrO_4^{-2}.

Solution: Total-system water-capacity is 700 gal + 200 gal = 900 gal. On the figure:

1. Connect 900 gal on the V-scale with 60% on the W-scale. Read the answer as 125 lb of polyphosphate cleaner on the P_p-scale.

2. Connect 900 gal on the V-scale with 35% on the W-scale. Read the answer as 21 lb of chromate product on the P_c-scale.

Equations used for the graph:

Polyphosphate required:

$$P_p = 8.35 \frac{V}{W}$$

where:

P_p = Amount of polyphosphate product required to attain 1% PO_4^{-2} solution, lb
V = Volume of system, gal
W = Weight percent PO_4^{-2} in product, %

Chromate required

$$P_c = 0.835 \frac{V}{W}$$

where:

P_c = Amount of chromate product required to attain 1,000-ppm CrO_4^{-2} solution, lb
V = Volume of system, gal
W = Weight percent CrO_4^{-2} in product, %

Section III
Design

Program for evaluation of shell-and-tube heat exchangers

With this program, it is possible to vary the design values for all the mechanical specifications and find out quickly if the designed exchanger will do the job.

Roger Crane and *Robert Gregg*, *University of South Florida*

☐ The program described here has been developed to perform the thermal and hydraulic design of standard shell-and-tube heat exchangers. Its applicability is limited to sensible heat exchange in units with normal segmented baffles [1]. Within these constraints, the program is quite general, accommodating a wide variety of shell and tube pass configurations.

The program (Table I) is written for the Hewlett-Packard HP-97 programmable calculator. It is designed to permit the rapid evaluation of a known heat-exchanger configuration, or to permit the rapid iteration of the design so as to obtain the optimum configuration. The user may vary the pass arrangement, tube array configurations and mechanical design parameters, and determine the effect of each on the relative thermal resistance and hydraulic performance.

Development

A suitable heat exchanger for a specific service will provide a sufficient overall heat-exchange coefficient, U, and area, A, such that the process heat load, Q, is transferred. Thus:

$$UA \, \Delta T_{eff} \geq Q \tag{1}$$

where: $\Delta T_{eff} = F_T (\text{LMTD})$

$$\text{Now} \quad \frac{1}{U} = \frac{1}{h(D_i/D_o)} + R_{fi} + \frac{D_o \ln(D_o/D_i)}{2k_w} + R_{fo} + \frac{1}{h_o}$$

[from 2, p. 86]

A rearrangement of these terms leads to the equation:

$$\frac{Q}{F_T(\text{LMTD})} \frac{1}{A} \times$$

$$\left[\frac{1}{h(D_i/D_o)} + R_{fi} + \frac{D_o \ln(D_o/D_i)}{2k_w} + R_{fo} + \frac{1}{h_o} \right] \leq 1 \tag{2}$$

The five terms on the left side of the equation may be viewed as fractional thermal resistances, each one corresponding to one of the five resistances to heat transfer (Table II). In a well-sized unit, the sum of these fractional resistances approaches, but is less than, 1.

The heat-transfer correlations used in the program are taken from Kern [2, p. 103, 107]:

$$\frac{h_i D}{k} = 0.027 \left(\frac{4m}{\pi D \mu} \right)^{0.8} \left(\frac{c_p \mu}{k} \right)^{1/3} \phi^{0.14} \tag{3}$$

$$\frac{h_o D_e}{k} = 0.36 \left(\frac{M D_e}{\mu A} \right)^{0.55} \left(\frac{c_p \mu}{k} \right)^{1/3} \phi^{0.14} \tag{4}$$

Introducing Eq. 3 and 4 into Eq. 2, the individual fractional resistances may be obtained as shown in Table II. Here, the viscosity ratio, ϕ, is omitted from the tubeside and shellside convective terms. For more-viscous fluids, this term should be reintroduced prior to final sizing for greater accuracy.

Once the thermal design is optimized, the pressure drops on the shellside and tubeside are checked to ensure that they are not excessive. The equations used for calculating pressure drops are shown below [2, pp. 836, 839]. Both equations include frictional losses through the bundle. The tubeside term also includes typical turning losses [2, p. 148].

$$\Delta P_s = \frac{f_s G_s^2 D_s \left(\frac{12L}{B} \right) N_s}{5.22 \times 10^{10} D_e S_o} \frac{1}{\phi^{0.14}} \tag{5}$$

$$\Delta P_t = \frac{f_t G_t^2 n L}{5.22 \times 10^{10} D_i S_i} \frac{1}{\phi^{0.14}} + \frac{2n V_i^2}{S_i g} \tag{6}$$

In the turbulent-flow region, the friction factors may be approximated (as in Ref. 2):

$$f_s = 0.0128 N_{Re}^{-0.1964}$$

(straight-line portion of correlation p. 839);

$$f_t = 0.0014 + 0.125 N_{Re}^{-0.32} \text{ (p. 53)}.$$

These correlations are incorporated in the program.

Originally published July 25, 1983

Program

There are three parts to the program, which may be loaded onto three separate cards (Table I). Program 1 calculates the log mean temperature difference (LMTD) and the correction factor, F_T, for either a one-shell pass/two-tube pass (1–2), or a 2–4 pass arrangement. For other than a 1–2 or 2–4 pass heat exchanger, the product $F_T \times$ (LMTD) may be stored in the I register for use with Program 2.

If $F_{T,1-2} < 0.75$, or an error message is displayed, the temperature cross is too great and the 1–2 pass arrangement is not appropriate. (Temperature cross is the difference between the outlet temperature of the cold fluid and the outlet temperature of the hot fluid, when the first is higher than the second.)

Program 2 evaluates the fractional temperature drops in Table II. Each product may be printed out prior to summation. This feature is highly useful in indicating areas where design improvements can be made. Design modifications will be most beneficial in the areas where fractional resistances are relatively large.

Program 3 calculates pressure drops on the shellside and tubeside, using Eq. 5 and 6 with turbulent friction factors. The program pauses to permit incorporation of the viscosity ratio effect.

Example 1

Evaluate a 2–6 heat exchanger that uses water on the tubeside to cool 33.5° API oil. (From *[2]*, Example 8.1.)

Process conditions:

Oil inlet temp.	= 358°F
Oil outlet temp.	= 100°F
Water inlet temp.	= 90°F
Water outlet temp.	= 120°F
Oil flowrate	= 49,600 lb/h
Water flowrate	= 233,000 lb/h
Combined fouling factor	= 0.004
Allowable ΔP	= 10 psi

Heat-exchanger parameters:

Shell I.D.	= 35 in.
Baffle spacing	= 7 in.
Shell passes	= 2
No. of tubes	= 454
Length of tubes	= 12 ft
Size of tubes	= 1 in. O.D., BWG, on 1¼-in. square pitch
No. of tube passes	= 6

The combined fouling factor is referenced to the outside of the tube and therefore is calculated in the R_{fo} term. In this case, the inside fouling factor is equal to zero. The calculated value for $F_{T,2-4}$ may be used for $F_{T,2-6}$ with very little error.

Procedure

1. Load both sides of Card 1. Routine A calculates the LMTD. Routines B and C determine the LMTD correction factor, F_T, for 1–2 and 2–4 pass arrangements.

a. Store T_1, T_2, t_1, t_2 in registers b, c, d and e.

b. Push A to calculate the LMTD (71.93°F).

c. Push B to find $F_{T,1-2}$; an error is displayed, indicating conditions will not permit the use of a 1–2 unit.

d. Push CLX, then push C to find $F_{T,2-4}$ (0.92). Push R/S to store F_T (LMTD) = 66.49°F in Register I.

2. Load both sides of Card 2.

a. Store the appropriate variables in Registers R_{s0}, R_{s1}, R_{s4}–R_{s9}, R_0–R_9, R_a (see Table III). In R_8, store the value for $N_{Pr,i}$.

b. Review the stack for proper values. Ensure that R_{s3} is set to 0. Then set the primary registers to active status.

Nomenclature

A	Area for heat transfer, ft^2
B	Baffle spacing, in.
C	Tube clearance, in.
c_p	Specific heat, Btu/(lb)(°F)
D	Dia., in.
f	Friction factor
F_T	Correction factor for LMTD, depending on type of flow in the heat exchanger, dimensionless
g	Gravity acceleration, ft/s^2
G	Weight flowrate per unit area, lb/(h)(ft^2)
h	Convective heat-transfer coefficient, Btu/(h)(ft^2)(°F)
k	Thermal conductivity, Btu/(h)(ft)(°F)
L	Length of tubes for heat transfer, ft
LMTD	Log mean temperature difference for counter flow, °F
m	Weight flowrate, cold fluid, lb/h
M	Weight flowrate, hot fluid, lb/h
n	Number of tube passes
N_{Pr}	Prandtl number, dimensionless
N_{Re}	Reynolds number, dimensionless
N_s	Number of shell passes
N_t	Number of tubes per pass
P	Pressure drop, psi
P_t	Tube pitch, in.
Q	Quantity of heat transferred, Btu/h
R_A, R_B, R_C, R_D, R_E	Fractional temperature drops, dimensionless
R_f	Fouling factor, (h)(ft^2)(°F)/Btu
S	Specific gravity, dimensionless
t	Cold-fluid temperature, °F
T	Hot-fluid temperature, °F
T_{eff}	Effective temperature difference, °F
U	Overall heat-exchanger conductance, Btu/(h)(ft^2)(°F)
V	Fluid velocity, ft/s
ϕ	Ratio of viscosity measured at bulk temperature to that at wall temperature
μ	Viscosity, cP

Subscripts

1	Inlet conditions
2	Outlet conditions
i	Inside the tube
o	Outside the tube
s	Shellside properties
e	Equivalent thermal or hydraulic diameter
w	Material properties of the tubewall
t	Tubeside properties

Vary the design specifications and find out quickly if the designed exchanger will do the job　Table I

Step	Key	Code	Step	Key	Code	Step	Key	Code	Step	Key	Code	Step	Key	Code	Step	Key	Code
	Program 1		076	R/S	51	152	1	01	047	8	08	123	RCL9	36 09	199	RCL8	36 08
001	*LBLA	21 11	077	RCL9	36 09	153	+	-55	048	GSB2	23 02	124	x	-35	200	.	-62
002	GSB1	23 01	078	x	-35	154	RTN	24	049	x	-35	125	RCL7	36 07	201	6	06
003	÷	-24	079	STOI	35 46	155	*LBL4	21 04	050	P≷S	16-51	126	÷	-24	202	7	07
004	STOI	35 46	080	RTN	24	156	X²	53	051	RCL8	36 08	127	RCLE	36 15	203	Yˣ	31
005	GSB1	23 01	081	*LBLC	21 13	157	1	01	052	÷	-24	128	÷	-24	204	RCLI	36 46
006	-	-45	082	2	02	158	+	-55	053	P≷S	16-51	129	.	-62	205	÷	-24
007	RCLI	36 46	083	GSB3	23 03	159	√X	54	054	GSB1	23 01	130	5	05	206	RCLE	36 15
008	LN	32	084	÷	-24	160	RTN	24	055	RCL2	36 02	131	5	05	207	RCLD	36 14
009	÷	-24	085	1	01	161	*LBL3	21 03	056	RCL1	36 01	132	Yˣ	31	208	-	-45
010	STO9	35 09	086	-	-45	162	RCLE	36 15	057	÷	-24	133	÷	-24	209	÷	-35
011	RTN	24	087	GSB2	23 02	163	RCLD	36 14	058	LN	32	134	GSB1	23 01	210	RTN	24
012	*LBL1	21 01	088	-	-45	164	-	-45	059	x	-35	135	GSB3	23 03	211	*LBL3	21 03
013	RCLB	36 12	089	STOI	35 46	165	RCLB	36 12	060	GSB3	23 03	136	RTN	24	212	PRTX	-14
014	RCLE	36 15	090	GSB5	23 05	166	RCLD	36 14	061	RTN	24	137	*LBL9	21 09	213	P≷S	16-51
015	-	-45	091	GSB6	23 06	167	-	-45	062	*LBLD	21 14	138	3	03	214	RCL3	36 03
016	RCLC	36 13	092	x	-35	168	÷	-24	063	1	01	139	.	-62	215	+	-55
017	RCLD	36 14	093	√X	54	169	RTN	24	064	.	-62	140	4	04	216	STO3	35 03
018	-	-45	094	2	02	170	*LBL2	21 02	065	5	05	141	4	04	217	P≷S	16-51
019	RTN	24	095	x	-35	171	RCLB	36 12	066	8	08	142	RCL7	36 07	218	RTN	24
020	*LBLB	21 12	096	GSB3	23 03	172	RCLC	36 13	067	GSB2	23 02	143	X²	53	219	R/S	51
021	GSB2	23 02	097	÷	-24	173	-	-45	068	x	-35	144	x	-35			
022	GSB4	23 04	098	RCLI	36 46	174	RCLE	36 15	069	P≷S	16-51	145	Pi	16-24		Program 3	
023	1	01	099	+	-55	175	RCLD	36 14	070	RCL9	36 09	146	RCL2	36 02	001	*LBLA	21 11
024	+	-55	100	STO1	35 01	176	-	-45	071	x	-35	147	X²	53	002	GSB2	23 02
025	GSB2	23 02	101	GSB2	23 02	177	÷	-24	072	P≷S	16-51	148	x	-35	003	STO8	35 08
026	+	-55	102	GSB4	23 04	178	RTN	24	073	RCL2	36 02	149	-	-45	004	GSB5	23 05
027	GSB3	23 03	103	STO2	35 02	179	R/S	51	074	÷	-24	150	Pi	16-24	005	RCL8	36 08
028	x	-35	104	RCL1	36 01				075	GSB1	23 01	151	÷	-24	006	x	-35
029	CHS	-22	105	+	-55		Program 2		076	GSB3	23 03	152	RCL2	36 02	007	P≷S	16-51
030	2	02	106	RCL1	36 01	001	*LBLA	21 11	077	RTN	24	153	÷	-24	008	RCL5	36 05
031	+	-55	107	RCL2	36 02	002	P≷S	16-51	078	*LBLE	21 15	154	RTN	24	009	P≷S	16-51
032	STOI	35 46	108	-	-45	003	RCL0	36 00	079	R/S	51	155	*LBL8	21 08	010	÷	-24
033	GSB2	23 02	109	÷	-24	004	RCL4	36 04	080	X=0?	16-43	156	Pi	16-24	011	.	-62
034	GSB4	23 04	110	LN	32	005	÷	-24	081	GTO7	22 07	157	RCL2	36 02	012	0	00
035	CHS	-22	111	1/X	52	006	P≷S	16-51	082	GTO6	22 06	158	X²	53	013	3	03
036	1	01	112	STOI	35 46	007	.	-62	083	*LBL7	21 07	159	x	-35	014	4	04
037	+	-55	113	GSB6	23 06	008	2	02	084	GSB9	23 09	160	4	04	015	4	04
038	GSB2	23 02	114	STO1	35 01	009	Yˣ	31	085	GTO5	22 05	161	÷	-24	016	x	-35
039	+	-55	115	GSB5	23 05	010	1	01	086	*LBL6	21 06	162	CHS	-22	017	.	-62
040	GSB3	23 03	116	1/X	52	011	.	-62	087	GSB8	23 08	163	RCL7	36 07	018	1	01
041	x	-35	117	RCL1	36 01	012	1	01	088	*LBL5	21 05	164	X²	53	019	9	09
042	CHS	-22	118	x	-35	013	2	02	089	.	-62	165	+	-55	020	6	06
043	2	02	119	LN	32	014	x	-35	090	4	04	166	4	04	021	4	04
044	+	-55	120	STO1	35 01	015	GSB0	23 00	091	5	05	167	x	-35	022	CHS	-22
045	RCLI	36 46	121	GSB2	23 02	016	x	-35	092	Yˣ	31	168	Pi	16-24	023	Yˣ	31
046	÷	-24	122	GSB4	23 04	017	GSB1	23 01	093	.	-62	169	÷	-24	024	.	-62
047	LN	32	123	RCL1	36 01	018	RCL1	36 01	094	1	01	170	RCL2	36 02	025	0	00
048	STOI	35 46	124	x	-35	019	RCL6	36 06	095	5	05	171	÷	-24	026	1	01
049	GSB2	23 02	125	2	02	020	x	-35	096	2	02	172	RTN	24	027	2	02
050	1	01	126	÷	-24	021	.	-62	097	x	-35	173	*LBL2	21 02	028	8	08
051	-	-45	127	STO1	35 01	022	8	08	098	P≷S	16-51	174	P≷S	16-51	029	x	-35
052	RCLI	36 46	128	GSB2	23 02	023	Yˣ	31	099	RCL1	36 01	175	RCL0	36 00	030	GSB5	23 05
053	x	-35	129	1	01	024	x	-35	100	RCL5	36 05	176	RCL4	36 04	031	X²	53
054	1/X	52	130	-	-45	025	GSB3	23 03	101	÷	-24	177	÷	-24	032	x	-35
055	STOI	35 46	131	1/X	52	026	RTN	24	102	.	-62	178	RCL6	36 06	033	RCL4	36 04
056	GSB2	23 02	132	RCL1	36 01	027	*LBLB	21 12	103	4	04	179	x	-35	034	x	-35
057	GSB4	23 04	133	x	-35	028	1	01	104	5	05	180	P≷S	16-51	035	RCLE	36 15
058	RCLI	36 46	134	RCLI	36 46	029	.	-62	105	Yˣ	31	181	RCL8	36 08	036	x	-35
059	x	-35	135	x	-35	030	5	05	106	x	-35	182	x	-35	037	RCL0	36 00
060	STOI	35 46	136	R/S	51	031	8	08	107	P≷S	16-51	183	RCLE	36 15	038	÷	-24
061	GSB2	23 02	137	RCL9	36 09	032	GSB2	23 02	108	RCL8	36 08	184	RCLD	36 14	039	RCL3	36 03
062	GSB3	23 03	138	x	-35	033	x	-35	109	.	-62	185	-	-45	040	x	-35
063	x	-35	139	STOI	35 46	034	RCLA	36 11	110	6	06	186	x	-35	041	5	05
064	CHS	-22	140	RTN	24	035	x	-35	111	7	07	187	RCLI	36 46	042	.	-62
065	1	01	141	*LBL6	21 06	036	RCL1	36 01	112	Yˣ	31	188	÷	-24	043	2	02
066	+	-55	142	GSB3	23 03	037	÷	-24	113	x	-35	189	RTN	24	044	2	02
067	1/X	52	143	CHS	-22	038	GSB1	23 01	114	RCLB	36 12	190	*LBL1	21 01	045	EEX	-23
068	GSB3	23 03	144	1	01	039	GSB3	23 03	115	RCLC	36 13	191	RCL4	36 04	046	1	01
069	CHS	-22	145	+	-55	040	RTN	24	116	-	-45	192	÷	-24	047	0	00
070	1	01	146	RTN	24	041	*LBLC	21 13	117	x	-35	193	RCL5	36 05	048	÷	-24
071	+	-55	147	*LBL5	21 05	042	.	-62	118	RCLI	36 46	194	÷	-24	049	RCL8	36 08
072	x	-35	148	GSB2	23 02	043	0	00	119	÷	-24	195	RCL6	36 06	050	1	01
073	LN	32	149	GSB3	23 03	044	0	00	120	RCL0	36 00	196	÷	-24	051	2	02
074	RCLI	36 46	150	x	-35	045	6	05	121	RCL3	36 03	197	RTN	24	052	÷	-24
075	x	-35	151	CHS	-22	046	6	06	122	x	-35	198	*LBL0	21 00	053	÷	-24

(Continued) Table I

Step	Key	Code	Step	Key	Code	Step	Key	Code	Step	Key	Code	Step	Key	Code	Step	Key	Code
054.	P⇄S	16-51	083	*LBL1	21 01	112	4	04	141	x	-35	170	.	-62	199	P⇄S	16-51
055	RCL3	36 03	084	GSB9	23 09	113	RCL7	36 07	142	2	02	171	0	00	200	RCL2	36 02
056	P⇄S	16-51	085	GTO0	22 00	114	X²	53	143	.	-62	172	2	02	201	P⇄S	16-51
057	÷	-24	086	*LBL4	21 04	115	x	-35	144	4	04	173	8	08	202	÷	-24
058	RTN	24	087	GSB8	23 08	116	Pi	16-24	145	2	02	174	2	02	203	4	04
059	*LBL5	21 05	088	*LBL0	21 00	117	RCL2	36 02	146	÷	-24	175	x	-35	204	.	-62
060	P⇄S	16-51	089	RTN	24	118	X²	53	147	1	01	176	RCL8	36 08	205	4	04
061	RCL1	36 01	090	*LBL8	21 08	119	x	-35	148	2	02	177	X²	53	206	5	05
062	P⇄S	16-51	091	Pi	16-24	120	-	-45	149	÷	-24	178	x	-35	207	EEX	-23
063	RCL7	36 07	092	RCL2	36 02	121	Pi	16-24	150	P⇄S	16-51	179	RCL4	36 04	208	6	06
064	x	-35	093	X²	53	122	÷	-24	151	RCL4	36 04	180	x	-35	209	CHS	-22
065	RCLE	36 15	094	x	-35	123	RCL2	36 02	152	P⇄S	16-51	181	RCL5	36 05	210	x	-35
066	x	-35	095	4	04	124	÷	-24	153	÷	-24	182	x	-35	211	X²	53
067	1	01	096	÷	-24	125	RTN	24	154	.	-62	183	5	05	212	.	-62
068	4	04	097	CHS	-22	126	*LBLC	21 13	155	3	03	184	.	-62	213	0	00
069	4	04	098	RCL7	36 07	127	P⇄S	16-51	156	2	02	185	2	02	214	2	02
070	x	-35	099	X²	53	128	RCL0	36 00	157	Y^x	31	186	2	02	215	7	07
071	RCL3	36 03	100	+	-55	129	P⇄S	16-51	158	1/X	52	187	EEX	-23	216	x	-35
072	÷	-24	101	4	04	130	1	01	159	.	-62	188	1	01	217	RCL5	36 05
073	RCL9	36 09	102	x	-35	131	8	08	160	1	01	189	0	00	218	x	-35
074	÷	-24	103	Pi	16-24	132	3	03	161	2	02	190	÷	-24	219	P⇄S	16-51
075	RCL0	36 00	104	÷	-24	133	x	-35	162	5	05	191	RCL1	36 01	220	RCL2	36 02
076	÷	-24	105	RCL2	36 02	134	RCL1	36 01	163	x	-35	192	1	01	221	P⇄S	16-51
077	RTN	24	106	÷	-24	135	X²	53	164	.	-62	193	2	02	222	÷	-24
078	*LBL2	21 02	107	RTN	24	136	÷	-24	165	0	00	194	÷	-24	223	+	-55
079	R/S	51	108	*LBL9	21 05	137	RCL6	36 06	166	0	00	195	÷	-24	224	RTN	24
080	X=0?	16-43	109	3	03	138	÷	-24	167	1	01	196	R/S	51			
081	GTO1	22 01	110	.	-62	139	STO8	35 08	168	4	04	197	÷	-24			
082	GTO4	22 04	111	4	04	140	RCL1	36 01	169	+	-55	198	RCL8	36 08			

c. Push A, B, C and then D in order, allowing time for the calculation to complete for each routine. Each fractional ΔT shown in Table II is printed out if the calculator is in the NORM position. $R_A = 0.0853$, $R_B = 0$, $R_C = 0.0034$, $R_D = 0.2955$. The sum of these products is accumulated and displayed.

d. Store the number of shell passes (2) in R_e. Store $N_{Pr,o}$ in R_8. Push E. Calculation will stop. Enter 0 for triangular tube array or 1 for square array. Push R/S. $R_E = 0.5393$ is printed out. The last number displayed is the sum of the products, $\Sigma R = 0.9235$.

This number may be used to calculate wall temperatures. The proportion of the temperature drop on the tubeside is 0.0853. The average tubeside temperature is $(90 + 120)/2 = 105°F$. The tubeside wall temperature will be higher than this by an amount $(0.0853/0.9235)$ $F_T(LMTD) = 6°F$, i.e., $T_w = 111°F$. The tubeside viscosity ratio is, therefore $\phi_t = \mu_{i,105°F}/\mu_{i,111°F}$. This is approximately equal to 1.

The shellside viscosity ratio, to the fouling layer, is obtained from the average temperature of the oil—$(358 + 100)/2 = 229°F$—and the temperature drop across the convective layer—$66.49(0.5393/0.9235) = 39°F$. Hence $\phi_s = \mu_{o,(229)}/\mu_{o,(190)}$.

These values of ϕ are calculated, and R_A and R_E are each divided by the relevant ratio raised to the 0.14 power (Eq. 3,4).

The correction on the water side is negligible, but the

Calculate thermal performance of each of the five regimes to see where resistance is highest   Table II

Resistance	Numerical factor	Driving potential	Materials properties	Mechanical design	Product
Inside tube, convective	3.169	$\dfrac{Q}{F_T(LMTD)}$	$\dfrac{\mu_i^{0.8}}{k_i m_i^{0.8}(N_{Pr,i})^{0.33}}$	$\dfrac{D_i^{0.8}}{N_t n L}$	R_A
Inside tube, fouling	$12/\pi$	$\dfrac{Q}{F_T(LMTD)}$	$R_{f,i}$	$\dfrac{1}{D_i L N_t n}$	R_B
Tubewall	$1/2\pi$	$\dfrac{Q}{F_T(LMTD)}$	$1/k_w$	$\dfrac{\ln(D_o/D_i)}{L N_t n}$	R_C
Outside tube, fouling	$12/\pi$	$\dfrac{Q}{F_T(LMTD)}$	$R_{f,o}$	$\dfrac{1}{D_o L N_t n}$	R_D
Outside tube, convective	0.367	$\dfrac{Q}{F_T(LMTD)}$	$\dfrac{\mu_o^{0.55}}{k_o m_o^{0.55}(N_{Pr,o})^{0.33}}$	$\dfrac{D_e^{0.45}(BD_sC)^{0.55}}{D_o L N_t n (P_t N_s)^{0.55}}$	R_E
				ΣProducts	ΣR

Contents of registers for Example 1					Table III
Secondary storage register	Variable	Numerical value	Primary storage register	Variable	Numerical value
R_{s0}	m_i	233,000 lb/h	R_0	B	7 in.
R_{s1}	m_o	49,600 lb/h	R_1	D_i	0.76 in.
R_{s2}	S_i	1.0	R_2	D_o	1.0 in.
R_{s3}	S_o	0.82	R_3	D_s	35 in.
R_{s4}	μ_i	0.73 cP	R_4	L	12 ft
R_{s5}	μ_o	1.12 cP	R_5	n	6
R_{s6}	k_i	0.37 Btu/(h)(ft)(°F)	R_6	N_t	$\frac{454}{6} = 75.67$
R_{s7}	k_o	0.076 Btu/(h)(ft)(°F)	R_7	P_t	1.25 in. (square array)
R_{s8}	k_w	25 Btu/(h)(ft)(°F)	R_8	$N_{Pr,i}(N_{Pr,o})$	4.78 (18.22)
R_{s9}	$R_{f,o}$	0.004	R_9	C	0.25 in.

Register	Variable	Value
R_a	$R_{f,i}$	0
R_b	T_1	358°F
R_c	T_2	100°F
R_d	t_1	90°F
R_e	$t_2, (N_s)$	120°F, (2)
I	F_t(LMTD)	Calculated by program

viscosity of 33.5° API oil is found to be 1.95 cP at 190°F. Therefore, $R_E = 0.5393/(1.12/1.95)^{0.14} = 0.5393/0.9253 = 0.5828$. The revised sum of products is 0.967, which is still less than 1, and hence the thermal design is fine.

If Σ products ≤ 1.0, continue to Step 3. Otherwise, the heat-exchanger design must be modified to obtain sufficient thermal performance.

3. Load both sides of Program 3.

a. Store S_i in R_{s2}, store S_o in R_{s3}. Return to primary registers.

b. Push A. When calculation stops, enter 0 for triangular array or 1 for square array. Push R/S to continue. $\Delta P_s = 6.21$ psi is displayed. $\Delta P_s < 10.0$ psi, which is acceptable. Divide by $\phi_s^{0.14}$. $\Delta P_s = 6.71$, still acceptable.

c. Push C. When the program stops, enter $\phi_t^{0.14}$ and push R/S. $\Delta P_t < 10.0$ psi, which is satisfactory.

Example 2

As a further example of the design procedure, again use the data of Example 1, but design for a fouling factor of $R_f = 0.003$ instead of 0.004. Any-size unit may be tried for the initial estimate. For convenience, start with the same unit used in Example 1.

1. The F_T and LMTD values remain unchanged; $F_{T,2-4} = 0.92$, LMTD = 71.93°F.

2. The uncorrected fractional temperature drops are: $R_A = 0.0853$, $R_B = 0$, $R_c = 0.0034$, $R_D = 0.2216$, $R_E = 0.5393$, $\Sigma R = 0.8496$.

a. The unit is oversized by about 15%. Size may be decreased by using a smaller-diameter shell or a shorter tube bundle. If the pressure drop permits, the smaller shell size will generally result in the lower-cost unit.

b. For the next iteration, try a 31-in. shell. Tube-count tables [2] suggest that up to 368 tubes may be in-

cluded in this size of shell.

Adjust the registers so that $R_3 = 31$ in. and $R_6 = 368/6$ (61.33). Repeating the calculations on Card 2, we find the following unadjusted fractional temperature drops: $R_A = 0.0890$, $R_B = 0$, $R_C = 0.0042$, $R_D = 0.2734$, $R_E = 0.6224$ and $\Sigma R = 0.9890$. The shellside wall temperature may be calculated at 187°F, for a wall viscosity of 2 cP. The corrected $\Sigma R = 1.038$.

The unit is now thermally too small. The major resistance is still on the shellside. Decreasing the baffle spacing would improve heat transfer by increasing turbulence. The minimum baffle spacing permitted by TEMA is 20% of the small diameter, or 6.2 in.

c. Set $R_0 = 6.2$, then rerun. Now $R_A = 0.0890$, $R_B = 0$, $R_C = 0.0042$, $R_D = 0.2734$, $R_E = 0.5822$. After applying the correction factor on viscosity, $R_E = 0.6314$ and $\Sigma R = 0.9980$. The unit is of adequate size.

3. Check the pressure drop on the new unit. $\Delta P_s = 10.39$, $\Delta P_t = 9.55$ psi. The shellside is slightly high but probably within the acceptable range.

For TI-58/59 users

The TI version closely follows the HP program. There are 3 TI programs (see listings in Tables IV, V and VI, respectively); and user instruction are offered in Table VII. A printout of the first example can be seen in Table VIII.

Listing for TI version—program A Table IV

Step	Code	Key	Step	Code	Key	Step	Code	Key
000	76	LBL	033	95	=	066	75	-
001	11	A	034	42	STO	067	43	RCL
002	43	RCL	035	29	29	068	23	23
003	21	21	036	99	PRT	069	54	)
004	75	-	037	53	(	070	95	=
005	43	RCL	038	43	RCL	071	42	STO
006	24	24	039	21	21	072	28	28
007	95	=	040	75	-	073	43	RCL
008	42	STO	041	43	RCL	074	29	29
009	25	25	042	22	22	075	91	R/S
010	43	RCL	043	54	)	076	76	LBL
011	22	22	044	55	÷	077	12	B
012	75	-	045	53	(	078	43	RCL
013	43	RCL	046	43	RCL	079	27	27
014	23	23	047	24	24	080	33	X²
015	95	=	048	75	-	081	85	+
016	42	STO	049	43	RCL	082	01	1
017	26	26	050	23	23	083	95	=
018	43	RCL	051	54	)	084	34	√X
019	25	25	052	95	=	085	85	+
020	75	-	053	42	STO	086	01	1
021	43	RCL	054	27	27	087	85	+
022	26	26	055	53	(	088	43	RCL
023	95	=	056	43	RCL	089	27	27
024	55	÷	057	24	24	090	95	=
025	53	(	058	75	-	091	65	×
026	43	RCL	059	43	RCL	092	43	RCL
027	25	25	060	23	23	093	28	28
028	55	÷	061	54	)	094	94	+/-
029	43	RCL	062	55	÷	095	85	+
030	26	26	063	53	(	096	02	2
031	54	)	064	43	RCL	097	95	=
032	23	LNX	065	21	21	098	42	STO

Step	Code	Key	Step	Code	Key	Step	Code	Key	Step	Code	Key	Step	Code	Key
099	30	30	143	85	+	187	76	LBL	231	33	X²	275	23	LNX
100	43	RCL	144	01	1	188	13	C	232	85	+	276	42	STO
101	27	27	145	95	=	189	02	2	233	01	1	277	59	59
102	33	X²	146	34	ΓX	190	55	÷	234	95	=	278	43	RCL
103	85	+	147	65	×	191	43	RCL	235	34	ΓX	279	27	27
104	01	1	148	43	RCL	192	28	28	236	42	STO	280	33	X²
105	95	=	149	30	30	193	75	-	237	58	58	281	85	+
106	34	ΓX	150	95	=	194	01	1	238	85	+	282	01	1
107	94	+/-	151	42	STO	195	75	-	239	43	RCL	283	95	=
108	85	+	152	30	30	196	43	RCL	240	59	59	284	34	ΓX
109	01	1	153	43	RCL	197	27	27	241	95	=	285	65	×
110	85	+	154	27	27	198	95	=	242	55	÷	286	43	RCL
111	43	RCL	155	65	×	199	42	STO	243	53	(	287	59	59
112	27	27	156	43	RCL	200	30	30	244	43	RCL	288	55	÷
113	95	=	157	28	28	201	01	1	245	59	59	289	02	2
114	65	×	158	94	+/-	202	75	-	246	75	-	290	95	=
115	43	RCL	159	85	+	203	43	RCL	247	43	RCL	291	42	STO
116	28	28	160	01	1	204	27	27	248	58	58	292	59	59
117	94	+/-	161	95	=	205	65	×	249	54	)	293	43	RCL
118	85	+	162	35	1/X	206	43	RCL	250	95	=	294	27	27
119	02	2	163	65	×	207	28	28	251	23	LNX	295	75	-
120	95	=	164	53	(	208	95	=	252	35	1/X	296	01	1
121	55	÷	165	43	RCL	209	65	×	253	42	STO	297	95	=
122	43	RCL	166	28	28	210	53	(	254	30	30	298	35	1/X
123	30	30	167	94	+/-	211	01	1	255	01	1	299	65	×
124	95	=	168	85	+	212	75	-	256	75	-	300	43	RCL
125	23	LNX	169	01	1	213	43	RCL	257	43	RCL	301	59	59
126	42	STO	170	54	)	214	28	28	258	28	28	302	65	×
127	30	30	171	95	=	215	54	)	259	95	=	303	43	RCL
128	43	RCL	172	23	LNX	216	95	=	260	42	STO	304	30	30
129	27	27	173	65	×	217	34	ΓX	261	59	59	305	95	=
130	75	-	174	43	RCL	218	65	×	262	01	1	306	99	PRT
131	01	1	175	30	30	219	02	2	263	75	-	307	91	R/S
132	95	=	176	95	=	220	55	÷	264	43	RCL	308	65	×
133	65	×	177	99	PRT	221	43	RCL	265	27	27	309	43	RCL
134	43	RCL	178	91	R/S	222	28	28	266	65	×	310	29	29
135	30	30	179	65	×	223	85	+	267	43	RCL	311	95	=
136	95	=	180	43	RCL	224	43	RCL	268	28	28	312	42	STO
137	35	1/X	181	29	29	225	30	30	269	95	=	313	30	30
138	42	STO	182	95	=	226	95	=	270	35	1/X	314	99	PRT
139	30	30	183	42	STO	227	42	STO	271	65	×	315	91	R/S
140	43	RCL	184	30	30	228	59	59	272	43	RCL			
141	27	27	185	99	PRT	229	43	RCL	273	59	59			
142	33	X²	186	91	R/S	230	27	27	274	95	=			

Listing for TI version—program B

Table V

Step	Code	Key	Step	Code	Key	Step	Code	Key	Step	Code	Key	Step	Code	Key
000	76	LBL	014	93	.	028	65	×	042	01	01	056	76	LBL
001	11	A	015	01	1	029	53	(	043	65	×	057	22	INV
002	43	RCL	016	02	2	030	43	RCL	044	43	RCL	058	55	÷
003	10	10	017	65	×	031	24	24	045	06	06	059	43	RCL
004	55	÷	018	53	(	032	75	-	046	54	)	060	04	04
005	43	RCL	019	43	RCL	033	43	RCL	047	45	Yˣ	061	55	÷
006	14	14	020	08	08	034	23	23	048	93	.	062	43	RCL
007	95	=	021	45	Yˣ	035	54	)	049	08	8	063	05	05
008	45	Yˣ	022	93	.	036	71	SBR	050	54	)	064	55	÷
009	93	.	023	06	6	037	22	INV	051	95	=	065	43	RCL
010	02	2	024	07	7	038	65	×	052	99	PRT	066	06	06
011	95	=	025	55	÷	039	53	(	053	44	SUM	067	92	RTN
012	65	×	026	43	RCL	040	53	(	054	33	33	068	76	LBL
013	01	1	027	30	30	041	43	RCL	055	91	R/S	069	12	B

(Continued) Table V

Step	Code	Key	Step	Code	Key	Step	Code	Key	Step	Code	Key	Step	Code	Key
070	01	1	117	93	.	164	95	=	211	43	RCL	258	07	7
071	93	.	118	00	0	165	99	PRT	212	07	07	259	65	×
072	05	5	119	00	0	166	44	SUM	213	33	X²	260	53	(
073	08	8	120	06	6	167	33	33	214	75	-	261	43	RCL
074	65	×	121	06	6	168	98	ADV	215	53	(	262	21	21
075	71	SBR	122	08	8	169	43	RCL	216	89	ñ	263	75	-
076	23	LNX	123	65	×	170	33	33	217	65	×	264	43	RCL
077	65	×	124	71	SBR	171	99	PRT	218	43	RCL	265	22	22
078	43	RCL	125	23	LNX	172	98	ADV	219	02	02	266	54	)
079	20	20	126	55	÷	173	91	R/S	220	33	X²	267	55	÷
080	55	÷	127	43	RCL	174	76	LBL	221	54	)	268	43	RCL
081	43	RCL	128	18	18	175	15	E	222	95	=	269	30	30
082	01	01	129	71	SBR	176	32	X:T	223	55	÷	270	65	×
083	71	SBR	130	22	INV	177	00	0	224	89	ñ	271	53	(
084	22	INV	131	65	×	178	67	EQ	225	55	÷	272	53	(
085	95	=	132	53	(	179	24	CE	226	43	RCL	273	43	RCL
086	99	PRT	133	53	(	180	89	ñ	227	02	02	274	00	00
087	44	SUM	134	43	RCL	181	65	×	228	95	=	275	65	×
088	33	33	135	02	02	182	43	RCL	229	76	LBL	276	43	RCL
089	91	R/S	136	55	÷	183	02	02	230	32	X:T	277	03	03
090	76	LBL	137	43	RCL	184	33	X²	231	45	Y×	278	65	×
091	23	LNX	138	01	01	185	55	÷	232	93	.	279	43	RCL
092	43	RCL	139	54	)	186	04	4	233	04	4	280	09	09
093	10	10	140	23	LNX	187	95	=	234	05	5	281	55	÷
094	55	÷	141	54	)	188	94	+/-	235	65	×	282	43	RCL
095	43	RCL	142	95	=	189	85	+	236	93	.	283	07	07
096	14	14	143	99	PRT	190	43	RCL	237	01	1	284	55	÷
097	65	×	144	44	SUM	191	07	07	238	05	5	285	43	RCL
098	43	RCL	145	33	33	192	33	X²	239	02	2	286	37	37
099	16	16	146	91	R/S	193	95	=	240	65	×	287	54	)
100	65	×	147	76	LBL	194	65	×	241	53	(	288	45	Y×
101	43	RCL	148	14	D	195	04	4	242	43	RCL	289	93	.
102	08	08	149	01	1	196	55	÷	243	11	11	290	05	5
103	55	÷	150	93	.	197	89	ñ	244	55	÷	291	05	5
104	43	RCL	151	05	5	198	55	÷	245	43	RCL	292	54	)
105	30	30	152	08	8	199	43	RCL	246	15	15	293	95	=
106	65	×	153	65	×	200	02	02	247	54	)	294	71	SBR
107	53	(	154	71	SBR	201	95	=	248	45	Y×	295	22	INV
108	43	RCL	155	23	LNX	202	61	GTO	249	93	.	296	95	=
109	24	24	156	65	×	203	32	X:T	250	04	4	297	99	PRT
110	75	-	157	43	RCL	204	76	LBL	251	05	5	298	44	SUM
111	43	RCL	158	19	19	205	24	CE	252	65	×	299	33	33
112	23	23	159	55	÷	206	03	3	253	43	RCL	300	98	ADV
113	54	)	160	43	RCL	207	93	.	254	38	38	301	43	RCL
114	92	RTN	161	02	02	208	04	4	255	45	Y×	302	33	33
115	76	LBL	162	71	SBR	209	04	4	256	93	.	303	99	PRT
116	13	C	163	22	INV	210	65	×	257	06	6	304	98	ADV
												305	91	R/S

Listing for TI version—program C   Table VI

Step	Code	Key	Step	Code	Key	Step	Code	Key	Step	Code	Key	Step	Code	Key
000	76	LBL	009	02	02	018	33	X²	027	95	=	036	04	4
001	11	A	010	33	X²	019	95	=	028	42	STO	037	04	4
002	32	X:T	011	55	÷	020	65	×	029	36	36	038	65	×
003	00	0	012	04	4	021	04	4	030	61	GTO	039	43	RCL
004	67	EQ	013	95	=	022	55	÷	031	23	LNX	040	07	07
005	22	INV	014	94	+/-	023	89	ñ	032	76	LBL	041	33	X²
006	89	ñ	015	85	+	024	55	÷	033	22	INV	042	75	-
007	65	×	016	43	RCL	025	43	RCL	034	03	3	043	53	(
008	43	RCL	017	07	07	026	02	02	035	93	.	044	89	ñ

(Continued) Table VI

Step	Code	Key	Step	Code	Key	Step	Code	Key	Step	Code	Key	Step	Code	Key
045	65	×	092	93	.	139	65	×	186	03	3	233	22	INV
046	43	RCL	093	00	0	140	01	1	187	02	2	234	52	EE
047	02	02	094	03	3	141	02	2	188	95	=	235	99	PRT
048	33	X²	095	04	4	142	55	÷	189	35	1/X	236	32	X:T
049	54	)	096	04	4	143	43	RCL	190	65	×	237	91	R/S
050	95	=	097	95	=	144	13	13	191	93	.	238	55	÷
051	55	÷	098	45	Y×	145	22	INV	192	01	1	239	32	X:T
052	89	π	099	93	.	146	52	EE	193	02	2	240	35	1/X
053	55	÷	100	01	1	147	95	=	194	05	5	241	85	+
054	43	RCL	101	09	9	148	99	PRT	195	85	+	242	53	(
055	02	02	102	06	6	149	98	ADV	196	93	.	243	93	.
056	95	=	103	04	4	150	91	R/S	197	00	0	244	00	0
057	42	STO	104	94	+/-	151	76	LBL	198	00	0	245	02	2
058	36	36	105	95	=	152	13	C	199	01	1	246	07	7
059	76	LBL	106	65	×	153	43	RCL	200	04	4	247	65	×
060	23	LNX	107	93	.	154	10	10	201	95	=	248	43	RCL
061	43	RCL	108	00	0	155	65	×	202	65	×	249	05	05
062	11	11	109	01	1	156	01	1	203	93	.	250	55	÷
063	65	×	110	02	2	157	08	8	204	00	0	251	43	RCL
064	43	RCL	111	08	8	158	03	3	205	02	2	252	12	12
065	07	07	112	65	×	159	55	÷	206	08	8	253	65	×
066	65	×	113	43	RCL	160	43	RCL	207	02	2	254	53	(
067	43	RCL	114	57	57	161	01	01	208	65	×	255	43	RCL
068	37	37	115	33	X²	162	33	X²	209	43	RCL	256	56	56
069	65	×	116	65	×	163	55	÷	210	56	56	257	55	÷
070	01	1	117	43	RCL	164	43	RCL	211	33	X²	258	43	RCL
071	04	4	118	04	04	165	06	06	212	65	×	259	12	12
072	04	4	119	65	×	166	95	=	213	43	RCL	260	65	×
073	55	÷	120	43	RCL	167	42	STO	214	04	04	261	04	4
074	43	RCL	121	37	37	168	56	56	215	65	×	262	93	.
075	03	03	122	55	÷	169	65	×	216	43	RCL	263	04	4
076	55	÷	123	43	RCL	170	43	RCL	217	05	05	264	05	5
077	43	RCL	124	00	00	171	01	01	218	55	÷	265	52	EE
078	09	09	125	65	×	172	55	÷	219	05	5	266	94	+/-
079	55	÷	126	43	RCL	173	02	2	220	93	.	267	06	6
080	43	RCL	127	03	03	174	93	.	221	02	2	268	54	)
081	00	00	128	55	÷	175	04	4	222	02	2	269	33	X²
082	95	=	129	05	5	176	02	2	223	52	EE	270	54	)
083	42	STO	130	93	.	177	55	÷	224	01	1	271	95	=
084	57	57	131	02	2	178	01	1	225	00	0	272	22	INV
085	65	×	132	02	2	179	02	2	226	55	÷	273	52	EE
086	43	RCL	133	52	EE	180	55	÷	227	43	RCL	274	99	PRT
087	36	36	134	01	1	181	43	RCL	228	01	01	275	98	ADV
088	55	÷	135	00	0	182	14	14	229	65	×	276	91	R/S
089	43	RCL	136	55	÷	183	95	=	230	01	1			
090	15	15	137	43	RCL	184	45	Y×	231	02	2			
091	65	×	138	36	36	185	93	.	232	95	=			

User instructions for TI version

Table VII

The TI version is in three parts, and must be run in order. However, all the data may be stored before any one of the parts is run, but not all the parts use all the data. The following tabulation shows the data storage areas and the data used by the separate programs.

(Note: The HP version calls for entering data between operation of the different parts of the program. This is because some of the HP storage areas are used for more than one value. The TI calculator has more storage capacity than the HP, so that all of the data may be entered before the start of program A, inasmuch as none of the storage areas are used for more than one value.)

Data	Register	Program Use A	B	C
Baffle spacing, in.	00		•	•
Tube I.D., in.	01		•	•
Tube O.D., in.	02		•	

Shell diameter, in.	03		•	•
Tube length, in.	04		•	•
No. of tube passes	05		•	
No. tubes per pass	05		•	•
Tube pitch, in.	07		•	•
Prandtl number, inside	08		•	
Prandtl number, outside	38		•	
Tube clearance	09		•	•
Weight flowrate, cold fluid, lb/h	10		•	•
Weight flowrate, hot fluid, lb/h	11		•	•
Specific gravity, cold fluid	12			•
Specific gravity, hot fluid	13			•
Viscosity, cold fluid, cP	14		•	•
Viscosity, hot fluid, cP	15		•	•
Thermal conductivity, cold	16		•	
Thermal conductivity, hot	17		•	
Thermal conductivity, tube	18		•	
Fouling factor, outside	19		•	
Fouling factor, inside	20		•	
Temperatures, °F				
Hot fluid, inlet	21	•	•	
Hot fluid, outlet	22	•	•	
Cold fluid, inlet	23	•	•	
Cold fluid, outlet	24	•	•	
Number of passes	37		•	•

Programs are run as follows:

Part A: Key **A** gives LMTD.
            Key **B** gives $F_{T,1-2}$ (if result is flashing, it indicates a temperature cross and inoperable condition. Use key **CLR**.)
            Key **C** gives $F_{T,2-4}$.
            Key **R/S** gives corrected LMTD for F_T.

Part B: Keys **A, B, C, D, E** give fractional temperature drops indicated in Table II.
            After key **D**, program prints both the D fraction (R_D) and the sum.
            Before key **E**, enter "0" for triangular tube array or "1" or square tube array.
            After key **E**, programs prints both the E fraction and the sum.

Part C: Enter "0" for triangular tube array or "1" for square tube array.
            Press key **A**
            Output will be shell-side pressure drop, psi.
            Press key **C**
            When program stops, enter $\phi + 0.14$ and key **R/S**.
            Output will be tube-side pressure drop, psi.

Printout for first example—TI Version   Table VII

Program A
```
71.9314240
.9585282826
0.924290356
66.40552224
```

Program B
```
0.08524324
        0.
.0034203053
.2954729306

.3842444759

.5393019662

.9235464421
```

Program C
```
6.205625639

3.386938169
6.403467672
```

References

1. Standards of Tubular Exchanger Manufacturers Assn. (TEMA), Sixth ed., New York, 1978.
2. Kern, D. Q., "Process Heat Transfer," McGraw-Hill Book Co., New York, 1950.
3. Gilmour, C. H., "Shortcut to heat exchanger design—I," *Chem. Eng.*, Oct. 1952.

The authors

Roger A. Crane is associate professor in the Dept. of Chemical and Mechanical Engineering at the University of South Florida, Tampa, FL 33620, tel: 813-974-2581. He received B.S. and M.S. degrees at the University of Missouri at Rolla and a Ph.D. at Auburn University. He was a senior engineer with Babcock and Wilcox and is a licensed professional engineer in Florida.

Robert Gregg is an instructor in mechanical engineering and machine design at the University of South Florida. He received B.S. and M.S. degrees there and has gained industrial experience as project design engineer for the construction and operation of plants producing cement and phosphate fertilizers. Gregg is a registered professional engineer in Florida.

Program solves airstream energy balances

This TI-59 calculator program includes correlations for the enthalpy of air and water vapor. Thus it needs little input to solve mixing and heat-transfer energy balances and predict outlet temperatures.

Calvin R. Brunner, Malcolm Pirnie, Inc.

☐ Predicting the final temperature of two airstreams that mix, or exchange heat without mixing, involves a trial-and-error procedure because both air and water-vapor enthalpies must be considered simultaneously. This TI-59 program solves such problems quickly, using correlations for the enthalpies of air and water vapor over the range 500°F–2,500°F. Though it is most accurate over this high-temperature range, the program can be used at lower temperatures with only moderate error.

Trial-and-error

The figure illustrates the problems that this program solves:

■ Heat exchange. Given the initial temperature, air flow and water flow for two airstreams, and the final temperature for one of the streams, predict the final temperature for the other stream.

■ Mixing. Given the initial conditions as above, predict the temperature of the mixed stream.

To see how the program solves these problems, we need to look at the energy balances for each case. First, the enthalpy flow of an airstream (H_i, Btu/h) is the sum of the dry-air and water-vapor enthalpy flows:

$$H_i = [M_a h_a + M_w h_w]_i \qquad (1)$$

where M is mass flowrate (1b/h), h is enthalpy (Btu/lb), a refers to air, and w refers to water.

Enthalpies h_a and h_w can be represented as functions of temperature (t, °F), based on least-squares correlations of the data shown in Table I:

$$h_a = 0.0805\ t^{1.1506} \qquad (2)$$

$$h_w = 0.56\ t + 937.8 \qquad (3)$$

Substituting these equations into Eq. (1) yields the enthalpy flow for a stream of known temperature and mass flowrates.

In the case of heat exchange, both the inlet and outlet

(Text continues on p. 111)

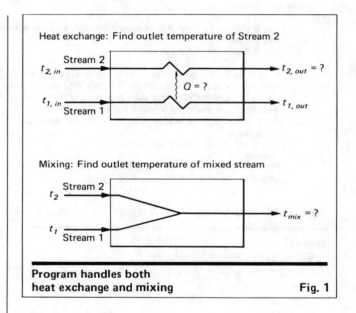

Heat exchange: Find outlet temperature of Stream 2

Mixing: Find outlet temperature of mixed stream

Program handles both heat exchange and mixing

Fig. 1

Correlations correspond closely to actual enthalpy values

Table I

	Temperature (t), °F				
	500	1,000	1,500	2,000	2,500
Air enthalpy (h_a), Btu/lb					
Actual*	102	230	365	505	650
$h_a = 0.0805\ t^{1.1506}$	103	228	363	506	654
Water-vapor enthalpy (h_w), Btu/lb					
Actual†	1,239	1,486	1,756	2,047	2,358
$h_w = 0.56\ t + 937.8$	1,218	1,497	1,777	2,057	2,337

*Keenan, J. H., and Kaye, J., "Gas Tables," John Wiley & Sons, New York, 1948.
†Keenan, J. H., and Keyes, F. G., "Thermodynamic Properties of Steam," John Wiley & Sons, New York, 1936.

Originally published November 16, 1981

Program listing for TI-59 calculator   Table II

Location	Code	Key	Location	Code	Key	Location	Code	Key	Location	Code	Key	Location	Code	Key
000	76	LBL	061	11	11	122	42	STO	183	01	1	244	01	1
001	24	CE	062	91	R/S	123	09	09	184	93	.	245	03	3
002	45	YX	063	76	LBL	124	43	RCL	185	01	1	246	02	2
003	01	1	064	18	C'	125	12	12	186	05	5	247	04	4
004	93	.	065	42	STO	126	71	SBR	187	35	1/X	248	03	3
005	01	1	066	12	12	127	24	CE	188	95	=	249	05	5
006	05	5	067	91	R/S	128	42	STO	189	42	STO	250	69	OP
007	00	0	068	76	LBL	129	13	13	190	18	18	251	04	04
008	06	6	069	19	D'	130	43	RCL	191	75	−	252	43	RCL
009	65	×	070	93	.	131	12	12	192	43	RCL	253	00	00
010	93	.	071	01	1	132	71	SBR	193	17	17	254	69	OP
011	00	0	072	32	X:T	133	23	LNX	194	95	=	255	06	06
012	08	8	073	43	RCL	134	42	STO	195	50	I×I	256	02	2
013	00	0	074	03	03	135	14	14	196	66	PAU	257	03	3
014	05	5	075	71	SBR	136	65	×	197	22	INV	258	07	7
015	95	=	076	24	CE	137	43	RCL	198	77	GE	259	00	0
016	92	RTN*	077	42	STO	138	11	11	199	02	02	260	03	3
017	76	LBL	078	04	04	139	85	+	200	06	06	261	02	2
018	23	LNX	079	43	RCL	140	43	RCL	201	43	RCL	262	69	OP
019	65	×	080	03	03	141	13	13	202	18	18	263	04	04
020	93	.	081	71	SBR	142	65	×	203	61	GTO	264	43	RCL
021	05	5	082	23	LNX	143	43	RCL	204	01	01	265	01	01
022	06	6	083	42	STO	144	10	10	205	53	53	266	69	OP
023	85	+	084	19	19	145	85	+	206	98	ADV	267	06	06
024	09	9	085	65	×	146	43	RCL	207	02	2	268	03	3
025	03	3	086	43	RCL	147	09	09	208	03	3	269	07	7
026	07	7	087	01	01	148	95	=	209	01	1	270	00	0
027	93	.	088	85	+	149	42	STO	210	07	7	271	00	0
028	08	8	089	43	RCL	150	16	16	211	01	1	272	02	2
029	01	1	090	04	04	151	43	RCL	212	03	3	273	04	4
030	04	4	091	65	×	152	12	12	213	69	OP	274	03	3
031	95	=	092	43	RCL	153	42	STO	214	02	02	275	01	1
032	92	RTN*	093	00	00	154	17	17	215	03	3	276	69	OP
033	76	LBL	094	95	=	155	65	×	216	07	7	277	04	04
034	11	A	095	42	STO	156	93	.	217	00	0	278	43	RCL
035	42	STO	096	05	05	157	05	5	218	00	0	279	02	02
036	00	00	097	43	RCL	158	06	6	219	03	3	280	69	OP
037	91	R/S	098	02	02	159	85	+	220	07	7	281	06	06
038	76	LBL	099	71	SBR	160	09	9	221	03	3	282	03	3
039	12	B	100	24	CE	161	03	3	222	05	5	283	07	7
040	42	STO	101	42	STO	162	08	8	223	01	1	284	03	3
041	01	01	102	06	06	163	95	=	224	03	3	285	02	2
042	91	R/S	103	43	RCL	164	65	×	225	69	OP	286	04	4
043	76	LBL	104	02	02	165	43	RCL	226	03	03	287	01	1
044	13	C	105	71	SBR	166	11	11	227	03	3	288	03	3
045	42	STO	106	23	LNX	167	94	+/−	228	01	1	289	07	7
046	02	02	107	42	STO	168	85	+	229	03	3	290	69	OP
047	91	R/S	108	07	07	169	43	RCL	230	06	6	291	04	04
048	76	LBL	109	65	×	170	16	16	231	02	2	292	43	RCL
049	14	D	110	43	RCL	171	95	=	232	01	1	293	03	03
050	42	STO	111	01	01	172	55	÷	233	01	1	294	69	OP
051	03	03	112	85	+	173	93	.	234	07	7	295	06	06
052	91	R/S	113	43	RCL	174	00	0	235	03	3	296	98	ADV
053	76	LBL	114	06	06	175	08	8	236	05	5	297	01	1
054	16	A'	115	65	×	176	00	0	237	69	OP	298	04	4
055	42	STO	116	43	RCL	177	05	5	238	04	04	299	06	6
056	10	10	117	00	00	178	55	÷	239	69	OP	300	03	3
057	91	R/S	118	75	−	179	43	RCL	240	05	05	301	02	2
058	76	LBL	119	43	RCL	180	10	10	241	98	ADV	302	03	3
059	17	B'	120	05	05	181	95	=	242	71	SBR	303	03	3
060	42	STO	121	95	=	182	45	YX	243	33	X²	304	05	5

Location	Code	Key	Location	Code	Key	Location	Code	Key	Location	Code	Key	Location	Code	Key
305	69	OP	366	98	ADV	427	03	03	488	02	02	549	08	8
306	04	04	367	71	SBR	428	03	3	489	71	SBR	550	54	)
307	43	RCL	368	35	1/X	429	07	7	490	24	CE	551	65	×
308	09	09	369	98	ADV	430	04	4	491	42	STO	552	53	(
309	69	OP	370	98	ADV	431	03	3	492	06	06	553	43	RCL
310	06	06	371	98	ADV	432	03	3	493	43	RCL	554	01	01
311	98	ADV	372	91	R/S	433	02	2	494	02	02	555	85	+
312	71	SBR	373	76	LBL	434	00	0	495	71	SBR	556	43	RCL
313	34	ГX	374	33	X²	435	00	0	496	23	LNX	557	11	11
314	01	1	375	03	3	436	00	0	497	42	STO	558	95	=
315	03	3	376	06	6	437	00	0	498	07	07	559	55	÷
316	02	2	377	03	3	438	69	OP	499	65	×	560	93	.
317	04	4	378	07	7	439	04	04	500	43	RCL	561	00	0
318	03	3	379	69	OP	440	69	OP	501	01	01	562	08	8
319	05	5	380	02	02	441	05	05	502	85	+	563	00	0
320	69	OP	381	03	3	442	92	RTN*	503	43	RCL	564	05	5
321	04	04	382	05	5	443	76	LBL	504	00	00	565	55	÷
322	43	RCL	383	01	1	444	35	1/X	505	65	×	566	53	(
323	10	10	384	07	7	445	05	5	506	43	RCL	567	43	RCL
324	69	OP	385	01	1	446	05	5	507	06	06	568	00	00
325	06	06	386	03	3	447	02	2	508	95	=	569	85	+
326	02	2	387	03	3	448	01	1	509	42	STO	570	43	RCL
327	03	3	388	00	0	449	02	2	510	08	08	571	10	10
328	07	7	389	00	0	450	07	7	511	43	RCL	572	95	=
329	00	0	390	00	0	451	03	3	512	12	12	573	45	Y^X
330	03	3	391	69	OP	452	02	2	513	71	SBR	574	01	1
331	02	2	392	03	03	453	04	4	514	24	CE	575	93	.
332	69	OP	393	03	3	454	03	3	515	42	STO	576	01	1
333	04	04	394	02	2	455	69	OP	516	13	13	577	05	5
334	43	RCL	395	03	3	456	02	02	517	43	RCL	578	35	1/X
335	11	11	396	01	1	457	02	2	518	12	12	579	95	=
336	69	OP	397	01	1	458	04	4	519	71	SBR	580	42	STO
337	06	06	398	07	7	459	03	3	520	23	LNX	581	15	15
338	03	3	399	00	0	460	01	1	521	42	STO	582	75	-
339	07	7	400	00	0	461	00	0	522	14	14	583	43	RCL
340	00	0	401	00	0	462	00	0	523	43	RCL	584	17	17
341	00	0	402	00	0	463	02	2	524	14	14	585	95	=
342	02	2	403	69	OP	464	07	7	525	65	×	586	50	I×I
343	04	4	404	04	04	465	69	OP	526	43	RCL	587	66	PAU
344	03	3	405	69	OP	466	03	03	527	11	11	588	22	INV
345	01	1	406	05	05	467	01	1	528	85	+	589	77	GE
346	69	OP	407	92	RTN*	468	04	4	529	43	RCL	590	05	05
347	04	04	408	76	LBL	469	06	6	530	10	10	591	99	99
348	43	RCL	409	34	ГX	470	03	3	531	65	×	592	43	RCL
349	12	12	410	03	3	471	02	2	532	43	RCL	593	15	15
350	69	OP	411	06	6	472	03	3	533	13	13	594	42	STO
351	06	06	412	03	3	473	03	3	534	85	+	595	17	17
352	03	3	413	07	7	474	05	5	535	43	RCL	596	61	GTO
353	07	7	414	69	OP	475	05	5	536	08	08	597	05	05
354	03	3	415	02	02	476	06	6	537	75	-	598	23	23
355	02	2	416	03	3	477	69	OP	538	53	(	599	98	ADV
356	04	4	417	05	5	478	04	04	539	53	(	600	01	1
357	01	1	418	01	1	479	69	OP	540	93	.	601	05	5
358	03	3	419	07	7	480	05	05	541	05	5	602	03	3
359	07	7	420	01	1	481	92	RTN*	542	06	6	603	02	2
360	69	OP	421	03	3	482	76	LBL	543	65	×	604	69	OP
361	04	04	422	03	3	483	15	E	544	43	RCL	605	01	01
362	43	RCL	423	00	0	484	93	.	545	17	17	606	03	3
363	17	17	424	00	0	485	01	1	546	85	+	607	00	0
364	69	OP	425	00	0	486	32	X:T	547	09	9	608	01	1
365	06	06	426	69	OP	487	43	RCL	548	03	3	609	04	4

Table II (Continued)

Location	Code	Key	Location	Code	Key	Location	Code	Key	Location	Code	Key	Location	Code	Key
610	02	2	648	03	3	686	03	3	724	04	4	762	00	00
611	04	4	649	02	2	687	05	5	725	69	OP	763	85	+
612	03	3	650	04	4	688	69	OP	726	02	02	764	43	RCL
613	01	1	651	03	3	689	04	04	727	03	3	765	10	10
614	01	1	652	05	5	690	43	RCL	728	01	1	766	95	=
615	07	7	653	69	OP	691	10	10	729	01	1	767	69	OP
616	69	OP	654	04	04	692	69	OP	730	07	7	768	06	06
617	02	02	655	43	RCL	693	06	06	731	01	1	769	02	2
618	01	1	656	00	00	694	02	2	732	06	6	770	03	3
619	06	6	657	69	OP	695	03	3	733	00	0	771	07	7
620	00	0	658	06	06	696	07	7	734	00	0	772	00	0
621	00	0	659	02	2	697	00	0	735	03	3	773	03	3
622	01	1	660	03	3	698	03	3	736	06	6	774	02	2
623	03	3	661	07	7	699	02	2	737	69	OP	775	69	OP
624	02	2	662	00	0	700	69	OP	738	03	03	776	04	04
625	04	4	663	03	3	701	04	04	739	03	3	777	43	RCL
626	03	3	664	02	2	702	43	RCL	740	07	7	778	01	01
627	05	5	665	69	OP	703	11	11	741	03	3	779	85	+
628	69	OP	666	04	04	704	69	OP	742	05	5	780	43	RCL
629	03	03	667	43	RCL	705	06	06	743	01	1	781	11	11
630	02	2	668	01	01	706	03	3	744	07	7	782	95	=
631	01	1	669	69	OP	707	07	7	745	01	1	783	69	OP
632	02	2	670	06	06	708	69	OP	746	03	3	784	06	06
633	07	7	671	03	3	709	04	04	747	03	3	785	03	3
634	03	3	672	07	7	710	43	RCL	748	00	0	786	07	7
635	02	2	673	69	OP	711	12	12	749	69	OP	787	69	OP
636	04	4	674	04	04	712	69	OP	750	04	04	788	04	04
637	03	3	675	43	RCL	713	06	06	751	69	OP	789	43	RCL
638	69	OP	676	02	02	714	98	ADV	752	05	05	790	17	17
639	04	04	677	69	OP	715	01	1	753	01	1	791	69	OP
640	69	OP	678	06	06	716	05	5	754	03	3	792	06	06
641	05	05	679	98	ADV	717	03	3	755	02	2	793	98	ADV
642	69	OP	680	71	SBR	718	02	2	756	04	4	794	71	SBR
643	00	00	681	34	√X	719	03	3	757	03	3	795	35	1/X
644	98	ADV	682	01	1	720	00	0	758	05	5	796	98	ADV
645	71	SBR	683	03	3	721	01	1	759	69	OP	797	98	ADV
646	33	X²	684	02	2	722	04	4	760	04	04	798	98	ADV
647	01	1	685	04	4	723	02	2	761	43	RCL	799	91	R/S

Notes: Calculator must be partitioned to 799.19 before entering the program (or reading the two program cards).

*The key shown as **RTN** must be entered as **INV SBR**. **RTN** appears at steps 016, 032, 407, 442 and 481.

temperatures are known for Stream 1, so $H_{1,in}$ and $H_{1,out}$ are known. From these, we can find the heat transferred to Stream 2 (Q, Btu/h):

$$Q = H_{1,in} - H_{1,out}$$
$$= [M_a (h_{a,in} - h_{a,out}) + M_w (h_{w,in} - h_{w,out})]_1 \quad (4)$$

Knowing Q, we can find the outlet enthalpy flow for Stream 2:

$$H_{2,out} = Q + H_{2,in} \quad (5)$$

The program calculates Q and $H_{2,in}$ from known conditions, using Eq. (1–4), then finds $H_{2,out}$ from Eq. (5). To find the outlet temperature of Stream 2, the program uses Eq. (1–3) again and solves for t:

Stream 2 outlet:

$$t^{1.1506} = \frac{H - (0.56\,t + 937.8)\,M_w}{0.0805\,M_a} \quad (6)$$

To solve Eq. (6), the program uses a trial-and-error procedure, with 0.1°F as the tolerable error.

1. Assume a temperature value, t', and use this to calculate the right-hand side of Eq. (6).

2. Solve for t on the left-hand side of Eq. (6).

3. If $|t - t'| < 0.1$, t' is considered equal to t, and the calculation is completed.

4. If $|t - t'| \geq 0.1$, the calculated value t is substituted for t' and the program returns to Step 1.

User instructions for TI-59 program   Table III

Step	Key	Comment
1. Partition	2 2nd OP 17	799.19
2. Enter program from Table II or from cards		
3. Enter essential data		
M_a, Stream 1	A	
M_w, Stream 1	B	
t, inlet, Stream 1	C	
M_a, Stream 2	2nd A'	
M_w, Stream 2	2nd B'	
t, inlet, Stream 2	2nd C'	
4. For mixing of two streams	E	Program runs, prints out for COMBINED FLOW
5. For heat transfer between streams		
t, outlet, Stream 1	D	
	2nd D'	Program runs, prints out for HEAT TRANSFER

Notes: While the program is running, the value $|t - t'|$ will flash in the display, decreasing each time until it is less than 0.1. At that time, the program is complete and will print out. If the printer is not used, the needed values can be recalled from the memory registers as listed in Table V.

Printouts for heat-exchange and mixing examples   Table IV

```
     HEAT TRANSFER              COMBINED AIR FLOW

        STREAM ONE                  STREAM ONE
   20000.      AIR            20000.      AIR
    1500.      H2O             1500.      H2O
     610.      T IN             610.      T
    2400.      TOUT

-11400179.83   B/HR              STREAM TWO
                               35000.      AIR
        STREAM TWO              2850.      H2O
   35000.      AIR             2450.      T
    2850.      H2O
    2450.      T IN          COMBINED STREAM
 1483.695621   TOUT           55000.      AIR
                               4350.      H2O
   (FLOW IN LB/HR)          1824.52377    T

                             (FLOW IN LB/HR)
```

Content of storage registers   Table V

Register	Content
00	M_a, Stream 1, lb/h
01	M_w, Stream 1, lb/h
02	t, inlet, Stream 1, °F
03	t, outlet, Stream 1, °F
04	h_a, outlet, Stream 1, Btu/lb
05	H, outlet, Stream 1, Btu/h
06	h_a, inlet, Stream 1, Btu/lb
07	h_w, inlet, Stream 1, Btu/lb
08	Used
09	Q, Btu/h
10	M_a, Stream 2, lb/h
11	M_w, Stream 2, lb/h
12	t, inlet, Stream 2, °F
13	h_a, inlet, Stream 2, Btu/lb
14	h_w, inlet, Stream 2, Btu/lb
15	t', °F for mixed-flow case
16	H, outlet, Stream 2, Btu/h
17	t', °F for heat-transfer case
18	t, °F, final result for either case
19	h_w, outlet, Stream 1, Btu/lb

In the case where two airstreams mix, the enthalpy and mass flowrates of the mixed stream are simply the sums of the individual enthalpy and mass flowrates of the two streams:

$$H_{mix} = H_1 + H_2 \qquad (7)$$
$$M_{a,mix} = M_{a,1} + M_{a,2} \qquad (8)$$
$$M_{w,mix} = M_{w,1} + M_{w,2} \qquad (9)$$

The program calculates these values from the given inputs, substitutes the H, M_a and M_w values for the mixed stream into Eq. (6), and solves for the temperature of the mixture by trial-and-error as above.

How to use the program

Table II lists the steps, and Table III the user instructions for the program. Note that the TI-59 calculator must be partitioned to 799.19 before entering the program (or before reading cards). If the program is used with the PC-100 printer, it prints out the relevant inputs and outputs as shown in Table IV. If the program is used without the printer, the results must be recalled from the data registers—Table V is the key to these. Register 18 holds the final result t for either mixing or heat transfer.

Example

Suppose that two airstreams are to exchange heat in a heat exchanger. The first stream is 20,000 lb/h of dry air, plus 1,500 lb/h of moisture, entering at 610°F. The second stream is 35,000 lb/h dry air, plus 2,850 lb/h moisture, entering at 2,450°F. What is the exit temperature of Stream 2 if the exit temperature of Stream 1 is assumed to be 2,400°F?

To solve this problem, partition the calculator and enter the program according to Table III. Then enter the data as shown, and press **2nd D'** to get the appropriate printout in about thirty seconds:

Airflow, Stream 1	20,000	A
Moisture flow, Stream 1	1,500	B
Inlet temperature, Stream 1	610	C
Airflow, Stream 2	35,000	2nd A'
Moisture flow, Stream 2	2,850	2nd B'
Inlet temperature, Stream 2	2,450	2nd C'
Exit temperature, Stream 1	2,400	D
		2nd D'

Table IV shows the resulting printout, under the heading HEAT TRANSFER. The outlet temperature of Stream 2 in this case is 1,484°F.

If the streams were to be mixed instead, the exit-temperature entry would have been unnecessary. Pressing **E** (instead of **2nd D'**) would get the program to run, and the printout would be as shown under COMBINED AIR FLOW in Table IV. In this case, the temperature of the mixed stream would be 1,824°F, as shown in Table IV.

For HP-67/97 users

The HP version closely follows the TI program. Table VI offers the HP program listing, and Table VII provides user instructions. Printouts for the examples are contained in Table VIII.

Program listing for HP version   Table VI

Step	Key	Code	Step	Key	Code	Step	Key	Code	Step	Key	Code	Step	Key	Code
001	*LBLA	21 11	029	RCL2	36 02	057	RCL9	36 09	085	−	−45	113	1	01
002	STO2	35 02	030	GSB2	23 02	058	+	−55	086	ABS	16 31	114	.	−62
003	R↓	−31	031	RCL1	36 01	059	STOE	35 15	087	,	−62	115	1	01
004	STO1	35 01	032	÷	−25	060	*LBLE	21 15	088	1	01	116	5	05
005	R↓	−31	033	ST+7	35-55 07	061	RCLE	36 15	089	X⤨Y	16-34	117	0	00
006	STO0	35 00	034	RCL3	36 03	062	RCLD	36 14	090	GTO7	22 07	118	6	06
007	R/S	51	035	GSB2	23 02	063	GSB2	23 02	091	RCLI	36 46	119	yx	31
008	*LBLd	21 16 14	036	RCL1	36 01	064	RCL5	36 05	092	STOD	35 14	120	.	−62
009	SF1	16 21 01	037	x	−35	065	y	−75	093	GTO6	22 06	121	0	00
010	*LBLD	21 14	038	ST+8	35-55 08	066	−	−45	094	*LBL7	21 07	122	8	08
011	STO3	35 03	039	RCL7	36 07	067	.	−62	095	RCLI	36 46	123	0	00
012	STOD	35 14	040	RCL8	36 08	068	0	00	096	PRTX	−14	124	5	05
013	R↓	−31	041	−	−45	069	8	08	097	RCLA	36 11	125	x	−35
014	STO4	35 04	042	STOA	35 11	070	0	00	098	PRTX	−14	126	RTN	24
015	R↓	−31	043	RCL4	36 04	071	5	05	099	CF1	16 22 01	127	*LBL3	21 03
016	STO5	35 05	044	GSB1	23 01	072	RCL6	36 06	100	SPC	16-11	128	.	−62
017	R↓	−31	045	RCL6	36 06	073	−	−35	101	R/S	51	129	5	05
018	STO6	35 06	046	x	−35	074	÷	−24	102	*LBL4	21 04	130	6	06
019	RCL2	36 02	047	STO3	35 03	075	1	01	103	RCL1	35 01	131	x	−35
020	GSB1	23 01	048	RCL4	36 04	076	.	−62	104	ST+5	35-55 05	132	9	09
021	RCL0	36 00	049	GSB2	23 02	077	1	01	105	RCL0	36 00	133	3	03
022	x	−35	050	RCL5	36 05	078	5	05	106	ST+6	35-55 06	134	7	07
023	STO7	35 07	051	x	−35	079	0	00	107	RCL7	36 07	135	.	−62
024	RCL3	36 03	052	ST+9	35-55 09	080	6	06	108	RCL9	36 09	136	8	08
025	GSB1	23 01	053	F1?	16 23 01	081	1/X	52	109	+	−55	137	+	−55
026	RCL9	36 09	054	GTO4	22 04	082	y^x	31	110	STOE	35 15	138	RTN	24
027	x	−35	055	*LBL3	21 03	083	STOI	35 46	111	GTO6	22 06	139	R/S	51
028	STO8	35 08	056	RCLA	36 11	084	RCLD	36 14	112	*LBL1	21 01			

User instructions for HP version   Table VII

Air flow, lb/h, stream 1	ENTER ↑
Moisture flow, lb/h, stream 1	ENTER ↑
Inlet temperature, °F, stream 1	Key A
Air flow, lb/h, stream 2	ENTER ↑
Moisture flow, lb/h, stream 2	ENTER ↑
Inlet temperature, °F, stream 2	ENTER ↑
Exit temperature, °F, stream 1	Key D for heat exchange case
	Key d for mixture case

With key D, answer is stream 2 outlet temperature, °F.
With key d, answer is mixture outlet temperature, °F.

Second number in both cases is heat transfered, Btu/h.

Note: HP answers for text example differ slightly from the text TI answers. The TI program uses 1.15 for the exponent constant in Eq. (2), but the HP program uses 1.1506 as shown in the equation. Also, the TI program uses two different values for the constant 937.814 of Eq. (3), 937.814 in one case—lines 024 to 030—and 938—lines 160 to 162—in the other). The HP program uses 937.814 in both cases.

Printouts for heat-exchange and mixing   Table VIII
examples—HP version

Heat-exchange example	Mixing example
20000.00 ENT↑	20000.00 ENT↑
1500.00 ENT↑	1500.00 ENT↑
610.00 GSBA	610.00 GSBA
35000.00 ENT↑	35000.00 ENT↑
2850.00 ENT↑	2850.00 ENT↑
2450.00 ENT↑	2450.00 ENT↑
3400.00 GSBD	2400.00 GSBd
1478.99 ***	1815.37 ***
−11400179.00 ***	−11400179.00 ***

The author

Calvin R. Brunner, P.E., is Chief Mechanical Engineer of Malcolm Pirnie, Inc., Consulting Environmental Engineers, Two Corporate Park Dr., White Plains, NY 10602. He is responsible for design, construction and operation of waste-incineration systems, as well as other combustion processes. Mr. Brunner holds a master's degree from Pennsylvania State University, and a bachelor's degree from City College of New York, both in mechanical engineering. He is licensed as a professional engineer in five states, and has written two texts on incineration, plus many technical articles.

Predicting the performance of a heat-exchanger train

Here is how to analyze the operation of each exchanger
in a series, predict performance in off-design cases,
and estimate modifications required for plant expansion.

Chyuan-Chung Chen, Hudson Engineering Corp.

☐ Heat-exchanger trains that have multiple shells are commonly used for applications involving large heat-transfer duties—for instance, for crude-feed preheat in petroleum refineries. The amount of heat transferred increases with the transfer area, but nonlinearly. When the system is sufficiently large, the impact of adding heat-transfer area becomes so small that doing so is not an economically feasible means of boosting heat transfer.

When the duty and outlet temperatures are unknown, trial-and-error procedures usually are required to solve a set of heat-balance and heat-transfer equations. The equations developed in this article can simplify matters.

Typical applications of these equations include:

■ Evaluating the incremental value of added heat-transfer area, to ensure that the exchanger system is economically sized.

■ Predicting performance at off-design conditions.

■ Estimating the modifications required for plant expansion.

■ Calculating the duty and temperature of each ex-
changer in a train to enable the proper design of piping and instrumentation.

Finding the overall duty of a train

The overall heat balance of hot and cold fluids in an exchanger system, as shown in Fig. 1, is:

$$Q = W_h C_h (T_N - T_o) \qquad (1)$$

$$Q = W_c C_c (t_N - t_o) \qquad (2)$$

where C_h and C_c are the specific heats of single-phase streams at average temperatures. Where condensation and/or vaporization can occur, the values of the average slopes of the heating and cooling curves can be used as C_h and C_c to take into account both sensible and latent heats.

The heat transferred in an exchanger system can be written as:

$$Q = UA[(\Delta T_N - \Delta T_o)/\ln (\Delta T_N/\Delta T_o)]F \qquad (3)$$

The parenthetical term is the logarithmic mean temperature difference, LMTD.

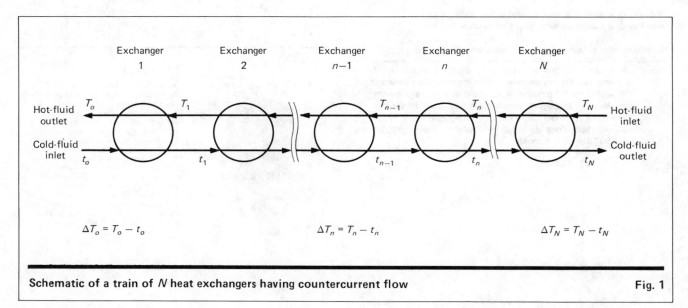

Exchanger 1   **Exchanger 2**   **Exchanger n−1**   **Exchanger n**   **Exchanger N**

Hot-fluid outlet T_o T_1 T_{n-1} T_n T_N Hot-fluid inlet

Cold-fluid inlet t_o t_1 t_{n-1} t_n t_N Cold-fluid outlet

$\Delta T_o = T_o - t_o$ $\Delta T_n = T_n - t_n$ $\Delta T_N = T_N - t_N$

Schematic of a train of *N* heat exchangers having countercurrent flow   **Fig. 1**

Originally published March 19, 1984

114

Nomenclature

A	Total heat-transfer area, ft²
A_n	Heat-transfer area of first n exchangers, ft²
C_c	Specific heat of cold fluid, Btu/(lb)(°F)
C_h	Specific heat of hot fluid, Btu/(lb)(°F)
F	LMTD correction factor, dimensionless
Q	Total heat duty, Btu/h
Q_n	Heat duty of first n exchangers, Btu/h
q_n	Heat duty of nth exchanger, Btu/h
T_N	Inlet temperature of hot fluid, °F
T_n	Temperature of hot fluid entering nth exchanger, °F
T_o	Outlet temperature of hot fluid, °F
t_N	Outlet temperature of cold fluid, °F
t_n	Temperature of cold fluid entering nth exchanger, °F
t_o	Inlet temperature of cold fluid, °F
U	Overall heat-transfer coefficient, Btu/(h)(ft²)(°F)
W_c	Flowrate of cold fluid, lb/h
W_h	Flowrate of hot fluid, lb/h
α_1	Difference in inlet temperatures, °F
α_2	Inverse of heat capacity of hot fluid, (h)(°F)/Btu
α_3	Inverse of heat capacity of cold fluid, (h)(°F)/Btu
α_4	Differential heat-transfer parameter, 1/ft²

Combining Eq. 1 – 3 to eliminate T_o and t_N yields:

$$Q = [\alpha_1(e^{\alpha_4 A} - 1)]/(\alpha_2 e^{\alpha_4 A} - \alpha_3) \qquad (4)$$

where

$$\alpha_1 = T_N - t_o \qquad (5)$$

$$\alpha_2 = 1/W_h C_h \qquad (6)$$

$$\alpha_3 = 1/W_c C_c \qquad (7)$$

$$\alpha_4 = UF(\alpha_2 - \alpha_3) \qquad (8)$$

The following relationships can be seen from Eq. 4:

1. The duty is proportional to the difference in inlet temperatures. Increasing the temperature of the heating medium or decreasing the temperature of the coolant, if possible, is the most effective way to increase the duty.

2. The area or transfer coefficient is related by an exponential function to the duty. The effect declines as either area or transfer coefficient rises. When either becomes infinitely large, the duty approaches a limiting value—which is the asymptotic solution of Eq. 4 when $\alpha_4 A \to \infty$:

$$Q = \alpha_1/\alpha_2 \text{ if } \alpha_4 > 0 \qquad (9a)$$

$$Q = \alpha_1/\alpha_3 \text{ if } \alpha_4 < 0 \qquad (9b)$$

Now let us substitute α_2 into Eq. 1, and rearrange:

$$T_o = T_N - \alpha_2 Q \qquad (1a)$$

By similar manipulation, we get:

$$t_N = t_o + \alpha_3 Q \qquad (2a)$$

Use Eq. 4 to calculate the duty. Then use Eq. 1a and 2a to determine the outlet temperatures.

The outlet temperature at the limiting condition can be obtained from Eq. 1a, 2a and 9a as:

$$T_o = t_o \qquad (1b)$$

$$t_N = t_o + \alpha_1 \alpha_3 / \alpha_2 \qquad (2b)$$

when $\alpha_4 > 0$, and from Eq. 1a, 2a and 9b as:

$$T_o = T_N - \alpha_1 \alpha_2 / \alpha_3 \qquad (1c)$$

$$t_N = T_N \qquad (2c)$$

when $\alpha_4 < 0$.

These equations indicate that if the heat capacity (WC) of the cold stream is greater than that of the hot stream ($\alpha_4 > 0$), the maximum heat transferable is the amount needed to cool the hot stream to the inlet temperature of the cold stream. If the heat capacity of the cold stream is less than that of the hot stream ($\alpha_4 < 0$), the transfer limit is that amount needed to heat the cold stream to the inlet temperature of the hot stream.

Duties of individual exchangers in a train

The heat balance and heat transfer in the first n exchangers in a train can be written in a manner similar to Eq. 1 to 3, namely:

$$Q_n = W_h C_h (T_n - T_o) \qquad (10)$$

$$Q_n = W_c C_c (t_n - t_o) \qquad (11)$$

$$Q_n = UA_n[(\Delta T_n - \Delta T_o)/\ln(\Delta T_n/\Delta T_o)]F \qquad (12)$$

These equations can be combined to give:

$$\Delta T_n / \Delta T_o = e^{\alpha_4 A_n} \qquad (13)$$

which, when $n = N$, becomes:

$$\Delta T_N / \Delta T_o = e^{\alpha_4 A} \qquad (13a)$$

Assuming that C_h, C_c, U and F are constant along the train, the following expression can be derived from Eq. 1, 2, 10, 11, 13 and 13a:

$$Q_n = Q(e^{\alpha_4 A_n} - 1)/(e^{\alpha_4 A} - 1) \qquad (14)$$

The duty of the nth exchanger can be calculated from:

$$q_n = Q_n - Q_{n-1} =$$
$$Q(e^{\alpha_4 A_n} - e^{\alpha_4 A_{n-1}})/(e^{\alpha_4 A} - 1) \qquad (15)$$

The temperatures T_n and t_n can be found via:

$$T_n = T_N - \alpha_2(Q - Q_n) \qquad (16)$$

$$t_n = t_o + \alpha_3 Q_n \qquad (17)$$

Using the technique

To illustrate how to use these equations, let us consider a train having seven identical exchangers (each with a transfer area of 2,213 ft²) for heating a cold feed against a hot product. The specifics are:

	Cold stream	Hot stream
Flowrate, lb/h	390,710	286,160
Inlet temperature, °F	282	626
Outlet temperature, °F	501	341
Specific heat, Btu/(lb)(°F)	0.61	0.64

We want to determine: (1) The duty and inlet/outlet temperatures of each exchanger in the train; (2) The changes in total duty and in outlet temperature when one exchanger is out of service for maintenance; (3)

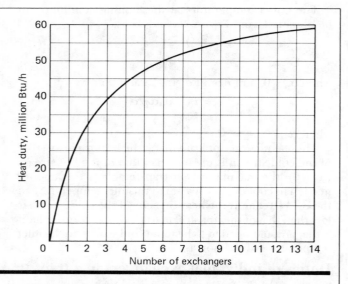

How transfer area affects heat duty   Fig. 2

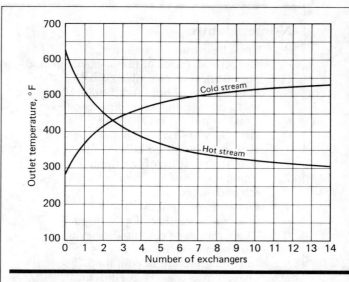

Outlet temperatures vary with transfer area   Fig. 3

Performance of the train when the flowrates of both streams are increased by 20% while other conditions remain the same; (4) The change in flowrate and/or temperature of the hot stream (if changeable) to heat the cold stream at 20% above the design flowrate from and to the same temperatures as in the design case.

Duty and inlet/outlet temperatures—First, the total duty of the entire train at design conditions can be calculated via Eq. 1:

$$Q = (286,160)(0.64)(626 - 341) = $$
$$52.196 \text{ million Btu/h}$$

Next, we calculate other necessary parameters:

$$A = 2,213 \times 7 = 15,491 \text{ ft}^2$$
$$\Delta T_o = 341 - 282 = 59°F$$
$$\Delta T_N = 626 - 501 = 125°F$$
$$\text{LMTD} = (125 - 59)/(\ln 125/59) = 87.91°F$$

Then, from Eq. 3:

$$UF = (52.196 \times 10^6)/(15,491)(87.91) = $$
$$38.33 \text{ Btu/(h)(ft}^2)(°F)$$

Similarly, we determine via Eq. 5, 6, 7 and 8:

$$\alpha_1 = 626 - 282 = 344$$
$$\alpha_2 = 1/(286,160)(0.64) = 5.460 \times 10^{-6}$$
$$\alpha_3 = 1/(390,710)(0.61) = 4.196 \times 10^{-6}$$
$$\alpha_4 = (38.33)(5.460 \times 10^{-6} - 4.196 \times 10^{-6}) = $$
$$4.845 \times 10^{-5}$$

The duty and temperature of each exchanger are then calculated from Eq. 15, 16 and 17. The results are summarized in Table I. They clearly show that although the exchangers are identical in area, the duties are different—with the highest duty nearly twice that of the lowest. The exchangers will have equal duty only when the hot and cold streams have the same heat capacity.

Impact of the number of exchangers in service—The total heat-transfer duty for different system sizes is deter-

mined via Eq. 4. For ease of approximation, the parameters α_2, α_3 and α_4 are assumed to be constant, regardless of the number of exchangers in service. (If more-accurate results are needed, the values of these parameters can be adjusted for the effect of temperature.) The outlet temperatures are calculated by Eq. 1a and 2a. The results for our example appear in Table II, and are plotted in Fig. 2 and 3. They show not only what happens if some of the exchangers are taken out of service but also the impact of adding more exchangers.

Table II illustrates that when one exchanger is removed for maintenance from our 7-exchanger train, the duty will decrease by about 2 million Btu/h, and the exiting cold stream will be approximately 9°F cooler than normal. This reflects some important facts, which bear amplification:

Each of the exchangers has a different duty (see Table I) during normal operation, but the duty loss will be the same regardless of which exchanger is removed from the system. The resulting 6-unit trains are identical in performance, no matter which exchanger is removed.

When one exchanger is out of service, the duty loss will *not* be the normal duty of the unit removed or one-

Duty and outlet temperatures of the exchangers in the example			Table I
Exchanger number	Exchanger duty (q_n), million Btu/h	Cold-stream temperature, (t_n), °F	Hot-stream temperature, (T_{n-1}), °F
1	5.283	304	341
2	5.881	329	370
3	6.547	356	402
4	7.288	387	438
5	8.113	421	478
6	9.031	459	522
7	10.053	501	571

Duty and outlet temperatures versus area of exchanger series				Table II
Number of exchangers	Total area (A), ft^2	Total duty (Q), million Btu/h	Cold-stream outlet temperature, (t_N), °F	Hot-stream outlet temperature, (T_o), °F
1	2,213	20.687	369	513
2	4,426	32.014	416	451
3	6,639	39.128	446	412
4	8,852	43.988	467	386
5	11,065	47.500	481	367
6	13,278	50.144	492	352
7	15,491	52.196	501	341
8	17,704	53.826	508	332
10	22,130	56.230	518	319
12	26,556	57.889	525	310
∞	∞	63.004	546	282

seventh of the total duty (since there are seven units). The actual loss will be less than either of these.

Performance under off-design conditions—When flowrates increase, the transfer coefficient goes up. Methods for correcting the transfer coefficient for off-design cases are available in most heat-transfer books, e.g., D. Q. Kern's "Process Heat Transfer" (McGraw-Hill).

Let us take our example system to be high-fouling, with a heat-transfer coefficient that will rise about 8% when both flowrates increase 20%. Assuming that C_h, C_c, and F are constant, the parameters in Eq. 4 can be calculated for the new situation:

$$\alpha_2 = 1/(286,160 \times 1.2)(0.64) =$$
$$(5.460 \times 10^{-6})/1.2 = 4.550 \times 10^{-6}$$

$$\alpha_3 = 1/(390,710 \times 1.2)(0.61) =$$
$$(4.196 \times 10^{-6})/1.2 = 3.497 \times 10^{-6}$$

$$\alpha_4 = (38.33)(4.550 \times 10^{-6} - 3.497 \times 10^{-6})(1.08) =$$
$$4.359 \times 10^{-5}$$

Plugging these values into Eq. 4 gives the duty:

$$Q = (344)(e^{(4.359 \times 10^{-5})(15,491)} - 1) \div$$
$$[(4.550 \times 10^{-6})(e^{(4.359 \times 10^{-5})(15,491)}) - 3.497 \times 10^{-6}] =$$
$$60.974 \text{ million Btu/h}$$

The outlet temperatures are determined via Eq. 1a and 2a:

$$T_o = 626 - (4.550 \times 10^{-6})(60.974 \times 10^6) = 349°F$$

$$t_N = 282 + (3.497 \times 10^{-6})(60.974 \times 10^6) = 495°F$$

The outlet temperatures of the original design case cannot be met unless the system is oversized to take into account the smaller increase in the coefficient compared with the flowrates. This requires an area oversizing for the original design of:

$$[(1.2/1.08) - 1] \times 100\% = 11\%$$

If the transfer coefficient is not corrected for the new flowrates, the calculated duty will be 59.703 millon Btu/h, or off by about 2%. So, if the change in operating

conditions is not very significant, U usually can be taken as a constant.

Modifications required for expansion—If additional duty is called for, this can be achieved either by increasing the hot-stream flowrate, the hot-stream temperature, or some combination of the two. The required changes are estimated as follows:

The necessary flowrate is determined via Eq. 4. In this case, we want a 20% increase in the cold-stream flowrate. We will assume a 1.08 correction factor for the heat-transfer coefficient, as we did previously, since both flowrates will be raised (even though we do not know as yet the amount for the hot stream). So, UFA becomes:

$$38.33 \times 1.08 \times 15,491 = 6.413 \times 10^5$$

Then:

$$(52.196 \times 10^6)(1.2) =$$
$$(344)(e^{(6.413 \times 10^5)(\alpha_2 - 3.497 \times 10^{-6})} - 1) \div$$
$$[(\alpha_2)(e^{(6.413 \times 10^5)(\alpha_2 - 3.497 \times 10^{-6})}) - 3.497 \times 10^{-6}]$$

Rearranging and solving:

$$\alpha_2 = 4.300 \times 10^{-6}$$

This, in turn, allows calculation of the flowrate:

$$W_h = 1/(4.300 \times 10^{-6})(0.64) = 363,372 \text{ lb/h}$$

Thus, the required increase in the hot-stream flowrate over the design rate is:

$$[(363,372/286,160) - 1] \times 100\% = 27\%$$

The required hot-stream inlet temperature is calculated by Eq. 4. In this case, we will use a correction factor of 1.05 for the heat-transfer coefficient, instead of 1.08, since we are taking only one flowrate as increased. Thus, $\alpha_4 A$ becomes:

$$38.33 \times 1.05 \times 15,491 \times$$
$$(5.460 \times 10^{-6} - 3.497 \times 10^{-6}) = 1.224$$

$$(52.196 \times 10^6)(1.2) =$$
$$(\alpha_1)(e^{1.224} - 1) \div [(5.460 \times 10^{-6})(e^{1.224}) - 3.497 \times 10^{-6}]$$

whose solution is:

$$\alpha_1 = 393$$

Then,

$$T_N = 282 + 393 = 675°F$$

and the required hot-stream temperature increase is:

$$675 - 626 = 49°F$$

The author

Chyuan-Chung Chen is Senior Section Leader for Hudson Engineering Corp., a McDermott company, 801 North Eldridge St., Houston, TX 77218; telephone: (713) 870-5000. He holds a B.S. from National Taiwan University, and M.S. and Ph.D. degrees from the University of Rochester, all in chemical engineering. A member of AIChE, he is a licensed Professional Engineer.

Simulating heat-exchanger performance

Programs for both the TI-59 and the HP-41CV predict heat-exchanger performance through a rigorous iteration method.

Shuh-Chih Chang, *Lawrence-Allison & Associates**

□ When evaluating the process conditions of an existing heat exchanger, the engineer is required to to calculate and determine the heat-exchanger duty, Q, and the outlet temperature, T_2, of the hot stream, and t_2 of the cold stream—given the inlet temperatures, T_1 and t_1, mass flowrates, W and w, and average specific heats, C and c.

For an existing heat exchanger, the overall heat-transfer coefficient, U, and transfer area, A, are known (or can be calculated from the individual heat-transfer coefficients, h, and the configuration of the heat exchanger, respectively). This technique of calculating the exit temperatures is especially useful when one deals with heat-exchanger networks [1].

Based on the data of W, C, T_1, w, c, t_1, U and A given above, the heat duty, Q, and the exit temperatures, T_2 and t_2, of the hot and cold streams can be calculated by the procedure discussed in this article. Program listings are included, so that a TI-59 or HP-41CV calculator can be compactly programmed for this task (although a computer or microcomputer with larger memory and faster computing speed is preferable).

The calculation output also provides the log mean temperature difference, [LMTD], true temperature difference, ΔT, temperature correction factor, F, temperature efficiency, P, heat-capacity-rate ratio, R, and the number of transfer units, [NTU]. Calculation selections are available for several types of heat exchangers:

1. True parallel-flow heat exchangers.
2. True countercurrent-flow heat exchangers.
3. (N shell pass)-($2N$ tube pass) heat exchangers.

Countercurrent and parallel-flow

The typical method used to solve the problem of predicting heat-exchanger performance employs an iterative procedure [2,3]. The basic heat-balance equations are:

$$Q = WC(T_1 - T_2) = wc(t_2 - t_1) \qquad (1)$$

and

$$Q = UA(\Delta T) \qquad (2)$$

*Since this article was written, the author has joined Summit CAD Corp. See "The author" on the last page of this article.

In fact, for true parallel-flow or true countercurrent-flow heat exchangers (double-pipe exchangers, for instance), Kern [4] gives the formula for straightforwardly evaluating the outlet temperature, T_2, of the hot fluid for both cases.

For countercurrent flow:

$$T_2 = \frac{(1 - R)T_1 + [1 - e^{(UA/wc)(R-1)}]Rt_1}{1 - Re^{(UA/wc)(R-1)}} \qquad (3)$$

For parallel flow:

$$T_2 = \frac{[R + e^{(UA/wc)(R+1)}]T_1 + [e^{(UA/wc)(R+1)} - 1]Rt_1}{(R+1)\,e^{(UA/wc)(R+1)}} \qquad (4)$$

Eq. (3) and (4) eliminate the need for a trial-and-error procedure. Once the outlet temperature, T_2, is solved, t_2 and Q can be calculated by Eq. (1). For a multipass shell-and-tube heat exchanger, however, the calculation is more complex and the iteration method is still required.

1-2 and 2-4 heat exchangers

Shell-and-tube heat exchangers with one shell pass and two tube passes, or two shell passes and four tube passes, are typically used when a large heat-transfer area is required. For this type of heat transfer, the flow pattern is neither countercurrent nor parallel. Therefore, the temperature correction factor, F, is introduced to obtain the actual temperature difference, i.e.:

$$\Delta T = F[LMTD] \qquad (5)$$

where [LMTD] is the log-mean temperature difference for true countercurrent flow.

The temperature correction factor, F, may be found graphically [5], or calculated from the proper equations if both inlet and outlet temperatures are known. Another method for determining F, by means of an outlet temperature gap, G, was recently proposed [6]. The formulas used in the program presented here are based on the conventional approach described below.

Underwood [7] derived a method, and Bowman, Mueller and Nagle [8] prepared the final forms of equations used to calculate the temperature correction factor, F, for N-$2N$ shell-and-tube exchangers. The equations

Originally published April 2, 1984

Program listing for TI-59 programmable calculator for a rigorous iteration method that simulates heat-exchanger performance

Table I continues

Input data

Location	Code	Key
000	76	LBL
001	11	A
002	42	STD
003	00	00
004	25	CLR
005	91	R/S
006	72	ST*
007	00	00
008	43	RCL
009	00	00
010	91	R/S
011	69	DP
012	20	20
013	61	GTD
014	00	00
015	06	06

Main program for N-2N and countercurrent exchanger

Location	Code	Key
016	76	LBL
017	12	B
018	42	STD
019	14	14
020	71	SBR
021	02	02
022	75	75
023	71	SBR
024	01	01
025	51	51
026	29	CP
027	43	RCL
028	14	14
029	67	EQ
030	00	00
031	45	45
032	87	IFF
033	01	01
034	00	00
035	42	42
036	71	SBR
037	03	03
038	84	84
039	61	GTD
040	00	00
041	45	45
042	71	SBR
043	05	05
044	91	91
045	71	SBR
046	03	03
047	47	47
048	43	RCL
049	03	03
050	75	-
051	43	RCL
052	22	22
053	95	=
054	42	STD
055	27	27
056	43	RCL
057	21	21
058	75	-
059	43	RCL
060	06	06
061	95	=
062	42	STD
063	29	29
064	71	SBR
065	01	01
066	29	29
067	65	×
068	43	RCL
069	13	13
070	95	=
071	42	STD
072	25	25

Output data

Location	Code	Key
073	43	RCL
074	23	23
075	91	R/S
076	43	RCL
077	21	21
078	91	R/S
079	43	RCL
080	22	22
081	91	R/S
082	43	RCL
083	24	24
084	91	R/S
085	43	RCL
086	25	25
087	91	R/S
088	43	RCL
089	13	13
090	91	R/S
091	73	RC*
092	09	09
093	91	R/S

Main program for parallel-flow exchanger

Location	Code	Key
094	76	LBL
095	13	C
096	42	STD
097	14	14
098	71	SBR
099	02	02
100	75	75
101	71	SBR
102	02	02
103	29	29
104	71	SBR
105	03	03
106	47	47
107	43	RCL
108	03	03
109	75	-
110	43	RCL
111	06	06
112	95	=
113	42	STD
114	27	27
115	43	RCL
116	21	21
117	75	-
118	43	RCL
119	22	22
120	95	=
121	42	STD
122	29	29
123	71	SBR
124	01	01
125	29	29
126	61	GTD
127	00	00
128	67	67

Calc. [LMTD]

Location	Code	Key
129	32	X:T
130	43	RCL
131	27	27
132	67	EQ
133	01	01
134	48	48
135	75	-
136	32	X:T
137	95	=
138	55	÷
139	53	(
140	43	RCL
141	27	27
142	55	÷
143	43	RCL
144	29	29
145	54	)
146	23	LNX
147	95	=
148	42	STD
149	24	24
150	92	RTN

Calc. T_2 (counter)

Location	Code	Key
151	43	RCL
152	11	11
153	65	×
154	43	RCL
155	13	13
156	65	×
157	42	STD
158	18	18
159	43	RCL
160	06	06
161	85	+
162	43	RCL
163	03	03
164	95	=
165	55	÷
166	53	(
167	01	1
168	85	+
169	43	RCL
170	18	18
171	54	)
172	95	=
173	42	STD
174	21	21
175	01	1
176	32	X:T
177	43	RCL
178	10	10
179	67	EQ
180	02	02
181	28	28
182	43	RCL
183	18	18
184	65	×
185	53	(
186	43	RCL
187	10	10
188	75	-
189	01	1
190	54	)
191	42	STD
192	18	18
193	95	=
194	22	INV
195	23	LNX
196	42	STD
197	12	12
198	94	+/-
199	85	+
200	01	1
201	95	=
202	65	×
203	43	RCL
204	10	10
205	65	×
206	73	RC*
207	00	00
208	75	-
209	43	RCL
210	03	03
211	65	×
212	43	RCL
213	18	18
214	95	=
215	55	÷
216	53	(
217	01	1
218	75	-
219	43	RCL
220	10	10
221	65	×
222	43	RCL
223	12	12
224	54	)
225	95	=
226	42	STD
227	21	21
228	92	RTN

Calc. T_2 (parallel)

Location	Code	Key
229	43	RCL
230	10	10
231	85	+
232	01	1
233	95	=
234	42	STD
235	18	18
236	65	×
237	43	RCL
238	11	11
239	95	=
240	22	INV
241	23	LNX
242	42	STD
243	12	12
244	85	+
245	43	RCL
246	10	10
247	95	=
248	65	×
249	43	RCL
250	03	03
251	85	+
252	53	(
253	43	RCL
254	12	12
255	75	-
256	01	1
257	54	)
258	65	×
259	43	RCL
260	10	10
261	65	×
262	73	RC*
263	00	00
264	95	=
265	55	÷
266	43	RCL
267	18	18
268	55	÷
269	43	RCL
270	12	12
271	95	=
272	42	STD
273	21	21
274	92	RTN

Initialize and calc. R and [NTU]

Location	Code	Key
275	01	1
276	42	STD
277	13	13
278	01	1
279	09	9
280	42	STD
281	09	09
282	06	6
283	42	STD
284	00	00
285	00	0
286	42	STD
287	19	19
288	42	STD
289	20	20
290	86	STF
291	08	08
292	22	INV
293	86	STF
294	01	01
295	43	RCL
296	14	14
297	32	X:T
298	02	2
299	77	GE
300	03	03
301	04	04
302	86	STF
303	01	01
304	43	RCL
305	04	04
306	65	×
307	43	RCL
308	05	05
309	55	÷
310	42	STD
311	18	18
312	43	RCL
313	01	01
314	55	÷
315	43	RCL
316	02	02
317	95	=
318	42	STD
319	10	10
320	33	X²
321	85	+
322	01	1
323	95	=
324	34	ΓX
325	42	STD
326	15	15
327	43	RCL
328	07	07
329	65	×
330	43	RCL
331	08	08
332	55	÷
333	43	RCL
334	18	18
335	22	INV
336	87	IFF
337	01	01
338	03	03
339	43	43
340	55	÷
341	43	RCL
342	14	14
343	95	=
344	42	STD
345	11	11
346	92	RTN

Calc. Q

Location	Code	Key
347	43	RCL
348	03	03
349	75	-
350	43	RCL
351	21	21
352	95	=
353	55	÷
354	43	RCL
355	10	10
356	85	+
357	43	RCL
358	06	06

Table I continued

Location	Code	Key	Location	Code	Key	Location	Code	Key	Location	Code	Key	Location	Code	Key	Location	Code	Key
359	95	=	419	03	03	481	16	16	543	85	+	603	75	-	665	65	×
360	42	STO	420	54	)	482	01	1	544	43	RCL	604	01	1	666	53	(
361	22	22	421	95	=	483	75	-	545	17	17	605	95	=	667	43	RCL
362	43	RCL	422	42	STO	484	43	RCL	546	85	+	606	42	STO	668	03	03
363	03	03	423	19	19	485	19	19	547	42	STO	607	27	27	669	75	-
364	75	-	424	55	÷	486	95	=	548	18	18	608	42	STO	670	43	RCL
365	43	RCL	425	53	(	487	65	×	549	43	RCL	609	26	26	671	06	06
366	21	21	426	01	1	488	53	(	550	15	15	610	02	2	672	54	)
367	95	=	427	75	-	489	01	1	551	95	=	611	06	6	673	85	+
368	65	×	428	43	RCL	490	75	-	552	55	÷	612	42	STO	674	43	RCL
369	43	RCL	429	19	19	491	43	RCL	553	53	(	613	00	00	675	06	06
370	01	01	430	54	)	492	19	19	554	43	RCL	614	43	RCL	676	75	-
371	65	×	431	95	=	493	65	×	555	18	18	615	26	26	677	42	STO
372	43	RCL	432	42	STO	494	43	RCL	556	75	-	616	85	+	678	22	22
373	02	02	433	28	28	495	10	10	557	43	RCL	617	43	RCL	679	43	RCL
374	55	÷	434	01	1	496	54	)	558	15	15	618	27	27	680	19	19
375	01	1	435	32	X:T	497	95	=	559	54	)	619	95	=	681	65	×
376	52	EE	436	43	RCL	498	34	⌐X	560	95	=	620	55	÷	682	53	(
377	06	6	437	10	10	499	65	×	561	23	LNX	621	02	2	683	43	RCL
378	95	=	438	67	EQ	500	43	RCL	562	35	1/X	622	95	=	684	03	03
379	22	INV	439	04	04	501	18	18	563	65	×	623	42	STO	685	75	-
380	52	EE	440	68	68	502	95	=	564	43	RCL	624	26	26	686	43	RCL
381	42	STO	441	01	1	503	42	STO	565	28	28	625	71	SBR	687	26	26
382	23	23	442	75	-	504	17	17	566	65	×	626	03	03	688	54	)
383	92	RTN	443	43	RCL	505	02	2	567	43	RCL	627	84	84	689	95	=
			444	19	19	506	32	X:T	568	15	15	628	01	1	690	42	STO

Iteration for 1-2 and 2-4 exchanger

Location	Code	Key	Location	Code	Key	Location	Code	Key	Location	Code	Key	Location	Code	Key	Location	Code	Key
384	43	RCL	445	95	=	507	43	RCL	569	55	÷	629	75	-	691	27	27
385	21	21	446	55	÷	508	14	14	570	02	2	630	43	RCL	692	75	-
386	42	STO	447	53	(	509	67	EQ	571	95	=	631	19	19	693	43	RCL
387	27	27	448	01	1	510	05	05	572	42	STO	632	65	×	694	26	26
388	87	IFF	449	75	-	511	41	41	573	13	13	633	43	RCL	695	95	=
389	01	01	450	43	RCL	512	43	RCL	574	71	SBR	634	10	10	696	50	I×I
390	03	03	451	19	19	513	16	16	575	01	01	635	95	=	697	32	X:T
391	95	95	452	65	×	514	85	+	576	51	51	636	55	÷	698	93	.
392	01	1	453	43	RCL	515	43	RCL	577	93	.	637	53	(	699	01	1
393	00	0	454	10	10	516	15	15	578	01	1	638	01	1	700	32	X:T
394	85	+	455	54	)	517	95	=	579	32	X:T	639	75	-	701	77	GE
395	43	RCL	456	95	=	518	55	÷	580	43	RCL	640	43	RCL	702	06	06
396	21	21	457	23	LNX	519	53	(	581	21	21	641	19	19	703	14	14
397	85	+	458	55	÷	520	43	RCL	582	75	-	642	54	)	704	43	RCL
398	43	RCL	459	53	(	521	16	16	583	43	RCL	643	95	=	705	06	06
399	27	27	460	43	RCL	522	75	-	584	27	27	644	45	Y^x	706	75	-
400	95	=	461	10	10	523	43	RCL	585	95	=	645	43	RCL	707	43	RCL
401	55	÷	462	75	-	524	15	15	586	50	I×I	646	14	14	708	22	22
402	02	2	463	01	1	525	54	)	587	77	GE	647	95	=	709	95	=
403	95	=	464	54	)	526	95	=	588	03	03	648	42	STO	710	65	×
404	42	STO	465	95	=	527	23	LNX	589	95	95	649	29	29	711	43	RCL
405	27	27	466	42	STO	528	35	1/X	590	92	RTN	650	94	+/-	712	10	10
406	75	-	467	28	28	529	65	×				651	85	+	713	85	+
407	43	RCL	468	02	2	530	43	RCL				652	01	1	714	43	RCL
408	03	03	469	55	÷	531	15	15	**Iteration for N-2N**			653	95	=	715	03	03
409	95	=	470	43	RCL	532	65	×	**exchanger where $N \geq 3$**			654	55	÷	716	95	=
410	55	÷	471	19	19	533	43	RCL	591	02	2	655	53	(	717	42	STO
411	43	RCL	472	75	-	534	28	28	592	00	0	656	43	RCL	718	21	21
412	10	10	473	42	STO	535	95	=	593	42	STO	657	10	10	719	92	RTN
413	55	÷	474	18	18	536	42	STO	594	09	09	658	75	-			
414	53	(	475	01	1	537	13	13	595	43	RCL	659	43	RCL			
415	73	RC*	476	75	-	538	61	GTO	596	06	06	660	29	29			
416	00	00	477	43	RCL	539	05	05	597	85	+	661	54	)			
417	75	-	478	10	10	540	74	74	598	43	RCL	662	95	=			
418	43	RCL	479	95	=	541	43	RCL	599	22	22	663	42	STO			
			480	42	STO	542	16	16	600	95	=	664	20	20			
									601	55	÷						
									602	02	2						

derived for N-$2N$ exchangers are briefly presented as follows. For a one-two exchanger:

$$F_{1,2} = \frac{mH}{\ln\left(\dfrac{a+m}{a-m}\right)} \qquad (6)$$

and for a two-four exchanger:

$$F_{2,4} = \frac{mH}{2\ln\left(\dfrac{a+b+m}{a+b-m}\right)} \qquad (7)$$

where:

$$a = (2/P) - 1 - R \qquad (8)$$

$$b = (2/P)[(1-P)(1-PR)]^{1/2} \qquad (9)$$

$$m = (R^2+1)^{1/2} \qquad (10)$$

$$H = \frac{1}{(R-1)} \ln\left(\frac{1-P}{1-PR}\right) \qquad (11)$$

When $R = 1$, l'Hospital's rule is applied, to obtain:

$$H = P/(1-P) \qquad (12)$$

Eq. (6) and (7) are based on the assumptions of Bowman, Mueller and Nagle:

1. The overall heat-transfer coefficient, U, is constant throughout the heat exchanger.

2. The rate of flow of each fluid is constant.

3. The specific heat of each fluid is constant.

Table I User's instructions, TI-59   Table II

Step	Procedure	Enter	Press	Display
1	Partition calculator	3	2nd OP 17	719.29
2	Enter learn code		LRN	000 00
3	Key in program			
4	Exit learn mode		LRN	719.29
5	Initialize to enter data starting with Input No. 1	1	A	0
6	Enter input data in sequence			
	Flowrate of hot fluid	W	R/S	1 (i=1)
	Specific heat of hot fluid	C	R/S	2 (i=2)
	Inlet temperature of hot fluid	T_1	R/S	3 (i=3)
	Flowrate of cold fluid	w	R/S	4 (i=4)
	Specific heat of cold fluid	c	R/S	5 (i=5)
	Inlet temperature of cold fluid	t_1	R/S	6 (i=6)
	Heat-transfer coefficient	U	R/S	7 (i=7)
	Heat-transfer area	A	R/S	8 (i=8)
7	To change or correct an entry error, simply enter input sequence number, i, and press A. Then reenter the correct data, using the R/S key.	i Correct ith input datum	A R/S	0 i
8	Enter number of shell passes, N, then choose the type of exchanger by pressing B or C.			
	a. True counterflow	0	B	Q
	b. N-$2N$ exchanger ($N \geqslant 1$)	N	B	Q
	c. True parallel flow	0	C	Q
9	Repeatedly press the R/S key to read the output data		R/S	T_2
			R/S	t_2
			R/S	$[LMTD]$
			R/S	$\triangle T$
			R/S	F
			R/S	P^*
			RCL 10	R
			RCL 11†	$[NTU]$
			RCL 26‡	t_i

*For true counter or parallel flow, P will display 0.
†For $N \geqslant 3$, the display value will be $[NTU]/N$.
‡For N-$2N$ exchanger where $N \geqslant 3$ only.

Data registers used for storing information   Table III

Data Register	Contents	Data Register	Contents
00	Input pointer	20	$P_{N,2N}$ for $N \geqslant 3$ @ $R \neq 1$
01	W	21	T_2
02	C	22	t_2
03	T_1	23	Q
04	w	24	$[LMTD]$
05	c	25	$\triangle T$
06	t_1	26	t_i
07	U	27	T_2 trial, t_i trial, $T_1 - t_2$, $T_1 - t_1$
08	A		
09	Pointer to recall P	28	H @ $R=1$ or $R \neq 1$
10	R	29	$((1-PR)/(1-P))^N$, T_2-t_1, T_2-t_2
11	$[NTU]$ or $[NTU]/N$ for $N \geqslant 3$		
12	$\exp((UAF/wc)(R \pm 1))$		
13	F	30	T_2 initial (HP only)
14	N		
15	m		
16	a		
17	b		
18	wc, FUA/wc, $R \pm 1$, $a + b$		
19	$P_{1,2}$		

Nomenclature

A	Heat-transfer surface, ft²
a	$(2/P) - R - 1$, dimensionless
b	$(2/P)[(1-P)(1-PR)]^{1/2}$, dimensionless
C	Hot-fluid heat capacity, Btu/(lb)(°F)
c	Cold-fluid heat capacity, Btu/(lb)(°F)
F	Temperature correction factor, dimensionless
G	Outlet temperature gap, $(T_2-t_2)/(T_1-t_1)$, dimensionless
H	$[1/(R-1)]\ln[(1-P)/(1-PR)]$, or $P/(1-P)$, dimensionless
$[LMTD]$	Log-mean temperature difference, °F
m	$(R^2+1)^{1/2}$, dimensionless
N	Number of shell passes
$[NTU]$	Number of transfer units, UA/wc, dimensionless
P	Temperature efficiency, $(t_2-t_1)/(T_1-t_1)$, dimensionless
Q	Heat flow, million Btu/h
R	Heat-capacity-rate ratio, wc/WC, dimensionless
T_1	Hot-fluid inlet temperature, °F
T_2	Hot-fluid outlet temperature, °F
ΔT	True temperature difference, $F[LMTD]$, °F
t_1	Cold-fluid inlet temperature, °F
t_2	Cold-fluid outlet temperature, °F
t_i	Inlet temperature of cold fluid in the first 1-2 exchanger of an N-$2N$ exchanger, °F
t_2'	Outlet temperature of cold fluid in an N-$2N$ exchanger, °F
t_2''	Outlet temperature of cold fluid in the first 1-2 exchanger of an N-$2N$ exchanger, °F.
U	Overall heat-transfer coefficient, Btu/(h)(ft²)(°F)
W	Hot-fluid flowrate, lb/h
w	Cold-fluid flowrate, lb/h

4. There is no condensation of vapor or boiling of liquid in any part of the exchanger.

5. Heat losses are negligible.

6. There is equal heat-transfer surface in each pass.

7. The temperature of the shell-side fluid in any shell-side pass is uniform over any cross-section.

In the above equations, the temperature efficiency, P, and heat-capacity-rate ratio, R, are defined as:

$$P = (t_2 - t_1)/(T_1 - t_1) \tag{13}$$

$$R = (T_1 - T_2)/(t_2 - t_1) \tag{14}$$

Note that R can also be expressed as:

$$R = wc/WC \tag{15}$$

and therefore R is determinable. In order to properly use Eq. (6) through (12) in the program, Eq. (13) and (14) are combined to obtain the more useful form:

$$P = \frac{(T_1 - T_2)}{R(T_1 - t_1)} \tag{16}$$

$$t_2 = t_1 + (T_1 - T_2)/R \tag{17}$$

$$T_2 = T_1 - R(t_2 - t_1) \tag{18}$$

The iterative procedure for one-two or two-four exchanger calculations introduces the correction factor F into Eq. (3). Following a similar procedure in Kern's

---

Problem 1—Standard-case run   Table IV

Input data*:

$W = 20{,}160$ lb/h	$c = 1.0$ Btu/(lb)(°F)
$C = 0.757$ Btu/(lb)(°F)	$t_1 = 68$°F
$T_1 = 150$°F	$U = 183$ Btu/(ft²)(h)(°F)
$w = 41{,}600$ lb/h	$A = 163$ ft²

Output data:

Entry	0	1	2	3	4	0	Kern's
Press	C	B	B	B	B	B	example
Type of exchanger	Parallel flow	1-2	2-4	3-6	4-8	Counter flow	1-2 exchanger
Executing time, min:sec	0:10	0:34	1:19	3:32	4:15	0:11	
Q, 10^6 Btu/h	0.852	0.915	0.974	0.985	0.990	0.994	0.915
T_2, °F	94.16	90.07	86.20	85.42	85.16	84.86	90.
t_2, °F	88.49	89.99	91.41	91.69	91.79	91.90	90.
$[LMTD]$, °F	28.57	37.93	34.55	33.85	33.61	33.33	37.9
$\triangle T$, °F	28.57	30.66	32.64	33.0	33.12	33.33	30.7
F	1.0	0.808	0.945	0.975	0.986	1.0	0.81
P		0.268	0.285	0.289	0.290		0.268

*Input data are taken from Kern's "Process Heat Transfer," [4], Example 7.6.

Problem 2—Case run for R = 1   Table V

Input data:

$W = 75{,}000$ lb/h	$c = 0.3$ Btu/(lb)(°F)
$C = 0.6$ Btu/(lb)(°F)	$t_1 = 100$°F
$T_1 = 300$°F	$U = 150$ Btu/(ft²)(h)(°F)
$w = 150{,}000$ lb/h	$A = 375$ ft²

Output data*:

Entry	0	0	1	2
Press	C	B	B	B
Type of exchanger	Parallel flow	Counter flow	1-2	2-4
Executing time min:sec	0:10	0:08	0:39	0:47
Q, 10^6 Btu/h	4.131	5.	4.503	4.862
T_2, °F	208.2	188.9	199.9	192.0
t_2, °F	191.8	211.1	200.1	208.0
$[LMTD]$, °F	73.4	88.9	99.9	92.0
$\triangle T$, °F	73.4	88.9	80.1	86.4
F	1.0	1.0	0.801	0.940
P			0.50	0.540

* Not available for $N \geqslant 3$.

Problem 3—Comparison of Ten Broeck chart method and the program calculation   Table VI

Input data*:

$W = 43{,}800$ lb/h	$c = 0.49$ Btu/(lb)(°F)
$C = 0.60$ Btu/(lb)(°F)	$t_1 = 100$°F
$T_1 = 390$°F	$U = 69.3$ Btu/(ft²)(h)(°F)
$w = 149{,}000$ lb/h	$A = 662$ ft²

Output-data comparison:

	Kern's Example 7.9	This work
Type of exchanger	1-2 Exchanger	1-2 Exchanger
Q, 10^6 Btu/h	5.6	5.411
T_2, °F	176.	184.11
t_2, °F	177.	174.11
$[LMTD]$, °F	132.9	139.80
$\triangle T$, °F	107.3	117.94
F	0.807	0.844
P	0.265†	0.256

Heat-balance check:

$WC(T_1 - T_2)$, Btu/h	5,623,920.	5,410,789.
$wc(t_2 - t_1)$, Btu/h	5,621,770.	5,410,771.
$UA \triangle T$, Btu/h	4,922,559.	5,410,686.

* "Process Heat Transfer," [4], p. 170.

† A misreading may result in a temperature cross, which is impractical in a 1-2 heat exchanger.

Problem 4—Case run where inoperative condition occurs   Table VII

Input data*:

$W = 60{,}000$ lb/h	$c = 0.51$ Btu/(lb)(°F)
$C = 0.57$ Btu/(lb)(°F)	$t_1 = 90$°F
$T_1 = 250$°F	$U = 58.8$ Btu/(ft²)(h)(°F)
$w = 168{,}000$ lb/h	$A = 2{,}540$ ft²

Output data:

Entry	0	0	1†	2	3‡	4
Press	C	B	B	B	B	B
Type of exchanger	Parallel flow	Counter flow	1-2	2-4	3-6	4-8
Executing time, min:sec	0:10	0:12	0:45	7:00	3:37	
Q, 10^6 Btu/h	3.902	5.226	N/A	4.999	5.129	5.172
T_2, °F	135.9	97.18	103.8	100.0	98.8	
t_2 °F	135.5	151.00	148.3	149.9	150.4	
$[LMTD]$, °F	26.13	35.0	44.0	39.2	37.4	
$\triangle T$, °F	26.13	35.0	33.5	34.3	34.6	
F	1.0	1.0	0.760	0.876	0.926	
P	-	-	0.365	0.374	0.377	

*Input data are taken from Kern's "Process Heat Transfer," Example 8.2.

†For TI program, a flashing display indicates the inoperative condition. For HP program, DATA ERROR will be displayed.

‡For TI, press GTO 603 LRN = R/S LRN. After pressing 3B, enter a smaller value than displayed (98, for example) to continue the program. For HP, press GTO .467 PRGM R/S, and follow same procedure as for TI.

HP program listing for heat-exchanger calculations (Before entering program, Execute SIZE 031). Program requires the HP-41CV or the HP41 with at least three memory modules

Table VIII continues

```
01♦LBL "HTEX"        63 RCL 21          125 "TEMP. EFF. P="    187 1              249 RCL 10          311 STO 11
02 "HOT RATE W, LB/" 64 X>Y?            126 ARCL IND 09        188 RCL 10          250 *               312 FS? 01
03 "├HR?"            65 GTO 04          127 AVIEW              189 X=Y?            251 RCL IND 00       313 GTO 11
04 PROMPT            66 X<>Y            128 STOP               190 GTO 03          252 *               314 RTN
05 STO 01            67 2              129 "RATIO R="          191 RCL 18          253 +               315♦LBL 11
06 "HOT CP, BTU/LB." 68 +              130 ARCL 10             192 RCL 10          254 RCL 18          316 RCL 14
07 "├F?"             69 STO 21          131 AVIEW              193 1              255 /               317 /
08 PROMPT            70♦LBL 04          132 STOP               194 -              256 RCL 12          318 STO 11
09 STO 02            71 FS? 01          133 2                 195 STO 18          257 /               319 RTN
10 "HOT TEMP. T1, F" 72 GTO 24          134 RCL 14             196 *              258 STO 21          320♦LBL 20
11 "├?"              73 XEQ 18          135 X>Y?               197 E↑X            259 0               321 RCL 03
12 PROMPT            74 GTO 23          136 ST* 11             198 STO 12          260 RCL 14          322 RCL 21
13 STO 03            75♦LBL 24          137 "NTU ="            199 CHS            261 X>Y?            323 -
14 "COLD RATE W, LB" 76 XEQ 19          138 ARCL 11            200 1              262 RTN             324 RCL 10
15 "├/HR?"           77♦LBL 23          139 AVIEW              201 +              263 RCL 03          325 /
16 PROMPT            78 XEQ 20          140 STOP               202 RCL 10          264 RCL 21          326 RCL 06
17 STO 04            79 RCL 03          141 "*** END ***"      203 *              265 -               327 +
18 "COLD CP, BTU/LB" 80 RCL 22          142 AVIEW              204 RCL IND 00      266 RCL 10          328 STO 22
19 "├.F?"            81 -              143 STOP               205 *              267 /               329 RCL 03
20 PROMPT            82 STO 27          144♦LBL B             206 RCL 03          268 RCL 03          330 RCL 21
21 STO 05            83 RCL 21          145 STO 14             207 RCL 18          269 RCL 06          331 -
22 "COLD TEMP. T1, " 84 RCL 06          146 XEQ 15             208 *              270 -               332 RCL 01
23 "├F?"             85 -              147 XEQ 17             209 -              271 /               333 *
24 PROMPT            86 STO 29          148 XEQ 20             210 1              272 STO 19          334 RCL 02
25 STO 06            87 XEQ 21          149 RCL 03             211 RCL 10          273 RTN             335 *
26 "COEFF. U, BTU/F" 88♦LBL 22          150 RCL 06             212 RCL 12          274♦LBL 15          336 1 E6
27 "├T2.F.HR?"       89 RCL 13          151 -                 213 *              275 1               337 /
28 PROMPT            90 *              152 STO 27             214 -              276 STO 13          338 STO 23
29 STO 07            91 STO 25          153 RCL 21             215 /              277 19              339 RTN
30 "AREA A, FT2?"    92 FIX 3           154 RCL 22             216 STO 21          278 STO 09          340♦LBL 18
31 PROMPT            93 CF 01           155 -                 217♦LBL 03          279 6               341 RCL 21
32 STO 08            94 "DUTY Q="       156 STO 29             218 0              280 STO 00          342 STO 27
33 SF 27             95 ARCL 23         157 XEQ 21             219 RCL 14          281 0               343 FS? 01
34 "SHELL N? PRESS " 96 "├MMBTU/HR"     158 GTO 22             220 X>Y?           282 STO 19          344 GTO 26
35 "├A OR B"         97 AVIEW           159♦LBL 21             221 RTN            283 STO 20          345 10
36 PROMPT            98 STOP            160 RCL 27             222 RCL 03          284 CF 01           346 ST+ 27
37♦LBL A             99 FIX 1           161 X=Y?               223 RCL 21          285 RCL 14          347♦LBL 26
38 STO 14            100 "HOT TEMP. T2=" 162 GTO 25            224 -              286 2               348 RCL 21
39 XEQ 15            101 ARCL 21        163 X<>Y               225 RCL 10          287 X<>Y            349 RCL 27
40 XEQ 16            102 "├F"           164 -                 226 /              288 X<=Y?            350 +
41 0                 103 AVIEW          165 RCL 27             227 RCL 03          289 GTO 10          351 2
42 RCL 14            104 STOP           166 RCL 29             228 RCL 06          290 SF 01           352 /
43 X=Y?              105 "COLD TEMP. T2=" 167 /                229 -              291♦LBL 10          353 STO 27
44 GTO 23            106 ARCL 22        168 LN                230 /              292 RCL 04          354 RCL 03
45 2                 107 "├F"           169 /                 231 STO 19          293 RCL 05          355 -
46 RCL 10            108 AVIEW          170♦LBL 25             232 RTN            294 *               356 RCL 10
47 *                 109 STOP           171 STO 24             233♦LBL 17          295 STO 18          357 /
48 RCL 03            110 "LMTD="        172 RTN                234 RCL 10          296 RCL 01          358 RCL IND 00
49 RCL 06            111 ARCL 24        173♦LBL 16             235 1              297 /               359 RCL 03
50 -                 112 "├F"           174 RCL 11             236 +              298 RCL 02          360 -
51 *                 113 AVIEW          175 RCL 13             237 STO 18          299 /               361 /
52 1                 114 STOP           176 *                 238 RCL 11          300 STO 10          362 STO 19
53 ENTER↑            115 "TRUE DT="     177 STO 18             239 *              301 X↑2             363 1
54 RCL 10            116 ARCL 25        178 RCL 06             240 E↑X            302 1               364 RCL 19
55 +                 117 "├F"           179 *                 241 STO 12          303 +               365 -
56 RCL 15            118 AVIEW          180 RCL 03             242 RCL 10          304 SQRT            366 /
57 +                 119 STOP           181 +                 243 +              305 STO 15          367 STO 28
58 /                 120 FIX 3          182 1                 244 RCL 03          306 RCL 07          368 1
59 CHS               121 "CORRECT. F="  183 RCL 18             245 *              307 RCL 08          369 RCL 10
60 RCL 03            122 ARCL 13        184 +                 246 RCL 12          308 *               370 X=Y?
61 +                 123 AVIEW          185 /                 247 1              309 RCL 18          371 GTO 09
62 STO 30            124 STOP           186 STO 21             248 -              310 /               372 1
```

Table VIII continued

373 RCL 19	403 *	433 STO 18	463 RCL 06	493 X=Y?	523 STO 22
374 -	404 -	434 RCL 15	464 RCL 22	494 GTO 05	524 RCL 19
375 1	405 *	435 +	465 +	495 1	525 RCL 03
376 RCL 19	406 SQRT	436 RCL 18	466 2	496 RCL 19	526 RCL 26
377 RCL 10	407 RCL 18	437 RCL 15	467 /	497 RCL 10	527 -
378 *	408 *	438 -	468 STO 26	498 *	528 *
379 -	409 STO 17	439 /	469 STO 27	499 -	529 -
380 /	410 2	440 LN	470 26	500 1	530 STO 27
381 LN	411 RCL 14	441 1/X	471 STO 00	501 RCL 19	531 RCL 26
382 RCL 10	412 X=Y?	442 RCL 28	472◆LBL 27	502 -	532 -
383 1	413 GTO 08	443 *	473 RCL 26	503 /	533 ABS
384 -	414 RCL 16	444 RCL 15	474 RCL 27	504 RCL 14	534 .1
385 /	415 RCL 15	445 *	475 +	505 Y↑X	535 X<=Y?
386 STO 28	416 +	446 2	476 2	506 STO 29	536 GTO 27
387◆LBL 09	417 RCL 16	447 /	477 /	507 CHS	537 RCL 06
388 2	418 RCL 15	448 STO 13	478 STO 26	508 1	538 RCL 22
389 RCL 19	419 -	449◆LBL 07	479 XEQ 18	509 +	539 -
390 /	420 /	450 XEQ 16	480 RCL 19	510 RCL 10	540 RCL 10
391 STO 18	421 LN	451 .1	481 RCL 14	511 RCL 29	541 *
392 1	422 1/X	452 RCL 21	482 *	512 -	542 RCL 03
393 -	423 RCL 15	453 RCL 27	483 STO 29	513 /	543 +
394 RCL 10	424 *	454 -	484 RCL 29	514 STO 20	544 STO 21
395 -	425 RCL 28	455 ABS	485 RCL 19	515◆LBL 05	545 .END.
396 STO 16	426 *	456 X<>Y	486 -	516 RCL 20	
397 1	427 STO 13	457 X=Y?	487 1	517 RCL 03	
398 RCL 19	428 GTO 07	458 GTO 26	488 +	518 RCL 06	
399 -	429◆LBL 08	459 RTN	489 /	519 -	
400 1	430 RCL 16	460◆LBL 19	490 STO 20	520 *	
401 RCL 19	431 RCL 17	461 20	491 1	521 RCL 06	
402 RCL 10	432 +	462 STO 09	492 RCL 10	522 +	

derivation [4], the iteration formula for T_2 is obtained:

$$T_2 = \frac{(1-R)T_1 + [1 - e^{(FUA/wc)(R-1)}]Rt_1}{1 - Re^{(FUA/wc)(R-1)}} \tag{19}$$

When $R=1$, Eq. (19) is reduced to:

$$T_2 = \frac{T_1 + t_1(FUA/wc)}{1 + FUA/wc} \tag{20}$$

Note that Eq. (19) and (20) are implicit forms of T_2 because the correction factor $F_{1,2}$ or $F_{2,4}$ can be converted to a function of T_2, which is the only unknown.

N-2N heat exchangers where $N \gtreqless 3$

The general equation for an N-$2N$ exchanger is [8]:

$$P_{N,2N} = \frac{1 - [(1-P_{1,2}R)/(1-P_{1,2})]^N}{R - [(1-P_{1,2}R)/(1-P_{1,2})]^N} \tag{21}$$

When $R=1$:

$$P_{N,2N} = NP_{1,2}/(NP_{1,2} - P_{1,2} + 1) \tag{22}$$

Fig. 1 shows the first one-two exchanger of an N-$2N$ exchanger. The temperature efficiency of the N-$2N$ exchanger, $P_{N,2N}$, and that of the first one-two exchanger for the N-$2N$ exchanger, $P_{1,2}$, are expressed as:

$$P_{N,2N} = (t_2' - t_1)/(T_1 - t_1) \tag{23}$$

$$P_{1,2} = (t_2'' - t_i)/(T_1 - t_i) \tag{24}$$

where t_2' and t_2'' are the outlet temperatures of the cold stream for the N-$2N$ exchanger and for the first one-two exchanger of the N-$2N$ exchanger, respectively. Rearrange Eq. (23) and (24) to obtain:

$$t_2' = P_{N,2N}(T_1 - t_1) + t_1 \tag{25}$$

$$t_2'' = P_{1,2}(T_1 - t_i) + t_i \tag{26}$$

Let $t_2' = t_2''$, then the iteration formula for t_i in terms of the inlet temperatures, T_1 and t_1, can be expressed as:

$$t_i = P_{N,2N}(T_1 - t_1) + t_1 - P_{1,2}(T_1 - t_i) \tag{27}$$

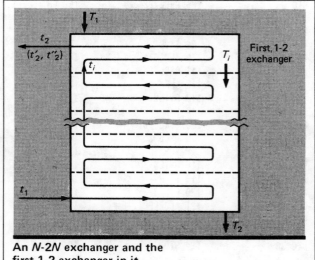

An N-2N exchanger and the first 1-2 exchanger in it

Calculator program development

The algorithm for a one-two or two-four exchanger calculation is:

1. Guess initial T_2.
2. Calculate P from Eq. (16).
3. Calculate F from Eq. (6) or (7).
4. Calculate the new T_2 from Eq. (19) or (20).
5. Adjust T_2 and repeat Steps 2, 3 and 4 until the T_2 value is converged.

For an N-$2N$ exchanger, where $N \leqq 3$, a double-iteration method is proposed as follows:

1. Guess initial t_i.
2. Use W, C, T_1, w, c, t_i, U and A/N to calculate t_2, T_2, $P_{1,2}$, $F_{1,2}$ by the previous iterative procedure for the one-two exchanger.
3. Calculate $P_{N,2N}$ from $P_{1,2}$ by Eq. (21) or (22).
4. Substitute $P_{1,2}$, $P_{N,2N}$ and t_i into Eq. (27) to obtain the new t_i.
5. Adjust t_i, and repeat Steps 2, 3 and 4 until the t_i value is converged.

Due to the many calculations involved in the different types of heat exchangers, the program requires a large number of memory steps, and the data registers are repeatedly used. All but one of the required equations have been included in the program, so that it can be run even for the special cases. Unfortunately, Eq. (22) is not coded in the program because of lack of memory space. Therefore, for N-$2N$ exchangers where $N \leqq 3$, the calculation at $R = 1$ is the only limited case not available. As mentioned before, with a larger program memory, it is no problem to include Eq. (22). The convergence tolerance for both T_2 and t_i is set at 0.1, which is accurate enough for process calculations.

Basically, the program is applicable for N-$2N$ heat exchangers. However, it should be noted that when the number of tube passes is even and the tube-shell pass ratio is greater than two, the results for N-$2N$ exchangers still may be used with little error.

How to use the program

The required input data are W, C, T_1, w, c, t_1, U and A. Load the program (Table I) and save it on magnetic cards. Then follow the user instructions in Table II.

Label A is used for data input, B is for true countercurrent and N-$2N$ exchanger calculation, and C is for a true parallel-flow exchanger. All the contents of the data registers used in the program are shown in Table III.

Some sample problems are listed in Tables IV to VII for various applications. Problem 1 is a standard-case run. Problem 2 is for the case where $R = 1$. Note that it is not applicable for the case where $N \leqq 3$ because of the absence of Eq. (22). Problem 3 illustrates the comparison between the Ten Broeck chart method [9] and this program. Because the value of P is quite sensitive to t_2, any tiny error in reading from the chart will result in a great difference for t_2, and therefore will predict an unreal exchanger performance. Problem 4 will show flashing data during the program run if the input-data combination is inoperative. Increasing the shell number, N, will cause the set of data to be operable. However, this case should be observed with care. The program-generated initial T_2 and t_i are adequate for most cases. Under some conditions, the initial generated value will not converge the calculation, and flashing-error data will be obtained. Increasing the initial generated T_2 value and/or decreasing the initial generated t_i value in the program will compensate for this problem. If a flashing error still exists, it indicates the inoperative case.

To generate a higher value of initial T_2, press GTO 392 LRN, then enter a two-digit value greater than 10 (20 for instance), and press LRN. To manually generate a lower value of initial t_i, press GTO 603 LRN=R/S LRN. After rerunning the program, enter a value smaller than the display but greater than t_1 (RCL 06) at the first stop, and then press R/S to continue the program. It may be tried for a couple of initial values to ensure that the input-data combination is indeed inoperative.

A calculator program for HP-41CV (or HP-41C with three additional memory modules) was also developed for this heat-exchanger calculation, based on the same algorithm presented in this article. The alphanumerical input and output in this HP calculator program are self-explanatory, and the program is listed in Table VIII. Note that one should key in XEQ ALPHA HTEX ALPHA to start the program, and that Keys A and B are equivalent to the TI Keys B and C, respectively.

To ensure that the initial T_2 is properly generated, a should be greater than m, and the following relation has been coded in the HP program:

$$T_2 > T_1 - \frac{2R(T_1 - t_1)}{(1 + R) + (R^2 + 1)^{1/2}} \qquad (28)$$

If the program-generated initial T_2 does not meet Eq. (28), a value of the right-hand side of Eq. (28) plus 2 is used as the initial T_2.

References

1. Al-Zakri, A. S., and Bell, K. J., Estimating Performance When Uncertainties Exist, *Chem. Eng. Prog.*, July 1981.
2. Spencer, Robert, A., Jr., Predicting Heat-Exchanger Performance by Successive Summation, *Chem. Eng.*, Dec. 4, 1978, p. 121.
3. Jagannath, S., "Calculator Programs for the Hydrocarbon Processing Industries," Vol. 1, p. 100, Gulf Pub. Co., Houston, Tex., 1980.
4. Kern, D. Q., "Process Heat Transfer," McGraw-Hill, New York, 1950.
5. "Standards of Tubular Exchanger Manufacturers Association," 6th ed., Tubular Exchangers Mfrs. Assn., New York, 1978.
6. Wales, Ronald E., Mean Temperature Difference in Heat Exchangers, *Chem. Eng.*, Feb. 23, 1981, p. 77.
7. Underwood, A. J. V., The Calculation of the Mean Temperature Difference in Multipass Heat Exchangers, *J. Inst. Pet. Technol.*, Vol. 20, pp. 145–158 (1934).
8. Bowman, R. A., others, Mean Temperature Difference in Design, *Trans. ASME*, May 1940, pp. 283–294.
9. Ten Broeck, H., Multipass Exchanger Calculations, *Ind. Eng. Chem.*, Vol. 30, pp. 1041–1042 (1938).

The author

Shuh-Chih Chang is a Software Engineer for Summit CAD Corp., 5222 FM 1960, Houston, TX 77069, tel. (713) 440-1468, where he is responsible for software design and development, system implementation, and hardware configuration for Computer-Aided Design (CAD). His prior experience has been mainly in process simulation, facilities design, engineering design and software development for oil and gas production and processing systems. He holds B.S. and M.S. degrees from the National Central University, Taiwan, and an M.S. from the University of Iowa, all in chemical engineering. He is a member of AIChE.

Heat-transfer coefficient depends on tubeside flowrate

The overall heat-transfer coefficient of a shell-and-tube heat exchanger is sensitive to the flow regime in the tubes, particularly if the tubeside fluid is a gas.

Christopher G. Lower, Gould Inc., Foil Div.

☐ It is well known that the heat transfer in a shell-and-tube heat exchanger depends on both the overall heat-transfer coefficient and the mass flowrate in the tubes. What is not commonly known, however, is the relationship between the two, and the precise way it affects the exit temperature of the process fluid.

In three examples below, we begin with a properly sized shell-and-tube heat exchanger. The flowrate of the process fluid is then reduced, and we will examine how this affects its outlet temperature. Each time the flowrate is reduced, a new overall heat-transfer coefficient is calculated. From this new coefficient and the changed flowrate, a new exit temperature is calculated from the energy balance:

$$wc(t_2 - t_1) = UA\Delta T_{LM} \tag{1}$$

which becomes, for counterflow exchangers:

$$T_2 = \frac{(1 - R)T_1 + [1 - e^{(UA/wc)(R - 1)}]Rt_1}{1 - R.e^{(UA/wc)(R - 1)}} \tag{2}$$

Here:

T_1, T_2 = Entrance and exit temperatures of process fluid, respectively, °F

t_1, t_2 = Entrance and exit temperatures of cooling (or heating) medium respectively, °F

U = Overall heat-transfer coefficient, Btu/(h)(ft²)(°F)

A = Area of heat exchanger, ft²

w = Mass flowrate of cooling (or heating) medium, lb/h

c = Specific heat of cooling (or heating) medium, Btu/(lb)(°F)

W = Mass flowrate of process fluid, lb/h

C = Specific heat of process fluid, Btu/(lb)(°F)

R = wc/WC

For a counterflow heat exchanger, t_2 is found by another energy balance:

$$t_2 = (T_1 - T_2)1/R + t_1 \tag{3}$$

Let us now look at three examples: One in which the process fluid is air, tubeside; another in which the fluid is acetic acid, tubeside; and the last in which it is a petroleum fluid on the shellside.

What we will find is that, with a gas (air) on the tubeside, the overall heat-transfer coefficient (and therefore the exit temperature of the gas) is dramatically affected by the flowrate in the transition region. The effect is noticeable but less dramatic with a liquid (acetic acid) on the tubeside, and apparently nonexistent with a liquid on the shellside. Therefore, in the rather unusual case that a gas is being run as the process fluid tubeside, this irregularity should be taken into account.

Air on the tubeside

In our first example, air passes through the tubes and is cooled by water on the shellside. The original design of the exchanger (Table I) is for 67,900 lb/h air (N_{RE} = 1.7×10^6), which is being cooled from 360 to 290°F. The mass flowrate and the inlet temperature of the cooling water are held constant throughout at 39,608 lb/h and 85°F, respectively.

The air is then throttled back. With the longer residence time in the exchanger, we should expect that the air would be cooled more thoroughly, and indeed, the exit temperature does drop as the Reynolds number decreases. But the curve is not smooth. There is a point of inflection, as shown in Fig. 1, which plots the outlet temperature of the air as a function of the Reynolds number inside the tubes. The irregularity occurs because the overall heat-transfer coefficient changes according to the flow regime inside the tube. These changes are examined in the next section.

Coefficient versus Reynolds number

The tubeside heat-transfer coefficient is determined by:

$$h_i = j_{Hi} k/D (C\mu/k)^{1/3} \phi_T^{0.14} \tag{4}$$

Originally published July 9, 1984

Heat exchanger specifications and fluid properties — Table I

Exchanger data

Tube length, ft	4
Tube size, in.	1
Number of tube passes	1
Tube pitch, in.	1.25
Shell dia., in.	15.75
Fouling factor, tubes	1.5×10^{-3}
Number of tubes	121
Tube I.D., in.	0.87
Shell passes	1
Type of pitch	Triangular
Baffle spacing, in.	4
Fouling factor, shell	2.0×10^{-3}

Physical properties	Shellside	Tubeside
Fluid	Cooling water	Air
Flowrate, lb/h	39,608	67,900
Specific gravity	1	1.2×10^{-3}
Inlet temperature, °F	85	360
Outlet temperature, °F	115	290
Viscosity, cP	0.682	2.4×10^{-3}
Specific heat, Btu/lb °F	1	0.25
Thermal conductivity, Btu/h ft² °F/ft	0.36	2.11×10^{-3}
Viscosity at wall, cP	0.6819996	2.40×10^{-3}

For each new flowrate a new overall heat-transfer coefficient is calculated — Table II

Air flow, lb/h	N_{RE}	U	T_2	t_2
67,900	1.70×10^6	41.89	289.9	115.0
40,000	1.00×10^6	29.49	277.2	105.9
20,000	5.00×10^5	18.03	261.1	97.5
10,000	2.50×10^5	10.75	245.7	92.2
5,000	1.25×10^5	6.32	230.7	89.1
2,000	5.00×10^4	3.08	211.3	86.9
900	2.25×10^4	1.64	194.4	85.9
600	1.50×10^4	1.19	186.0	85.7
400	1.00×10^4	0.82	182.0	85.4
300	7.50×10^3	0.64	178.2	85.3
250	6.25×10^3	0.54	177.5	85.3
200	5.00×10^3	0.43	178.3	85.2
150	3.75×10^3	0.31	182.2	85.2
100	2.50×10^3	0.18	193.2	85.1
90	2.25×10^3	0.16	197.3	85.1
80	2.00×10^3	0.14	200.1	85.1
70	1.75×10^3	0.13	191.2	85.1
60	1.50×10^3	0.125	180.8	85.1
50	1.25×10^3	0.118	168.5	85.1
30	750	0.099	136.5	85.0
15	375	0.079	104.2	85.0
10	250	0.068	93.4	85.0
3	75	0.046	85.1	85.0

Here:

j_{Hi} = Inside Colburn factor
k = Thermal conductivity, Btu/h ft² ft
D = Inside diameter of the tubes, ft
μ = Viscosity, cP
ϕ_T = Wall viscosity correction
$\quad = \mu/\mu_{wall}$

The original design flowrate in our first example is in the turbulent region. Therefore, we would expect both conductive and convective heat transfer within the medium. The Colburn factor, j_{Hi}, under these conditions ($N_{RE} > 10,000$) is described by:

$$j_{Hi} = 0.023 [1 + (D/L)^{0.7}] (N_{RE})^{0.8} \qquad (5)$$

Here:

$\quad L$ = Length of tubes, ft

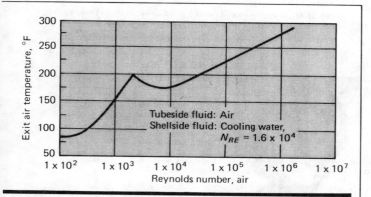

Tubeside fluid: Air
Shellside fluid: Cooling water,
$N_{RE} = 1.6 \times 10^4$

With air in the tubes, exit temperature versus Reynolds number shows inflection — Fig. 1

Throughout the turbulent region, the decreasing mass flowrate produces a consistently decreasing exit temperature. This occurs because the mass is reduced at a greater rate than is the heat-transfer coefficient.

The decline in temperature continues to a Reynolds number of approximately 10,000.

At this point, the flow enters the transition region, where the flow displays characteristics of both turbulent and laminar patterns. The Colburn factor is not easily described in the transition region, but it can be approximated by fitting experimental data to a curve. Assume that, within the region $2,100 \leq N_{RE} \leq 10,000$, the curve (Fig. 24, Ref. 2) follows the form:

$$y = A + Bx + Cx^2 \qquad (6)$$

Here:

$y = \ln j_{Hi}$, $x = \ln N_{RE}$, and A, B and C are constants

To determine A, B and C, three points must be described:

1. Lower turbulent region, ($N_{RE} = 10,000$), Eq. (5).
2. Slope at the point where $N_{RE} = 10,000$.
3. Upper laminar region, ($N_{RE} = 2,100$), use Eq. (7).

The resulting constants are:

$$A = Z - 9.21B - 84.82 \, C$$
$$B = 0.819 - 18.42 \, C$$
$$C = 1.82 + 0.137 \ln (D/L) - (Z/2.44)$$

Here:

$$Z = \ln 36.45[1 + (D/L)^{0.7}]$$

It is within the transition region that the decrease in the heat-transfer coefficient is greater than the decrease in mass. The result is a rise in the exit temperature, rather than a drop. As the Reynolds number continues to

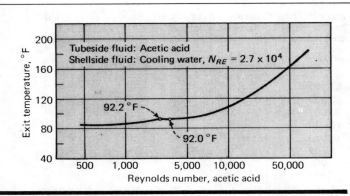

With acetic acid in tubes, point of inflection in transition region is much less marked   Fig. 2

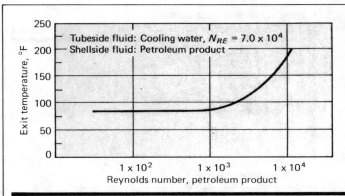

With petroleum product on shellside, there is no point of inflection in transition region   Fig. 3

decrease through the transition region, the rising temperature breaks sharply at about $N_{RE} = 2,100$, as the flow declines into the laminar region.

In laminar flow ($N_{RE} < 2,100$), the controlling source of heat transfer is via conduction. The Colburn factor for laminar flow is described by:

$$j_{Hi} = 1.86 \ (D/L)^{1/3} \ (N_{RE})^{1/3} \qquad (7)$$

Once again, as in the turbulent region, the decrease in mass is greater than the decrease in heat-transfer coefficient. This will allow the exit temperature to fall asymptotically to a temperature equal to that of the incoming cooling water.

Hence it may be clearly seen that the heat transfer coefficient for the tubeside fluid is a function of the tubeside flowrate (i.e., Reynolds number). Further, an inflection in the cooling curve can be expected in the transition region, with a peak at $N_{RE} = 2,100$. The degree to which the inflection occurs (the magnitude of the peak) will depend on the fluid properties in each case.

Table II represents the results of a calculation, via computer program, of the overall heat-transfer coefficient as a function of the reduced flowrates used in plotting Fig. 1.

Acetic acid tubeside

Fig. 2 is a cooling curve for acetic acid running on the tubeside—exit temperature versus Reynolds number. The initial design duty is to lower the temperature of 150,000 lb/h of acetic acid from 220 to 175°F. The mass flowrate and inlet temperature of the shellside cooling water are held constant at 117,450 lb/h and 85°F, respectively, as the acetic acid flowrate is reduced. It is found that the peak at $N_{RE} = 2,100$ is only 0.2°F higher than the lowest transition-region temperature of 92.0°F. However, the curve itself is consistent with Fig. 1.

Petroleum product shellside

Fig. 3 represents the exit temperature of the shellside fluid (a petroleum product) as a function of the shellside Reynolds number. The exchanger was designed for a mass flowrate of 75,000 lb/h, cooling from 390 to 200°F. The mass flowrate and inlet temperature of the cooling

water were again held constant, this time at 280,250 lb/h and 85°F, respectively.

The shellside coefficient is determined by:

$$h_o = j_{Ho} \, k/De \ (C\mu/k)^{1/3} \ \phi_s^{0.14} \qquad (8)$$

Here:

j_{Ho} = Shellside Colburn factor
De = Equivalent diameter, ft
    = 4 × Free area/wetted perimeter
ϕ_s = Wall viscosity correction = μ/μ_{wall}

The Colburn factor for the shellside fluid is given by:

$$j_{Ho} = \exp \left[0.3077 \ (\ln N_{RE})^{1.175} - 0.1779\right]$$

and is valid for $10 < N_{RE} < 1 \times 10^6$.

The results obtained by throttling the shellside fluid are different from those obtained by throttling the tubeside fluid. The exit temperature drops through the turbulent, transitional and laminar regions to the asymptotic temperature of 85°F, with no inflections. This is due to the nature of the shellside flow patterns. The fluid is always changing direction between baffles and through the tube bundle; this induces eddies and creates friction. Therefore, both conductive and convective heat transfer apply over the entire curve, and no discontinuity occurs at the point where $N_{RE} = 2,100$.

References

1. McCabe, W. L., and Smith, J. C., "Unit Operations of Chemical Engineering," McGraw-Hill, New York, 1976.
2. Kern, D. Q., "Process Heat Transfer," McGraw-Hill, New York, 1950.
3. Holman, J. P., "Heat Transfer," McGraw-Hill, New York, 1976.

The author

Christopher G. Lower was working with The H. K. Ferguson Co. when this article was written. His responsibilities now include overseeing the design, installation and startup of new process technology for Gould, Inc., Foil Div., 5045 North State, Route N.W., McConnelsville, OH 43756. Tel: (614) 962-5252. Lower graduated from Ohio University with a B.S. in chemical engineering.

Thermal and hydraulic design of hairpin and finned-bundle exchangers

Select new exchangers and re-rate old ones via this method for predicting the performance of longitudinal-flow hairpin and finned-bundle heat exchangers in single-phase heat transfer.

G. P. Purohit, Fluor Corp.

☐ Hairpin (or double-pipe) heat exchangers find application in a variety of services requiring small heat-transfer surfaces, particularly when one stream is a gas or viscous liquid, or its flowrate is small. Economics favor hairpin exchangers for high-pressure service because of their small shell diameter.

These exchangers are suited for handling dirty streams because of the ease of cleaning and maintenance. They are flexible, with their heat-transfer surfaces readily shifted for changing process conditions and heat loads. Spare parts are interchangeable, and available off-the-shelf (as is original equipment).

Longitudinally-finned-bundle exchangers, which are similar to the hairpin exchanger, are also available as standard designs. They are used extensively for heating viscous liquids, and can serve in applications involving fluids having significantly different heat-transfer characteristics and fouling resistances.

Basic features

A hairpin exchanger basically consists of concentric pipes (Fig. 1). One fluid flows through the inner pipe (tubeside fluid), the other through the annular space (shellside fluid). The inner pipes are connected by U-shaped return bends enclosed in a return-bend housing. Because the inlets and outlets are close together at one end, these exchangers can easily be stacked.

Hairpin exchangers are available in two basic types: single-tube and multitube. The first consists of a pipe within a pipe, and is generally called a double-pipe exchanger (Fig. 1). In the multitube, a bundle of small-diameter tubes replaces the inner pipe. In either case,

the tubes may be bare or have longitudinal fins (Fig. 2a).

Fins increase heat-transfer surface per unit length (up to as much as 15 times), and reduce the size and number of exchangers required for a given service. Fins are usually welded to the tube but may be integrally formed (Fig. 2b). They may be made of carbon steel or of a variety of alloys, and may be cut and twisted to promote turbulent flow, or perforated for special applications. This discussion is limited to plain, continuous (uncut), longitudinal fins.

Hairpin exchangers have shell diameters of 50–410 mm (2–16 in.) at nominal lengths of 1.5–12.0 m (5–40 ft). Surface area ranges from about 0.25–150 m² (3–1,615 ft²) per section. Major dimensions are summarized in Table I. Combinations of these dimensions yield a variety of design configurations available as standard equipment.

Fluid flow in hairpin exchangers

Flows in hairpin exchangers are pure countercurrent, which eliminates the *F* correction factor for the logarithmic mean temperature difference (LMTD) of the mixed countercurrent and cocurrent flows common in shell-and-tube exchangers. This results in up to 20% higher energy efficiency, correspondingly reducing required heat-transfer surface. Pure countercurrent flows in a shell-and-tube exchanger would require a single-shell/single-tube pass configuration or a double-shell/double-tube pass configuration, at higher cost and greater complexity.

Pure countercurrent flow allows close temperature approaches and temperature crossovers. With conven-

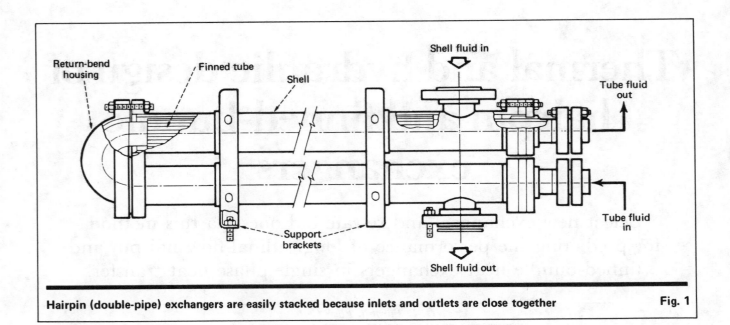

Hairpin (double-pipe) exchangers are easily stacked because inlets and outlets are close together   **Fig. 1**

tional shell-and-tube exchangers of single-shell pass and multiple-tube passes, multiple shells in series would be required to eliminate crossovers.

In the shell-and-tube design, long temperature ranges (which usually result in temperature crosses) require either multiple shells in series or a single pass on both shellside and tubeside. Using a single shell with a longitudinal baffle would result in high thermal stresses in the channel, tubesheet and shell, and in warpage of the baffle, causing leakage across it. An advantage of the hairpin exchanger is that its inner pipe can expand comparatively freely; however, care must be taken to avoid severe thermal stresses, particularly repetitive or cyclically reversing ones.

The longitudinally-finned-bundle exchanger

In this type, a bundle of finned tubes connected by U-bends and mounted in a shell provides a single shell pass and two tube passes. Available in shell diameters up to about 1,200 mm, this exchanger is used extensively to heat viscous liquids, such as asphalts and fuel oils, in order to reduce viscosity for easier pumping.

Two examples of this type of exchanger are tank-suction and inline heaters. The open end of the first extends into the tank (Fig. 3a). The second is installed in pipelines (Fig. 3b). In both, the viscous fluid flows on the finned side, and the heating medium through the tubes.

Thermal and hydraulic design

A flowchart of the steps involved in the thermal and hydraulic design of longitudinal-flow hairpin and finned-bundle heat exchangers is presented in Fig. 4. Published correlations for flow inside circular tubes are used in heat-transfer and pressure-drop calculations.

Although both exchangers are used with single- and two-phase (especially condensing) fluids, this article is limited to single-phase ones. It also only considers exchangers having continuous, longitudinal finned tubes, and not those having baffles and tube supports.

The equivalent-diameter approach is used to calculate heat transfer and pressure drop in the annulus, because these are in theory similar to those inside the tube, although the annulus has no dimensions corresponding to those of the tube (its geometry being distinctly dissimilar), and only a portion of its wetted surface is heated or cooled.

The simplest approach to applying inside-tube heat-transfer correlations to the annulus is to substitute equivalent diameter of the annulus for the inner-tube diameter:

$$d_e = 4\,r_h = \frac{4 \times \text{Net free-flow area}}{\text{Wetted (or heated or cooled) perimeter}} \quad (1)$$

This approximation is accepted for heat-transfer and pressure-drop calculations in annuli of small equivalent diameters. The validity of the equivalent-diameter approach has been substantiated by the results of experiments with finned annuli [1].

The *total wetted perimeter* of the annulus of a longitudinally finned heat exchanger is given by:

$$P_W = \pi(D_i + d_o N_T) + 2(Y N_F N_T) \quad (2)$$

Nomenclature

A	Heat-transfer surface, m^2
A_{NF}	Net free area, m^2
a_f	Cross-section of fin, m^2
C	Numerical constant, Eq. (40)
C_p	Specific heat, kJ/(kg)(K)
cosh	Hyperbolic cosine
D	Characteristic diameter in heat-transfer and pressure-drop correlations, m
D_i	Inside dia. of shell (outer pipe), m
d	Dia. of tube (inner pipe), m
d_e	Equivalent dia. of annulus, m
E_T	Tip temperature factor, dimensionless
F	Correction factor for LMTD, dimensionless
f	Fanning friction factor, dimensionless
G	Mean mass velocity, $kg/(m^2)(s)$
g	Acceleration due to gravity, 9.81 m/s^2
H	Difference in elevation between inlet and outlet nozzles, m
h	Heat-transfer coefficient, $W/(m^2)(K)$
K	Velocity-head coefficient, dimensionless
k	Thermal conductivity, W/(m)(K)
k_f	Thermal conductivity of fin, W/(m)(K)
L	Nominal length of exchanger section, also length of longitudinal fin, m
M	Parameter of fin efficiency, Eq. (18), m^{-1}
m	Exponent of correction factor ψ for pressure-drop correlations
N	Number of exchanger sections in series
N_F	Number of fins per tube
N_{Gr}	Grashof No., $(\rho^2\, g\, \beta\, \Delta T\, D^3)/\mu^2$, dimensionless
N_{Gz}	Graetz No., $W\, C_p/k\, L = [N_{Re}\, N_{Pr}(D/L)](A_{NF}/D^2)$, dimensionless
N_{Nu}	Nusselt No., hD/k, dimensionless
N_{Pr}	Prandtl No., $\mu\, C_p/k$, dimensionless
N_{Re}	Reynolds No., $G\, D/\mu$, dimensionless
N_T	Number of tubes in one leg of hairpin exchanger; number of tube holes in tubesheet of bundle exchanger (twice the number of U-bends), Eq. (2), (3) and (4); or number of U-tubes in bundle exchanger, Eq. (7), (8), (9) and (10)
n	Exponent for correction factor ϕ in heat-transfer correlations, Eq. (12)
n_1,n_2	Number of parallel cold and hot streams, Eq. (31) and (32)
P	Perimeter, m
P_1,P_2	Temperature ratios, Eq. (31) and (32)
ΔP	Pressure drop, kPa
Q	Heat load or duty, W
R	Thermal resistance, $(m^2)(K)/W$
R_1,R_2	Temperature ratios, Eq. (31) and (32)
r	Fouling resistance, $(m^2)(K)/W$
r_h	Hydraulic radius, m
S	Dimensionless parameter for true temperature difference of series-parallel arrangement, Eq. (31)
T	Absolute temperature, K
ΔT	Temperature difference, K
tanh	Hyperbolic tangent
U	Overall heat-transfer coefficient based on total external surface, $W/(m^2)(K)$
v	Mean stream velocity, m/s
W	Mass flowrate, kg/s
X	Fin thickness, m
Y	Fin height, m
β	Temperature coefficient of volume expansion or contraction for N_{Gr}, $(1/\rho_{out} - 1/\rho_{in})/(1/\rho_b)(T_{out} - T_{in})$, K^{-1}
ϵ	Fin effectiveness
η	Fin efficiency
μ	Dynamic viscosity, mPa
ρ	Mass density, kg/m^3
ϕ	Property-ratio correction factor for nonisothermal flow in heat-transfer correlations, Eq. (11)
ψ	Property ratio correction factor for nonisothermal flow in pressure-drop correlations, Eq. (34)

Subscripts

allow	Allowable
b	Bulk
C	Cold
calc	Calculated
F	Frictional
f	Fins or finside
H	Hot
HT	Based on heat-transfer perimeter, Eq. (3) and (6)
h	Hydraulic
i	Inside of tubes
in	Inlet
max	Maximum
m	Momentum
n	Nozzles
o	Outside of bare tubes
out	Outlet
S	Static head
s	Shellside or finside
s-p	Series-parallel arrangement
T	Fin tip
t	Total external or tubeside coefficient based on total external surface
U	Equivalent of U-bend
W	Wetted, Eq. (2) and (5)
w	Wall of bare tube
wtd	Weighted

And the *heat-transfer perimeter* of the annulus by:

$$P_{HT} = \pi d_o N_T + 2(Y N_F N_T) \qquad (3)$$

The only difference between Eq. (2) and (3) is the term D_i in Eq. (2), which represents the inside diameter of the shell. This difference is due to the shell I.D. resistance to fluid flowing in the annulus. Such, however, is not the case for the heat-transfer perimeter, because heat transfer takes place from the total external surface and does not depend on the shell I.D.

The *net free area* in the annulus of a longitudinally finned exchanger is given by:

$$A_{NF} = (\pi/4)(D_i^2 - d_o^2 N_T) - (X\, Y N_F\, N_T) \qquad (4)$$

Equivalent diameter based on wetted perimeter (also known as *hydraulic diameter*) is:

$$(d_e)_W = 4 A_{NF}/P_W \qquad (5)$$

Equivalent diameter based on heat-transfer perimeter is:

$$(d_e)_{HT} = 4 A_{NF}/P_{HT} \qquad (6)$$

Calculate Reynolds number, Graetz number and the ratio D/L with the Eq. (5) diameter. Use the Eq. (6) diameter to calculate the heat-transfer coefficient from the Nusselt number, and in evaluating the Grashof number. Slightly higher coefficients result from using the equivalent diameter based on wetted perimeter, Eq. (5), for heat-transfer calculations.

Total outside heat-transfer surface is the sum of the bare-tube and fin surfaces:

$$A_o = 2 N_T (\pi d_o L - N_F L X) \qquad (7)$$
$$A_f = 2 N_T N_F L (2Y + X) \qquad (8)$$
$$A_t = A_o + A_f = 2 N_T L (\pi d_o + 2 N_F Y) \qquad (9)$$

For bare tubes, $A_f = 0$ and $A_t = A_o$.

The *inside heat-transfer surface* (based on tube I.D.) is:

$$A_i = 2(\pi d_i L N_T) \qquad (10)$$

Area ratios A_f/A_t and A_o/A_t, or $1 - (A_f/A_t)$, are needed for calculating weighted fin efficiency, and A_t/A_i for basing inside (tubeside) heat-transfer coefficient on the total outside heat-transfer surface.

For standard equipment, calculated values for Eq. (2) through (10) may be found tabulated in manufacturers' catalogs.

Heat-transfer correlations

For *turbulent flow* ($N_{Re} > 10,000$), with $L/D > 60$, and $0.67 < N_{Pr} < 120$, calculate N_{Nu} with the Sieder-Tate correlation [2]:

$$N_{Nu} = 0.023 N_{Re}^{0.8} N_{Pr}^{0.33} \phi \qquad (11)$$

It is convenient to evaluate all fluid properties at the bulk temperature and apply a correction factor ϕ to account for the effects of property variations with respect to temperature. A common way of doing this is to lump the variant effects into a ratio of the dominant property evaluated at the surface temperature to that property evaluated at the bulk temperature.

For liquids, the dominant property is viscosity:

$$\phi = (\mu_b/\mu_w)^n \qquad (12)$$

For both heating and cooling, $n = 0.14$, per experimental results [2]. Values of the correction factor n have also been derived analytically [3].

For gases, the key variable is absolute temperature:

$$\phi = (T_b/T_w)^n \qquad (13)$$

For heating, $n = 0.5$, $T_b/T_w < 1$. For cooling, $n = 0$, $T_b/T_w > 1$. These values of n are based on gases for which temperature-dependent properties are similar to those of air between 310–1,920 K [3].

For *laminar flow* ($N_{Re} < 2,100$), data are frequently correlated via three dimensionless numbers: N_{Nu}, based on arithmetic or logarithmic mean temperature difference; N_{Gz}, for the contribution due to forced convec-

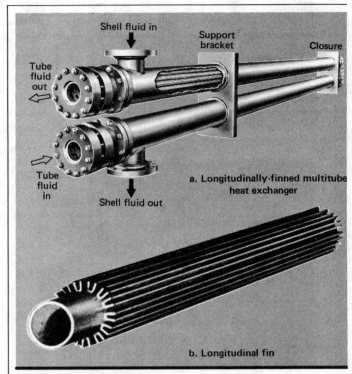

a. Longitudinally-finned multitube heat exchanger

b. Longitudinal fin

Multitube exchanger with longitudinally finned tubes   **Fig. 2**

tion; and N_{Gr}, for the contribution due to natural convection. At laminar flow, inside-tube heat transfer chiefly involves moderately to highly viscous liquids.

The Sieder-Tate correlation based on constant wall temperature can be used to calculate N_{Nu} for small tubes (I.D. up to 76 mm) and LMTDs (up to 315 K), and $N_{Gz} > 100$ [2,4]:

$$N_{Nu} = 1.86[(N_{Re})(N_{Pr})(D/L)]^{0.33} \phi \qquad (14)$$

Eq. (15) is a more general correlation, covering all diameters and ΔTs:

$$N_{Nu} = 1.86[(N_{Re})(N_{Pr})(D/L)]^{0.33} \phi + 0.87[1 + 0.015(N_{Gr})^{0.33}] \qquad (15)$$

For $N_{Gz} < 100$, use Hausen's correlation, developed on the basis of constant wall temperature [5]:

$$N_{Nu} = 3.66 + \left\{ \frac{0.085[(N_{Re})(N_{Pr})(D/L)(A_{NF}/D^2)]}{1 + 0.047[(N_{Re})(N_{Pr})(D/L)(A_{NF}/D^2)]^{0.67}} \right\} \phi \qquad (16)$$

With liquids, ϕ is obtained as for turbulent flow, via Eq. (12), for both heating and cooling [2]. With gases, the variation of fluid properties has only minor influence on inside-tube laminar-flow convective heat transfer, according to several investigations. For the temperature-ratio range of 0.5–2, $n = 0$ ($\phi = 1$; T_w/T_b for heating, T_b/T_w for cooling) [6].

For *transition flow* ($2,100 < N_{Re} < 10,000$), heat transfer is unstable and difficult to define. Graphical methods relating laminar and turbulent correlations have been used. Hausen's equation is recommended [7]:

$$N_{Nu} = 0.116(N_{Re}^{0.67} - 125) N_{Pr}^{0.33} [1 + (D/L)^{0.67}] \phi \qquad (17)$$

Typical dimensions of hairpin heat exchangers	Table I
Parameter	**Dimension range**
Inside dia. of shell (outer pipe)	50-400 mm (2-16 in.)
Nominal length of section	1.5-12.2 m (5-40 ft)
Outside dia. of tube (inner pipe)	19-100 mm (0.75-4 in.)
Number of tubes	1-208
Number of fins/tube	0-72
Fin height	5.3-25 mm (0.21-1 in.)
Fin thickness	0.5-0.9 mm (0.02-0.035 in.)
Surface-area/section	0.25-150 m2 (3-1,600 ft2)

The correction factor, ϕ, is determined via Eq. (12).

Correlations for transition flow are discussed by Afgan and Schlunder [8].

Fin performance

Fins present a special heat-transfer problem because ΔT cannot be specified independently. It is determined via thermal properties, fin geometry and heat-transfer coefficient, with the last being itself dependent on ΔT. Therefore, analyzing fins is difficult without resorting to simplified theory. Fin performance is gauged by fin efficiency, η_f, and fin effectiveness, ϵ_f.

Fin efficiency, η_f, is defined as the actual heat transferred by the fin divided by the heat that ideally would be transferred if the entire fin were at the base temperature. Based on simplifying assumptions, Eq. (18) gives the efficiency of rectangular continuous longitudinal fins [9]:

$$\eta_f = \tanh{(M Y)}/M Y \qquad (18)$$

Here, $M = (h_f P_f/k_f a_f)^{0.5}$; $a_f = X L$; and $P_f = 2 (L + X) \simeq 2L$. Substituting these values of a_f and P_f yields $M = (2 h_f/k_f X)^{0.5}$.

Fin efficiency is inversely proportional to the finside film coefficient, h_f; for this reason, it is always advantageous to place the fluid having the lower heat-transfer coefficient or the higher fouling resistance on the finside of the exchanger. Eq. (18) shows that fin efficiency decreases as fin height, Y, increases. Although efficiency provides some insight into how well a fin is contrived, it is not possible to design fins for a particular value of η_f.

Fin effectiveness, ϵ_f, is defined as the heat transferred through the fin, divided by the heat transferred through the same base surface having no fins. Effectiveness can be related to efficiency via: $\epsilon_f = \eta_f$ (surface area of fin/cross-sectional area of fin). This indicates that increasing the fin height boosts fin effectiveness. Normally, the design objective is to achieve as high a fin effectiveness as possible. However, as noted previously, increasing the fin height adversely affects η_f, and thus, ϵ_f. Values of η_f and ϵ_f are useful in characterizing fins. Optimum fin performance depends on a tradeoff between the two.

Fin efficiency applies only to the finned portion of the total heat-transfer surface; the bare portion is regarded as having an efficiency of unity. Thus, efficiency of the total heat-transfer surface is expressed as a *weighted fin efficiency*:

$$(\eta_f)_{wtd} = \eta_f(A_f/A_t) + A_o/A_t \qquad (19)$$

Here, $A_o/A_t = (1 - A_f/A_t)$.

Sometimes, it is necessary to know the *tip temperature* of the fin. The relationship between the fin-base and fin-tip temperatures is expressed by a tip temperature factor, E_T, defined as the ratio of the temperature driving force at the tip of the fin to that at its base [10]:

$$E_T = \frac{(T_b)_f - T_T}{(T_b)_f - T_w} = \frac{1}{\cosh{(M Y)}} \qquad (20)$$

Here, $(T_b)_f$ is the bulk temperature of the fluid on the finside, and T_w is the wall temperature at the bare surface (i.e., the fin base).

Temperature variation across the fin can sometimes be used to advantage in heating and cooling heat-sensitive fluids. In heating oil or asphalt with steam, for example, the average metal temperature on the finside is lower than at the bare surface. Therefore, heating takes place at a relatively low temperature, which reduces the tendency of the oil or asphalt to coke. If the fluid to be cooled is on the finside, cooling takes place at the higher temperature, and this helps prevent the solidification of viscous liquids at the tube wall.

Overall heat-transfer coefficient

Based on total external surface, the overall heat-transfer coefficient is given by:

$$U = 1/(R_s + R_t + R_w) \qquad (21)$$

The shellside (finside) thermal resistance is:

$$R_s = [(1/h_f) + r_f]/(\eta_f)_{wtd} \qquad (22)$$

The tubeside thermal resistance based on the inside tube surface is:

$$R_i = (1/h_i) + r_i \qquad (23)$$

The foregoing resistance is referred to the total external surface as:

$$R_t = R_i(A_t/A_i) = [(1/h_i) + r_i](A_t/A_i) \qquad (24)$$

The tube wall resistance, R_w, equals the wall thickness divided by the thermal conductivity of the tube material; it is calculated only for bare tubes. In the case of finned tubes, the wall resistance is built into the fin efficiency, because the latter is based on uniform heat flux at the inside tube surface [9].

Tube wall temperature

As was explained previously, the tube wall temperature must be known to evaluate fluid properties at the surface temperature in order to calculate ϕ, the property-ratio correction factor. In the case of viscous liquid and laminar flow, the values of heat-transfer coefficients and pressure drops are greatly influenced by the tube wall temperature. Turbulent flow is also affected, but to a lesser extent.

For the purposes of this article, fluids of $\mu_b > 1$ mPa (1 cP) are defined as viscous; for those of $\mu_b < 1$ mPa, μ_b/μ_w can be assumed to be unity and the effect of ϕ can be ignored.

The tube wall temperature depends on the bulk temperature of the two fluids and their film and fouling resistances. In calculating the wall temperature of bare tubes, the temperature difference across the metal wall is neglected, and the entire tube is considered to be at the temperature of the outside surface of the wall.

When the cold fluid is on the shellside, the wall temperature is given by:

$$\frac{T_w - (T_b)_C}{(T_b)_H - (T_b)_C} = \frac{R_s}{R_s + R_t} \qquad (25)$$

When the cold fluid is on the tubeside, calculate the wall temperature via:

$$\frac{T_w - (T_b)_C}{(T_b)_H - (T_b)_C} = \frac{R_t}{R_s + R_t} \qquad (26)$$

In Eq. (25) and (26), the caloric temperatures of the hot and cold fluids may be used, instead of their bulk temperatures [4]. However, doing this involves additional calculations and does not significantly improve final results.

Tube wall temperature is computed via iteration. First, a wall temperature is assumed and the heat-transfer coefficients calculated. With these, the tube wall temperature is calculated by means of Eq. (25) or (26). This temperature is used to recalculate the heat-transfer coefficients, with which another temperature is computed. The process is repeated until a fixed value for wall temperature is attained. This is shown in Fig. 4 by a converging loop for tube wall temperature.

For finned tubes, no simple method is available for computing mean wall temperature for the combination of fins and base surfaces. A simplified approach is suggested for estimating the mean temperature of a finned tube as the arithmetic average of the fin-base (bare tube surface) and fin-tip temperatures [given by Eq. (20), (25) and (26)]:

$$(T_w)_f = (T_w + T_T)/2 \qquad (27)$$

Eq. (27) will yield only approximate results, which may differ from those of more-precise methods, such as finite-difference schemes and the method in Ref. 4. Discrepancies will be greater for high-fin tubes.

LMTD and true temperature difference

When the heat-transfer curve is linear with respect to temperature, LMTD is calculated conventionally. When the curve is not linear, it can be divided into zones that are approximately linear, and a weighted LMTD calculated via:

$$(\text{LMTD})_{wtd} = Q \bigg/ \sum_{i=1}^{i=n} [Q_i/(\text{LMTD})_i] \qquad (28)$$

Pure countercurrent flow offers the maximum thermal potential for heat transfer. When a large number of hairpin exchangers are required in a single service, it may not always be possible to connect both the shellsides and tubesides in series for pure countercurrent flow. An example would be when a large quantity of one fluid is to undergo a minor temperature change and a small quantity of the other a major change.

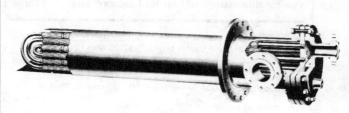

a. Tank suction heater

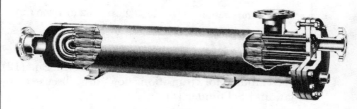

b. Inline heater

Suction and inline heaters are often used to heat viscous fluids   Fig. 3

It may not be possible to circulate the large flow through the required number of hairpins connected in series, with the available pressure drop. In such a circumstance, the large flow may be manifolded into a number of parallel streams, and the small flow manifolded in series, in a series-parallel arrangement. The true overall mean temperature difference for such an arrangement is given by [9]:

$$(\Delta T)_{s\text{-}p} = S(T_{H_{in}} - T_{C_{in}}) \qquad (29)$$

The dimensionless parameter S is defined as:

$$S = [W(C_p)_H(T_{H_{in}} - T_{H_{out}})]/UA(T_{H_{in}} - T_{C_{in}}) \qquad (30)$$

The value of S depends on the number of hot and cold streams and their series-parallel arrangement:

For 1-series hot fluid and n_1-parallel cold streams:

$$S = \frac{(1 - P_1)}{\left(\dfrac{n_1 R_1}{R_1 - 1}\right)\ln\left[\left(\dfrac{R_1 - 1}{R_1}\right)\left(\dfrac{1}{P_1}\right)^{1/n_1} + \dfrac{1}{R_1}\right]} \qquad (31)$$

Here, $P_1 = (T_{H_{out}} - T_{C_{in}})/(T_{H_{in}} - T_{C_{in}})$, and $R_1 = (T_{H_{in}} - T_{H_{out}})/n_1(T_{C_{out}} - T_{C_{in}})$.

For 1-series cold stream and n_2-parallel hot streams:

$$S = \frac{(1 - P_2)}{\left(\dfrac{n_2}{1 - R_2}\right)\ln\left[(1 - R_2)\left(\dfrac{1}{P_2}\right)^{1/n_2} + R_2\right]} \qquad (32)$$

Here, $P_2 = (T_{H_{in}} - T_{C_{out}})/(T_{H_{in}} - T_{C_{in}})$, and $R_2 = n_2(T_{H_{in}} - T_{H_{out}})/(T_{C_{out}} - T_{C_{in}})$.

The F correction factor must be applied to LMTD in the case of the longitudinal-flow bundle-type exchanger because the flow is not purely countercurrent. It is calculated using curves for single-shell pass, and for two, or

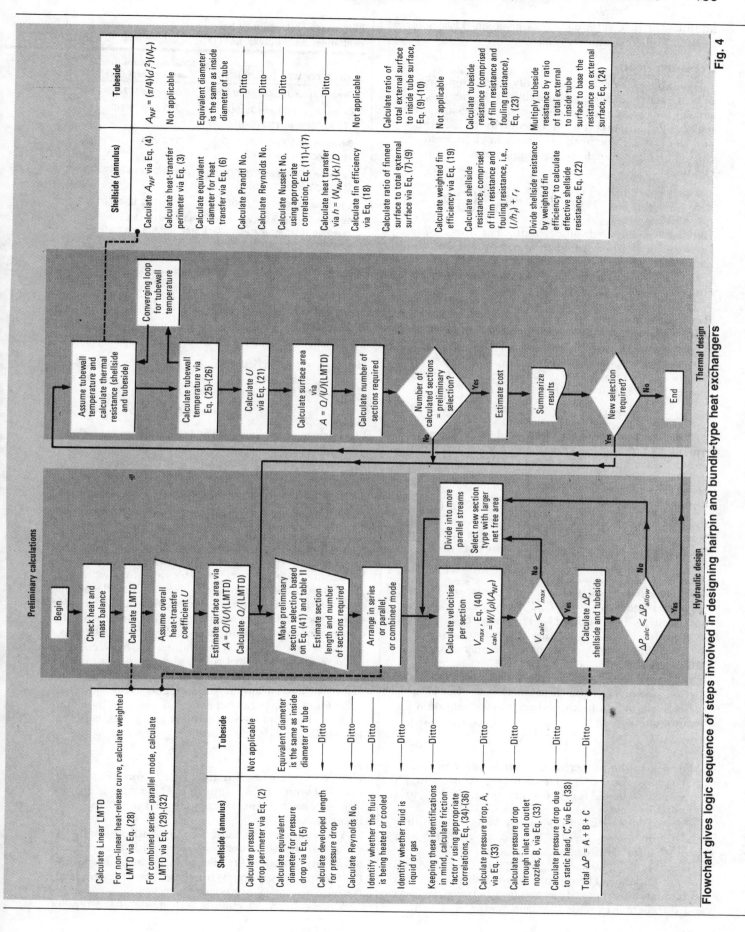

Flowchart gives logic sequence of steps involved in designing hairpin and bundle-type heat exchangers

Fig. 4

multiples of two, tube passes, as given in TEMA for shell-and-tube exchangers [11].

Pressure drop

Available pressure drop is usually a governing factor in heat-exchanger design. Normally, the higher it is, the smaller the exchanger size and cost. Total pressure drop includes fluid friction (the largest contributor), entrance and exit losses, static head and moment-change effects.

Frictional pressure drop is calculated via the Fanning equation:

$$(\Delta P)_F = 4f\left(\frac{\rho v^2}{2}\right)\left(\frac{L_h}{D}\right)N = 4f\left(\frac{G^2}{2\rho}\right)\left(\frac{L_h}{D}\right)N \quad (33)$$

Here, L_h is the hydraulic length. For the *shellside*, $L_h = 2L$ for hairpin exchangers, and $L_h = L$ for bundle-type exchangers. For the *tubeside*, $L_h = 2L + L_U$ for both hairpin and bundle exchangers. Here, L_U is the equivalent straight length of the U-bend.

The friction factor, f, in Eq. (33) depends on the Reynolds number.

In *turbulent flow* ($N_{Re} > 3,000$), f is strongly related to tube roughness:

With smooth pipe and $3,000 < N_{Re} < 3,000,000$, the empirical equation of Koo (with isothermal representation of heating and cooling) is used [12,13]:

$$f = 0.0014 + (0.125/N_{Re}^{0.32})\psi \quad (34)$$

With clean commercial iron and steel pipes, the equation of Wilson et al. (with isothermal representation of heating and cooling) is used [14]:

$$f = 0.0035 + (0.264/N_{Re}^{0.42})\psi \quad (35)$$

For *liquids,* the viscosity correction factor in Eq. (34) and (35) is taken to be: $\psi = (\mu_b/\mu_w)^m$, with $m = 0.25$ for both heating and cooling [15]. Values of m are also derived analytically and given as a function of N_{Pr} by Kays [3]. For non-viscous liquids, $\mu_b < 1$ mPa, and the effect of ψ can be ignored.

For *gases,* data on friction coefficients are limited. Kays recommends: $\psi = (T_b/T_w)^m$, with $m = 0.1$ for both heating and cooling [3]. This value of m is based on a gas for which temperature dependence on properties is similar to that of air in the temperature range 310–1,920 K.

With fully developed *laminar flow* ($N_{Re} < 2,000$), the friction factor can be calculated via the Hagen-Poiseuille equation (with isothermal representation for heating and cooling):

$$f = (16/N_{Re})\psi \quad (36)$$

For *liquids,* the properties correction factor is used as analytically derived by Diessler [16]: $\psi = (\mu_b/\mu_w)^m$, with $m = -0.58$ for heating $(\mu_b/\mu_w > 1)$, and $m = -0.5$ $(\mu_b/\mu_w < 1)$.

For *gases,* the value of ψ given by Worsoe-Schmidt is used in the range 277–1,666 K: $\psi = (T_b/T_w)^m$, with $m = -1.0$ for both heating and cooling [17].

Transition flow ($2,000 < N_{Re} < 3,000$) is unstable and difficult to define. The pressure drop in this regime can be estimated by linear interpolation between laminar and turbulent Reynolds numbers.

Preliminary selection of hairpin exchanger based on Q/LMTD	Table II

Q/LMTD, W/K [Btu/(h)(°F)]	Shell dia. of multitube section, mm (in.)*
Greater than 79,000 [150,000]	305-406 (12-16)
Between 53,000 and 79,000 [100,000 and 150,000]	203-406 (8-16)
Between 26,500 and 53,000 [50,000 and 100,000]	152-254 (6-10)
Between 10,500 and 26,500 [20,000 and 50,000]	102-203 (4-8)†
Less than 10,500 [20,000]	51-102 (2-4)†

*Length and number of sections required can be calculated based on total estimated surface area and known surface area per unit length of section

†Also consider double-pipe (single tube) unit.

Nozzle pressure drop is estimated by simply calculating kinetic energy:

$$(\Delta P)_n = K(\rho v_n^2/2) = K(G_n^2/2\rho) \quad (37)$$

Here, K is the velocity head, 1.0 at inlet and 0.5 at outlet.

Static head is calculated as:

$$(\Delta P)_S = \rho H \quad (38)$$

For horizontal shells, H is the total difference in elevation between the inlet and outlet nozzles.

Momentum-change pressure drop is calculated as:

$$(\Delta P)_m = G^2(1/\rho_{out} - 1/\rho_{in}) \quad (39)$$

With single-phase liquids and gases, the momentum-change pressure drop can be ignored.

Preliminary selection criteria

Maximum velocities in heat exchangers are limited by the effects of erosion, pressure drop and vibration. These velocities are sometimes used as a guide during preliminary calculations for designing an exchanger within specified limits of pressure drop. Because of the effects of erosion, pressure drop and vibration, the maximum velocity of water is limited to between 2.4–3.0 m/s for carbon steel tubes, and up to 4–5 m/s for some alloy materials. Copper alloy tubes permit lower velocities and titanium tubes higher velocities than these limits. A generalized equation based on these limits is used by some manufacturers to set maximum velocities for liquids and gases:

$$v_{max} = C/\rho^{1/2} \quad (40)$$

Here, $C = 122$ with SI units (v in m/s and ρ in kg/m³); $C = 100$ with English units (v in ft/s and ρ in lb/ft³), although some manufacturers assume this C to be 1.6 times higher.

A preliminary evaluation begins with assuming (1) an overall heat-transfer coefficient, with U dependent on fluid properties; (2) allowable pressure drops; and (3) fouling resistances. With these, the required heat-transfer surface is estimated.

A preliminary selection is next made of the type of

Cost breakdown of hairpin heat exchangers				Table III
Component	Percent of total exchanger cost			
	Double-pipe		Multitube	
	Bare	Finned	Bare	Finned
Tubes	15-25	20-40	30-40	35-60
Shell	40-55	40-50	35-50	25-40
Closure	20-35	20-35	15-30	15-30

hairpin section, depending on the ratio of duty/LMTD (also known as the UA product):

$$Q/(\text{LMTD}) = UA \cong 53,000 - 106,000 \text{ W/K} \quad (41)$$

This ratio is used as a guide for limiting the size of the exchanger. However, other factors influence this limiting. Criteria for the preliminary selection of the size of hairpin exchangers on the basis of this ratio are summarized in Table II.

Having selected the section type and determined the necessary heat-transfer surface, the number of sections required for the chosen length can be calculated and arranged in series or in parallel, or in a combination—depending on flowrates, available pressure drops, and other considerations.

In the selection of section type, the choice of finned tubes over bare tubes depends on several factors. For the optimum design, the shellside and tubeside effective film-coefficients referred to the same surfaces should be as high as possible, and be approximately of the same magnitude, within specified limits of pressure drops and fouling resistances. However, this is not always possible; the fluid film coefficients may differ significantly because of one or more of the following reasons:

1. The physical properties (density, viscosity, specific heat and thermal conductivity) of the two fluids may differ greatly (as when one is a viscous liquid and the other is water).

2. The physical properties of the fluids may be similar, but the allowable pressure drops may differ greatly (i.e., one gas is at high and the other at low pressure).

3. Flowrates may differ greatly.

4. The fouling resistance of one of the fluids is high.

When the foregoing limitations are applicable, finned tubes are generally preferred. The fluid having the lower heat-transfer coefficient is placed on the fin side to keep the UA product of Eq. (41) about equal by combining a high A with a low U.

Evaluation involves the following steps:

1. Calculate the velocities per section.

2. If the calculated velocity is larger than the maximum velocity [Eq. (40)], divide the flow into parallel streams, or select a section type having a greater net free area.

3. When $v_{calc} \leq v_{max}$, calculate the pressure drop. If the calculated ΔP is greater than the allowable ΔP, further split the flow into parallel streams, or select a section type having a larger net free area.

4. When $\Delta P_{calc} \leq \Delta P_{max}$, proceed with the thermal calculations to determine the heat-transfer coefficients and the required surface area.

5. Depending on the percent over- or underdesign, and on the difference between the calculated and allowable pressure drops, vary the exchanger length, number of sections and other parameters, to select a section type that will give the optimum design.

Costs of hairpin heat exchangers

For cost estimating purposes, hairpin exchangers can be broken down into three major components: shell, tubes and closure. The cost of the shell and tubes increases with longer exchanger length, whereas that of the closure remains fixed. This makes longer exchangers more economical.

In Table III, cost breakdowns in percent of total exchanger cost are given for carbon steel at design pressures to 4,000 kPa (600 psi) and temperatures to 600 K, and for shell diameters to 200 mm in nominal lengths of 1.5-9.1 m. These percentages may vary with manufacturers due to differences in design patents and mechanical standards.

References

1. De Lorenzo, B., and Anderson, E. D., Heat Transfer and Pressure Drop of Liquids in Double Pipe Fintube Exchangers, *Trans. ASME,* Vol. 67, 1945, p. 697.
2. Sieder, E. N., and Tate, G. E., Heat Transfer and Pressure Drop of Liquids in Tubes, *Ind. and Eng. Chem.,* Vol. 28, No. 12, 1936, pp. 1429–1434.
3. Kays, W. M., "Convective Heat and Mass Transfer," McGraw-Hill, Inc., New York, 1966.
4. Kern, D. Q., and Kraus, A. D., "Extended Surfaces Heat Transfer," McGraw-Hill, Inc., New York, 1972.
5. Perry, R. H., and Chilton, C. H., "Chemical Engineers' Handbook," 5th ed., McGraw-Hill, Inc., New York, 1973.
6. Kays, W. M., and Nicoll, W. B., *Trans. ASME, J. of Heat Transfer,* Vol. 85, 1963, pp. 329–338.
7. Hausen, H. Z., *V. D. K. Beiheft Verfahrenstechnik,* Vol. 4, 1943, pp. 91–98.
8. Afgan, N., and Schlunder, E. U., "Heat Exchangers: Design and Theory Source Book," McGraw-Hill, Inc., New York, 1974.
9. Kern, D. Q., "Process Heat Transfer," McGraw-Hill, Inc., New York, 1950.
10. Lienhard, J. H., "A Heat Transfer Textbook," Prentice-Hall, Inc., Englewood Cliffs, N.J., 1981.
11. "Standards of Tubular Exchanger Manufacturers Assn.," TEMA, Inc., New York, 1978.
12. Koo, E. C., Sc.D. thesis in chemical engineering., Mass. Inst. of Technology, 1932.
13. Drew, T. B., Koo, E. C., and McAdams, W. H., The Friction Factor for Clean Round Pipe, *Trans. AIChE,* Vol. 28, 1932, pp. 56–72.
14. Wilson, R. E., McAdams, W. H., and Seltzer, M., The Flow of Fluids Through Commercial Pipelines, *Ind. and Eng. Chem.,* Vol. 14, 1922, p. 105.
15. Rohsenow, W. M., and Hartnett, J. P., "Handbook of Heat Transfer," McGraw-Hill, Inc., New York, 1973.
16. Diessler, R. C., Natl. Advisory Committee for Aeronautics, TN 2410, Washington, D.C., July 1941.
17. Worsoe-Schmidt, P. M., Technical Report No. 247-8, Thermosciences Div., Dept. of Mech. Eng., Stanford University, Stanford, Calif., Nov. 1964.

The author

G. P. Purohit, a consultant in the design and analysis of energy systems (3801 Parkview Lane, No. 6C, Irvine, CA 92715; telephone 714-857-4762), has until recently been a heat-transfer engineer with Fluor Corp. Previously, he had been a systems development engineer with the Jet Propulsion Laboratory of the California Institute of Technology. He holds an M.S. degree in engineering from the University of California at Los Angeles (UCLA) and a B.E. degree in mechanical engineering from Birla Vishvakarma Mahavidyalaya of Sardar Patel University, Vallabh Vidyanagar, India.

Predict storage-tank heat transfer precisely

Use this procedure to determine the rate of heat transfer from a vertical storage tank when shortcut methods are inadequate.

Jimmy D. Kumana and Samir P. Kothari, Henningson, Durham and Richardson, Inc.

☐ Heating or cooling storage tanks can be a major energy expense at plants and tankfarms. Though many procedures for calculating such heat-transfer requirements have been published [1,3,5,7,8,10], the simplifying assumptions that they use can lead to significant errors in computed heat-transfer rates. This is of concern because efficient sizing of tanks, insulation, heaters and coolers depends on accurate estimates of heat transfer to and from the various tank surfaces. And the ultimate value of being accurate increases as energy costs continue to rise.

The procedure presented here determines the heat transfer to or from a vertical-cylindrical storage tank seated on the ground—like the one in Fig. 1. It includes the effects of tank configuration, liquid level, ambient temperature and wind speed, as well as temperature variations within the tank and between air and ground. A partially worked example shows how to use the technique, and how to do the calculations on a computer.

The theory

Storage tanks come in many different shapes and sizes. Horizontal-cylindrical and spherical tanks are used for storage of liquids under pressure; atmospheric tanks tend to be vertical-cylindrical, with flat bottoms and conical roofs as shown in Fig. 1. The example presented here is for the latter configuration, but the procedure applies to any tank for which reliable heat-transfer correlations are available.

For the sake of simplicity, we assume that the tank contents are warmer than the ambient air, and that we are concerned with heat loss from the tank rather than heat gain. But the method may, of course, be applied to either case.

Consider, then, the categories of surfaces from which heat may be transferred across the tank boundaries: wet or dry sidewalls, tank bottom, and roof. In the context used here, "wet" refers to the portion of the wall submerged under the liquid surface, whereas "dry" refers to the portion of the wall in the vapor space, above the liquid surface.

In general, the heating coils would be located near the bottom of the tank, in the form of flat "pancakes." Therefore, the temperature of the air (or vapor) space

above the liquid level may be expected to be lower than the liquid itself. Experience has shown that the average bulk temperatures of the liquid and vapor space may be significantly (i.e., more than 5°F) different, and they are treated accordingly in our procedure. Use of different liquid and vapor temperatures is an important departure from the traditional approach, which assumes the same value for both.

Our basic approach is to develop equations for calculating the heat loss from each of the four categories of surfaces, and then add the individual heat losses to get the total heat loss. Thus:

For dry sidewall $\quad q_d = U_d A_d (T_V - T_A) \quad$ (1)

For wet sidewall $\quad q_w = U_w A_w (T_L - T_A) \quad$ (2)

For tank bottom $\quad q_b = U_b A_b (T_L - T_G) \quad$ (3)

For tank roof $\quad q_r = U_r A_r (T_V - T_A) \quad$ (4)

Total $\quad Q = q_d + q_w + q_b + q_r \quad$ (5)

When using these equations in design or rating problems, we either assume the various temperatures for typ-

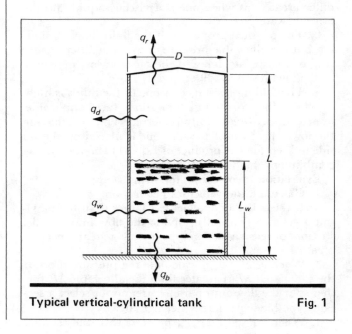

Typical vertical-cylindrical tank   **Fig. 1**

Originally published March 22, 1982

Individual film heat-transfer coefficients — Table I

Type/surface	Dry wall	Wet wall	Roof	Bottom
Inside	h_{Vw}	h_{Lw}	h_{Vr}	h_{Lb}
Wall conduction	$\left(\dfrac{t_M}{k_M}+\dfrac{t_I}{k_I}\right)^{-1}$	$\left(\dfrac{t_M}{k_M}+\dfrac{t_I}{k_I}\right)^{-1}$	$\left(\dfrac{t_M}{k_M}\right)^{-1}$	$\left(\dfrac{t_M}{k_M}\right)^{-1}$
Outside	$W_f h'_{Aw}+h_{Rd}$	$W_f h'_{Aw}+h_{Rw}$	$W_f h'_{Ar}+h_{Rr}$	h_G
Fouling	h_{Fd}	h_{Fw}	h_{Fr}	h_{Fb}

Note: Tank roof and bottom are uninsulated.

ical conditions or determine them by measurement. The area values are also easy to obtain:

$$A_d = \pi D (L - L_w) \quad (6)$$
$$A_w = \pi D L_w \quad (7)$$
$$A_b = \pi D^2/4 \quad (8)$$
$$A_r = (\pi D/2)(D^2/4 + d^2)^{1/2} \quad (9)$$

The complications arise when we try to estimate the overall heat-transfer coefficients U_d, U_w, U_b and U_r, for the four surfaces of the tank. For the tank geometry chosen, these can fortunately be calculated from the individual film heat-transfer coefficients in the conventional manner, using published correlations.

The overall coefficients

Table I shows the component coefficients for each surface. The overall heat-transfer coefficient for the dry sidewall of the tank (U_d) is calculated as the sum of the resistances of vapor film, fouling, metal wall, insulation (if any), and outside air (convection plus radiation).

The outside-air heat-transfer coefficient (h_{Aw}) is a function of wind velocity as well as temperature gradi-

ent. Data on the effect of wind velocity and ΔT have been presented by Stuhlbarg [10] and Boyen [2]. With a little bit of manipulation, their data were replotted, yielding the "wind enhancement factor" (W_f) in Fig. 2. By definition:

$$W_f = h_{Aw}/h'_{Aw} = h_{Ar}/h'_{Ar} \quad (10)$$

Therefore, once the outside-air coefficient for *still* air (h'_{Aw}) is known, the overall dry-sidewall coefficient at various wind velocities can be computed as:

$$1/U_d = 1/h_{Vw} + t_M/k_M + t_I/k_I + \\ 1/(W_f h'_{Aw} + h_{Rd}) + 1/h_{Fd} \quad (11)$$

Similarly, the overall coefficients for the wet sidewall, bottom and roof surfaces are:

$$1/U_w = 1/h_{Lw} + t_M/k_M + t_I/k_I + \\ 1/(W_f h'_{Aw} + h_{Rw}) + 1/h_{Fw} \quad (12)$$
$$1/U_b = 1/h_{Lb} + t_M/k_M + 1/h_G + 1/h_{Fb} \quad (13)$$
$$1/U_r = 1/h_{Vr} + t_M/k_M + \\ 1/(W_f h'_{Ar} + h_{Rr}) + 1/h_{Fd} \quad (14)$$

Eq. 13 and 14 assume that the roof and bottom are not insulated, which is generally the case in temperate climates. We shall now review correlations for the individual heat-transfer coefficients needed to obtain the overall coefficients.

Individual film heat-transfer coefficients

The film heat-transfer coefficients may be divided into four categories: convection from vertical walls, convection from horizontal surfaces, pure conduction, and radiative heat transfer. Within each category, correlations are presented for several flow regimes.

Vertical-wall film coefficients. These apply to the inside wall (wet or dry) and the outside wall (still air). For vertical plates and cylinders, Kato et al. [6] recommend the following for liquids and vapors:

$$N_{Nu} = 0.138 N_{Gr}^{0.36}(N_{Pr}^{0.175} - 0.55) \quad (15)$$

where $0.1 < N_{Pr} < 40$ and $N_{Gr} > 10^9$.

For isothermal vertical plates, Ede [4] reported the following for liquids:

$$N_{Nu} = 0.495(N_{Gr}N_{Pr})^{0.25} \quad (16)$$

where $N_{Pr} > 100$ and $10^4 < (N_{Gr}N_{Pr}) < 10^9$, and for gases:

$$N_{Nu} = 0.0295 N_{Gr}^{0.40} N_{Pr}^{0.47} (1 + 0.5 N_{Pr}^{0.67})^{-0.40} \quad (17)$$

where $N_{Pr} \approx 5$ and $(N_{Gr}N_{Pr}) > 10^9$.

For vertical plates taller than 3 ft, Stuhlbarg [10] recommends:

$$h = 0.45 k L^{-0.75} (N_{Gr} N_{Pr})^{0.25} \quad (18)$$

where $10^4 < (N_{Gr} N_{Pr}) < 10^9$.

Horizontal-surface heat-transfer coefficients. These coefficients apply to the roof and inside-bottom surfaces of the tank. The bottom is assumed to be flat. For surfaces facing up [8]:

$$N_{Nu} = 0.14 (N_{Gr} N_{Pr})^{0.33} \quad (19)$$

For surfaces facing down:

$$N_{Nu} = 0.27 (N_{Gr} N_{Pr})^{0.25} \quad (20)$$

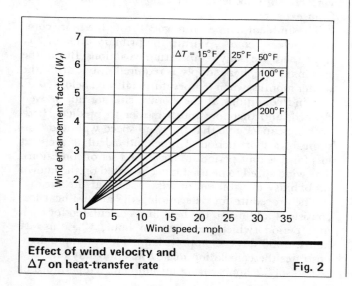

Effect of wind velocity and ΔT on heat-transfer rate — Fig. 2

Nomenclature

A	Area of heat-transfer surface, ft^2; A_b for bottom, A_d for dry wall, A_w for wet wall, A_r for roof
c_p	Specific heat at constant pressure, Btu/lb-°F
D	Diameter of tank, ft
d	Height of conical roof at center, ft
g	Acceleration due to gravity, 4.17×10^8 ft/h^2
h	Individual film coefficient of heat transfer, Btu/ft^{2}h-°F; h_{Aw} for air outside the walls, h_{Ar} for air above the roof, h'_{Aw} and h'_{Ar} for still air, h_{Lw} for liquid between the walls, h_{Lb} for liquid near the bottom, h_{Vw} for vapor near the walls, h_{Vr} for vapor near the roof
h_F	Fouling coefficient, Btu/ft^{2}h-°F; h_{Fw} for liquid at the walls, h_{Fb} for liquid at the bottom, h_{Fv} for vapor at the walls or the roof
h_G	Heat-transfer coefficient for ground, Btu/ft^{2}h-°F
h_I	Heat-transfer coefficient for insulation, Btu/ft^{2}h-°F
h_M	Heat-transfer coefficient for metal, Btu/ft^{2}h-°F
h_R	Heat-transfer coefficient for radiation, Btu/ft^{2}h-°F; h_{Rb} for bottom, h_{Rd} for dry wall, h_{Rw} for wet wall, h_{Rr} for roof
k	Thermal conductivity, Btu/ft-h-°F; k_G for ground, k_I for insulation, k_M for metal wall
L	Total length for heat-transfer surface, ft
L_w	Total length for wetted surface, ft
N_{Gr}	Grashof number, $L^3\rho^2 g\beta\Delta T/\mu^2$
N_{Nu}	Nusselt number, hD/k or hL/k
N_{Pr}	Prandtl number, $c_p\mu/k$
Q	Rate of heat transfer, Btu/h
q	Individual rate of heat transfer, Btu/h; q_b for bottom, q_d for dry wall, q_w for wet wall, q_r for roof
T	Temperature, °F; T_A for ambient air, T_L for bulk liquid, T_V for vapor, T_G for ground, T_w for inside wall, T_{ws} for outside wall
ΔT	Temperature difference, °F
t	Surface thickness, ft; t_I for insulation, t_M for metal
U	Overall heat-transfer coefficient, Btu/ft^{2}h-°F; U_b for bottom, U_d for dry wall, U_w for wet wall, U_r for roof
W_f	Wind enhancement factor
β	Volumetric coefficient for thermal expansion, °F^{-1}
μ	Viscosity of fluid, lb/ft-h
ρ	Density of fluid, lb/ft^3
ε	Emissivity

Both equations apply in the range $2 \times 10^7 < (N_{Gr}N_{Pr}) < 3 \times 10^{10}$.

Equivalent coefficients for conductive heat transfer. The wall and insulation coefficients are derived from the thermal conductivities:

$$h_M = k_M/t_M \qquad (21)$$

$$h_I = k_I/t_I \qquad (22)$$

The coefficient for heat transfer to and from the ground is the coefficient for heat conduction from a semi-infinite solid [9]:

$$h_G = 8\,k_G/\pi D \qquad (23)$$

Fouling coefficients. The coefficients h_{Fd}, h_{Fw} and h_{Fb} apply to the vapor and liquid at the wall, and the liquid at the bottom of the tank, respectively. These are empirical, and depend on the type of fluid and other factors such as tank cleaning. Generally, h_{Fd} is the greatest of the three, and h_{Fb} the least, indicating that the greatest fouling resistance is at the bottom of the tank.

Equivalent coefficient for radiative heat transfer. The coefficient for sidewalls and roof depends on the emissivity of these surfaces, and is given by [8]:

$$h_R = \frac{0.1713\varepsilon}{(T_{ws} - T_A)}\left[\left(\frac{T_{ws} + 460}{100}\right)^4 - \left(\frac{T_A + 460}{100}\right)^4\right] \qquad (24)$$

With these relationships, we now have the tools to calculate heat transfer to or from the tank.

Example

ABC Chemical Corp. has a single manufacturing plant in the U.S., and exports a high-viscosity specialty oil product to Europe. The oil is offloaded in Port City, and stored in a flat-bottom, conical-roof tank rented from XYZ Terminal Co. Ltd. The tank is located outdoors and rests on the ground. It is equipped with pancake-type steam-heating coils because the oil must be maintained above 50°F in order to preserve its fluidity. Other pertinent data are: tank diameter is 20 ft; tank height is 48 ft (to the edge of the roof); roof incline is $\frac{3}{4}$ in. per foot; tank sidewalls are $\frac{3}{16}$-in. carbon steel; insulation is 1½-in. fiberglass, on the sidewall only.

XYZ Terminal Co. does not have metering stations on the steam supply to individual tanks, and proposes to charge ABC Chemical for tank heating on the basis of calculated heat losses, using the conventional tables [1], and assuming a tank wall temperature of 50°F. The project engineer from ABC Chemical decided to investigate how XYZ's estimate would compare with the more elaborate one described in this article.

First, the engineer collected basic data on storage and climate. Oil shipments from the U.S. arrive at Port City approximately once a month, in 100,000-gal batches. Deliveries to local customers are made in 8,000-gal tanktrucks, three times a week on average. The typical variation in tank level over a 30-day period is known from experience.

The ambient temperature goes through a more complex cycle, of course. Within the primary cycle of 365 days, there are daily temperature variations. But in the seasonal cycle, heat supply is required only during the winter months, when temperatures fall well below 50°F.

Wind conditions at the storage site are not as well defined, and therefore much harder to predict. However, we can assume that the wind speed will hold constant for a short period of time, and calculate the heat loss for this unit period under a fixed set of conditions. The wind speed to be used must be based on the known probability distribution of wind speeds at the site.

The procedure for determining the annual heat loss consists of adding up the heat losses calculated for each unit period (which could be an hour, 12 hours, 24 hours, or 30 days, as appropriate). This example demonstrates the calculation of heat loss for only one unit period, of 12 hours, using an ambient temperature of

Data for ABC Chemical Co. example — Table II

Physical properties	Liquid	Air	Vapor*
Density, lb/ft³	4.68	0.08	0.08
Specific heat, Btu/lb-°F	0.6	0.25	0.25
Viscosity, cP	40	0.007	0.007
Thermal conductivity, Btu/ft-h-°F	0.12	0.0151	0.015
Coefficient of volumetric expansion per °F	1 × 10⁻⁶	0.002	0.002
Assumed fouling coefficients			
Dry wall	1,000 Btu/ft²h-°F		
Wet wall	800		
Roof	1,000		
Bottom	500		
Thermal conductivities			
Metal walls	10 Btu/ft-h-°F		
Insulation	0.028		
Ground	0.80		
Surface emissivity			
Wall and roof	0.9		
Temperatures			
Vapor in tank	50°F		
Liquid in tank	55		
Outside air	35		
Ground	40		

*Since the liquid has low volatility, the vapor space is assumed to be mostly air.

Heat-transfer coefficients after first iteration — Table III

Coefficient	Dry wall	Wet wall	Roof	Bottom
h_{Vw}	0.5815	—	—	—
h_{Lw}	—	1.415	—	—
h_{Vr}	—	—	0.1537	—
h_{Lb}	—	—	—	1.105
h_G	—	—	—	0.102
h'_{Ar}	—	—	0.6635	—
$h_{Ar}*$	—	—	2.057	—
h'_{Aw}	0.51	0.51	—	—
$h_{Aw}*$	1.683	1.683	—	—
h_M	640	640	640	640
h_I	0.224	0.224	—	—
h_F	1,000	800	1,000	500
h_R	0.7565	0.7594	0.7651	—
$U*$	0.1516	0.1828	0.1457	0.0933

*For 10-mph wind

Second iteration yields closer temperature estimates — Table IV

Temperature	Iteration	Dry wall	Wet wall	Roof	Bottom
T_w (inside), °F	2	46.0	52.7	35.75	53.7
	1	42.5	45	42.5	47.5
T_{ws} (outside), °F	2	35.9	36.5	35.75	—
	1	38.75	40	42.5	—

35°F, a wind velocity of 10 mph, and a liquid level of 50%. The other data required are given in Table II. Note that the liquid temperature is controlled at 55°F to provide a 5°F margin of safety.

Since the Prandtl and Grashof numbers occur repeatedly in the film heat-transfer coefficient equations, and remain relatively unchanged for all the conditions of interest, let us first calculate their values. Thus, for the liquid phase:

$$N_{Gr} = L^3\rho^2 g\beta\,\Delta T/\mu^2 = 97.5\,L^3\,\Delta T$$
$$N_{Pr} = C_p\mu/k = 484$$

Similarly, for the vapor phase, $N_{Gr} = 1.90 \times 10^7 L^3\,\Delta T$, and $N_{Pr} = 0.28$. We can now calculate the individual film heat-transfer coefficients, using the appropriate L and ΔT values in the Grashof-number equations. This is an iterative process that requires initial estimates for wall and ground temperatures, plus wall temperatures.

Coefficient for vapor at wall (h_{Vw}). As an initial approximation, assume that the wall temperature is the average of the vapor and outside-air temperatures: $T_w = (50 + 35)/2 = 42.5°F$. Then find the Grashof number:

$$N_{Gr} = 1.90 \times 10^7(L - L_w)^3(T_V - T_w)$$
$$= 1.90 \times 10^7(24)^3(7.5)$$
$$= 1.97 \times 10^{12}$$

Employing Eq. 15, find the Nusselt number and then the coefficient ($k = 0.0151$, $L = 48$ ft, $L_w = 24$ ft):

$$N_{Nu} = 0.138(N_{Gr})^{0.36}(N_{Pr}^{0.175} - 0.55) = 921.1$$
$$h_{Vw} = (921.1)(k)/(L - L_w) = 0.581 \text{ Btu/ft}^2\text{h-°F}$$

Coefficient for liquid at the wall (h_{Lw}). Here, neither N_{Pr} nor ($N_{Gr} N_{Pr}$) falls within the range of the applicable correlations (Eq. 16,18). Let us try both, again using an average for T_w.

$$T_w = (T_L + T_A)/2 = 45°F$$
$$N_{Gr} = 97.47L^3(T_L - T_w) = 1.35 \times 10^7$$

Using Eq. 16 and 18, we get two estimates for the heat-transfer coefficient ($k = 0.12$, $N_{Pr} = 484$):

$$h_{Lw} = (0.495k/L_w)(N_{Gr} N_{Pr})^{0.25} = 0.704 \text{ Btu/ft}^2\text{h-°F}$$
$$h_{Lw} = (0.45k/L_w^{0.75})(N_{Gr} N_{Pr})^{0.25}$$
$$= 1.415 \text{ Btu/ft}^2\text{h-°F}$$

To be conservative, we use the higher value: $h_{Lw} = 1.415$ Btu/ft²h-°F.

Coefficient for vapor at roof (h_{Vr}). We consider this a flat plate, with a diameter of 20 ft, and use Eq. 20, again with an average T_w of 42.5°F ($k = 0.0151$):

$$N_{Gr} = 1.9 \times 10^7 D^3(T_V - T_w) = 1.14 \times 10^{12}$$
$$h_{Vr} = (0.27k/D)(N_{Gr} N_{Pr})^{0.25} = 0.154 \text{ Btu/ft}^2\text{h-°F}$$

Coefficient for liquid at tank bottom (h_{Lb}). Assume that the ground temperature (T_G) is 5°F above ambient, and use an average of liquid and ground temperatures as a first approximation for the tank-bottom temperature:

$$T_w = (T_L + T_G)/2 = (T_L + T_A + 5)/2 = 47.5°F$$

Then, figure the Grashof number, and use Eq. 19 to get the coefficient:

$$N_{Gr} = 97.47D^3(T_L - T_w) = 5.85 \times 10^6$$
$$N_{Gr}N_{Pr} = 2.83 \times 10^9$$
$$h_{Lb} = 1.105 \text{ Btu/ft}^2\text{h-}°\text{F}$$

Coefficient for outside air at roof (h'_{Ar}). Assume $T_{ws} = T_w$ since the roof is uninsulated, and get the coefficient for still air from Eq. 19:

$$N_{Gr} = 1.9 \times 10^7D^3(T_{ws} - T_A) = 1.14 \times 10^{12}$$
$$h'_{Ar} = 0.663 \text{ Btu/ft}^2\text{h-}°\text{F}$$

Coefficient for outside air at wall (h'_{Aw}). Assume that the temperature drop across the film is one-fourth of the drop from the inside fluid to the outside air (averaged for the wet and dry walls), and use Eq. 15 to find the coefficient:

$$\Delta T = 17.5/4 = 4.375°\text{F}$$
$$N_{Gr} = 1.9 \times 10^7L^3 \Delta T = 9.19 \times 10^{12}$$
$$h'_{Aw} = 0.51 \text{ Btu/ft}^2\text{h-}°\text{F}$$

Conduction coefficients for ground, metal wall, and insulation (h_G, h_M and h_I). These are straightforward, from Eq. 21-23:

$$h_M = k_M/t_M = 640 \text{ Btu/ft}^2\text{h-}°\text{F}$$
$$h_I = k_I/t_I = 0.224 \text{ Btu/ft}^2\text{h-}°\text{F}$$
$$h_G = 8 k_G/\pi D = 0.102 \text{ Btu/ft}^2\text{h-}°\text{F}$$

Radiation coefficients for dry and wet sidewall, and roof (h_{Rd}, h_{Rw}, h_{Rr}). As for the outside-air film coefficients, assume that $T_{ws} = T_A + 0.25 \ (T_{bulk} - T_A)$, where T_{bulk} is the temperature of the liquid or vapor inside the tank, if the surface is insulated. For the uninsulated roof, assume that $T_{ws} = T_A + 0.5(T_V - T_A)$. Then $T_{ws} = 38.75°\text{F}$ for the (insulated) dry sidewall, $T_{ws} = 40°\text{F}$ for the wet sidewall, and $T_{ws} = 42.5°\text{F}$ for the roof. Using Eq. 24, find the coefficient for each of the three cases:

$$h_{Rd} = 0.757 \text{ Btu/ft}^2\text{h-}°\text{F}$$
$$h_{Rw} = 0.759 \text{ Btu/ft}^2\text{h-}°\text{F}$$
$$h_{Rr} = 0.765 \text{ Btu/ft}^2\text{h-}°\text{F}$$

Closing in on results

Table III summarizes the heat-transfer coefficients just calculated, including the corrections for wind—h'_{Aw} and h'_{Ar} are multiplied by 3.3 and 3.1, respectively, based on data for 10-mph wind in Fig. 2. Substituting these individual coefficients in Eq. 11-14, we obtain the U values listed in Table III.

What remains to be done? When we began the calculations, we assumed that the outside-wall temperatures were related to the bulk-fluid temperatures by:

$$T_w = T_A + 0.5(T_{bulk} - T_A) \text{ for uninsulated surfaces}$$
$$T_{ws} = T_A + 0.25(T_{bulk} - T_A) \text{ for insulated surfaces}$$

In order to calculate accurate coefficients for heat transfer, we must now obtain better estimates of these wall temperatures. This requires an iterative procedure that can be programmed and run on a computer.

Revised coefficients after second iteration				Table V
Coefficient	Dry wall	Wet wall	Roof	Bottom
hV_w	0.463	—	—	—
h_{Lw}	—	0.98	—	—
hV_r	—	—	0.181	—
h_{Lb}	—	—	—	0.619
h_G	—	—	—	0.102
h'_{Ar}	—	—	0.31	—
$h_{Ar}*$	—	—	0.96	—
h'_{Aw}	0.317	0.317	—	—
$h_{Aw}*$	1.047	1.047	—	—
h_M	640	640	640	640
h_I	0.224	0.224	—	—
h_F	1,000	800	1,000	500
h_R	0.7500	0.7514	0.7500	—
$U*$	0.1392	0.1655	0.1636	0.0875

*For 10-mph wind

For dry wall, the rate of heat loss is given by all three of the following:

$$q_d = U_d A_d(T_V - T_A) \quad (25)$$
$$= h_{Vw} A_d(T_V - T_w) \quad (26)$$
$$= (h_{Rd} + h_{Aw})A_d(T_{ws} - T_A) \quad (27)$$

Solving Eq. 25 and 27 for T_{ws} yields:

$$T_{ws} = (U_d/(h_{Rd} + h_{Aw}))(T_V - T_A) + T_A \quad (28)$$

Similarly, solving Eq. 25 and 26 for T_w yields:

$$T_w = T_V - (U_d/h_{Vw})(T_V - T_A) \quad (29)$$

Using the same approach, now calculate T_w and T_{ws} for the wet wall, and T_w for the roof and bottom of the tank.

To find the correct wall temperatures, use the initial estimates of U and h values in Eq. 28 and 29 (and in the parallel equations for the other surfaces) to get new T_w and T_{ws} values. Table IV shows these temperatures after a second iteration. Using these new temperatures, recompute Grashof numbers, individual heat-transfer coefficients and overall coefficients, and then iterate again to get a new set of T_w and T_{ws} values. When the current and previous iteration's temperature estimates are the same (within a specified tolerance), the iteration is completed.

Table V lists the individual and overall coefficients after the second iteration. Although it is clear that additional iterations are needed, let us accept these values as sufficiently accurate for the present purpose. Then we can obtain the total heat-transfer rate (Q) by using the U values in Eq. 1-5 and summing. Table VI shows the calculated heat-transfer rates through each boundary, and the total rate. Note that the roof and bottom of the tank account for only slight heat loss, despite being uninsulated.

This, of course, is for the unit period of time, when wind speed is 10 mph, the tank is half full, and the air is 35°F. Table VII shows how the results of unit-period

Rate of heat transfer during unit period				Table VI
Surface	U, Btu/ft^{2}h-°F	Area, ft^2	ΔT, °F	q, Btu/h
Dry wall	0.1392	1,508	15	3,148.7
Wet wall	0.1655	1,508	20	4,991.5
Roof	0.1636	315	15	773.0
Bottom	0.0875	314	15	412.1
Total		3,645		9,325.3

Note: Total for 12-h period is 111,904 Btu

Summing losses for unit periods yields heat loss for 30 days				Table VII
Period	Liquid level, %	T_A, °F	Wind speed, mph	Heat loss, Btu
1	50	35	10	111,904
2	50	27	5	392,407
3	43	42	0	42,591
.	.	.	.	
.	.	.	.	
.	.	.	.	
42	93	55	30	0
.	.	.	.	
.	.	.	.	
59	56	48	20	12,368
60	49	60	15	0
Total for 30-day period				8,389,050

heat losses can be tabulated and added to get the cumulative heat loss for a month or year. Of course, this requires climatic data and tank-level estimates for the overall time-period.

Comparison with other methods

Aerstin and Street [1] offer a very simple method for calculating heat loss from tanks. For a tank with 1.5 in. of sidewall insulation, and a wind speed of 10 mph, the recommended overall U (based on $k = 0.019$ for the insulation) is 0.14 for $\Delta T = 60$°F and 0.14 for $\Delta T = 100$°F. Adjusting these values for $k = 0.028$ and $\Delta T = 17$°F, as in our example, yields an overall U of 0.206 Btu/ft^{2}h-°F. The total exposed surface is 3,331 ft^2 (tank bottom not included), and thus the overall rate of heat transfer by their method is:

$$Q = 0.206 \times 3,331 \times 17 = 11,666 \text{ Btu/h}$$

This compares with a heat loss of 8,913 Btu/h (for the exposed surface) calculated by the procedure of this article—see Table VI. Thus their method yields a result 31% too high in this case.

Stuhlbarg [10] takes an approach similar to that proposed here, but his method differs in how the outside tankwall film coefficient is computed. Stuhlbarg recommends the use of a manufacturer's data table, and does not explicitly distinguish between the bulk liquid temperature and the outside-wall surface temperature in calculating the proper heat-transfer coefficient.

The algebraic method of Hughes and Deumaga [5] resembles the one presented in this article in many ways. But it does not recognize differences between liquid and vapor temperatures inside the tank, nor does it account for the interaction between ΔT and wind speed in calculating a wind-enhancement factor. Finally, even though their procedure requires iteration, the focus of the iterative efforts is to get better estimates of fluid properties, not tankwall temperatures.

Conclusions

Our engineer at ABC Chemical was able to negotiate a significant reduction in the heating charges proposed by the XYZ Terminal Co., which had used a shortcut method for its estimate, because the procedure presented here is rational and defensible. A rigorous solution of the iterations can easily be reached on a digital computer or even a programmable calculator, and the effort pays off in better design or operation criteria.

References

1. Aerstin, F., and Street, G., "Applied Chemical Process Design," Plenum Press, New York, 1978, p. 121.
2. Boyen, J. L., "Thermal Energy Recovery," John Wiley & Sons, New York, 1978, p. 285.
3. Cordero, R., The cost of missing pipe insulation, *Chem. Eng.*, Feb. 14, 1977, p. 77.
4. Ede, A. J., "Advances in Heat Transfer," Vol. 4, Academic Press, New York, 1967, p. 1.
5. Hughes, R., and Deumaga, V., Insulation saves energy, *Chem. Eng.*, May 27, 1974, p. 95.
6. Kato, Nishiwaki, and Hirata, *Intl. J. of Heat and Mass Transfer*, Vol. 11 (1968), p. 1117.
7. Kern, D. Q., "Process Heat Transfer," McGraw-Hill, New York, 1950, p. 217.
8. Perry, R. H., and Chilton, C. H., "Chemical Engineers' Handbook," 5th ed., McGraw-Hill, New York, 1973, p. 10–17.
9. Rohsenow, W. M., and Hartnett, J. P., "Handbook of Heat Transfer," McGraw-Hill, New York, 1973, p. 3-120.
10. Stuhlbarg, D., How to Design Tank Heating Coils, *Pet. Refiner*, Vol. 38, No. 4 (1959), p. 143.

The authors

Jimmy D. Kumana is Chief Process Engineer at Henningson, Durham and Richardson, Inc., P.O. Box 12744, Pensacola, FL 32575, where he is involved in all aspects of engineering design. He holds a B. Tech. degree from the Indian Institute of Technology, and an M.S. from the University of Cincinnati, both in chemical engineering. Mr. Kumana is registered as a professional engineer in four states, and is active in AIChE as chairman of his local section. He holds a patent on distillation of fuel-grade ethanol.

Samir P. Kothari was a process engineer with Henningson, Durham and Richardson, Inc. when this article was written. He recently joined Hoffmann-La Roche Inc., Nutley, NJ 07110, where he is engaged in designing chemical-process plants. Mr. Kothari holds a B.S. degree from M. Sayajirao University (India) and an M.S. from the University of Cincinnati, both in chemical engineering. He belongs to AIChE and has written several technical articles.

Modeling heat-transfer systems

The equations that model heat-transfer systems in batch or continuous equipment must allow for isothermal or nonisothermal heat exchange between process fluids and heat-transfer media.

John L. Guy, Dynamod Enterprises

☐ The dynamic modeling of heat-transfer equipment involves energy flow and temperature, both varying with time at a fixed point in space. We will look at two basic modeling-concepts: (1) the process is intended to operate in the manner so described, i.e., a batch-type process, and (2) the process is *not* intended to operate with transient variations, i.e., a continuous process undergoing upset.

Batch processes

The reasons, as listed by Kern,* for choosing a batch heat-transfer operation rather than a continuous one are that the:

■ Liquid being processed for the product is not continuously available.

■ Heating or cooling medium to the equipment is not continuously available.

■ Reaction-time or treating-time requirements necessitate holdup.

■ Economics of intermittently processing a large batch justify the accumulation of a small continuous stream.

■ Cleaning or regeneration procedure takes up a significant part of the total operating period.

■ Simplified operation of most batch processes is advantageous.

Let us review some of the common batch operations considered by Kern for which manual calculations are suitable. The heat-transfer coefficients for these are considered constant, and the physical properties are calculated at some average temperature. These examples are all concerned with the heating or cooling of liquid batches. For liquid batches, we will consider the use of isothermal and nonisothermal exchange media for jacketed vessels and for external heat-exchangers. Such pro-

*Kern, D. Q., "Process Heat Transfer," McGraw-Hill, New York, 1950.

cedures applied to inherently batch processes are also applicable to the startup of continuous processes.

Jacketed vessel: isothermal medium

The liquid contents of a jacketed tank are being heated by steam condensing in the jacket, as shown in Fig. 1a. The relationships that will be developed are also valid for a heating or cooling coil inside the tank. We can use the same equations for cooling the tank contents via an isothermal boiling fluid—the only change being a negative heat duty that indicates heat transfer is occurring in the opposite direction.

Letting q Btu equal the energy transferred to M lb of liquid having a heat capacity C_p Btu/(lb)(°F), we have:

$$\frac{dq}{d\theta} = MC_p\left(\frac{dt}{d\theta}\right) \qquad (1)$$

where t is temperature of liquid, °F, and θ is time, h.

For an agitated batch, we can assume that the tank temperature varies with time but not with position. For a given time, θ, we can therefore write:

$$dq/d\theta = UA(T - t) \qquad (2)$$

where U is the overall heat-transfer coefficient, Btu/(h)(ft²)(°F); A is heat-transfer surface, ft²; and T is temperature in the jacket, °F.

By combining Eq. (1) and (2), we can solve for the time θ to heat the liquid from t_1 to t_2:

$$MC_p(dt/d\theta) = UA(T - t) \qquad (3)$$

$$\int_0^\theta d\theta = \int_{t_1}^{t_2} \frac{MC_p}{UA} \frac{dt}{(T - t)} \qquad (4)$$

$$\left. \begin{aligned} \theta &= -\frac{MC_p}{UA} \ln\left[\frac{T - t_2}{T - t_1}\right] \\ \theta &= \frac{MC_p}{UA} \ln\left[\frac{T - t_1}{T - t_2}\right] \end{aligned} \right\} \qquad (5)$$

Originally published May 3, 1982

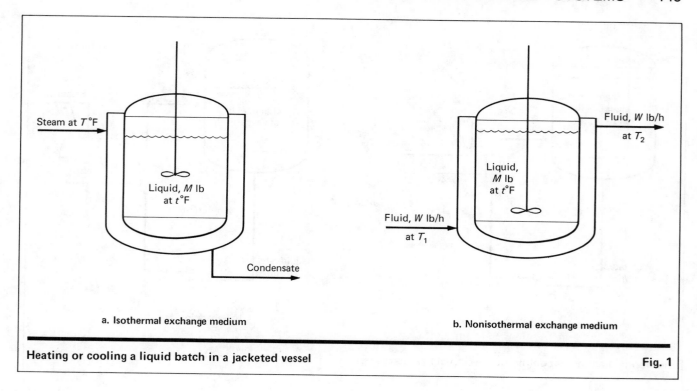

Heating or cooling a liquid batch in a jacketed vessel

a. Isothermal exchange medium

b. Nonisothermal exchange medium

Fig. 1

Example 1—Let us calculate the time required to cool 10,000 lb of liquid water from 300°F to 250°F by generating atmospheric steam at 212°F—given $U = 100$ Btu/(h)(ft^2)(°F), and $A = 100$ ft^2, and $C_p = 1$ Btu/(lb)(°F).

Substituting the appropriate quantities into Eq. (5), we find:

$$\theta = \frac{(10,000)(1)}{(100)(100)} \ln\left[\frac{212 - 300}{212 - 250}\right] = 0.84 \text{ h}$$

We can also calculate the total heat transferred, q, the total steam generated, V, maximum heat transferred, q_{max}, and maximum steam rate, V_{max}, as:

$$q = 10,000(1)(300 - 250) = 500,000 \text{ Btu}$$
$$V = 500,000/970.3 = 515 \text{ lb}$$
$$q_{max} = (100)(100)(300 - 212) = 880,000 \text{ Btu/h}$$
$$V_{max} = 880,000/970.3 = 907 \text{ lb/h}$$

In these computations, we use 970.3 Btu/lb as the heat of vaporization for steam at 212°F.

Jacketed vessel: nonisothermal medium

The contents of a jacketed tank are being heated or cooled by a circulating fluid (undergoing no phase change) in the jacket, as shown in Fig. 1b. Again, the relationships can be applied to a heating or cooling coil inside a tank. For a constant circulating flowrate, the outlet temperature, T_2, varies with time. A heat balance for the liquid batch and the heat-transfer medium yields:

$$\frac{dq}{d\theta} = MC_{pi}\left(\frac{dt}{d\theta}\right) \tag{6}$$

$$dq/d\theta = W_o C_{po}(T_2 - T_1) \tag{7}$$

For a uniform temperature, t, in the tank, we can write:

$$dq/d\theta = UA(LMTD)$$
$$\frac{dq}{d\theta} = UA\left[\frac{(T_1 - t) - (T_2 - t)}{\ln\left(\dfrac{T_1 - t}{T_2 - t}\right)}\right] \tag{8}$$

We can eliminate the variable T_2 by equating Eq. (7) and (8) to get:

$$T_2 = t + \frac{T_1 - t}{\exp(UA/W_o C_{po})} \tag{9}$$

Equating Eq. (6) and (7), substituting the expression for T_2 from Eq. (9), performing the necessary integration, and rearranging, we can obtain the time, θ, that is necessary to heat or cool M lb of liquid from t_1 to t_2 as:

$$\theta = \frac{MC_{pi}}{WC_{po}}\left[\frac{\exp(UA/WC_{po}) - 1}{\exp(UA/WC_{po})}\right]\ln\left[\frac{T_1 - t_1}{T_1 - t_2}\right] \tag{10}$$

Example 2—Let us calculate the time required to cool the same batch as in Example 1, from 250°F to 120°F, by using cooling water at 90°F and at a circulation rate of 50 gpm (equivalent to 25,000 lb/h). For this example, let $U = 75$ Btu/(h)(ft^2)(°F).

We solve the following terms from Eq. (10):

$$\frac{UA}{WC_{po}} = \frac{75(100)}{25,000(1)} = 0.3$$
$$\exp(UA/WC_{po}) = \exp(0.3) = 1.350$$

Using these values and the appropriate data for the example, we substitute them into Eq. (10):

$$\theta = \left(\frac{10,000(1)}{25,000(1)}\right)\left(\frac{1.35 - 1}{1.35}\right)\ln\left(\frac{90 - 250}{90 - 120}\right) = 0.174 \text{ h}$$

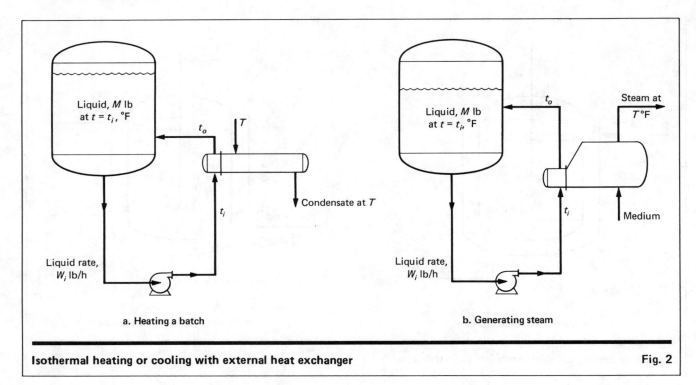

a. Heating a batch

b. Generating steam

Isothermal heating or cooling with external heat exchanger   **Fig. 2**

Next, we obtain the maximum outlet temperature for the cooling water at time θ equal to zero, from Eq. (9):

$$T_2 = 250 + \frac{(90 - 250)}{1.35} = 131.5°F$$

Based on this outlet temperature, the total annual operating time and the materials of construction, we can now decide whether we need a higher cooling-water circulation rate to reduce the initial temperature rise of the cooling water.

External exchanger: isothermal medium

A batch of liquid in a tank is being heated in an external heat exchanger by a fluid condensing on the shellside of the exchanger, as shown in Fig. 2a. We have now added these variables: the circulation rate of the fluid being heated, W_i, lb/h; and t_i and t_o, the inlet and outlet temperatures to and from the exchanger. The variable t_i is also the temperature in the tank.

We will again start with a heat balance, eliminate the variable t_o, integrate the resulting equations, and rearrange to solve for the time, θ, needed to heat the fluid from t_1 to t_2. The results are:

$$\left.\begin{array}{l} \dfrac{dq}{d\theta} = MC_{pi}\dfrac{dt}{d\theta} = W_iC_{pi}(t_o - t_i) \\[2ex] \dfrac{dq}{d\theta} = UA(LMTD) = UA\left[\dfrac{(T - t_o) - (T - t_i)}{\ln\left[\dfrac{T - t_o}{T - t_i}\right]}\right] \end{array}\right\} \quad (11)$$

$$t_o = T - \frac{(T - t_i)}{\exp(UA/W_iC_{pi})} \quad (12)$$

$$\theta = \frac{M}{W_i}\left[\frac{\exp(UA/W_iC_{pi})}{\exp(UA/W_iC_{pi}) - 1}\right]\ln\left[\frac{T - t_1}{T - t_2}\right] \quad (13)$$

Example 3—We will rework Example 1 for the process shown in Fig. 2b. Let the circulation rate equal 10 gpm (5,000 lb/h), $U = 200$ Btu/(h)(ft²)(°F), and $A = 100$ ft².

We begin by solving the following terms of Eq. (13):

$$\frac{UA}{W_iC_{pi}} = \frac{(200)(100)}{(5,000)(1)} = 4$$

$$\exp(UA/W_iC_{pi}) = \exp(4) = 54.6$$

Now, by substituting the numerical values for these terms and other applicable data into Eq. (13), we calculate the required time, θ, initial shellside fluid temperature, $t_{o(init)}$, maximum heat transferred, q_{max}, and maximum steam rate, V_{max}, as:

$$\theta = \frac{10,000}{5,000}\left(\frac{54.6}{53.6}\right)\ln\left[\frac{212 - 300}{212 - 250}\right] = 1.71 \text{ h}$$

$$t_{o(init)} = 212 - \frac{(212 - 300)}{54.6} = 213.6°F$$

$$q_{max} = W_iC_{pi}(t_o - t_i)$$

$$q_{max} = 5,000(1)(300 - 213.6) = 432,000 \text{ Btu/h}$$

$$V_{max} = 432,000/970.3 = 445 \text{ lb/h}$$

If we double the circulation rate, the time to accomplish the total cooling is about the same as for Example 1. (Note: This example does not prove that jacketed vessels are superior to external exchangers.)

External exchanger: nonisothermal medium

For external exchangers using a nonisothermal medium, we will consider two situations: a counterflow arrangement, and a multiple-pass one.

Counterflow—The equipment and its arrangement that we will now simulate is shown in Fig. 3a. The equations that model the system follow. Unfortunately,

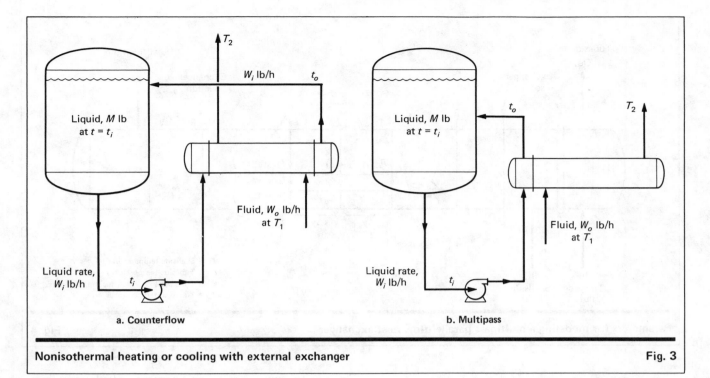

Nonisothermal heating or cooling with external exchanger   **Fig. 3**

a. Counterflow

b. Multipass

we will have to calculate the outlet temperatures, t_o and T_2, by trial and error from Eq. (15) and (16).

$$\frac{dq}{d\theta} = MC_{pi}\frac{dt}{d\theta} = W_iC_{pi}(t_o - t_i) = W_oC_{po}(T_1 - T_2)$$

$$\frac{dq}{d\theta} = UA\left[\frac{(T_2 - t_i) - (T_1 - t_o)}{\ln[(T_2 - t_i)/(T_1 - t_o)]}\right] \quad (14)$$

$$\frac{W_iC_{pi}}{UA}(t_o - t_i) = \frac{(T_2 - t_i) - (T_1 - t_o)}{\ln[(T_2 - t_i)/(T_1 - t_o)]} \quad (15)$$

$$T_2 = T_1 - \frac{W_iC_{pi}}{W_oC_{pi}}(t_o - t_i) \quad (16)$$

$$\theta = \frac{M}{W_iW_oC_{po}} \times$$

$$\left[\frac{W_iC_{pi}\left\{\exp\left[UA\left(\frac{1}{W_iC_{pi}} - \frac{1}{W_oC_{po}}\right)\right]\right\} - W_oC_{po}}{\exp\left\{UA\left[\left(\frac{1}{W_iC_{pi}}\right) - \left(\frac{1}{W_oC_{po}}\right)\right]\right\} - 1}\right] \times$$

$$\ln[(T_1 - t_1)/T_1 - t_2)] \quad (17)$$

Example 4—Let us solve Example 2 for an external exchanger having a counterflow arrangement. We will assume $U = 150$ Btu/(h)(ft²)(°F), $A = 100$ ft², $W_i = 10$ gpm (5,000 lb/h), and $W_o = 50$ gpm (25,000 lb/h). We calculate the following terms in Eq. (17):

$$UA\left[\frac{1}{W_iC_{pi}} - \frac{1}{W_oC_{po}}\right] =$$

$$150(100)\left[\frac{1}{5,000(1)} - \frac{1}{25,000(1)}\right] = 2.4$$

$$\exp(2.4) = 11.023$$

We make the necessary substitutions into Eq. (17) to find the required cooling time:

$$\theta = \frac{10,000}{5,000(25,000)(1)} \times$$

$$\left[\frac{5,000(1)(11.023) - 25,000(1)}{11.023 - 1}\right]\ln\left(\frac{90 - 250}{90 - 120}\right)$$

$$\theta = 0.402 \text{ h}$$

Solving Eq. (15) and (16) by trial and error, we find the initial temperatures $T_2 \cong 120°$F, and $t_o \cong 102°$F.

Multipass exchanger—The equipment and its arrangement that we will model is shown in Fig. 3b. We must now add a temperature-correction factor, F, to our calculations (this makes the algebra more tedious). However, we can calculate the time, θ, directly. For the initial exchanger conditions, the solution is again trial and error. We can add the F-factor from charts rather than algebraically to perhaps save some calculation work. The procedure for calculating time, θ, is:

$$\left.\begin{array}{l}\dfrac{dq}{d\theta} = MC_{pi}\dfrac{dt}{d\theta} = W_iC_{pi}(t_o - t_i) \\[2mm] = W_oC_{po}(T_1 - T_2) \\[2mm] = UAF\left[\dfrac{(T_2 - t_i) - (T_1 - t_o)}{\ln\left(\dfrac{T_2 - t_i}{T_1 - t_o}\right)}\right]\end{array}\right\} \quad (18)$$

$$R = W_iC_{pi}/W_oC_{po} \quad (19)$$

$$K = \exp\left[\frac{UA}{W_iC_{pi}}\sqrt{R^2 + 1}\right] \quad (20)$$

$$S = \frac{2(K - 1)}{K(R + 1 + \sqrt{R^2 + 1}) - (R + 1 - \sqrt{R^2 + 1})} \quad (21)$$

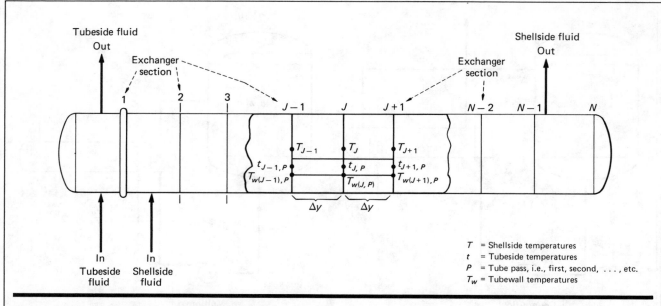

Parameters for modeling a multipass parallel-flow heat exchanger   **Fig. 4**

$$F = \frac{\sqrt{R^2+1}\,\ln\left(\dfrac{1-S}{1-RS}\right)}{(R-1)\ln\left\{\dfrac{2-S[(R+1)-\sqrt{R^2-1}]}{2-S[(R+1)+\sqrt{R^2+1}]}\right\}} \quad (22)$$

$$\theta = \frac{M}{SW_i}\ln\left(\frac{T_1-t_1}{T_1-t_2}\right) \quad (23)$$

Example 5—Let us solve Example 4, using a multipass exchanger that increases U to 200 Btu/(h)(ft²)(°F).

We calculate the values for R, K and S by substituting appropriate values into Eq. (19), (20) and (21), respectively; and get:

$$R = 0.2, \; K = 59.1, \; S = 0.887$$

We can now make the appropriate numerical substitutions into Eq. (23) to calculate the time for cooling:

$$\theta = \frac{10,000}{0.887(5,000)}\ln\left[\frac{90-250}{90-120}\right] = 3.77 \text{ h}$$

In this example, the increased value for the heat-transfer coefficient, U, does not compensate for the temperature correction factor.

Continuous heat-transfer processes

In *Chem. Eng.*, June 29, 1981, p. 77, (Part 1), we derived expressions for a heat-exchanger tube with a sensible liquid being heated on the tubeside by condensing steam on the shellside of the exchanger. The constant temperature of the shellside fluid simplified the derivation of the model because there was one less differential equation than in the sensible-sensible case. If we add a multipass exchanger to the sensible-sensible case, we have increased the computational difficulty even more. The model derived in Part 1 contains two partial-differential equations and is very difficult, if not impossible, to solve manually without creating more complications.

With computers to perform the calculations, however, we can solve such problems by numerical methods. Using the same derivation as in Part 1, and depending on the service conditions, we can set up the following equations:

■ Tubeside: sensible (no phase change)

$$\frac{\pi}{4}\frac{\rho_i d_i^2}{144}C_p\frac{\partial t}{\partial\theta} = \frac{h_i\pi d_i}{12}(T_w-t) - \rho_i v_i\frac{\frac{\pi}{4}d_i^2}{144}C_p\frac{\partial t}{\partial Y} \quad (24)$$

■ Tubeside: isothermal condensing or boiling

$$-\lambda\frac{dV}{dY} + h_i\frac{\pi d_i}{12}(T_w-t) = 0 \quad (25)$$

■ Shellside: sensible (no phase change)

$$\rho_o A_o C_{po}\frac{\partial T}{\partial\theta} = -\rho_o v_o A_o C_{po}\frac{\partial T}{\partial Y} - \frac{nh_o\pi d_o}{12}(T-T_w) \quad (26)$$

■ Shellside: isothermal condensing or boiling

$$\lambda\frac{dV}{dY} + h_o\frac{\pi d_o}{12}(T-T_w) = 0 \quad (27)$$

■ Tubewall

$$\rho_w\frac{(d_o^2-d_i^2)}{4\times144}C_{pw}\frac{dT_w}{d\theta} =$$
$$\frac{h_o d_o}{12}(T-T_w) - \frac{h_i d_i}{12}(T_w-t) \quad (28)$$

where A_o = shellside flow area, ft²; C_p = heat capacity, Btu/(lb)(°F); d = diameter, in.; h = heat-transfer coefficient, Btu/(h)(ft²)(°F); t = temperature of tubeside fluid, °F; T = temperature of shellside fluid, °F; T_w = temperature of the tubewall, °F; v = fluid velocity, ft/h; V = vapor generated, lb/h; Y = tube length, ft; ρ = density, lb/ft³; λ = heat of vaporization, Btu/lb; and θ = time, h. And where the subscript i = inside or tubeside, o = outside or shellside, and w = tubewall.

The convention adopted here is that heat is flowing from the shellside to the tubeside. We have also assumed parallel flow in the shellside of the exchanger, and are not modeling the tortuous path that the fluid actually makes in a cross-baffled exchanger. We can extend Eq. (25) and (27) to condensing or boiling a mixture of components. This will cause us to consider changes in the temperature and the individual coefficients as a function of tube length.

The only difficulty with Eq. (24) to (28) is that the steady-state solutions (at $\theta = \infty$) do not agree exactly with the generally accepted design equations. We have omitted the resistance due to the tubewall. While this resistance is not a large percentage of the total resistance to heat transfer, it would be appropriate to remove this slight discrepancy. The exact procedure is to calculate a heat balance across a differential segment of the tubewall. The resulting equation is given by:

$$\frac{dT_w}{d\theta} = \frac{k_w}{C_{pw}\rho_w}\left[\frac{\partial^2 T_w}{\partial R^2} + \frac{1}{R}\frac{\partial T_w}{\partial R}\right] \quad (29)$$

However, this procedure involves a lot more work in setting up the computer program. In all but the most precise work, the following approach is recommended.

As was indicated in Part 1 of this series, the inside and outside coefficients include a resistance due to fouling. In the same manner, we can include the resistance of the tubewall. If we add half of this resistance to both the inside and outside coefficients, we can calculate (manually or by machine) composite coefficients that will agree with the standard solutions calculated by steady-state methods.

The composite coefficients are given by:

$$h_i = \left[\frac{1}{h_i} + R_{di} + \frac{d_i(d_o - d_i)}{12k_w(d_o + d_i)}\right]^{-1} \quad (30)$$

$$h_o = \left[\frac{1}{h_o} + R_{do} + \frac{d_o(d_o - d_i)}{12k_w(d_o + d_i)}\right]^{-1} \quad (31)$$

The most common approach in solving partial-differential equations is to convert the variables, other than time, to finite-difference equations. The resulting equations are then solved by one of several numerical-integration methods such as those discussed in *Chem. Eng.*, Mar. 8, 1982, pp. 98–101.

Let the exchanger shown in Fig. 4 be partitioned into N sections. For each section, we can write the equivalent form of Eq. (24) to (28). For convenience in writing these equations, we will use the following notation:

$$A_i = (\pi d_i^2/4 \times 144)\rho_i C_{pi}$$
$$B_i = h_i\pi d_i/12$$
$$C_i = (\rho_i v_i\pi d_i^2/4 \times 144)C_{pi}$$
P = Pass, i.e., first, second, third, . . . , number of tube passes

$$\frac{dt_{J,P}}{d\theta} = \frac{B_i(T_{w(J,P)} - t_{J,P})}{A_i} +$$
$$\frac{(-1)^P C_i}{A_i}\left[\frac{t_{J+1,P} - t_{J-1,P}}{2\Delta Y}\right] \quad (32)$$

$$\frac{dv_i}{dY} = \frac{h_i\pi d_i}{12\lambda}(T_{w(J,P)} - t_{J,P}) \quad (33)$$

$$\frac{dT_J}{d\theta} = \sum_{P=1}^{N_{Pt}}\left\{\frac{B_o}{A_o}\left[T_J - T_{w(J,P)}\right]n\right\} +$$
$$\frac{C_o}{A_o}\left[\frac{T_{J+1} - T_{J-1}}{2\Delta Y}\right] \quad (34)$$

where N_{Pt} = number of tube passes/shell, n = number of tubes/pass, and:

$$A_o = \rho_o A_o C_{po}$$
$$B_o = -(h_o\pi d_o/12)$$
$$C_o = -\rho_o v_o A_o C_{po}$$

$$\frac{dv_o}{dY} = \sum_{P=1}^{N_{Pt}}\frac{h_o\pi d_o}{12\lambda}\left[T_J - T_{w(J,P)}\right]n \quad (35)$$

$$\frac{dT_{w(J,P)}}{d\theta} = \frac{B_w}{A_w}(T_J - T_{w(J,P)}) - \frac{C_w}{A_w}(T_{w(J,P)} - t_{J,P}) \quad (36)$$

where:

$$A_w = [\rho_w C_{pw}(d_o^2 - d_i^2)]/4 \times 144$$
$$B_w = h_o d_o/12$$
$$C_w = h_i d_i/12$$

In the last section (N in Fig. 4) of the exchanger, we need to replace Eq. (32) and (34) with the forward difference for the derivative instead of the central difference. These equations become:

$$\frac{dt_{N,P}}{d\theta} = \frac{B_i(T_{w(N,P)} - t_{N,P})}{A_i} +$$
$$\frac{(-1)^P C_i}{A_i}\left(\frac{t_{N,P} - t_{N-1,P}}{\Delta Y}\right) \quad (37)$$

$$\frac{dT_N}{d\theta} = \sum_{P=1}^{N_{Pt}}\left\{\frac{B_o}{A_o}\left[T_N - T_{w(N,P)}\right]n\right\} +$$
$$\frac{C_o}{A_o}\left(\frac{T_N - T_{N-1}}{\Delta Y}\right) \quad (38)$$

A computer program was written to simulate Example 5. This program handled the numerous trial-and-error computations for solving the model equations, and illustrates the method that we have described for continuous systems.

The author

John L. Guy founded Dynamod Enterprises, P.O. Box 240, Swarthmore, PA 19081 [Phone: (215) 328-4545] to market a rigorous dynamic multicomponent-distillation computer program and to provide consulting services to the petrochemical industry. He has had over ten years of industrial experience with Union Carbide Corp. and Atlantic Richfield Co. He has a B.S. in chemical engineering from Ohio State University and an M.S. in chemical engineering from West Virginia University. He is a member of AIChE, and a part-time instructor at Widener University.

Solving problems of varying heat-transfer areas in batch processes

Here are two methods—one graphical and one using an HP-41 calculator program—for solving batch-process heat-transfer problems, where transfer area varies with time.

Maurizio Marzi, Recordati S.p.A.

☐ Batch processes often involve an evaporation step, either to concentrate an initially dilute solution, or for solvent recovery. Another often-used process is a neutralization operation, with the aim of salt formation, or the precipitation of crystals from a mother liquor.

Heating and cooling problems are often hard to solve in these batch processes, particularly because the change in volume causes an associated change in heat-transfer area.

In this article we will show how to find the relation between the time of the operation and the volume of evaporated liquid, by the use of mathematical relations and, where that is not possible, by graphs.

Such operations are normally conducted in a vertical, stirred, jacketed vessel. Heat is added or removed by a heating or cooling medium (steam, water, brine, heat-transfer fluid, etc.).

Fig. 1 shows the described equipment. During the operation, normally, the physical characteristics of the liquid in the vessel (density, viscosity, thermal conductivity, etc.) change. This makes it almost impossible to work out a strictly theoretical treatment, but often it is possible to arrive at a sufficiently accurate solution by means of simplifying assumptions.

Simplifying assumptions

■ Fluid properties do not change during the operation, so that it will be possible to assume a constant overall heat-transfer coefficient.

■ At the beginning of the operation the liquid temperature is at the boiling point.

■ Liquid-level change in the equipment occurs only in the cylindrical part of the vessel.

■ The heat-transfer surface is assumed to be coincident with the wetted wall, when the stirrer is stopped.

■ The condition of the fluid entering the jacket is constant, during the operation.

■ Heat losses are negligible.

First of all, we will consider the problem of the concentration. The results will then be extended to the neutralization operation.

Concentration of a solution

Let us consider the instant θ of the operation.

If U is the constant overall heat transfer:

$$Q_\theta = UA_\theta \, \Delta t_m \qquad (1)$$

where
$$A_\theta = A_f + \pi D L_\theta \qquad (2)$$

then
$$\Delta t_m = T_m - t_e = 0.5(T_i + T_o) - t_e \qquad (3)$$

From a heat balance:

$$Q_\theta = WC(T_i - T_o) \qquad (4)$$

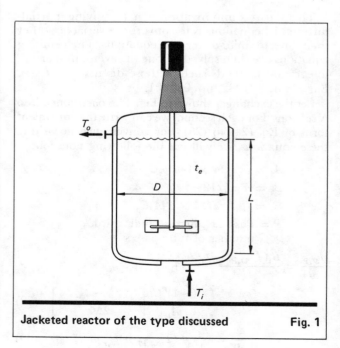

Jacketed reactor of the type discussed   **Fig. 1**

Originally published May 17, 1982

Combining Eq. (3) and (4):

$$\Delta t_m = \Delta t_i - \frac{Q_\theta}{2WC} \quad (5)$$

where $\Delta t_i = T_i - t_e$.

Substituting Eq. (2) and (5) in Eq. (1):

$$Q_\theta = U(A_f + \pi D L_\theta)\left(\Delta t_i - \frac{Q_\theta}{2WC}\right)$$

which is the thermal flow, at instant θ, through the wet-ted wall. Due to the assumptions, the only variable with time is L_θ:

$$L_\theta = \frac{mQ_\theta}{n - Q_\theta} - \frac{A_f}{\pi D} \quad (6)$$

where

$$m = \frac{2WC}{\pi UD}; \qquad n = 2WC(\Delta t_i)$$

From Eq. (6), differentiating:

$$dL_\theta = \frac{mn}{(n - Q_\theta)^2} dQ_\theta \quad (6a)$$

The transferred heat during the interval $d\theta$ is related to the mass of evaporated liquid and to the liquid level in the vessel, following the equation:

$$Q_\theta d\theta = -\lambda dM_\theta = -\lambda \rho_1 \frac{\pi}{4} D^2 dL_\theta$$

Substituting the preceding relation in Eq. (6a), we have

$$\frac{dQ_\theta}{Q_\theta(n - Q_\theta)^2} = -K d\theta$$

where

$$\frac{1}{K} = \frac{\lambda \rho_1 D(WC)^2}{U}\Delta t_i$$

Integrating the differential equation from 0 to θ:

$$\frac{n(Q^0 - Q_\theta)}{(n - Q_\theta)(n - Q^0)} + \ln\frac{(n - Q_\theta)}{(n - Q^0)}\frac{Q^0}{Q_\theta} = Kn^2\theta \quad (7)$$

Thermal flow at time 0 can be found by means of a thermal balance:

Heat flow through the wetted wall at time 0:

$$Q^0 = UA^0(T_m - t_e) = UA^0\left(\frac{T_i + T_o^0}{2} - t_e\right)$$

$$T_o^0 = 2\left(\frac{Q^0}{UA^0} + t_e\right) - T_i \quad (8)$$

Heat given up from heating fluid at time 0:

$$Q^0 = WC(T_i - T_o^0)$$

$$T_o^0 = T_i - \frac{Q^0}{WC} \quad (9)$$

Equating Eq. (8) and (9):

$$Q^0 = \frac{\Delta t_i}{\dfrac{1}{UA^0} + \dfrac{1}{2WC}} \quad (10)$$

Nomenclature

A_f	Bottom reactor exchanging surface, m²
A_θ	Total exchanging surface at time θ, m²
A^0	Total exchanging surface at time 0, m²
C,c	Jacket-fluid specific heat, J/kg°C
D	Reactor diameter, m
K	System constant
L_θ	Wetted cylindrical height at time θ, m
M_θ	Mass of the liquid in the reactor at time θ, kg
m	System constant
n	System constant
Q_θ	Thermal flow at time θ, W
Q^0	Thermal flow at time 0, W
T_i, t_i	Fluid inlet temperature in the jacket, °C
T_m, t_m	Average jacket temperature at time θ, °C
T_o, t_o	Jacket-fluid outlet temperature at time θ, °C
T_r, t_e	Constant temperature of the fluid in the reactor, °C
T_o^0, t_o^0	Jacket-fluid outlet temperature at time 0, °C
U	Overall heat-transfer coefficient, W/m²°C
V	Liquid volume in the reactor, m³
V_i	Initial liquid volume in reactor, m³
V_f	Final liquid volume in reactor, m³
V_{ij}	Liquid volume variation during the interval θ_{ij}, m³
W,w	Jacket fluid flowrate, kg/s
λ	Latent heat of evaporation, J/kg
ρ_1	Process-fluid density, kg/m³
θ	Time, s
θ', θ''	Time to complete the operation, s
θ_{ij}	Time interval expressed as $(\theta_j - \theta_i)$, s
ΔH	Heat of reaction, J/kg
ΔT_i	Temperature difference expressed as $(T_r - t_i)$, °C
Δt_i	Temperature difference expressed as $(T_i - t_e)$, °C
Δt_m	Temperature difference expressed as $(T_m - t_e)$, °C
$\Delta V', \Delta V''$	Evaporated or added volume until time θ' or θ'', m³

Eq. (7) can be put in the form:

$$\frac{(Q^0/n) - (Q_\theta/n)}{[1 - (Q^0/n)][1 - (Q_\theta/n)]} +$$

$$\ln\left[\frac{1 - (Q_\theta/n)}{1 - (Q^0/n)}\frac{Q^0/n}{Q_\theta/n}\right] = Kn^2\theta \quad (7a)$$

where, remembering the expression for K, m and n:

$$Kn^2 = \frac{4U\Delta t_i}{\lambda \rho_1 D} \quad \text{and} \quad \frac{n}{Q^0} = 1 + \frac{2WC}{UA^0} \quad (11)$$

The relation (7a) has been plotted (Fig. 2) with Q_θ/n vs. $Kn^2\theta$, for different values of the constant n/Q^0.

Thus, it will be possible to find, graphically, the value of Q_θ for every time θ, knowing the system constants, Eq. (11).

A recurrent problem is to calculate the time needed to concentrate an initial solution of volume V_i to the final volume V_f. It is practically impossible to solve the problem analytically, because it would be necessary to integrate the relation:

$$dV = -\frac{Q_\theta}{\lambda \rho_1} d\theta \qquad (12)$$

where Q_θ should be made explicit from Eq. (7a). It is easier instead to answer the question graphically in the following way:

■ Assuming we know the characteristics of the system (i.e., U, A_o, T_i, t_e, W, D) and the physical properties of the involved fluids (C, λ, ρ_1), it is possible to compute the constants Q^0, Kn^2 and n/Q^0 by Eq. (10) and (11).

■ Assuming a time interval $\Delta\theta$ and starting from $\theta = 0$, let us find by means of Fig. 2 the value of Q_θ at $\theta = \Delta\theta$, $\theta = 2\Delta\theta$ . . . . , until we go beyond a realistic time for the operation.

■ Plot $Q_\theta/\lambda\rho_1$ vs. θ and make a graphical integration of the curve or do a numerical integration by a simplified method (i.e., Simpson's).

■ The integration must be stopped when the value $\Delta V' = V_i - V_f$ has been reached.

A very easy case occurs when it is possible to add a further simplification—that the temperature of the heating medium is constant (i.e., when using steam). This is also the case of a fluid for which $(T_i - T_o) \ll (T_m - t_e)$. Adopting this added simplification, and repeating the logic used to arrive at Eq. (7a), it is easy to show:

$$Q_\theta = Q^0 \exp[-Kn^2\theta] \qquad (13)$$

where $Q^0 = UA^0 \Delta t_i$

It is possible to arrive at Eq. (13) by the following considerations:

If the heating-medium temperature is assumed to be constant, this is the same as supposing that its thermal capacity is infinite. In that case $n \gg Q$ and Q/n tends to zero, so that Eq. (7a) becomes Eq. (13). Introducing Q_θ, in the form of Eq. (13), in the differential equation (12):

$$dV = -\frac{Q^0}{\lambda\rho_1} \exp[-Kn^2\theta] d\theta$$

By integration,

$$\Delta V' = \frac{Q^0 D}{4U \Delta t_i}[1 - \exp[-Kn^2\theta']] \qquad (14)$$

or

$$\Delta V' = \frac{A^0 D}{4}[1 - \exp[-Kn^2\theta']] \qquad (14a)$$

or rearranging,

$$\theta' = \frac{1}{Kn^2} \ln\left[1 - \frac{4\,\Delta V'}{A^0 D}\right]^{-1} \qquad (14b)$$

where $\Delta V' = V_i - V_f$

Eq. (14), (14a), and (14b) are very easy to solve.

Application examples

In the two following examples we will be comparing two different fluids as heating media: superheated water and saturated steam. In order to compare the more rigorous method with the simplified one, we will assume the same inlet temperature and the same overall heat-transfer coefficient for both fluids.

Common data

Fluid to be evaporated	Ethanol
Evaporation temperature, t_e	78.4°C
Initial volume of liquid in the vessel, V_i	3.0 m³
Final volume of liquid in the vessel, V_f	1.5 m³
Internal diameter of the vessel, D	1.4 m
Initial wetted surface, A^0	9.5 m²
Overall heat-transfer coefficient, U	580 W/m²°C
Density of the evaporating liquid, ρ_1	730 kg/m³
Latent heat of vaporization, λ	878 kJ/kg
Inlet temperature of the heating medium, t_i	110°C

System constants

From Eq. (11):

$$Kn^2 = \frac{4 \times 580 \times (110 - 78.4)}{878,000 \times 730 \times 1.4} = 8.17 \times 10^{-5}$$

$$n/Q^0 = 1 + \frac{2 \times W \times C}{580 \times 9.5}$$

Example 1

Heating medium: saturated steam at 110°C. Find the necessary time for concentration.

$$\theta' = \frac{1}{8.75 \times 10^{-5}} \ln\left[1 - \frac{4 \times 1.5}{9.5 \times 1.4}\right]^{-1}$$

$$= 6,856 \text{ s } (1^h 54' 16'')$$

Example 2

Heating medium: superheated water at 110°C; water flowrate in the jacket, W, 20 m³/h = 5.56 kg/s; specific heat of heating medium, C, 4,222 J/kg°C.

therefore, $n/Q^0 = 1 + \dfrac{2 \times 5.56 \times 4,222}{580 \times 9.5} = 9.52$

Using the system constants calculated before, we can tabulate Q_θ vs. θ, solving relation (7a) by the use of the diagram of Fig. 2 (see Table I).

The same calculation can be made by means of a pocket calculator program. In Table II you will find a program that can be used with the HP-41.

The liquid volume that evaporates in any interval of time θ_{ij} is computed by Eq. (12), which for finite intervals becomes:

$$V_{ij} = \frac{Q_i + Q_j}{2\lambda\rho_1}\theta_{ij} \qquad (12a)$$

where $\theta_{ij} = \theta_j - \theta_i$.

Substituting Eq. (12) by (12a) does not introduce sensible errors, due to the regularity of $Q_\theta = f(\theta)$.

The cumulative volumes of the evaporated liquid are shown in Fig. 3 (Curve B). By graphic interpolation it is easy to deduce the evaporation time of 1,500 L of ethanol, for the examined case: 2^{h}13'20''

Curve A represents the evaporated liquid volume, computed by applying the simplified Eq. (14), valid for

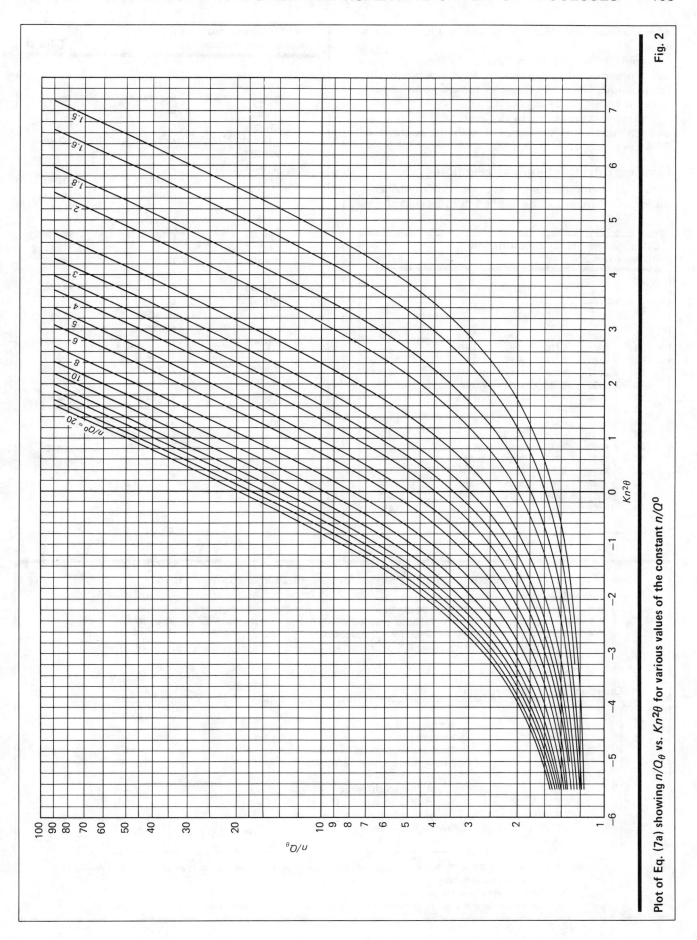

Plot of Eq. (7a) showing n/Q_θ vs. $Kn^2\theta$ for various values of the constant n/Q^0 Fig. 2

an isothermal fluid. The error introduced in this case is about 9%.

It tends to rise with fluids having low thermal capacity (diathermic oils), or when the fluid flowrate is low. For example, Curve C represents the case of a heating medium with $C = 2,000 J/kg°C$.

The evaporation time, in this case, is $2^h25'50''$ with an increment of 19% compared with the isothermal-fluid operation.

Cooling during chemical reaction

The same considerations developed for the concentration of solutions can be applied in other operations,

Calculated results for concentration example No. 2   Table I

θ, h	θ, s	$Kn^2\theta$	n/Q_θ	Q_θ, W	V_{ij}, m^3	V_t, m^3
0.0	0	0.000	9.52	155,827		0.000
					0.086	
0.1	360	0.029	9.75	152,190		0.086
					0.085	
0.2	720	0.059	9.98	148,617		0.171
					0.163	
0.4	1,440	0.118	10.47	141,671		0.334
					0.156	
0.6	2,160	0.176	10.99	134,981		0.489
					0.148	
0.8	2,880	0.235	11.54	128,547		0.638
					0.141	
1.0	3,600	0.294	12.12	122,363		0.779
					0.134	
1.2	4,320	0.353	12.74	116,425		0.913
					0.128	
1.4	5,040	0.412	13.40	110,729		1.041
					0.121	
1.6	5,760	0.471	14.09	105,268		1.162
					0.115	
1.8	6,840	0.559	14.83	97,506		1,277
					0.109	
2.0	7,200	0.588	15.61	95,030		1.387
					0.104	
2.2	7,920	0.647	16.44	90,241		1.491
					0.099	
2.4	8,640	0.706	17.32	85,664		1.590
					0.094	
2.6	9,360	0.765	18.25	81,291		1.684
					0.089	
2.8	10,080	0.824	19.24	77,117		1.773
					0.084	
3.0	10,800	0.882	20.28	73,135		1.857

HP-41 Calculator program for both concentration and cooling   Table II

```
01 LBL SOLVE        /               STO 06
   SF 00            "DELTA T ?"      RCL 04
   0.001            PROMPT           -
   STO 00           ST x 03          ABS
   "1 OR 2?"        x                RCL 00
   PROMPT           4             70 X↔Y
   GTO IND X        x                X≤Y?
   LBL 01        40 "LAMBDA ?"       GTO 03
   CF 00            PROMPT           RCL 06
10 LBL 02           /                STO 04
   "D?"             "DENS. ?"        1
   PROMPT           PROMPT           ST - 06
   "A?"             /                RCL 06
   PROMPT           FS ? 00          RCL 01
   "U ?"            CHS              -
   PROMPT           STO 02        80 RCL 06
   STO 07           LBL 04           /
   x             50 "TIME ?"         RCL 01
   "W ?"           PROMPT            /
20 PROMPT           1                CHS
   "C ?"            STO 04           RCL 05
   PROMPT           X↔Y              GTO 05
   x                STO 05           LBL 03
   2                LBL 05           RCL 06
   x                RCL 02           1/X
   STO 03           x             90 RCL 03
   /                +                x
   1/X           60 eˣ               PSE
   STO 01           RCL 01           PSE
30 CLX              x                GTO 04
   RCL 07           1             95 END
   X↔Y              +
```

Instructions for running program for concentration or cooling on HP-41

- **XEQ SOLVE**
- Answer the questions put by the program, using the units specified in the nomenclature, below.
- When requested, introduce the TIME (in seconds) at which you need the value for Q_θ. Press **R/S**.
- The program will show you the result for a few seconds.
  You can hold the result by pressing **R/S** or, if you like, by substituting instruction 92 and 93 (PSE) with **R/S** (one time).
- Then the program will request "TIME?" again for the next point.

Questions put by the program

		Symbol	Units
"1 or 2?"	See note below		
"D ?"	Reactor diameter	D	m
"A ?"	Total exchanging surface at time 0	A^0	m^2
"U ?"	Overall heat transfer coefficient	U	W/m^2°C
"W ?"	Jacket fluid flowrate	W, w	kg/sec
"C ?"	Jacket fluid specific heat	C, c	J/kg°C
"DELTA T ?"	Temp. diff. expressed as $(T - t)$	$\Delta T_j, \Delta t_j$	°C
"LAMBDA ?"	Lat. heat or Heat of reaction	λ	J/kg
"DENS. ?"	Process fluid density	ρ_1	kg/m^3
"TIME ?"	Time	θ	s

Note: The question "1 OR 2 ?" that appears at the beginning of the run refers to the type of calculation requested:
1. Concentration of solutions.
2. Cooling during chemical reaction.

Input the number and press **R/S**. After inputting each item of data, press **R/S**.

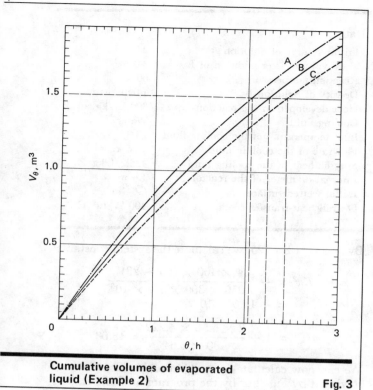

Cumulative volumes of evaporated liquid (Example 2)   **Fig. 3**

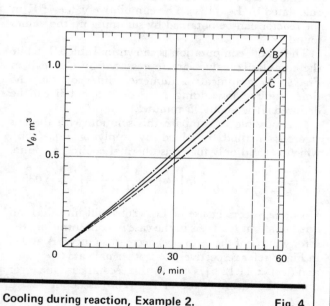

Cooling during reaction, Example 2.   **Fig. 4**

Calculated results for cooling example No. 2   Table III

θ, min	θ, s	$Kn^2\theta$	n/Q_θ	Q_θ, W	V_{ij}, m³	V_t, m³
0.0	0	0.000	18.68	80,441		0.000
10.0	600	−0.070	17.55	85,602	0.158	0.158
20.0	1,200	−0.139	16.50	91,051	0.168	0.326
30.0	1,800	−0.209	15.52	96,799	0.179	0.505
40.0	2,400	−0.279	14.61	102,858	1.190	0.695
50.0	3,000	−0.348	13.75	109,238	0.202	0.897
60.0	3,600	−0.418	12.96	115,948	0.214	1.111

such as liquid-phase reactions obtained by addition of one liquid reagent to another.

The discontinuous operation is conducted by pouring a second reagent into a reactor that contains the first reagent. During pouring, the liquid temperature must be maintained at a fixed value, T_r, subtracting heat by means of a cooling medium that flows through the reactor jacket.

Let us consider the heat flow through the wetted wall:

$$Q_\theta = UA_\theta \Delta t_m \tag{15}$$

where

$$A_\theta = A_f + \pi D L_\theta \tag{16}$$

$$\Delta t_m = T_r - t_m = T_r - 0.5(t_i + t_o) \tag{17}$$

From a heat balance:

$$Q_\theta = wc(t_o - t_i) \tag{18}$$

Combining Eq. (17) and (18):

$$\Delta t_m = \Delta T_i - \frac{Q_\theta}{2wc} \tag{19}$$

where $\Delta T_i = (T_r - t_i)$.

Heat flow at time θ through the wetted wall can be put in the form:

$$Q_\theta = U(A_f + \pi D L_\theta)\left(\Delta T_i - \frac{Q_\theta}{2wc}\right)$$

and following the same procedures as for Eq. (6) and (6a):

$$dL_\theta = \frac{mn}{(n - Q_\theta)^2} dQ_\theta \tag{20}$$

where,

$$m = \frac{2wc}{\pi UD}; \quad \text{and} \quad n = 2we(\Delta T_i)$$

Let ΔH be the heat of reaction, referred to the unit of mass of the liquid reagent poured into the reactor. The heat balance during the interval $d\theta$ is:

$$Q_\theta d\theta = \Delta H dM_\theta = \Delta H \rho_1 \frac{\pi}{4} D^2 dL_\theta$$

Combining this relation with Eq. (20) and putting:

$$K = -\frac{U}{\rho_1 \Delta HD(wc)^2 \Delta T_i}$$

we obtain:

$$\frac{dQ_\theta}{Q_\theta(n - Q_\theta)^2} = -Kd\theta \tag{21}$$

Integrating between $\theta = 0$ and $\theta = \theta$ we will again find Eq. (7).

Initial conditions (i.e., heat flow at time 0) can be found in the same way as for concentration, reaching by means of the two equations, (22) and (23):

$$t_o^0 = 2\left[T_r - \frac{Q^0}{UA^0}\right] - t_i \tag{22}$$

$$t_o^0 = \frac{Q^0}{wc} + t_i \tag{23}$$

the expression:

$$Q^0 = \frac{\Delta T_i}{\dfrac{1}{UA^0} + \dfrac{1}{2wc}} \qquad (24)$$

Defining, as for concentration, the two system constants:

$$Kn^2 = -\frac{4U\,\Delta T_i}{\rho_1\,\Delta H D} \qquad \frac{n}{Q^0} = 1 + \frac{2wc}{UA^0} \qquad (25)$$

it is again possible to arrive at Eq. (7a), which is therefore valid for cooling during chemical reaction, as well as for evaporation.

We can also use the same diagram (Fig. 2) to solve the problem of finding the minimum time necessary for the reaction, due to the system geometry.

Obviously, the system constant should be determined before, by Eq. (25). Two situations are again possible. The most common one is the cooling by a nonisothermal fluid. In that case we must apply the graphic method, or use a program for the programmable calculator (Table II) in order to compute Q_θ vs. θ

Then, using Eq. (12a), we will be able to calculate the volume of solution to be added in each time interval $\Delta\theta$. Summing up these values we will obtain the total cumulative volume.

It is also possible to proceed with a graphic integration, but the more precise solution does not justify the complexity of the computation.

A less common situation arises with an isothermal cooling fluid, for example a pure boiling liquid. The same consideration made for concentration can be applied, arriving at equations like Eq. (14), Eq. (14a) and Eq. (14b).

Still, in order to have only positive values for $\Delta V'$, it will be more convenient to modify the definition

$$\Delta V'' = V_f - V_i$$

Then relations (14), (14a) and (14b) become:

$$\Delta V'' = \frac{Q^0 D}{4U\,\Delta T_i}[\exp(-Kn^2\theta'') - 1] \qquad (26)$$

$$\Delta V'' = \frac{A^0 D}{4}[\exp(-Kn^2\theta'') - 1] \qquad (26a)$$

$$\theta'' = \frac{1}{Kn^2}\ln\left[1 + \frac{4\,\Delta V''}{A^0 D}\right]^{-1} \qquad (26b)$$

Application example

Since the case involving isothermal fluid is easy to solve, using Eq. (26) to (26b), it is better to examine the case with a nonisothermal fluid.

Let us consider a reactor in which, at the beginning of the operation, only the Reagent A is present (in liquid phase or in solution). It is then necessary to add a volume $\Delta V''$ of Reagent B to produce the desired chemical reaction.

Let us assume that the initial temperature is exactly the preselected reaction temperature, and that it must be kept constant throughout the operation. (No effect of volume change is considered for the substances present in the system.)

Data:

Initial volume of Solution A	2.0 m³
Total volume after addition of B	3.0 m³
Reaction temperature	60°C
Density of Solution B	1,050 kg/m³
Heat developed in the reaction	300 kJ/kg sol. B
Cooling fluid	Water
Inlet temperature of the cooling fluid	28°C
Flowrate of the cooling fluid	5.56 kg/s
Specific heat of the cooling fluid	4,222 J/kg°C
Internal diameter of the reactor	1.4 m
Initial wetted surface	6.64 m²
Overall heat-transfer coefficient	400 W/m²°C

By means of Eq. (25) we calculate the system constants:

$$Kn^2 = -\frac{4 \times 400 \times (60 - 28)}{1{,}050 \times 300 \times 1.4 \times 10^3}$$
$$= -1.16 \times 10^{-4}$$

$$n/Q^0 = 1 + \frac{2 \times 5.56 \times 4{,}222}{400 \times 6.64} = 18.68$$

We can now calculate Q_θ vs. θ, applying Eq. (7a) and solving it by Fig. 2 or by the program of Table II. The volume of Reagent B added in every interval θ_{ij} can be calculated by Eq. (12a). The cumulative value of V_t in the reactor can be obtained by summing up the values found, V_{ij}.

The whole computation is shown in Table III, while the cumulative values are represented by Curve B in Fig. 4. By graphical or numerical interpolation, the time required for adding 1 m³ of Reagent B can be deduced. It is about 55 minutes.

It is possible to compare this solution with the less rigorous method, obtained by applying Eq. (26b), which is valid only for an isothermal cooling medium.

$$\theta'' = \frac{1}{-1.16 \times 10^{-4}}\ln\left[1 + \frac{4 \times 1.0}{6.64 \times 1.4}\right]^{-1} = 3{,}085 \text{ s}$$

However, incorrect use of Eq. (26b) will introduce an error of about 6.5%. As in the case of concentration, the error will rise if the term (wc) decreases. Curves A and C in Fig. 4 refer respectively to isothermal case $(wc = \infty)$ and $(wc) = 11.12$ kJ/s °C. In this second case the error due to the incorrect use of Eq. (26b) is about 12.5%.

The author

Maurizio Marzi, Via Divisione Torino, 7, 00143—Rome, Italy, is a senior process and project engineer for Recordati S.p.A. (a chemical and pharmaceutical company), Campoverde di Aprilia, Latina, Italy, where he works as a project and technological manager on jobs involving process and project engineering, plant erection and startup. He holds a degree in chemical engineering from the Universita degli Studi di Roma, and is registered as a professional engineer in Italy.

Designing a helical-coil heat exchanger

An HCHE offers advantages over a double-pipe heat exchanger in some situations. Here are a few cases where you might want to consider using one, and a simple procedure for designing it.

Ramachandra K. Patil, *Rathi Industrial Equipment Co.*;
B. W. Shende, *Polychem Ltd.*; and Prasanta K. Ghosh, *Hindustan Antibiotics Ltd.*

☐ The double-pipe heat exchanger would normally be used for many continuous systems having small to medium heat duties. However, the helical-coil heat exchanger (HCHE) might be a better choice in some cases:

■ Where space is limited, so that not enough straight pipe can be laid.

■ Under conditions of laminar flow or low flowrates, where a shell-and-tube heat exchanger would become uneconomical because of the resulting low heat-transfer coefficients.

■ Where the pressure drop of one fluid is limited (for example, because of flow through other process equipment). By setting the velocity of the annulus fluid in an HCHE at about 1 m/s, the pressure drop will be low.

An HCHE consists of a helical coil fabricated out of a metal pipe that is fitted in the annular portion of two concentric cylinders, as shown in Fig. 1. The fluids flow inside the coil and the annulus, with heat transfer taking place across the coil wall. The dimensions of both cylinders are determined by the velocity of the fluid in the annulus needed to meet heat-transfer requirements.

Fig. 2 is a schematic cutaway view of the HCHE. The minimum clearances between the annulus walls and the coil and between two consecutive turns of the coil must be equal. In this case, both clearances are taken as $d_o/2$. The pitch, p, which is the spacing between consecutive coil turns (measured from center to center), is $1.5d_o$. Assuming that the average fluid velocity is uniform, the mass velocity of the fluid, G_s, is computed based on the minimum clearance between the helix and the cylinder wall.

Design procedure

Here is a simple procedure for designing an HCHE:
Determine the heat-transfer coefficients. To calculate the heat-transfer coefficients in the coil and the annulus, the following parameters must be known:

1. The length of coil, L, needed to make N turns:

$$L = N\sqrt{(2\pi r)^2 + p^2} \qquad (1)$$

2. The volume occupied by the coil, V_c:

$$V_c = (\pi/4)d_o^2 L \qquad (2)$$

3. The volume of the annulus, V_a:

$$V_a = (\pi/4)(C^2 - B^2)pN \qquad (3)$$

4. The volume available for the flow of fluid in the annulus, V_f:

$$V_f = V_a - V_c \qquad (4)$$

5. The shell-side equivalent diameter of the coiled tube, D_e:

$$D_e = 4V_f/\pi d_o L \qquad (5)$$

The heat-transfer coefficient in the annulus, h_o, can

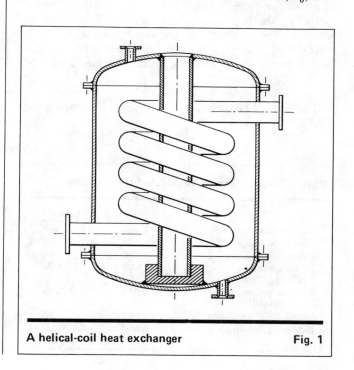

A helical-coil heat exchanger   **Fig. 1**

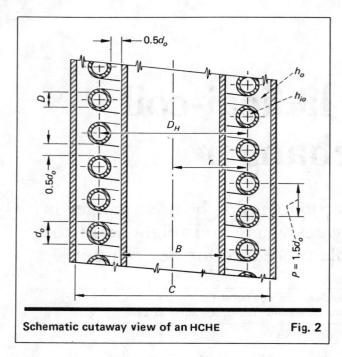

Schematic cutaway view of an HCHE   **Fig. 2**

now be calculated using one of the following two equations. For Reynolds numbers, N_{Re}, in the range of 50–10,000, Eq. (6) [3] is recommended:

$$h_o D_e/k = 0.6 N_{Re}^{0.5} N_{Pr}^{0.31} \qquad (6)$$

For N_{Re} over 10,000, Eq. (7) [4] should be used:

$$h_o D_e/k = 0.36 N_{Re}^{0.55} N_{Pr}^{1/3}(\mu/\mu_w)^{0.14} \qquad (7)$$

The heat-transfer coefficient of the fluid flowing inside the coil, h_{io}, can be determined using conventional methods, such as described in Ref. [4]. The heat-transfer coefficient based on the inside coil diameter, h_i, is obtained using a method for a straight tube—either one of the Sieder-Tate relationships, or a plot of the Colburn factor, j_H, vs. N_{Re}, such as Fig. 3. That must then be corrected for a coiled tube by multiplying h_i by $[1 + 3.5(D/D_H)]$ to get h_{ic}. The coefficient based on the outside diameter of the coil, h_{io}, is then obtained by:

$$h_{io} = h_{ic}(D/d_o) \qquad (8)$$

The overall heat-transfer coefficient, U, is given by:

$$1/U = 1/h_o + 1/h_{io} + x/k_c + R_t + R_a \qquad (9)$$

Physical properties and other data for the example

	Liquid A	Liquid B
Mass flowrate, M, kg/h	1,350	2,141
Inlet temperature, °C	127	30
Outlet temperature, °C	100	47
Heat capacity, c_p, kcal/(kg)(°C)	1.00	1.00
Thermal conductivity, k, kcal/(h)(m)(°C)	0.419	0.4075
Viscosity, μ, kg/(m)(h)	1.89	5.76
Density, ρ, kg/m³	870.0	935.0

Nomenclature

A	Area for heat transfer, m²
A_a	Area for fluid flow in annulus, $(\pi/4)[(C^2 - B^2) - (D_{H2}^2 - D_{H1}^2)]$, m²
A_f	Cross-sectional area of coil, $\pi D/4$, m²
B	Outside dia. of inner cylinder, m
C	Inside dia. of outer cylinder, m
c_p	Fluid heat capacity, kcal/(kg)(°C)
D	Inside dia. of coil, m
D_e	Shell-side equivalent dia. of coil, m
D_H	Average dia. of helix, m
D_{H1}	Inside dia. of helix, m
D_{H2}	Outside dia. of helix, m
d_o	Outside dia. of coil, m
G_s	Mass velocity of fluid, $M/[(\pi/4)((C^2 - B^2) - (D_{H2}^2 - D_{H1}^2))]$, kg/(m²)(h)
H	Height of cylinder, m
h_i	Heat-transfer coefficient inside straight tube, based on inside dia., kcal/(h)(m²)(°C)
h_{ic}	Heat-transfer coefficient inside coiled tube (h_i corrected for coil), based on inside dia., kcal/(h)(m²)(°C)
h_{io}	Heat-transfer coefficient inside coil, based on outside dia. of coil, kcal/(h)(m²)(°C)
h_o	Heat-transfer coefficient outside coil, kcal/(h)(m²)(°C)
j_H	Colburn factor for heat transfer, $(h_i D/k)(N_{Pr})^{-1/3}(\mu/\mu_w)^{-0.14}$, dimensionless
k	Thermal conductivity of fluid, kcal/(h)(m)(°C)
k_c	Thermal conductivity of coil wall, kcal/(h)(m)(°C)
L	Length of helical coil needed to form N turns, m
M	Mass flowrate of fluid, kg/h
N	Theoretical number of turns of helical coil
n	Actual number of turns of coil needed for given process heat duty (N rounded to the next highest integer)
N_{Pr}	Prandtl number, $c_p\mu/k$, dimensionless
N_{Re}	Reynolds number, $Du\rho/\mu$ or DG/μ, dimensionless
Q	Heat load, kcal/h
q	Volumetric flowrate of fluid, m³/h
r	Average radius of helical coil, taken from the centerline of the helix to the centerline of the coil, m
R_a	Shell-side fouling factor, (h)(m²)(°C)/kcal
R_t	Tube-side fouling factor, (h)(m²)(°C)/kcal
Δt_c	Corrected log-mean-temperature-difference, °C
Δt_{lm}	Log-mean-temperature-difference, °C
u	Fluid velocity, m/h
U	Overall heat-transfer coefficient, kcal/(h)(m²)(°C)
V_a	Volume of annulus, m³
V_c	Volume occupied by N turns of coil, m³
V_f	Volume available for fluid flow in the annulus, m³
x	Thickness of coil wall, m
μ	Fluid viscosity at mean bulk-fluid temperature, kg/(m)(h)
μ_w	Fluid viscosity at pipe-wall temperature, kg/(m)(h)
ρ	Fluid density, kg/m³

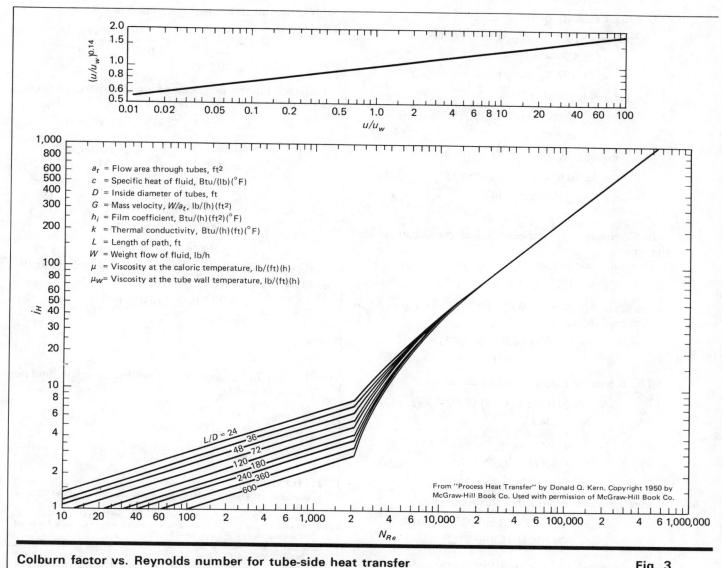

a_t = Flow area through tubes, ft^2
c = Specific heat of fluid, Btu/(lb)($^\circ$F)
D = Inside diameter of tubes, ft
G = Mass velocity, W/a_t, lb/(h)(ft^2)
h_i = Film coefficient, Btu/(h)(ft^2)($^\circ$F)
k = Thermal conductivity, Btu/(h)(ft)($^\circ$F)
L = Length of path, ft
W = Weight flow of fluid, lb/h
μ = Viscosity at the caloric temperature, lb/(ft)(h)
μ_w= Viscosity at the tube wall temperature, lb/(ft)(h)

$L/D = 24$ 36
48 72
120 180
240 360
600

From "Process Heat Transfer" by Donald Q. Kern. Copyright 1950 by McGraw-Hill Book Co. Used with permission of McGraw-Hill Book Co.

Colburn factor vs. Reynolds number for tube-side heat transfer   **Fig. 3**

Determine the required area. The area needed for heat transfer is determined by:

$$A = Q/U\Delta t_c \qquad (10)$$

The log-mean-temperature-difference, Δt_{lm}, must be corrected to take into account the fact that the fluids are flowing perpendicular to each other, which is done by applying the standard correction factor for perpendicular flow [4].

Determine the number of turns of coil. Since $A = \pi d_o L$, and L is expressed in terms of N, the number of turns of coil needed can be calculated by:

$$N = A/(\pi d_o(L/N)) \qquad (11)$$

The actual number of coil turns needed, n, is simply N rounded to the next highest integer.

An example

Liquid A flows inside a 316 stainless-steel pipe coil; Liquid B, in the annulus. The flowrates, the inlet and outlet temperatures, and the physical properties of the fluids are given in the table. The geometry of the HCHE is that shown in Fig. 2, where $B = 0.340$ m; $C = 0.460$ m; $D = 0.025$ m; $D_H = 0.400$ m; $d_o = 0.03$ m; and $p = 0.045$ m.

A. Calculate the shell-side heat-transfer coefficient, h_o.

From Eq. (1), the length of coil needed is:

$$L = N\sqrt{(2\pi(0.2)^2 + (0.045)^2}$$
$$= 1.257N$$

Using Eq. (2-4), the volume available for fluid flow in the annulus, V_f, is:

$$V_f = [(\pi/4)(0.46^2 - 0.34^2)(0.045)N] -$$
$$[(\pi/4)(0.03)^2(1.257N)]$$
$$= 2.504 \times 10^{-3}N$$

The shell-side equivalent diameter is:

$$D_e = (4)(2.504 \times 10^{-3}N)/(\pi)(0.03)(1.257N)$$
$$= 0.0845 \text{ m}$$

The mass velocity of the fluid is:

$$G_s = (2{,}141)/[(\pi/4)((0.46^2 - 0.34^2) - (0.43^2 - 0.37^2))]$$
$$= 56{,}792 \text{ kg/(m}^2)(\text{h})$$

The Reynolds number is:

$$N_{Re} = (0.0845)(56{,}792)/(5.76)$$
$$= 833$$

Using Eq. (6), we get:

$$h_o = (0.6)(0.4075/0.0845)(833)^{0.5}$$
$$((1)(5.76)/(0.4075))^{0.31}$$
$$= 190$$

B. Compute h_{io}, the heat-transfer coefficient inside the coil.

The fluid velocity is:

$$u = q/A_f$$

where $A_f = \pi D^2/4 = 4.909 \times 10^{-4} \text{ m}^2$ and $q = M/\rho = 1.552 \text{ m}^3/\text{h}$, so that:

$$u = 1.552/(4.909 \times 10^{-4})$$
$$= 3{,}161.5 \text{ m/h}$$

The Reynolds number (tube-side) is then:

$$N_{Re} = (0.025)(3{,}161.5)(870)/(1.89)$$
$$= 36{,}383$$

From Fig. 3, j_H (for $N_{Re} = 36{,}383$) is 110, and:

$$h_i = j_H(k/D)(N_{Pr})^{1/3}$$
$$= 3{,}046 \text{ kcal/(h)(m}^2)(^\circ\text{C})$$

Corrected for a coiled tube, this becomes:

$$h_{ic} = (3{,}046)[1 + 3.5(0.025/0.400)]$$
$$= 3{,}712 \text{ kcal/(h)(m}^2)(^\circ\text{C})$$

The heat-transfer coefficient based on the outside diameter of the coil is:

$$h_{io} = (3{,}712)(0.025/0.03)$$
$$= 3{,}093 \text{ kcal/(h)(m}^2)(^\circ\text{C})$$

C. Calculate the overall heat-transfer coefficient, U. The coil-wall thickness, x, is:

$$x = (d_o - D)/2$$
$$= 0.0025 \text{ m}$$

The fouling factors, R_t and R_a, depend on the nature of the liquids, the presence of suspended matter in the liquids, the operating temperatures, and the velocities of the fluids. In this case, both R_a and R_t are 8.2×10^{-4} (h)(m^2)($^\circ$C)/kcal. The thermal conductivity of stainless steel is $k_c = 14$ kcal/(h)(m)($^\circ$C).

Using Eq. (9):

$$1/U = 1/190 + 1/3{,}093 + 0.0025/14 + 0.00082 + 0.00082$$
$$U = 135 \text{ kcal/(h)(m}^2)(^\circ\text{C})$$

D. Determine the required area.

The log-mean-temperature-difference is:

$$\Delta t_{lm} = \frac{[(127 - 30) - (100 - 47)]}{\ln((127 - 30)/(100 - 47))}$$
$$= 72.8^\circ\text{C}$$

To account for perpendicular flow, the correction factor [6] is 0.99, so that:

$$\Delta t_c = (0.99)(72.8) = 72.1^\circ\text{C}$$

The heat load is:

$$Q = (1{,}350)(1.0)(127 - 100)$$
$$= 36{,}450 \text{ kcal/h}$$

Using Eq. (10), the required area is:

$$A = (36{,}450)/(135)(72.1)$$
$$= 3.745 \text{ m}^2$$

E. Calculate the number of turns of coil required. From Eq. (11):

$$N = (3.745)/(\pi)(0.03)(1.257)$$
$$= 31.6$$

and:

$$n = 32$$

The height of the cylinder needed to accommodate 32 turns of coil is:

$$H = (32)(0.045) + (0.03)$$
$$= 1.470 \text{ m}$$

References

1. Dimopłon, W., Finding the length of helical coils, *Chem. Eng.*, Oct. 23, 1978, p. 177.
2. Edwards, M. F., others, Heat Transfer and Pressure Drop Characteristics of a Plate Exchanger Using Newtonian and Non-Newtonian Liquids, *The Chemical Engineer*, May 1974, pp. 286–288, 293.
3. Coates, J., and Pressburg, B. S., Heat Transfer to Moving Fluids, *Chem. Eng.*, Dec. 28, 1959, pp. 67–72.
4. Kern, D. Q., "Process Heat Transfer," pp. 103, 137, 549, 721, 834, McGraw-Hill, New York, 1950.

The authors

R. K. Patil   B. W. Shende   P. K. Ghosh

Ramachandra K. Patil is Superintendent, Process Engineering at Rathi Industrial Equipment Co., Pune 411 001, India. He is involved in the design of pneumatic solids-handling and other process equipment. He obtained a bachelor's degree in chemical engineering in 1970 from Bombay University.

B. W. Shende was Technical Executive at Kirloskar Consultants Pvt. Ltd. when this article was written.. He is now a Production Officer at Polychem Ltd., Pune 412 102, India. He obtained master's and Ph.D. degrees in chemical engineering from Bombay University.

Prasanta K. Ghosh is Corporate Planning Manager at Hindustan Antibiotics Ltd., Pimpri, Pune 411 018, India. He previously worked for the government of India and was involved in the planning and development of the Indian drug industry. He received a master of technology degree in chemical engineering in 1963 and a Ph.D. in chemical kinetics in 1968, both from Calcutta University.

Mean temperature difference in heat exchangers

Correction factors for log mean temperature difference are related to the ratio of outlet temperature gap to inlet temperature difference, instead of to the Bowman-Mueller-Nagle parameters.

Ronald E. Wales, Texaco Inc.

☐ When making heat-exchanger calculations, engineers generally rely on the familiar formula:

$$Q = UA\Delta T_M \qquad (1)$$

The mean temperature difference is usually calculated as:

$$\Delta T_M = F\Delta T_{LM} \qquad (2)$$

The countercurrent log mean temperature difference is computed via:

$$\Delta T_{LM} = \frac{(T_1 - t_2) - (T_2 - t_1)}{\ln\left[(T_1 - t_2)/(T_2 - t_1)\right]} \qquad (3)$$

The term F, the correction factor for ΔT_{LM}, typically depends on the four terminal temperatures and the configuration of the heat exchanger.

Bowman-Mueller-Nagle F relationship

Many researchers have attempted to determine the proper mean temperature differences for the numerous types of heat exchangers. In 1940, Bowman, Mueller and Nagle published ΔT_{LM} correction factors for many common types of exchangers [1]. For typical shell-and-tube exchangers, they expressed F as a function of two dimensionless parameters:

$$F = f_1(P, R) \qquad (4)$$

Here, P is the cold-fluid temperature efficiency:

$$P = (t_2 - t_1)/(T_1 - t_1) \qquad (5)$$

And R is the hot-fluid-to-cold-fluid temperature-change ratio:

$$R = (T_1 - T_2)/(t_2 - t_1) \qquad (6)$$

Bowman, Mueller and Nagle represented $F = f_1(P, R)$ by means of equations and graphs for different types of exchangers.

F as a function of outlet gap

This article presents another method for determining ΔT_{LM} correction factors. For typical shell-and-tube exchangers, it shows that:

$$F = f_2(G, R) \qquad (7)$$

For some common situations, it further demonstrates that:

$$F \cong f_3(G) \qquad (8)$$

In Eq. (7), and in the approximate relationship, Eq. (8), G is defined as the ratio of the outlet temperature gap (or approach) to the inlet temperature difference:

$$G = (T_2 - t_2)/(T_1 - t_1) \qquad (9)$$

Values of G can range from $+1$ (no heat exchange) to -1 (highest heat exchange). When outlet temperatures are equal ($T_2 = t_2$), $G = 0$. A negative G value represents a fluid-temperature cross (i.e., temperature t_2 exceeds temperature T_2).

Verifying Eq. (7)

Bowman, Mueller and Nagle gave the following equations for ΔT_{LM} correction factors of one-two and two-four shell-and-tube exchangers:

One-pass shellside, two-pass tubeside

$$F_{1,2} = \frac{\left[\dfrac{(R^2+1)^{1/2}}{R-1}\right]\ln\left(\dfrac{1-P}{1-PR}\right)}{\ln\left[\dfrac{(2/P)-1-R+(R^2+1)^{1/2}}{(2/P)-1-R-(R^2+1)^{1/2}}\right]} \qquad (10)$$

Two-pass shellside, four-pass tubeside

$$F_{2,4} = \frac{\left[\dfrac{(R^2+1)^{1/2}}{2(R-1)}\right]\ln\left(\dfrac{1-P}{1-PR}\right)}{\ln\left\{\dfrac{A+B+(R^2+1)^{1/2}}{A+B-(R^2+1)^{1/2}}\right\}} \qquad (11)$$

In Eq. (11), $A = (2/P) - 1 - R$, and $B = (2/P)[(1-P)(1-PR)]^{1/2}$.

The numerators of Eq. (10) and (11) become indeterminate when $R = 1$. The usual treatment of such indeterminates is:

$$[1/(R-1)]\ln[(1-P)/(1-PR)] = P/(1-P) \qquad (12)$$

The graphical representations of Eq. (10) and (11) are familiar to engineers who make heat-exchanger calculations.

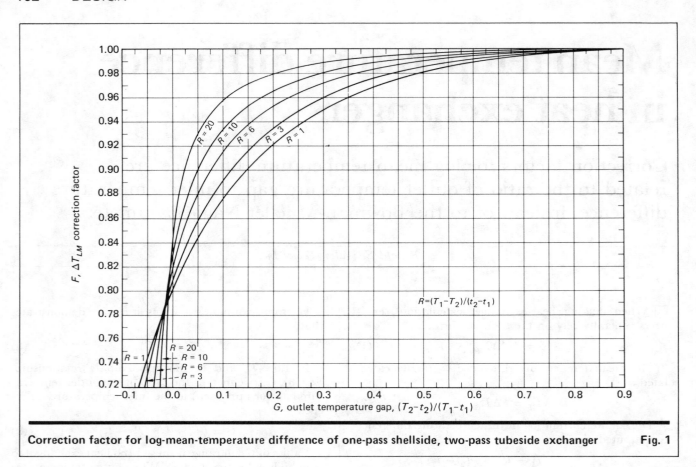

Correction factor for log-mean-temperature difference of one-pass shellside, two-pass tubeside exchanger   Fig. 1

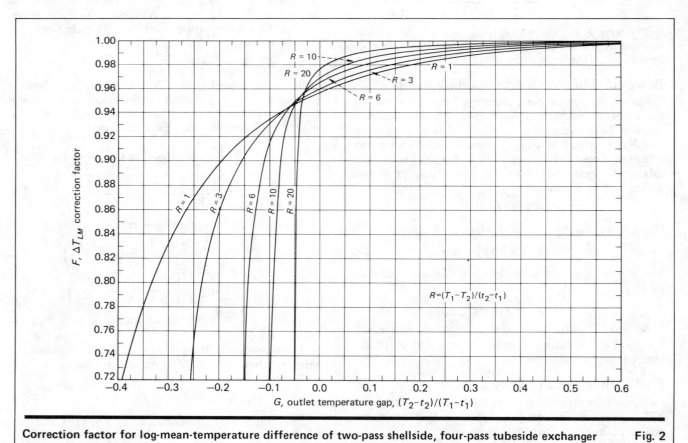

Correction factor for log-mean-temperature difference of two-pass shellside, four-pass tubeside exchanger   Fig. 2

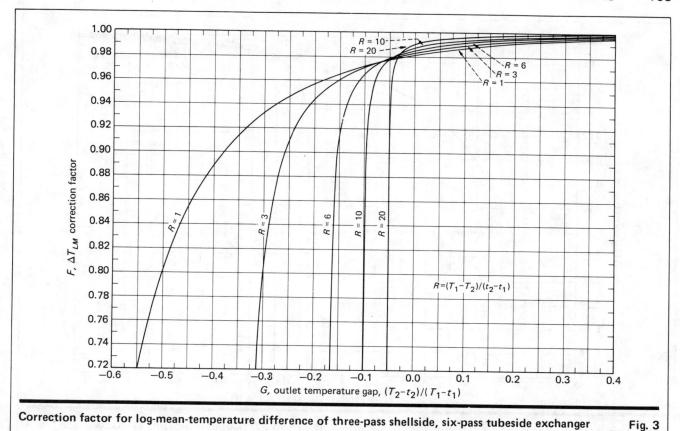

Correction factor for log-mean-temperature difference of three-pass shellside, six-pass tubeside exchanger   Fig. 3

Eq. (10) and (11) are based on the following assumptions of Bowman, Mueller and Nagle:

1. The overall heat-transfer coefficient is constant throughout the exchanger.
2. The flowrate of each fluid is constant.
3. The specific heat of each fluid is constant.
4. There is no condensation or boiling.
5. Heat losses are negligible.
6. The heat-transfer surface of each pass is equal.
7. The temperature of the shellside fluid in any shellside pass is uniform over any cross section.

Examination of the definitions of P, R and G in, respectively, Eq. (5), (6) and (9) will show that they are related algebraically by:

$$P = (1 - G)/(1 + R) \qquad (13)$$

Eq. (4) and (13), when combined, establish the validity of Eq. (7).

F in terms of G

Substituting the cold-fluid temperature efficiency term, P, from Eq. (13) into Eq. (10) and (11), and simplifying, yields the following equations for the ΔT_{LM} correction factors of one-two and two-four shell-and-tube exchangers:

One-pass shellside, two-pass tubeside

$$F_{1,2} = \frac{\left[\dfrac{(R^2 + 1)^{1/2}}{R - 1} \right] \ln\left(\dfrac{R + G}{1 + GR} \right)}{\ln\left\{ \dfrac{C + D}{C - D} \right\}} \qquad (14)$$

Nomenclature

A	Heat-transfer surface area, ft²	T_1	Hot-fluid inlet temperature, °F
F	ΔT_{LM} correction factor, dimensionless	T_2	Hot-fluid outlet temperature, °F
G	Outlet temperature gap, $(T_2 - t_2)/(T_1 - t_1)$, dimensionless	t_1	Cold-fluid inlet temperature, °F
		t_2	Cold-fluid outlet temperature, °F
N	Number of shellside passes	U	Overall heat-transfer coefficient, Btu/(h)(ft²) (°F)
P	Cold-fluid temperature efficiency, $(t_2 - t_1)/(T_1 - t_1)$, dimensionless		
Q	Heat-transfer rate, Btu/h	ΔT_{LM}	Countercurrent log mean temperature difference, °F
R	Hot-fluid-to-cold-fluid temperature-change ratio, $(T_1 - T_2)/(t_2 - t_1)$, dimensionless	ΔT_M	Mean temperature difference, °F

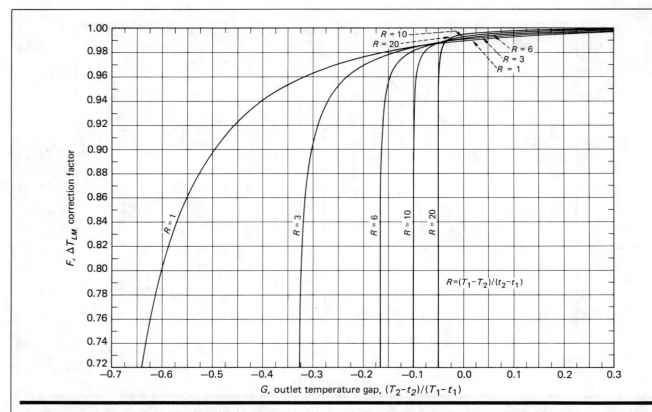

Correction factor for log-mean-temperature difference of four-pass shellside, eight-pass tubeside exchanger   **Fig. 4**

Two-pass shellside, four-pass tubeside

$$F_{2,4} = \frac{\left[\dfrac{(R^2 + 1)^{1/2}}{2(R - 1)}\right] \ln\left(\dfrac{R + G}{1 + GR}\right)}{\ln\left\{\dfrac{C + 2[(R + G)(1 + GR)]^{1/2} + D}{C + 2[(R + G)(1 + GR)]^{1/2} - D}\right\}} \quad (15)$$

In Eq. (14) and (15), $C = (R + 1)(1 + G)$, and $D = [(1 - G)(R^2 + 1)]^{1/2}$.

The numerators in Eq. (14) and (15) become indeterminate when $R = 1$. By the usual treatment:

$$[1/(R - 1)] \ln [(R + G)/(1 + GR)]$$
$$= (1 - G)/(1 + G) \quad (16)$$

The same seven assumptions of Bowman, Mueller and Nagle for Eq. (10) and (11) also apply to Eq. (14) and (15).

Advantage of revised equations

Eq. (14) and (15) are mathematically equivalent to Eq. (10) and (11). The only advantage of Eq. (14) and (15) is revealed by their graphic representations (Fig. 1 and 2, respectively), which show that G is a more powerful parameter than either P or R, because G, by itself, provides more information about the value of F than does either P or R alone.

That the correction factor F is the same for R and $1/R$ is apparent from Eq. (14) and (15). Substituting $1/R$ for R results in no change after simplification.

The correction factor is the same whether the shellside fluid enters at the fixed- or floating-head end of the shell (as noted by Bowman, Mueller and Nagle). This is the case with the one-two exchanger, as well as with the two-four exchanger made up of two shells. Therefore, in the latter case, the nozzles directly connecting two stacked shells can be located at either end of the shells, with the piping nozzles of the exchanger placed at the other ends.

The correction factor F is also the same whether the hot fluid is on the shellside or tubeside. However, the placement of the hot fluid could make a significant difference in the overall heat-transfer coefficient, U, or the fluid pressure drops, or both.

Fig. 1 makes possible a practical generalization about the design of one-two shell-and-tube exchangers, namely: Outlet temperatures may be allowed to be equal, but fluid temperatures should not be permitted to cross, because the ΔT_{LM} correction factor becomes unsatisfactorily low.

Factors for more-complex exchangers

Bowman, Mueller and Nagle also presented a general method for calculating F factors for three-six, four-eight, five-ten and six-twelve exchangers, from the F factors of one-two exchangers. They stated that, at any particular values of F and R, the value of P for an exchanger having N shellside passes and $2N$ tubeside passes is related to P of a one-two exchanger by the equation:

$$P_{N,2N} = \frac{1 - [(1 - P_{1,2}R)/(1 - P_{1,2})]^N}{R - [(1 - P_{1,2}R)/(1 - P_{1,2})]^N} \quad (17)$$

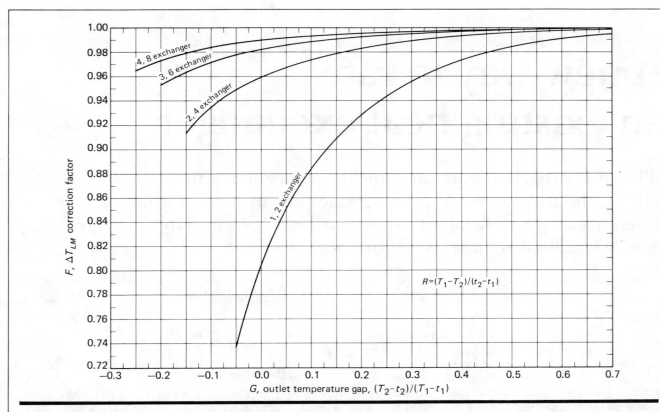

Correction factor for log-mean-temperature difference when R is greater than 0.33 but less than 3.00   Fig. 5

For the special case of $R = 1$, Eq. (17) becomes indeterminate, and the usual treatment of such indeterminates yields:

$$P_{N,2N} = (P_{1,2}N)/(P_{1,2}N - P_{1,2} + 1) \qquad (18)$$

Eq. (13), (17) and (18) may be combined to arrive at similar equations in terms of G. Therefore, for any particular values of F and R, the value of G for an exchanger having N shellside passes and $2N$ tubeside passes is related to G for a one-two exchanger by:

$$G_{N,2N} = 1 - (1 + R)\left\{ \frac{[(1 + RG_{1,2})/(R + G_{1,2})]^N - 1}{[(1 + RG_{1,2})/(R + G_{1,2})]^N - R} \right\} \qquad (19)$$

For the special case of $R = 1$, Eq. (19) becomes indeterminate, and the usual treatment gives:

$$G_{N,2N} = \frac{G_{1,2} + 1 - N(1 - G_{1,2})}{G_{1,2} + 1 + N(1 - G_{1,2})} \qquad (20)$$

Values of F vs. G are plotted in Fig. 3 and 4 for, respectively, three-six and four-eight exchangers.

Examination of Fig. 1, 2, 3 and 4 show that F is primarily related to G, with relatively little error, for the common situation of no temperature cross ($G \geqq 0$) and R between 0.33 and 3.00. Therefore, the approximate-relationship Eq. (8) must apply to the foregoing situation, as was stated earlier.

Eq. (8) may be determined mathematically for one-two shell-and-tube exchangers by setting $R = 2.15$ in

Eq. (14). When $G \geqq -0.05$, and $0.33 < R < 3.00$:

$$F_{1,2} \cong \frac{2.0619 \ln\left[(2.15 + G)/(1 + 2.15G)\right]}{\ln\left(\dfrac{5.5212 + 0.7788\,G}{0.7788 + 5.5212\,G}\right)} \qquad (21)$$

Eq. (21) is accurate to within 1%. (Note that $G < -0.05$ is not a normal design situation because F would be less than 0.75.)

Eq. (21) can be combined with Eq. (19) to yield approximate values of F for two-four, three-six and four-eight exchangers. These are plotted in Fig. 5.

Reference

1. Bowman, R. A., Mueller, A. C., and Nagle, W. M., Mean Temperature Difference in Design, *Trans. of ASME*, May 1940, pp. 283–293.

The author

Ronald E. Wales is a technologist with Texaco Inc., Engineering Dept., P. O. Box 52332, Houston, TX 77052. During 23 years with Texaco, he has been engaged in refinery technical service for more than half that time, with the rest being spent in refinery process design. He has served as Chief Process Engineer at Texaco's Los Angeles plant, and been involved in several unit startups. Holder of a B.S.Ch.E. from the University of Illinois and an M.S.Ch.E. from Purdue University, he is a registered engineer in Texas.

A new way to rate an existing heat exchanger

With this program, you can quickly determine what will be the unknown temperatures, log-mean temperature-difference correction factor, and heat load for any job that an exchanger may be required to do.

Rogério G. Herkenhoff, *Petróleo Brasileiro S.A.*

☐ Prediction of the thermal performance of an existing multipass exchanger is usually carried out by an iterative procedure in the following way:

1. Compute the overall heat-transfer coefficient, based on the most recent information about fluid temperatures. (In the first trial, using an arbitrary value may be advantageous.)
2. Calculate the unknown temperatures.
3. If necessary, return to Step 1.

In the second step, one unknown temperature is calculated from a dimensionless group (P), and the other is usually computed via a thermal balance.

The value of P can be obtained graphically [1,2] if the number of shell passes does not exceed two, or by means of a trial-and-error solution involving the calculation of the log-mean temperature-difference (LMTD) correction factor. This time-consuming method can be avoided by using a noniterative general solution.

Developing a general solution

The assumptions are the same as those made for the derivation of the LMTD correction factor [3]. In addition, the number of tube passes must be a multiple of the number of shell passes.

For an exchanger having N shell passes, Bowman [4] developed a general solution for the correction factor:

$$F = \frac{\sqrt{R^2 + 1}\, \ln\left[(1 - X)/(1 - RX)\right]}{(R - 1) \ln \dfrac{2 - X(R + 1 - \sqrt{R^2 + 1})}{2 - X(R + 1 + \sqrt{R^2 + 1})}} \tag{1}$$

where:

$$X = \frac{1 - \left(\dfrac{1 - RP}{1 - P}\right)^{1/N}}{R - \left(\dfrac{1 - RP}{1 - P}\right)^{1/N}} \tag{2}$$

$$R = \frac{T_1 - T_2}{t_2 - t_1} = \frac{wc}{WC} \tag{3}$$

$$P = \frac{t_2 - t_1}{T_1 - t_1} \tag{4}$$

From the original definition:

$$F = \frac{(UA/wc)_{cc}}{(UA/wc)} = \frac{\ln\left[(1 - P)/(1 - RP)\right]}{(R - 1)(UA/wc)}$$

Substituting F in Eq. (1):

$$\frac{\ln\left[(1 - P)/(1 - RP)\right]}{(UA/wc)} = \frac{\sqrt{R^2 + 1}\, \ln\left[(1 - X)/(1 - RX)\right]}{\ln \dfrac{2 - X(R + 1 - \sqrt{R^2 + 1})}{2 - X(R + 1 + \sqrt{R^2 + 1})}} \tag{5}$$

From Eq. (2), it is possible to prove that:

$$\ln\left[(1 - P)/(1 - RP)\right] = N \ln\left[(1 - X)/(1 - RX)\right]$$

Substituting in Eq. (5), and rearranging:

$$X = \frac{2E - 2}{(R + 1 + \sqrt{R^2 + 1})E - (R + 1 - \sqrt{R^2 + 1})} \tag{6}$$

where:

$$E = e^{(UA/wc)\sqrt{R^2 + 1}/N}$$

From Eq. (2):

$$P = \frac{1 - \left(\dfrac{1 - RX}{1 - X}\right)^N}{R - \left(\dfrac{1 - RX}{1 - X}\right)^N} \qquad \text{when } (R \neq 1) \tag{7}$$

When $R = 1$, part of the equation becomes indeterminate, but Eq. (6) is still valid. However, Eq. (7) must be replaced by:

$$P = \frac{NX}{NX - X + 1} \qquad \text{when } (R = 1) \tag{7a}$$

Originally published March 23, 1981

Program to rate an existing exchanger | Table I

Step	Key	Code	Step	Key	Code	Step	Key	Code	Step	Key	Code	Step	Key	Code	Step	Key	Code	Step	Key	Code
001	*LBLa	21 16 11	039	STOI	35 46	077	SPC	16-11	115	P≷S	16-51	153	RCL8	36 08	191	STOA	35 11
002	STOB	35 12	040	GSB1	23 01	078	2	02	116	GSB9	23 09	154	×	-35	192	RCLE	36 15
003	X≷Y	-41	041	*LBL1	21 01	079	GSB0	23 00	117	RCLi	36 45	155	RCLC	36 13	193	*LBL2	21 02
004	STOD	35 14	042	GSB5	23 05	080	*LBL2	21 02	118	-	-45	156	÷	-24	194	-	-45
005	RTN	24	043	RCLE	36 15	081	1	01	119	×	-35	157	RCLB	36 12	195	RCLA	36 11
006	*LBLb	21 16 12	044	×	-35	082	P≷S	16-51	120	RCLi	36 45	158	÷	-24	196	*LBL6	21 06
007	P≷S	16-51	045	STO0	35 00	083	GSB4	23 04	121	+	-55	159	e^x	33	197	1	01
008	GSBc	23 16 13	046	RCLi	36 45	084	P≷S	16-51	122	GSB9	23 09	160	STO0	35 00	198	-	-45
009	P≷S	16-51	047	×	-35	085	GT01	22 01	123	STOi	35 45	161	RCL9	36 09	199	÷	-24
010	RTN	24	048	RCLi	36 45	086	*LBL7	21 07	124	GT07	22 07	162	×	-35	200	RTN	24
011	*LBLc	21 16 13	049	-	-45	087	PRTX	-14	125	*LBL5	21 05	163	RCL7	36 07	201	*LBL9	21 09
012	STO3	35 03	050	GSB9	23 09	088	*LBL1	21 01	126	GSB8	23 08	164	-	-45	202	DSZI	16 25 46
013	R↓	-31	051	P≷S	16-51	089	RTN	24	127	STOC	35 13	165	2	02	203	RTN	24
014	STO4	35 04	052	RCLi	36 45	090	RCLA	36 11	128	P≷S	16-51	166	RCL0	36 00	204	ISZI	16 26 46
015	R↓	-31	053	P≷S	16-51	091	1/X	52	129	GSB8	23 08	167	×	-35	205	ISZI	16 26 46
016	STO5	35 05	054	RCLA	36 11	092	GSB3	23 03	130	P≷S	16-51	168	2	02	206	RTN	24
017	RTN	24	055	ST+0	35-55 00	093	1/X	52	131	÷	-24	169	-	-45	207	*LBL8	21 08
018	*LBLd	21 16 14	056	×	-35	094	LN	32	132	1	01	170	÷	-24	208	RCL1	36 01
019	X≷Y	-41	057	+	-55	095	RCLE	36 15	133	STO7	35 07	171	GSB3	23 03	209	RCL2	36 02
020	P≷S	16-51	058	RCL0	36 00	096	GSB6	23 06	134	STO9	35 09	172	RCLB	36 12	210	+	-55
021	GSBe	23 16 15	059	GSB5	23 06	097	RCLD	36 14	135	X≷Y	-41	173	Y^X	31	211	2	02
022	P≷S	16-51	060	STOi	35 45	098	÷	-24	136	X≠Y?	16-32	174	GSB3	23 03	212	÷	-24
023	X≷Y	-41	061	P≷S	16-51	099	RCL8	36 08	137	GT01	22 01	175	1/X	52	213	RCL3	36 03
024	RTN	24	062	GT07	22 07	100	÷	-24	138	EEX	-23	176	STOA	35 11	214	×	-35
025	*LBLe	21 16 15	063	*LBLC	21 13	101	RCLC	36 13	139	5	05	177	RTN	24	215	RCL4	36 04
026	STO2	35 02	064	SPC	16-11	102	×	-35	140	CHS	-22	178	*LBL0	21 00	216	+	-55
027	X≷Y	-41	065	GSB1	23 01	103	R/S	51	141	+	-55	179	STOI	35 46	217	STO6	35 06
028	STO1	35 01	066	GT02	22 02	104	GSB8	23 08	142	*LBL1	21 01	180	GSB5	23 05	218	RCL5	36 05
029	RTN	24	067	*LBLD	21 14	105	RCL2	36 02	143	STOE	35 15	181	P≷S	16-51	219	×	-35
030	*LBLA	21 11	068	SPC	16-11	106	RCL1	36 01	144	ST+7	35-55 07	182	RCLi	36 45	220	RTN	24
031	STO8	35 08	069	1	01	107	-	-45	145	ST+9	35-55 09	183	P≷S	16-51	221	R/S	51
032	P≷S	16-51	070	P≷S	16-51	108	×	-35	146	X²	53	184	×	-35			
033	STO8	35 08	071	GSB0	23 00	109	RTN	24	147	+	-55	185	RCLi	36 45			
034	P≷S	16-51	072	P≷S	16-51	110	*LBL4	21 04	148	√X	54	186	GSB2	23 02			
035	RTN	24	073	*LBL1	21 01	111	STOI	35 46	149	ST-7	35-45 07	187	GSB3	23 03			
036	*LBLB	21 12	074	2	02	112	GSB5	23 05	150	ST+9	35-55 09	188	STOi	35 45			
037	SPC	16-11	075	GT04	22 04	113	P≷S	16-51	151	RCLD	36 14	189	GT07	22 07			
038	2	02	076	*LBLE	21 15	114	RCLi	36 45	152	×	-35	190	*LBL3	21 03			

User's instructions | Table II

Necessary input (in any order):

1 — $A \uparrow N$ .... f [a]
2 — $W \uparrow \beta \uparrow \alpha$ (hot fluid) .... f [b]
3 — $w \uparrow \beta \uparrow \alpha$ (cold fluid) .... f [c]
4 — $T_1 \uparrow T_2$ .... f [d]
5 — $t_1 \uparrow t_2$ .... f [e]
6 — U .... [A]

Output:

1 — To print t_1 and T_1 .... [B]
2 — To print t_2 and T_2 .... [C]
3 — To print T_1 and t_2 .... [D]
4 — To print t_1 and T_2 .... [E]
5 — After pressing B, C, D or E:
   (a) To display F .... [R/S]
   (b) To display q .... [R/S]

Registers:

0 Used	1 t_1	2 t_2	3 α_c	4 β_c	5 w	6 c	7 Used	8 U	9 Used
S0 Used	S1 T_2	S2 T_1	S3 α_h	S4 β_h	S5 W	S6 C	S7 Used	S8 U	S9 Used
A Used	B N	C wc or WC	D A	E R	I Used				

where: $c = \beta_c + \alpha_c t_m$
$C = \beta_h + \alpha_h T_m$

When the cold fluid is isothermally vaporized ($R = \infty$), or when in Eq. (4) only one temperature is known, the definitions of R and P [Eq. (3) and (4)] may be changed to:

$$R = \frac{t_2 - t_1}{T_1 - T_2} = \frac{WC}{wc} \qquad (3a)$$

$$P = \frac{T_1 - T_2}{T_1 - t_1} \qquad (4a)$$

In this case, in Eq. (6), E must be replaced by:

$$E = e^{(UA/WC)\sqrt{R^2 + 1}/N}$$

Program description

The HP-97/67/41-C program calculates any combination of two terminal temperatures in an existing multipass exchanger, except if both refer to the same fluid.

The specific heats are calculated as a linear function of the arithmetic mean temperatures of the fluids, using the last temperatures stored. Thus, unless the desired temperatures had been preliminarily stored, or the specific heats had been considered to be constants ($\alpha = 0$), the temperatures will not be precisely calculated.

So, an iterative procedure must be used to yield a consistent result. The desired calculation is repeated (by pressing B, C, D or E) until the calculated temperatures remain approximately constant. An initial estimate of the desired temperatures may be stored to avoid using the remaining values of any previous operation and to improve the convergence.

Also, the overall heat-transfer coefficient is often seriously affected by the variation of the physical properties of the fluids and cannot be precisely evaluated without knowledge of both inlet and outlet fluid temperatures. Therefore, the iterative procedure must usually include the replacement of the previous estimate of the overall coefficient with a new one, computed for the latest temperature conditions.

After the convergence, the program can calculate the LMTD correction factor and the heat load, based on the cold-fluid conditions.

Any system of units may be used, as long as UA/wc remains dimensionless.

The program can also accommodate a pure counterflow exchanger, assuming a number of shell passes sufficiently large to make the LMTD correction factor equal to 1.

When $R = 1$, the program assumes $R = 1.00001$ to permit the solution without using Eq. (7a).

The second unknown temperature is calculated from a second value of P, which is evaluated by using the first calculated temperature to compute the specific heat.

Rating an exchanger for heating crude oil

It is desired to heat 700,000 lb/h of a 36.4° API gravity crude ($k = 11.5$) at 433°F, using 400,000 lb/h of a 38° API gravity oil ($k = 12.1$) at 556°F. Available for this service is a 3:6 exchanger having a heat-transfer surface area of 15,000 ft². Assuming a total dirt factor of 0.005, and that the individual heat-transfer coefficients may be estimated by the following equations, what will the outlet temperatures be?

$$h_c = 25 + 0.35\,t_m \quad \text{and} \quad h_h = 160 + 0.30\,T_m$$

Nomenclature

A	Heat-transfer surface area
c	Cold-fluid specific heat
C	Hot-fluid specific heat
E	Dimensionless group
F	LMTD correction factor
h	Individual heat-transfer coefficient
k	Characterization factor
N	Total number of shell passes
P	Dimensionless group
q	Heat load
R	Dimensionless group
R_d	Fouling factor
t	Cold-fluid temperature
T	Hot-fluid temperature
U	Overall heat-transfer coefficient
w	Cold-fluid mass flowrate
W	Hot-fluid mass flowrate
X	Dimensionless group
α	Angular coefficient for specific-heat calculation
β	Linear coefficient for specific-heat calculation

Subscripts

c	Cold fluid
cc	Countercurrent
h	Hot fluid
m	Arithmetic mean
1	Inlet
2	Outlet

Using the equation presented in Ref. 2, p. 149:

$$\beta_c = 0.4142$$
$$\alpha_c = 0.0005474$$
$$\beta_h = 0.4306$$
$$\alpha_h = 0.0005682$$

In the first trial, let us assume that $U = 50$, $t_2 = t_1$, and $T_2 = T_1$.

Preliminary input:

1—15,000↑3	f [a]	
2—400,000↑.4306↑.0005682	f [b]	
3—700,000↑.4142↑.0005474	f [c]	
4—556↑	f [d]	
5—433↑	f [e]	

1st trial:

1—50	[A]	
2—(Compute t_2 and T_2)	[C]	($t_2 = 496.01$, $T_2 = 459.29$)

2nd trial:

1—Compute the overall coefficient ($U = 73.89$)		
2—73.89	[A]	
3—Compute t_2 and T_2	[C]	($t_2 = 498.77$, $T_2 = 448.98$)

3rd trial:

1—Compute the overall coefficient ($U = 73.88$)		
2—73.88	[A]	
3—Compute t_2 and T_2	[C]	($t_2 = 498.53$, $T_2 = 448.83$)

4th trial:

1—Compute the overall coefficient ($U = 73.87$)		
2—73.87	[A]	
3—Compute t_2 and T_2	[C]	($t_2 = 498.53$, $T_2 = 448.83$)
4—Compute F	[R/S]	($F = 0.85792$)
5—Compute q	[R/S]	($q = 3.06972 \times 10^7$)

For TI-58/59 users

The TI programs closely follow the HP version. However, program A (see Table III listing) supplies output 1 and 2 of Table II. (i.e., output 1 is inlet temperatures, and output 2 is outlet temperatures). Program B (listing in Table IV) offers output 3 and 4. Table V provides user instructions.

Listing for TI version—program A   **Table III**

Step	Code	Key	Step	Code	Key	Step	Code	Key
000	76	LBL	012	42	STD	024	65	×
001	11	A	013	25	25	025	43	RCL
002	42	STD	014	71	SBR	026	24	24
003	08	08	015	22	INV	027	95	=
004	42	STD	016	71	SBR	028	42	STD
005	18	18	017	23	LNX	029	00	00
006	01	1	018	76	LBL	030	65	×
007	32	X:T	019	22	INV	031	73	RC*
008	91	R/S	020	71	SBR	032	25	25
009	76	LBL	021	24	CE	033	75	-
010	12	B	022	43	RCL	034	73	RC*
011	02	2	023	20	20	035	25	25

Step	Code	Key	Step	Code	Key	Step	Code	Key	Step	Code	Key	Step	Code	Key
036	95	=	100	61	GTO	164	25	25	228	85	+	292	43	RCL
037	42	STO	101	34	ΓX	165	67	EQ	229	01	1	293	24	24
038	40	40	102	76	LBL	166	44	SUM	230	95	=	294	95	=
039	01	1	103	33	X²	167	02	2	231	34	ΓX	295	55	÷
040	22	INV	104	02	2	168	42	STO	232	22	INV	296	53	(
041	44	SUM	105	42	STO	169	25	25	233	44	SUM	297	43	RCL
042	25	25	106	25	25	170	76	LBL	234	07	07	298	20	20
043	43	RCL	107	61	GTO	171	44	SUM	235	44	SUM	299	75	-
044	25	25	108	42	STO	172	43	RCL	236	09	09	300	01	1
045	67	EQ	109	76	LBL	173	43	43	237	65	×	301	54	)
046	25	CLR	110	34	ΓX	174	72	ST*	238	43	RCL	302	95	=
047	02	2	111	71	SBR	175	25	25	239	23	23	303	35	1/X
048	42	STO	112	23	LNX	176	99	PRT	240	65	×	304	42	STO
049	25	25	113	01	1	177	61	GTO	241	43	RCL	305	20	20
050	76	LBL	114	42	STO	178	45	Y×	242	08	08	306	92	RTN
051	25	CLR	115	25	25	179	76	LBL	243	55	÷	307	76	LBL
052	43	RCL	116	71	SBR	180	24	CE	244	43	RCL	308	52	EE
053	20	20	117	42	STO	181	71	SBR	245	22	22	309	53	(
054	44	SUM	118	76	LBL	182	52	EE	246	55	÷	310	43	RCL
055	00	00	119	42	STO	183	43	RCL	247	43	RCL	311	01	01
056	71	SBR	120	71	SBR	184	44	44	248	21	21	312	85	+
057	23	LNX	121	24	CE	185	42	STO	249	95	=	313	43	RCL
058	73	RC*	122	71	SBR	186	22	22	250	22	INV	314	02	02
059	25	25	123	23	LNX	187	71	SBR	251	23	LNX	315	54	)
060	65	×	124	73	RC*	188	23	LNX	252	42	STO	316	55	÷
061	43	RCL	125	25	25	189	71	SBR	253	00	00	317	02	2
062	20	20	126	42	STO	190	52	EE	254	65	×	318	65	×
063	85	+	127	42	42	191	71	SBR	255	43	RCL	319	43	RCL
064	43	RCL	128	71	SBR	192	23	LNX	256	09	09	320	03	03
065	40	40	129	23	LNX	193	43	RCL	257	75	-	321	85	+
066	95	=	130	01	1	194	22	22	258	43	RCL	322	43	RCL
067	42	STO	131	22	INV	195	55	÷	259	07	07	323	04	04
068	41	41	132	44	SUM	196	43	RCL	260	95	=	324	95	=
069	71	SBR	133	25	25	197	44	44	261	55	÷	325	42	STO
070	23	LNX	134	43	RCL	198	95	=	262	53	(	326	06	06
071	43	RCL	135	25	25	199	42	STO	263	02	2	327	65	×
072	41	41	136	67	EQ	200	45	45	264	65	×	328	43	RCL
073	55	÷	137	43	RCL	201	01	1	265	43	RCL	329	05	05
074	53	(	138	02	2	202	42	STO	266	00	00	330	95	=
075	43	RCL	139	42	STO	203	07	07	267	75	-	331	42	STO
076	00	00	140	25	25	204	42	STO	268	02	2	332	44	44
077	75	-	141	76	LBL	205	09	09	269	54	)	333	92	RTN
078	01	1	142	43	RCL	206	43	RCL	270	95	=	334	76	LBL
079	54	)	143	53	(	207	45	45	271	42	STO	335	23	LNX
080	95	=	144	43	RCL	208	22	INV	272	20	20	336	00	0
081	72	ST*	145	42	42	209	67	EQ	273	75	-	337	42	STO
082	25	25	146	75	-	210	53	(	274	43	RCL	338	49	49
083	76	LBL	147	73	RC*	211	01	1	275	24	24	339	76	LBL
084	55	÷	148	25	25	212	52	EE	276	95	=	340	61	GTO
085	73	RC*	149	54	)	213	94	+/-	277	55	÷	341	73	RC*
086	25	25	150	65	×	214	05	5	278	53	(	342	49	49
087	99	PRT	151	43	RCL	215	44	SUM	279	43	RCL	343	32	X:T
088	76	LBL	152	20	20	216	45	45	280	20	20	344	01	1
089	45	Y×	153	85	+	217	76	LBL	281	75	-	345	00	0
090	87	IFF	154	73	RC*	218	53	(	282	01	1	346	44	SUM
091	01	01	155	25	25	219	43	RCL	283	54	)	347	49	49
092	32	X:T	156	95	=	220	45	45	284	95	=	348	73	RC*
093	86	STF	157	42	STO	221	42	STO	285	45	Y×	349	49	49
094	01	01	158	43	43	222	24	24	286	43	RCL	350	32	X:T
095	92	RTN	159	01	1	223	44	SUM	287	21	21	351	72	ST*
096	76	LBL	160	22	INV	224	07	07	288	95	=	352	49	49
097	13	C	161	44	SUM	225	44	SUM	289	42	STO	353	01	1
098	71	SBR	162	25	25	226	09	09	290	20	20	354	00	0
099	33	X²	163	43	RCL	227	33	X²	291	75	-	355	22	INV

Step	Code	Key	Step	Code	Key	Step	Code	Key	Step	Code	Key	Step	Code	Key
356	44	SUM	376	32	X:T	396	75	–	416	43	RCL	436	42	STO
357	49	49	377	92	RTN	397	01	1	417	08	08	437	06	06
358	32	X:T	378	76	LBL	398	54	)	418	95	=	438	71	SBR
359	72	ST*	379	32	X:T	399	95	=	419	99	PRT	439	23	LNX
360	49	49	380	22	INV	400	35	1/X	420	53	(	440	43	RCL
361	01	1	381	86	STF	401	23	LNX	421	43	RCL	441	06	06
362	44	SUM	382	01	01	402	65	×	422	01	01	442	65	×
363	49	49	383	43	RCL	403	43	RCL	423	85	+	443	43	RCL
364	01	1	384	20	20	404	22	22	424	43	RCL	444	05	05
365	00	0	385	35	1/X	405	55	÷	425	02	02	445	65	×
366	32	X:T	386	42	STO	406	53	(	426	54	)	446	53	(
367	43	RCL	387	20	20	407	43	RCL	427	55	÷	447	43	RCL
368	49	49	388	75	–	408	24	24	428	02	2	448	02	02
369	67	EQ	389	43	RCL	409	75	–	429	65	×	449	75	–
370	65	×	390	24	24	410	01	1	430	43	RCL	450	43	RCL
371	61	GTO	391	95	=	411	54	)	431	03	03	451	01	01
372	61	GTO	392	55	÷	412	55	÷	432	85	+	452	54	)
373	76	LBL	393	53	(	413	43	RCL	433	43	RCL	453	95	=
374	65	×	394	43	RCL	414	23	23	434	04	04	454	99	PRT
375	01	1	395	20	20	415	55	÷	435	95	=	455	91	R/S

Listing for VI version—program B

Table IV

Step	Code	Key	Step	Code	Key	Step	Code	Key	Step	Code	Key	Step	Code	Key
000	76	LBL	036	42	STO	072	43	RCL	108	76	LBL	144	43	RCL
001	11	A	037	25	25	073	25	25	109	44	SUM	145	45	45
002	42	STO	038	61	GTO	074	67	EQ	110	43	RCL	146	22	INV
003	08	08	039	42	STO	075	43	RCL	111	43	43	147	67	EQ
004	42	STO	040	76	LBL	076	02	2	112	72	ST*	148	53	(
005	18	18	041	15	E	077	42	STO	113	25	25	149	01	1
006	01	1	042	02	2	078	25	25	114	99	PRT	150	52	EE
007	32	X:T	043	42	STO	079	76	LBL	115	61	GTO	151	94	+/–
008	91	R/S	044	25	25	080	43	RCL	116	45	Y×	152	05	5
009	76	LBL	045	71	SBR	081	53	(	117	76	LBL	153	44	SUM
010	55	÷	046	35	1/X	082	43	RCL	118	24	CE	154	45	45
011	73	RC*	047	76	LBL	083	42	42	119	71	SBR	155	76	LBL
012	25	25	048	34	ΓX	084	75	–	120	52	EE	156	53	(
013	99	PRT	049	71	SBR	085	73	RC*	121	43	RCL	157	43	RCL
014	76	LBL	050	23	LNX	086	25	25	122	44	44	158	45	45
015	45	Y×	051	01	1	087	54	)	123	42	STO	159	42	STO
016	87	IFF	052	42	STO	088	65	×	124	22	22	160	24	24
017	01	01	053	25	25	089	43	RCL	125	71	SBR	161	44	SUM
018	32	X:T	054	71	SBR	090	20	20	126	23	LNX	162	07	07
019	86	STF	055	42	STO	091	85	+	127	71	SBR	163	44	SUM
020	01	01	056	76	LBL	092	73	RC*	128	52	EE	164	09	09
021	92	RTN	057	42	STO	093	25	25	129	71	SBR	165	33	X²
022	76	LBL	058	71	SBR	094	95	=	130	23	LNX	166	85	+
023	14	D	059	24	CE	095	42	STO	131	43	RCL	167	01	1
024	01	1	060	71	SBR	096	43	43	132	22	22	168	95	=
025	42	STO	061	23	LNX	097	01	1	133	55	÷	169	34	ΓX
026	25	25	062	73	RC*	098	22	INV	134	43	RCL	170	22	INV
027	71	SBR	063	25	25	099	44	SUM	135	44	44	171	44	SUM
028	23	LNX	064	42	STO	100	25	25	136	95	=	172	07	07
029	71	SBR	065	42	42	101	43	RCL	137	42	STO	173	44	SUM
030	35	1/X	066	71	SBR	102	25	25	138	45	45	174	09	09
031	71	SBR	067	23	LNX	103	67	EQ	139	01	1	175	65	×
032	23	LNX	068	01	1	104	44	SUM	140	42	STO	176	43	RCL
033	76	LBL	069	22	INV	105	02	2	141	07	07	177	23	23
034	33	X²	070	44	SUM	106	42	STO	142	42	STO	178	65	×
035	02	2	071	25	25	107	25	25	143	09	09	179	43	RCL

Step	Code	Key	Step	Code	Key	Step	Code	Key	Step	Code	Key	Step	Code	Key
180	08	08	233	55	÷	286	42	STO	339	32	X:T	392	43	RCL
181	55	÷	234	53	(	287	25	25	340	72	ST*	393	22	22
182	43	RCL	235	43	RCL	288	76	LBL	341	49	49	394	55	÷
183	22	22	236	20	20	289	54	)	342	01	1	395	53	(
184	55	÷	237	75	-	290	43	RCL	343	00	0	396	43	RCL
185	43	RCL	238	01	1	291	48	48	344	22	INV	397	24	24
186	21	21	239	54	)	292	72	ST*	345	44	SUM	398	75	-
187	95	=	240	95	=	293	25	25	346	49	49	399	01	1
188	22	INV	241	35	1/X	294	61	GTO	347	32	X:T	400	54	)
189	23	LNX	242	42	STO	295	55	÷	348	72	ST*	401	55	÷
190	42	STO	243	20	20	296	76	LBL	349	49	49	402	43	RCL
191	00	00	244	92	RTN	297	52	EE	350	01	1	403	23	23
192	65	×	245	76	LBL	298	53	(	351	44	SUM	404	55	÷
193	43	RCL	246	35	1/X	299	43	RCL	352	49	49	405	43	RCL
194	09	09	247	71	SBR	300	01	01	353	01	1	406	08	08
195	75	-	248	24	CE	301	85	+	354	00	0	407	95	=
196	43	RCL	249	71	SBR	302	43	RCL	355	32	X:T	408	99	PRT
197	07	07	250	23	LNX	303	02	02	356	43	RCL	409	53	(
198	95	=	251	43	RCL	304	54	)	357	49	49	410	43	RCL
199	55	÷	252	20	20	305	55	÷	358	67	EQ	411	01	01
200	53	(	253	65	×	306	02	2	359	65	×	412	85	+
201	02	2	254	73	RC*	307	65	×	360	61	GTO	413	43	RCL
202	65	×	255	25	25	308	43	RCL	361	61	GTO	414	02	02
203	43	RCL	256	95	=	309	03	03	362	76	LBL	415	54	)
204	00	00	257	42	STO	310	85	+	363	65	×	416	55	÷
205	75	-	258	47	47	311	43	RCL	364	01	1	417	02	2
206	02	2	259	71	SBR	312	04	04	365	32	X:T	418	65	×
207	54	)	260	23	LNX	313	95	=	366	92	RTN	419	43	RCL
208	95	=	261	43	RCL	314	42	STO	367	76	LBL	420	03	03
209	42	STO	262	47	47	315	06	06	368	32	X:T	421	85	+
210	20	20	263	75	-	316	65	×	369	22	INV	422	43	RCL
211	75	-	264	73	RC*	317	43	RCL	370	86	STF	423	04	04
212	43	RCL	265	25	25	318	05	05	371	01	01	424	95	=
213	24	24	266	95	=	319	95	=	372	43	RCL	425	42	STO
214	95	=	267	55	÷	320	42	STO	373	20	20	426	06	06
215	55	÷	268	53	(	321	44	44	374	35	1/X	427	71	SBR
216	53	(	269	43	RCL	322	92	RTN	375	42	STO	428	23	LNX
217	43	RCL	270	20	20	323	76	LBL	376	20	20	429	43	RCL
218	20	20	271	75	-	324	23	LNX	377	75	-	430	06	06
219	75	-	272	01	1	325	00	0	378	43	RCL	431	65	×
220	01	1	273	54	)	326	42	STO	379	24	24	432	43	RCL
221	54	)	274	95	=	327	49	49	380	95	=	433	05	05
222	95	=	275	42	STO	328	76	LBL	381	55	÷	434	65	×
223	45	Yˣ	276	48	48	329	61	GTO	382	53	(	435	53	(
224	43	RCL	277	01	1	330	73	RC*	383	43	RCL	436	43	RCL
225	21	21	278	22	INV	331	49	49	384	20	20	437	02	02
226	95	=	279	44	SUM	332	32	X:T	385	75	-	438	75	-
227	42	STO	280	25	25	333	01	1	386	01	1	439	43	RCL
228	20	20	281	43	RCL	334	00	0	387	54	)	440	01	01
229	75	-	282	25	25	335	44	SUM	388	95	=	441	54	)
230	43	RCL	283	67	EQ	336	49	49	389	35	1/X	442	95	=
231	24	24	284	54	)	337	73	RC*	390	23	LNX	443	99	PRT
232	95	=	285	02	2	338	49	49	391	65	×	444	91	R/S

User instructions for TI version   Table V

Two separate programs are given. They are run in similar manner.

For both programs data are entered as follows:

A, heat-transfer surface area	STO 23
N, number of passes	STO 21

Hot side
W, mass flowrate	STO 15
α factor	STO 13
β factor	STO 14
T_1, temperature in	STO 12
T_2, temperature out	STO 11

Cold side
w, mass flowrate	STO 05
α factor	STO 03
β factor	STO 04
t_1, temperature in	STO 01
t_2, temperature out	STO 02

Enter estimate of overall heat-transfer coefficient U, then press key **A**

With the program A:

Key **B** gives inlet temperatures, cold side (t_1) and hot side (T_1)
Key **C** gives outlet temperatures, cold side (t_2) and hot side (T_2)

With program B:

Key **D** gives hot side inlet (T_1) and cold side outlet (t_2) temperatures
Key **E** gives cold side inlet (t_1) and hot side outlet (T_2) temperatures

Calculation will take a few minutes. In all cases, programs also give LMTD correction factor F, and heat load q.

With further estimate of overall heat-transfer coefficient, key **A**, additional calculations of the same two temperatures are made until satisfactory convergence is obtained.

(Note: Any consistent set of units may be used as long as UA/wc remains dimensionless.)

All output are printed in the order indicated in the explanation (i.e., temperatures, LMTD correction factor F, and heat load q.).

References

1. Ten Broeck, H., Multipass Exchanger Calculations, *Ind. & Eng. Chem.,* Vol. 30, No. 9, pp. 1,041–1,042 (1938).
2. "Standards of Tubular Exchanger Manufacturers' Assn.," pp. 138–139, New York (1968).
3. Kern, D. Q., "Process Heat Transfer," pp. 140, 176, McGraw-Hill, New York (1950).
4. Bowman, R. A., "Mean Temperature Difference Correction in Multipass Exchangers," *Ind. & Eng. Chem.,* Vol. 28, No. 5, pp. 541–544 (1936).

The author

Rogério Geaquinto Herkenhoff is a process engineer for Petróleo Brasileiro S.A. (Petrobrás), Pç. Mahatma Gandhi 14/804, Centro, CEP 20031, Rio de Janeiro, Brazil. He holds a B.S. degree in chemical engineering from Universidade Federal Rural do Rio de Janeiro. He lectures on heat-exchanger design in Petrobrás courses for process engineers.

Calculating the corrected LMTD in shell-and-tube heat exchangers

The programs described here, one for the Hewlett-Packard system and one for the Texas Instruments system, will allow the engineer to calculate the corrected logarithmic mean temperature-difference for shell-and-tube heat exchangers in series.

W. Wayne Blackwell, *Ford, Bacon & Davis, Texas Inc.*, and **Larry Haydu**, *Kennecott Corp.*

☐ Heat transfer is more efficient when there is countercurrent flow rather than cocurrent flow. When multipass shell-and-tube exchangers are used in series, the flow more closely approximates countercurrent when a lot of shells are used. But this means greater cost. So it is desirable to use the minimum number of shells that will achieve an acceptable level of efficiency.

The programs described here will determine that minimum number of shells, and will calculate a corrected mean temperature-difference for the system chosen. This eliminates the need for laborious calculations with charts and graphs that are normally used in such designs.

Heat-exchanger design

The thermal design of heat-exchange equipment often requires the calculation of the logarithmic mean temperature-difference (LMTD). This is defined by the following equation:

$$\text{LMTD} = \frac{\Delta t_1 - \Delta t_2}{\ln \dfrac{\Delta t_1}{\Delta t_2}}$$

Where, for countercurrent flow:
Δt_1 = the larger terminal difference, $T_1 - t_2$, and
Δt_2 = the smaller terminal difference, $T_2 - t_1$.

For cocurrent flow,
$\Delta t_1 = T_1 - t_1$, and
$\Delta t_2 = T_2 - t_2$.

Temperatures:
T_1 = hot-fluid inlet temperature, °F,
T_2 = hot-fluid exit temperature, °F,
t_1 = cold-fluid inlet temperature, °F,
t_2 = cold-fluid exit temperature, °F.

Fig. 1 shows a typical temperature profile of two fluids in true countercurrent flow through a 1-1 exchanger (one shell pass, one tube pass).

The 1-1 exchanger is very simple but has its limitations. In the majority of industrial operations, higher velocities, shorter tubes, and a more economical exchanger can be found using multipass design. In a multipass exchanger such as shown in Fig. 2, the flow is part countercurrent and part cocurrent. As a result, the mean temperature difference lies somewhere between the countercurrent and cocurrent LMTDs.

In this situation, a correction factor, F, is defined so that, when it is multiplied by the LMTD, the product is the corrected mean temperature-difference (CMTD).

$$\text{CMTD} = F \times \text{LMTD}$$

Thus, for pure countercurrent flow, $F = 1$. As more cocurrent flow is introduced, F is reduced and the efficiency of the exchanger drops. The lower limit of prac-

Originally published August 24, 1981

Hewlett-Packard program listing   Table I

Step	Key	Code	Step	Key	Code	Step	Key	Code	Step	Key	Code	Step	Key	Code	Step	Key	Code
001	*LBLA	21 11	038	RCL1	36 01	075	÷	-24	112	X<0?	16-45	149	GTOB	22 12	186	X≠Y	-41
002	1	01	039	RCL2	36 02	076	STO9	35 09	113	GTOC	22 13	150	*LBLD	21 14	187	X=Y?	16-33
003	STCI	35 46	040	-	-45	077	2	02	114	1	01	151	RCLI	36 46	188	GTOC	22 13
004	P≠S	16-51	041	RCL4	36 04	078	X≠Y	-41	115	X≠Y	-41	152	RCL5	36 05	189	LN	32
005	.	-62	042	RCL3	36 03	079	÷	-24	116	X=Y?	16-33	153	x	-35	190	P≠S	16-51
006	8	08	043	-	-45	080	1	01	117	GTOC	22 13	154	CHS	-22	191	STO1	35 01
007	STO2	35 02	044	X=0?	16-43	081	-	-45	118	LN	32	155	RCL5	36 05	192	P≠S	16-51
008	P≠S	16-51	045	GTOd	22 16 14	082	RCL6	36 06	119	RCL7	36 07	156	+	-55	193	RCL9	36 09
009	RCL1	36 01	046	÷	-24	083	-	-45	120	x	-35	157	RCLI	36 46	194	RCL7	36 07
010	RCL4	36 04	047	STO6	35 06	084	P≠S	16-51	121	RCL6	36 06	158	+	-55	195	x	-35
011	-	-45	048	X²	53	085	STO0	35 00	122	1	01	159	RCL5	36 05	196	1	01
012	ABS	16 31	049	1	01	086	P≠S	16-51	123	-	-45	160	X≠Y	-41	197	RCL9	36 09
013	STO5	35 05	050	+	-55	087	RCL7	36 07	124	÷	-24	161	÷	-24	198	-	-45
014	RCL2	36 02	051	√X	54	088	+	-55	125	P≠S	16-51	162	STO9	35 09	199	÷	-24
015	RCL3	36 03	052	STO7	35 07	089	P≠S	16-51	126	RCL1	36 01	163	2	02	200	P≠S	16-51
016	-	-45	053	1	01	090	RCL0	36 00	127	÷	-24	164	X≠Y	-41	201	RCL1	36 01
017	ABS	16 31	054	RCL6	36 06	091	P≠S	16-51	128	P≠S	16-51	165	÷	-24	202	÷	-24
018	STO6	35 06	055	X=Y?	16-33	092	RCL7	36 07	129	*LBLE	21 15	166	1	01	203	P≠S	16-51
019	-	-45	056	GTOD	22 14	093	-	-45	130	STOD	35 14	167	-	-45	204	GTOE	22 15
020	X=0?	16-43	057	RCL5	36 05	094	X=0?	16-43	131	RCL0	36 00	168	RCL6	36 06	205	*LBLb	21 16 12
021	GTOb	22 16 12	058	x	-35	095	GTOC	22 13	132	X=0?	16-43	169	-	-45	206	RCL5	36 05
022	RCL5	36 05	059	1	01	096	÷	-24	133	GSBc	23 16 13	170	P≠S	16-51	207	STOB	35 12
023	RCL6	36 06	060	-	-45	097	X<0?	16-45	134	X>Y?	16-34	171	STO0	35 00	208	GTOa	22 16 11
024	÷	-24	061	RCL5	36 05	098	GTOC	22 13	135	GTOC	22 13	172	P≠S	16-51	209	*LBLd	21 16 14
025	LN	32	062	1	01	099	LN	32	136	RCLI	36 46	173	RCL7	36 07	210	RCLB	36 12
026	-	-24	063	-	-45	100	P≠S	16-51	137	STOA	35 11	174	+	-55	211	STOC	35 13
027	*LBLa	21 16 11	064	÷	-24	101	STO1	35 01	138	SPC	16-11	175	P≠S	16-51	212	1	01
028	STOB	35 12	065	RCLI	36 46	102	P≠S	16-51	139	PRTX	-14	176	RCL0	36 00	213	GTOE	22 15
029	*LBLB	21 12	066	1/X	52	103	1	01	140	RCLB	36 12	177	P≠S	16-51	214	*LBLc	21 16 13
030	RCL4	36 04	067	Yˣ	31	104	RCL9	36 09	141	PRTX	-14	178	RCL7	36 07	215	R↓	-31
031	RCL3	36 03	068	STO8	35 08	105	-	-45	142	RCLD	36 14	179	-	-45	216	P≠S	16-51
032	-	-45	069	1	01	106	1	01	143	x	-35	180	X=0?	16-43	217	RCL2	36 02
033	RCL1	36 01	070	X≠Y	-41	107	RCL9	36 09	144	STOC	35 13	181	GTOC	22 13	218	P≠S	16-51
034	RCL3	36 03	071	-	-45	108	RCL6	36 06	145	PRTX	-14	182	÷	-24	219	RTN	24
035	-	-45	072	RCL6	36 06	109	x	-35	146	RTN	24	183	X<0?	16-45	220	R/S	51
036	÷	-24	073	RCL8	36 08	110	-	-45	147	*LBLC	21 13	184	GTOC	22 13			
037	STO5	35 05	074	-	-45	111	÷	-24	148	ISZI	16 26 46	185	1	01			

Texas Instruments program listing   Table II

Step	Code	Key	Step	Code	Key	Step	Code	Key	Step	Code	Key	Step	Code	Key
000	76	LBL	022	43	RCL	044	43	RCL	066	42	STO	088	69	OP
001	11	A	023	25	25	045	20	20	067	02	02	089	02	02
002	04	4	024	69	OP	046	69	OP	068	43	RCL	090	69	OP
003	69	OP	025	03	03	047	02	02	069	20	20	091	05	05
004	17	17	026	69	OP	048	69	OP	070	69	OP	092	43	RCL
005	25	CLR	027	05	05	049	05	05	071	02	02	093	24	24
006	69	OP	028	69	OP	050	91	R/S	072	69	OP	094	91	R/S
007	00	00	029	00	00	051	42	STO	073	05	05	095	42	STO
008	43	RCL	030	43	RCL	052	01	01	074	91	R/S	096	04	04
009	16	16	031	18	18	053	99	PRT	075	42	STO	097	99	PRT
010	69	OP	032	69	OP	054	43	RCL	076	03	03	098	53	(
011	02	02	033	01	01	055	21	21	077	99	PRT	099	43	RCL
012	43	RCL	034	43	RCL	056	69	OP	078	76	LBL	100	00	00
013	17	17	035	19	19	057	01	01	079	12	B	101	75	-
014	69	OP	036	69	OP	058	43	RCL	080	69	OP	102	43	RCL
015	03	03	037	02	02	059	19	19	081	00	00	103	01	01
016	69	OP	038	69	OP	060	69	OP	082	43	RCL	104	54	)
017	05	05	039	05	05	061	02	02	083	22	22	105	55	÷
018	43	RCL	040	91	R/S	062	69	OP	084	69	OP	106	53	(
019	25	25	041	42	STO	063	05	05	085	01	01	107	43	RCL
020	69	OP	042	00	00	064	91	R/S	086	43	RCL	108	03	03
021	02	02	043	99	PRT	065	99	PRT	087	23	23	109	75	-

Texas Instruments program listing   Table II (continued)

Step	Code	Key	Step	Code	Key	Step	Code	Key	Step	Code	Key	Step	Code	Key	Step	Code	Key
110	43	RCL	176	02	2	242	75	-	308	77	GE	374	03	03	440	31	31
111	02	02	177	85	+	243	01	1	309	18	C'	375	43	RCL	441	69	OP
112	54	)	178	02	2	244	95	=	310	28	LOG	376	27	27	442	02	02
113	95	=	179	34	ΓX	245	55	÷	311	42	STO	377	69	OP	443	43	RCL
114	42	STO	180	95	=	246	53	(	312	10	10	378	04	04	444	32	32
115	06	06	181	42	STO	247	43	RCL	313	01	1	379	69	OP	445	69	OP
116	53	(	182	10	10	248	07	07	314	75	-	380	05	05	446	03	03
117	43	RCL	183	02	2	249	75	-	315	43	RCL	381	43	RCL	447	69	OP
118	03	03	184	55	÷	250	01	1	316	05	05	382	04	04	448	05	05
119	75	-	185	43	RCL	251	54	)	317	95	=	383	99	PRT	449	43	RCL
120	43	RCL	186	08	08	252	95	=	318	55	÷	384	61	GTO	450	14	14
121	02	02	187	75	-	253	45	Y×	319	53	(	385	34	ΓX	451	65	×
122	54	)	188	02	2	254	43	RCL	320	01	1	386	76	LBL	452	43	RCL
123	55	÷	189	75	-	255	04	04	321	75	-	387	16	A'	453	11	11
124	53	(	190	02	2	256	35	1/X	322	43	RCL	388	25	CLR	454	95	=
125	43	RCL	191	34	ΓX	257	95	=	323	05	05	389	32	X:T	455	99	PRT
126	00	00	192	95	=	258	42	STO	324	65	×	390	43	RCL	456	98	ADV
127	75	-	193	22	INV	259	15	15	325	43	RCL	391	01	01	457	98	ADV
128	43	RCL	194	49	PRD	260	94	+/-	326	06	06	392	75	-	458	98	ADV
129	02	02	195	10	10	261	85	+	327	54	)	393	43	RCL	459	91	R/S
130	54	)	196	25	CLR	262	01	1	328	95	=	394	02	02	460	76	LBL
131	95	=	197	32	X:T	263	95	=	329	28	LOG	395	95	=	461	17	B'
132	42	STO	198	43	RCL	264	55	÷	330	65	×	396	42	STO	462	43	RCL
133	07	07	199	10	10	265	53	(	331	43	RCL	397	12	12	463	12	12
134	43	RCL	200	22	INV	266	43	RCL	332	09	09	398	43	RCL	464	42	STO
135	06	06	201	77	GE	267	06	06	333	55	÷	399	00	00	465	14	14
136	33	X²	202	18	C'	268	75	-	334	53	(	400	75	-	466	61	GTO
137	85	+	203	28	LOG	269	43	RCL	335	43	RCL	401	43	RCL	467	15	E
138	01	1	204	42	STO	270	15	15	336	06	06	402	03	03	468	76	LBL
139	95	=	205	10	10	271	54	)	337	75	-	403	95	=	469	45	Y×
140	34	ΓX	206	43	RCL	272	95	=	338	01	1	404	42	STO	470	69	OP
141	42	STO	207	09	09	273	42	STO	339	54	)	405	13	13	471	00	00
142	09	09	208	65	×	274	05	05	340	95	=	406	75	-	472	43	RCL
143	76	LBL	209	43	RCL	275	35	1/X	341	55	÷	407	43	RCL	473	28	28
144	34	ΓX	210	08	08	276	65	×	342	43	RCL	408	12	12	474	69	OP
145	01	1	211	55	÷	277	02	2	343	10	10	409	95	=	475	02	02
146	32	X:T	212	53	(	278	75	-	344	95	=	410	67	EQ	476	43	RCL
147	43	RCL	213	01	1	279	01	1	345	42	STO	411	17	B'	477	29	29
148	06	06	214	00	0	280	75	-	346	11	11	412	55	÷	478	69	OP
149	22	INV	215	23	LNX	281	43	RCL	347	76	LBL	413	53	(	479	03	03
150	67	EQ	216	65	×	282	06	06	348	14	D	414	43	RCL	480	69	OP
151	13	C	217	53	(	283	85	+	349	71	SBR	415	13	13	481	05	05
152	43	RCL	218	01	1	284	43	RCL	350	45	Y×	416	55	÷	482	43	RCL
153	07	07	219	75	-	285	09	09	351	43	RCL	417	43	RCL	483	11	11
154	55	÷	220	43	RCL	286	95	=	352	33	33	418	12	12	484	99	PRT
155	53	(	221	08	08	287	42	STO	353	32	X:T	419	54	)	485	92	RTN
156	43	RCL	222	54	)	288	10	10	354	43	RCL	420	23	LNX	486	76	LBL
157	04	04	223	95	=	289	02	2	355	11	11	421	95	=	487	19	D'
158	75	-	224	55	÷	290	55	÷	356	77	GE	422	42	STO	488	03	3
159	43	RCL	225	43	RCL	291	43	RCL	357	16	A'	423	14	14	489	05	5
160	07	07	226	10	10	292	05	05	358	76	LBL	424	76	LBL	490	69	OP
161	65	×	227	95	=	293	75	-	359	18	C'	425	15	E	491	04	04
162	53	(	228	42	STO	294	01	1	360	01	1	426	69	OP	492	43	RCL
163	43	RCL	229	11	11	295	75	-	361	44	SUM	427	00	00	493	06	06
164	04	04	230	61	GTO	296	43	RCL	362	04	04	428	43	RCL	494	69	OP
165	75	-	231	14	D	297	06	06	363	43	RCL	429	30	·30	495	06	06
166	01	1	232	76	LBL	298	75	-	364	22	22	430	69	OP	496	03	3
167	54	)	233	13	C	299	43	RCL	365	69	OP	431	02	02	497	03	3
168	95	=	234	25	CLR	300	09	09	366	01	01	432	69	OP	498	69	OP
169	42	STO	235	32	X:T	301	95	=	367	43	RCL	433	05	05	499	04	04
170	08	08	236	43	RCL	302	22	INV	368	23	23	434	43	RCL	500	43	RCL
171	35	1/X	237	07	07	303	49	PRD	369	69	OP	435	14	14	501	07	07
172	65	×	238	65	×	304	10	10	370	02	02	436	99	PRT	502	69	OP
173	02	2	239	43	RCL	305	43	RCL	371	43	RCL	437	69	OP	503	06	06
174	95	=	240	06	06	306	10	10	372	26	26	438	00	00	504	98	ADV
175	75	-	241	95	=	307	22	INV	373	69	OP	439	43	RCL	505	91	R/S

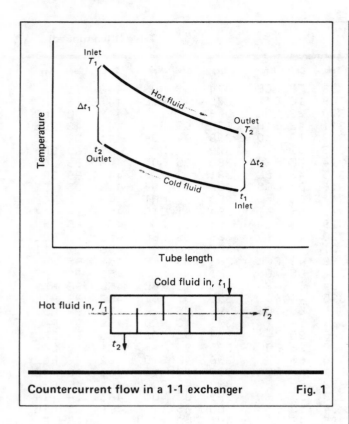

Countercurrent flow in a 1-1 exchanger   **Fig. 1**

tical efficiency is $F = 0.75$ to 0.80 [1].

When designing shell-and-tube heat exchangers in series, the lowest F value is for one shell. This value is raised as the number of shells increases, and the flow more nearly resembles countercurrent flow. The object of design is to find the minimum number of shells that will raise the F value above the chosen minimum of 0.75 to 0.80.

Determining the F factor

An article in the May 1940 issue of *The Transactions of the ASME* greatly simplified the calculations required to determine the mean temperature differences in shell-and-tube exchangers [2]. The equations supplied by that article have been adapted for calculator use.

The general equation, valid for any number of passes, is [3,4]:

$$F = \left(\frac{\sqrt{R^2 + 1}}{R - 1} \right) \frac{\ln\left[(1 - P_x)/(1 - RP_x)\right]}{\ln\left[\dfrac{(2/P_x) - 1 - R + \sqrt{R^2 + 1}}{(2/P_x) - 1 - R - \sqrt{R^2 + 1}} \right]}$$

where

$$P_x = \frac{1 - \left[\dfrac{RP - 1}{P - 1} \right]^{1/N}}{R - \left[\dfrac{RP - 1}{P - 1} \right]^{1/N}}$$

and

$$P = (t_2 - t_1)/(T_1 - t_1)$$
$$R = (T_1 - T_2)/(t_2 - t_1)$$

N is the total number of shell passes, i.e., the product of shell passes per shell and the number of units in series. Solving for N by repetitive trial and error with a minimum desired F, the minimum required number of shell passes can be determined.

If $R = 1$, the equation becomes indeterminate, and an alternate solution applies:

$$F = \frac{P_x \sqrt{R^2 + 1}/(1 - P_x)}{\ln\left[\dfrac{(2/P_x) - 1 - R + \sqrt{R^2 + 1}}{(2/P_x) - 1 - R - \sqrt{R^2 + 1}} \right]}$$

and: $P_x = P/(N - NP + P)$

The equations presented are based on certain assumptions: the overall heat-transfer coefficient, U, is constant throughout the heat exchanger; the flowrate of each fluid is constant; the specific heat of each fluid is constant; there is no condensation of vapor or boiling of liquid in any part of the exchanger; heat losses are negligible; the heat-transfer surface in each pass is equal; the temperature of the shell-side fluid in any shell-side pass is uniform over any cross section.

Hewlett-Packard program
Larry Haydu

The program in Table I is written for the Hewlett-Packard HP-67/97 programmable calculators. It is simple and efficient.

The inlet and outlet temperatures of the hot and cold fluids are fed into computer memory at the start of the program. The program will select the proper number of shells in series so that the value of F is greater than or equal to 0.8. If a cutoff value for F other than 0.8 is desired, a value can be registered in the memory.

Input: T_1 [STO] 1
T_2 [STO] 2
t_1 [STO] 3
t_2 [STO] 4
F [STO] 0 (if different than 0.8)

Begin computations—Press [A]

Output is stored on the HP-67 (printed on the HP-97) and can be recalled from storage registers.

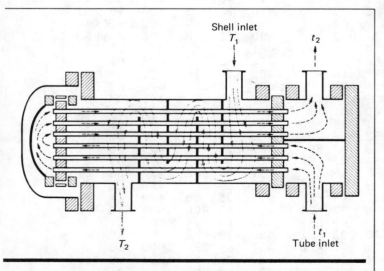

Flow patterns in a multipass exchanger   **Fig. 2**

TI user instructions — Table III

Step	Procedure	Enter	Press	Display
1.	Read in both magnetic cards, sides 1,2,3 and 4		CLR	1,2,3,4
2.	Begin computations		A	37002431
3.	Key-in hot inlet temp.	T_1	R/S	3700324137
4.	Key-in hot outlet temp.	T_2	R/S	37002431
5.	Key-in cold inlet temp.	t_1	R/S	3700324137
6.	Key-in cold outlet temp.	t_2	R/S	N
7.	Key-in number of shells (or use default value displayed)	N	R/S	Corr. LMTD
8.	Option: To calculate corrected LMTD for alternate number of shells, press B and continue with Step 7		B	N
9.	Option: Press D' to print R and P		D'	P

Output: Number of shell passes in series   RCL A

Logarithmic mean temperature-difference (LMTD)   RCL B

Corrected mean temperature-difference (CMTD)   RCL C

Correction factor calculated (CMTD/LMTD)   RCL D

Example

Acetone at 250°F is to be sent to storage at 100°F. The heat will be received by 100% acetic acid coming from storage at 90°F, and will raise its temperature to 150°F. Calculate the number of shell passes required, the LMTD, and the CMTD.

Answer: 3 shell passes in series required

39.09°F LMTD $(F = 0.87)$

34.20°F CMTD

TI data registers — Table IV

0.	Hot inlet temp.--T_1	17.	1513271536
1.	Hot outlet temp.--T_2	18.	23323700
2.	Cold inlet temp.--t_1	19.	37002431
3.	Cold outlet temp.--t_2	20.	3700324137
4.	Number of shells--N	21.	1532271600
5.	P'	22.	3132400036
6.	R	23.	2317272736
7.	P	24.	1 (optional value)
8.	P''	25.	5151515151
9.	$\sqrt{R^2+1}$	26.	24311535
10.	log term	27.	1713361716
11.	F	28.	2100211315
12.	Δt_1	29.	3732350000
13.	Δt_2	30.	27303716
14.	LMTD (uncorrected)	31.	2150552730
15.	$[(PR\text{-}1)/(P\text{-}1)]^{1/N}$	32.	3716560000
16.	2730371600	33.	0.75 (optional value)

Note: The numbers in registers 16-33 are all print codes and default values that must be stored in data registers before the program is recorded on magnetic cards. The program should be recorded on magnetic cards while it is in standard partitioning.

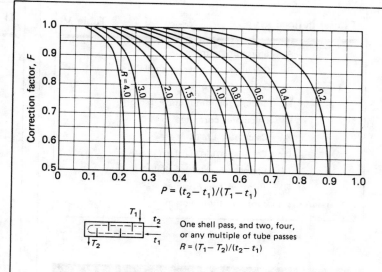

One shell pass, and two, four, or any multiple of tube passes
$R = (T_1 - T_2)/(t_2 - t_1)$

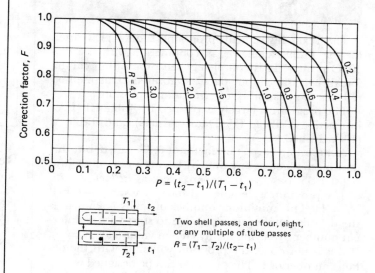

Two shell passes, and four, eight, or any multiple of tube passes
$R = (T_1 - T_2)/(t_2 - t_1)$

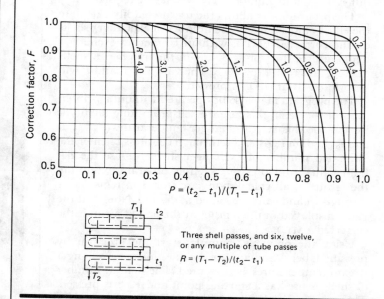

Three shell passes, and six, twelve, or any multiple of tube passes
$R = (T_1 - T_2)/(t_2 - t_1)$

Correction factor plot for multipass exchangers   Fig. 3

TI user-defined keys		Table V
A	Starts program	
B	Calculates number of shells*	
C	Solves *F* factor equations*	
D	Checks for minimum *F* factor*	
E	Calculates corrected LMTD	
A'	Calculates LMTD*	
B'	Calculates LMTD if $\Delta t_1 = \Delta t_2$*	
C'	Changes number of shells*	
D'	Prints *R* and *P* values	

* These keys all continue the program

Texas Instruments program
W. Wayne Blackwell

The Texas Instruments program (Table II) is written for the TI-59 calculator, in conjunction with the PC-100C printer. It is completely self-prompting, calling for all necessary input data as required. This method reduces the possibility of error and confusion.

The program is largely explained by Table III (User instructions), Table IV (Data registers), and Table V (User-defined keys).

A minimum correction factor of 0.75 is used, but this may be changed at any time by keying a new value into register 33. Base −10 logs are used, as in Ref. 2.

The program is designed to start by testing one shell ($N = 1$). This minimum number of shells is known as the default number. It is possible to start with a different number by keying in any chosen value (say, $N = 3$). If this is not done, the program will start with the default number of 1. The program can be designed with a different default number (say $N = 2$) in register 24.

The program also allows the user to readily calculate the effect of a change in the number of exchanger shells on the corrected LMTD, without reentering exchanger temperatures. Just press B, key in the number of shells desired, and then press R/S. The values of *R* and *P* will be printed if label D' is pressed.

The program may also be operated without a printer, (but, of course, in this case the self-prompting facility is lost). To do this, store exchanger temperatures (T_1, T_2, t_1 and t_2) in data registers 00 through 03, press B, enter the estimated number of exchanger shells (or use the default value) and press R/S. After program execution, the results of all calculations are stored in registers 04 through 15 and can be recalled as desired. Note that the value displayed in the register at the end of a run is the corrected LMTD.

Using the previous example, start the program by pressing label A and answer the questions on exchanger stream temperatures as presented. The default number of shells is used as a starting point and R/S is pressed. The printout is shown in Table VI. The *F* value for $N = 1$ isn't a viable solution. It does not print out.

Both programs give the same result, of course.

TI example printout	Table VI

```
HOT   T  IN
       250.
HOT  T  OUT
       100.
COLD   T  IN
        90.
COLD  T  OUT
       150.
NO.  SHELLS
         1.
NO.  SHELLS  INCREASED
         2.
    F  FACTOR
  .6160321966
NO.  SHELLS  INCREASED
         3.
    F  FACTOR
  .8748644663
      LMTD
  39.08650337
    F×(LMTD)
  34.19539291
```

References

1. Morton, D. S., Thermal Design of Heat Exchangers, *Ind. and Eng. Chem.*, Vol. 52, No. 6, 1960.
2. Bowman, R. A., Mueller, A. C., and Nagle, W. M., Mean Temperature Difference in Design, *Trans. ASME*, Vol. 62, 1940, pp. 283–294.
3. Taborek, J. J., Organizing Heat Exchanger Programs on Digital Computers, *Chem. Eng. Prog.*, Vol. 55, No. 10, 1959.
4. Gulley, D. L., Use Computers to Select Exchangers, *Pet. Refiner*, Vol. 39, No. 7, 1960.

The authors

W. Wayne Blackwell is a Senior Process Engineer with Ford, Bacon & Davis, Texas Inc., P.O. Box 38209, Dallas, TX 75238, tel: 214-278-8121. He holds a B.S. in Ch.E. from Texas Technological University, and is a registered professional engineer in Texas. Mr. Blackwell has 22 years of experience in gas process design, and has published numerous technical articles. He is presently working on a book "Process Design With a Programmable Calculator," to be published by McGraw-Hill.

Larry J. Haydu is a Senior Engineer with Kennecott Corp., Process Equipment Division, 31935 Aurora Rd., Solon, OH 44139, tel: 216-248-7100. He graduated from Cleveland State University with a bachelor's degree in chemical engineering, then worked for the city of Cleveland, Div. of Air Pollution, as an air pollution control engineer. Mr. Haydu is a licensed professional engineer in the state of Ohio.

Program calculates heat transfer through composite walls

This program computes the heat-transfer rate through walls composes of any two materials for which coefficients are known. It also determines the average and surface temperatures of the walls.

Calvin R. Brunner, Malcolm Pirnie, Inc.

☐ The heat transferring through each wall of a composite wall via conduction, $q' = (K/X)\Delta t$, is identical to the heat being lost at the outside surface via convection and radiation, $q' = h\Delta t$.

From the figure:

$$q' = (K_1/X_1)(t_i - t_1) = (K_2/X_2)(t_1 - t_2)$$
$$= h(t_2 - t_a) \quad (1)$$

Solving for t_1, t_2 and q':

$$t_1 = t_2 + (X_2/K_2)\ h(t_2 - t_a) \quad (2)$$

$$t_2 = t_a + (q'/h) \quad (3)$$

$$q' = (t_i - t_a)/[(X_1/K_1) + (X_2/K_2) + (1/h)] \quad (4)$$

Via regression analysis of the film coefficient data in Table II, an equation for h is derived:

$$h = 1.535 + 0.00582\ (|t_2 - t_a|) \quad (5)$$

The correlation coefficient for Eq. (5) is 0.9998.

From the thermal conductivity data in Table II, a value for K_1 is similarly determined:

$$K_1 = 4680/[(t_1 + t_i)/2] + 0.00425 \times$$
$$[(t_i + t_1)/2] \quad (6)$$

The correlation coefficient for Eq. (6) is 0.9992.

An equation for K_2 is also derived from data in Table II:

$$K_2 = 0.988\ e^{\ 0.00033\ [(t_1 + t_2)/2]} \quad (7)$$

The correlation coefficient for Eq. (7) is 0.9885.

How the program converges

A value for t_2 is assumed, and with t_i and t_a known, h is calculated via Eq. (5), and K_2 via Eq. (7). Initially, t_1 is taken as zero; however, a value for t_1 is inserted during the first calculation sequence.

With the values for h_1, K_2 and t_2, t_1' is calculated by

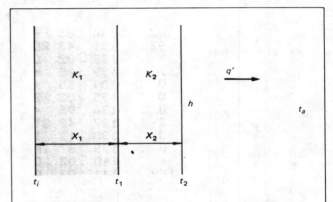

K Coefficient of thermal conductivity, Btu/(h)(ft²)(°F)
h Combined film-radiation coefficient, Btu/(h)(ft²)(°F)
q' Heat transfer, Btu/ft²
t Temperature, °F
X Length, in.

Variables involved in heat transfer through a composite wall

means of Eq. (2). This value of t_1' is compared to the initial t_1. If the absolute value of $t_1 - t_1' \geq 0.5$, the program repeats the calculation for t_1', setting $t_1 = t_1'$. The program converges when the absolute value of $t_1 - t_1' < 0.5$. Then the program continues, to test t_2.

With the assumed value for t_2, K_1 is calculated by means of Eq. (6), and q' by Eq. (4).

With t_a known, and q' and h now having been determined, t_2' is calculated. If the absolute value of $t_2' - t_2 < 0.5$, the program concludes with $t_2 = t_2'$ and $t_1 = t_1'$.

However, if the absolute value of $t_2' - t_2 \geq 0.5$, the program returns to its beginning, with $t_2 = t_2'$, and t_1 is calculated again.

Eventually, the program will converge, with the

(text continues on p. 182)

Originally published June 16, 1980

Program for calculating heat transfer through composite walls   **Table I**

Location	Code	Key	Location	Code	Key	Location	Code	Key	Location	Code	Key	Location	Code	Key	Location	Code	Key
000	76	LBL	061	95	=	122	08	08	183	75	−	244	91	R/S	462	06	6
001	11	A	062	55	÷	123	42	STO	184	43	RCL	245	00	0	463	02	2
002	42	STO	063	02	2	124	06	06	185	03	03				464	04	4
003	00	00	064	95	=	125	61	GTO	186	95	=	Space for			465	03	3
004	91	R/S	065	42	STO	126	00	00	187	42	STO	Alternate h equation			466	07	7
005	76	LBL	066	12	12	127	56	56	188	11	11				467	01	1
006	12	B	067	87	IFF	128	43	RCL	189	55	÷	283	00	0	468	07	7
007	42	STO	068	01	01	129	06	06	190	43	RCL	284	61	GTO	469	69	OP
008	01	01	069	02	02	130	85	+	191	05	05	285	00	00	470	03	03
009	91	R/S	070	92	92	131	43	RCL	192	85	+	286	47	47	471	04	4
010	76	LBL	071	65	×	132	02	02	193	43	RCL	287	76	LBL	472	03	3
011	13	C	072	93	.	133	95	=	194	03	03	288	18	C'	473	01	1
012	42	STO	073	00	0	134	55	÷	195	95	=	289	86	STF	474	03	3
013	02	02	074	00	0	135	02	2	196	42	STO	290	01	01	475	02	2
014	91	R/S	075	00	0	136	95	=	197	10	10	291	91	R/S	476	07	7
015	76	LBL	076	03	3	137	42	STO	198	75	−	292	00	0	477	02	2
016	14	D	077	03	3	138	13	13	199	43	RCL				478	07	7
017	42	STO	078	00	0	139	87	IFF	200	04	04	Space for			479	69	OP
018	03	03	079	95	=	140	02	02	201	95	=	Alternate K_2 equation			480	04	04
019	91	R/S	080	22	INV	141	03	03	202	50	IxI				481	69	OP
020	76	LBL	081	23	LNX	142	65	65	203	66	PAU	356	00	0	482	05	05
021	15	E	082	65	×	143	43	RCL	204	22	INV	357	61	GTO	483	98	ADV
022	42	STO	083	93	.	144	13	13	205	77	GE	358	00	00	484	04	4
023	04	04	084	09	9	145	65	×	206	04	04	359	88	88	485	04	4
024	87	IFF	085	08	8	146	93	.	207	35	35	360	76	LBL	486	00	0
025	00	00	086	08	8	147	00	0	208	43	RCL	361	19	D'	487	00	0
026	02	02	087	95	=	148	00	0	209	10	10	362	86	STF	488	00	0
027	45	45	088	42	STO	149	04	4	210	42	STO	363	02	02	489	02	2
028	75	−	089	07	07	150	02	2	211	04	04	364	91	R/S	490	69	OP
029	43	RCL	090	35	1/X	151	05	5	212	61	GTO	365	00	0	491	04	04
030	03	03	091	65	×	152	85	+	213	00	00				492	43	RCL
031	95	=	092	43	RCL	153	04	4	214	24	24	Space for			493	00	00
032	50	IxI	093	01	01	154	06	6	215	68	NOP	Alternate K_1 equation			494	69	OP
033	65	×	094	65	×	155	08	8	216	68	NOP				495	06	06
034	93	.	095	43	RCL	156	00	0	217	68	NOP	430	00	0	496	04	4
035	00	0	096	05	05	157	55	÷	218	91	R/S	431	61	GTO	497	04	4
036	00	0	097	65	×	158	43	RCL	219	76	LBL	432	01	01	498	00	0
037	05	5	098	53	(	159	13	13	220	16	A'	433	61	61	499	00	0
038	08	8	099	43	RCL	160	95	=	221	42	STO	434	00	0	500	00	0
039	02	2	100	04	04	161	42	STO	222	14	14	435	76	LBL	501	03	3
040	85	+	101	75	−	162	09	09	223	86	STF	436	10	E'	502	69	OP
041	01	1	102	43	RCL	163	35	1/X	224	03	03	437	02	2	503	04	04
042	93	.	103	03	03	164	65	×	225	91	R/S	438	03	3	504	43	RCL
043	05	5	104	95	=	165	43	RCL	226	43	RCL	439	06	6	505	01	01
044	03	3	105	85	+	166	00	00	227	14	14	440	03	3	506	69	OP
045	05	5	106	43	RCL	167	85	+	228	61	GTO	441	03	3	507	06	06
046	95	=	107	04	04	168	43	RCL	229	00	00	442	07	7	508	98	ADV
047	42	STO	108	95	=	169	01	01	230	55	55	443	00	0	509	03	3
048	05	05	109	42	STO	170	55	÷	231	00	0	444	00	0	510	07	7
049	87	IFF	110	08	08	171	43	RCL	232	00	0	445	02	2	511	00	0
050	03	03	111	75	−	172	07	07	233	00	0	446	00	0	512	00	0
051	02	02	112	43	RCL	173	85	+	234	00	0	447	69	OP	513	02	2
052	26	26	113	06	06	174	43	RCL	235	00	0	448	01	01	514	04	4
053	93	.	114	95	=	175	05	05	236	00	0	449	01	1	515	03	3
054	05	5	115	50	IxI	176	35	1/X	237	00	0	450	05	5	516	01	1
055	32	X:T	116	66	PAU	177	95	=	238	00	0	451	03	3	517	69	OP
056	43	RCL	117	22	INV	178	35	1/X	239	00	0	452	02	2	518	04	04
057	04	04	118	77	GE	179	65	×	240	76	LBL	453	03	3	519	43	RCL
058	85	+	119	01	01	180	53	(	241	17	B'	454	00	0	520	02	02
059	43	RCL	120	28	28	181	43	RCL	242	86	STF	455	03	3	521	69	OP
060	06	06	121	43	RCL	182	02	02	243	00	00	456	03	3	522	06	06
												457	69	OP			
												458	02	02			
												459	03	3			
												460	02	2			
												461	03	3			

Program for calculating heat transfer through composite walls (continued)   **Table I**

Location	Code	Key	Location	Code	Key	Location	Code	Key	Location	Code	Key	Location	Code	Key	Location	Code	Key
523	03	3	584	03	3	645	07	7	706	07	07	739	02	2	772	00	0
524	07	7	585	69	OP	646	00	0	707	30	30	740	03	3	773	00	0
525	01	1	586	04	04	647	69	OP	708	69	OP	741	01	1	774	00	0
526	03	3	587	43	RCL	648	04	04	709	00	00	742	07	7	775	00	0
527	03	3	588	12	12	649	43	RCL	710	01	1	743	01	1	776	69	OP
528	00	0	589	69	OP	650	11	11	711	05	5	744	05	5	777	03	03
529	01	1	590	06	06	651	69	OP	712	02	2	745	02	2	778	69	OP
530	04	4	591	02	2	652	06	06	713	03	3	746	06	6	779	05	05
531	69	OP	592	03	3	653	05	5	714	01	1	747	69	OP	780	43	RCL
532	04	04	593	00	0	654	04	4	715	07	7	748	03	03	781	11	11
533	43	RCL	594	00	0	655	00	0	716	01	1	749	02	2	782	98	ADV
534	03	03	595	00	0	656	00	0	717	05	5	750	06	6	783	98	ADV
535	69	OP	596	00	0	657	69	OP	718	02	2	751	00	0	784	98	ADV
536	06	06	597	69	OP	658	04	04	719	06	6	752	03	3	785	91	R/S
537	98	ADV	598	04	04	659	98	ADV	720	69	OP	753	69	OP	786	00	0
538	03	3	599	43	RCL	660	87	IFF	721	03	03	754	04	04	787	00	0
539	07	7	600	05	05	661	03	03	722	02	2	755	69	OP			
540	00	0	601	69	OP	662	06	06	723	06	6	756	05	05			
541	00	0	602	06	06	663	71	71	724	00	0	757	98	ADV			
542	00	0	603	02	2	664	93	.	725	02	2	758	69	OP	**Label addresses**		
543	02	2	604	06	6	665	05	5	726	69	OP	759	00	00			
544	69	OP	605	00	0	666	69	OP	727	04	04	760	02	2	001	11	A
545	04	04	606	00	0	667	06	06	728	69	OP	761	00	0	006	12	B
546	43	RCL	607	00	0	668	61	GTO	729	05	05	762	01	1	011	13	C
547	08	08	608	02	2	669	06	06	730	22	INV	763	07	7	016	14	D
548	69	OP	609	69	OP	670	75	75	731	87	IFF	764	69	OP	021	15	E
549	06	06	610	04	04	671	43	RCL	732	01	01	765	02	02	220	16	A'
550	03	3	611	43	RCL	672	14	14	733	07	07	766	03	3	241	17	B'
551	07	7	612	09	09	673	69	OP	734	57	57	767	01	1	288	18	C'
552	00	0	613	69	OP	674	06	06	735	69	OP	768	01	1	361	19	D'
553	00	0	614	06	06	675	22	INV	736	00	00	769	06	6	436	10	E'
554	00	0	615	02	2	676	87	IFF	737	01	1	770	02	2			
555	03	3	616	06	6	677	00	00	738	05	5	771	00	0			
556	69	OP	617	00	0	678	07	07									
557	04	04	618	00	0	679	03	03									
558	43	RCL	619	00	0	680	68	NOP									
559	10	10	620	03	3	681	69	OP									
560	69	OP	621	69	OP	682	00	00									
561	06	06	622	04	04	683	01	1									
562	98	ADV	623	43	RCL	684	05	5									
563	01	1	624	07	07	685	02	2									
564	03	3	625	69	OP	686	03	3									
565	04	4	626	06	06	687	01	1									
566	02	2	627	98	ADV	688	07	7									
567	02	2	628	61	GTO	689	01	1									
568	02	2	629	06	06	690	05	5									
569	00	0	630	39	39	691	02	2									
570	02	2	631	00	0	692	06	6									
571	69	OP	632	00	0	693	69	OP									
572	04	04	633	00	0	694	03	03									
573	43	RCL	634	00	0	695	02	2									
574	13	13	635	00	0	696	03	3									
575	69	OP	636	00	0	697	00	0									
576	06	06	637	00	0	698	00	0									
577	01	1	638	00	0	699	69	OP									
578	03	3	639	01	1	700	04	04									
579	04	4	640	04	4	701	69	OP									
580	02	2	641	06	6	702	05	05									
581	02	2	642	03	3	703	22	INV									
582	02	2	643	02	2	704	87	IFF									
583	00	0	644	01	1	705	02	02									

Data for derivation of Eq. (5), (6) and (7)   **Table II**

Eq. (5), film coefficient (h) for vertical flat plate:

$|t_2 - t_a|$, °F	50	100	150	200	250
Handbook h^{*}	1.82	2.13	2.40	2.70	2.99
Derived h	1.83	2.12	2.41	2.70	2.99

Eq. (6), thermal conductivity (K_1) for fireclay brick:

$\frac{1}{2}(t_i + t_1)$, °F	1,000	1,400	1,800	2,200	2,600
Catalog $K_1{}^{\dagger}$	9.2	9.4	10.0	11.3	13.0
Derived K_1	8.9	9.3	10.3	11.5	12.9

Eq. (7), thermal conductivity (K_2) for lightweight castable:

$\frac{1}{2}(t_i + t_1)$, °F	200	600	1,000	1,400
Catalog $K_2{}^{**}$	1.1	1.2	1.3	1.5
Derived K_2	1.1	1.2	1.4	1.6

*From Baumeister, T. and Marks, L., "Standard Handbook for Mechanical Engineers," 7th ed. McGraw-Hill, Inc., 1967, p. 4-106.

†From "Harbison-Walker Superduty Fireclay Brick," Harbison-Walker Refractories Div. of Dresser Industries, Inc.

**From "Harbison-Walker H-W Lightweight Castable 22," Harbison-Walker Refractories Div. of Dresser Industries, Inc.

Printouts of initial and alternate programs   Table III

Initial program		Alternate E, h, K, K_2:	
H/T - COMPOSITE WALL		H/T - COMPOSITE WALL	
2.	X 1	2.	X 1
4.5	X 2	4.5	X 2
1300.	T IN	1300.	T IN
80.	TAMB	80.	TAMB
1238.619241	T 1	1227.593374	T 1
206.3966218	T 2	987.8677006	T 2
1269.370596	AVG1	1263.796687	AVG1
722.6703142	AVG2	1107.730537	AVG2
2.271808718	H	0.65	H
9.081691591	K 1	16.3	K 1
1.254088009	K 2	11.07729928	K 2
287.1489473	B/F²	590.1140054	B/F²
0.5	e	0.001	e
		CHECK	H
-END-		CHECK	K1
		CHECK	K2
		-END-	

Alternate k_1:	
H/T - COMPOSITE WALL	
2.	X 1
4.5	X 2
1300.	T IN
80.	TAMB
1263.365891	T 1
209.1345196	T 2
1281.778161	AVG1
736.3231529	AVG2
2.28630371	H
16.3	K 1
1.259750971	K 2
295.2407311	B/F²
0.5	e
CHECK	K1
-END-	

Initial and alternate program operation, and storage information   Table IV

Initial operation:
Partitioning—799.19

Enter	Press	Comment
X_1	A	—
X_2	B	—
t_i	C	—
t_a	D	—
—	E	Program runs; q' is displayed.

Alternate operation:

Enter	Press	Comment
Alternate E	A'	First key in error limit
Alternate h	B', LRN	Insert alternate h
Alternate K_2	C', LRN	Insert alternate K_2
Alternate K_1	D', LRN	Insert alternate K_1
—	E	Program runs; q' is displayed.

Storage information

Location	Data
00	X_1
01	X_2
02	t_i
03	t_a
04	t_2
05	h_1
06	t_1
07	K_2
08	t_1
09	K_1
10	t_2
11	q'
12	½ $(t'_1 + t'_2)$
13	½ $(t_i + t_1)$
14	E

differences between t_1 and t_1' and t_2 and t_2' reduced to less than the 0.5 difference allowed by the calculator program (Table I).

Making the program more general

This program can be adapted in four ways:

1. The convergence point of 0.5 can be changed by means of the **A'** key to any value larger than zero. Of course, the smaller the error limit, the longer the program will run. However, it will always converge.

2. An alternate equation for Eq. (5), or value for h, can be inserted into the program by first pressing **B'**, then **LRN**. The alternate equation will be automatically flagged; that is, the main program will seek the alternate h.

3. An alternate equation for Eq. (6), or value for K_1, can be inserted into the program by first pressing **D'**, then **LRN**, and keying-in the equation. Again, the main program will seek the alternate K_1.

4. An alternate equation for Eq. (7), or value for K_2, can be inserted into the program by pressing **C'**, then **LRN**, and keying it in. Again, the main program is automatically flagged to seek the alternate K_2.

If the program is run without the printer, it will display q' at its end. Values for t, h and K can be obtained from the memory. With the printer, the program will print results as in Table III.

For an alternate K_1, the program will print the results as in Table III, indicating by the line "CHECK K1" that an alternate value for K_1 has been incorporated into the calculator program.

To demonstrate the flexibility of the program, the following alternate values for variables have been inserted: $t = 0.001$; $h = 0.65$; $K_1 = 16.3$; and $K_2 = [(t_1 + t_2)/2]$ (0.01).

The resulting printout is also listed in Table III.

For HP-67/97 users

The HP version of the program follows the instructions given in Table IV. Calculator output is in the same order indicated in Table III, but without the alphabetic information. The HP program contains the same arbitrary formulas given in the TI version. They are programmed as subroutines and may be readily changed:

E is in subroutine **Lbl a** at lines 184 to 188
h is in subroutine **Lbl b** at lines 165 to 183
K_2 is in subroutine **Lbl c** at lines 141 to 164
K_1 is in subroutine **Lbl d** at lines 117 to 142

A listing of the HP version of the program is provided in Table V.

Program listing for HP version

Table V

Subroutines for calculation of E, h, K_2, and K_1 (begins at step 017)

Step	Key	Code	Step	Key	Code	Step	Key	Code	Step	Key	Code	Step	Key	Code
001	*LBLA	21 11	036	RCL5	36 05	075	STOI	35 46	114	PRTX	-14	153	3	03
002	STO0	35 00	037	x	-35	076	RCL4	36 04	115	SPC	16-11	154	3	03
003	PRTX	-14	038	RCL7	36 07	077	-	-45	116	R/S	51	155	x	-35
004	R/S	51	039	÷	-24	078	ABS	16 31	117	*LBLd	21 16 14	156	e^x	33
005	*LBL5	21 12	040	RCL1	36 01	079	RCL8	36 08	118	4	04	157	.	-62
006	STO1	35 01	041	x	-35	080	X>Y?	16-34	119	6	06	158	9	09
007	PRTX	-14	042	RCL4	36 04	081	GTO3	22 03	120	8	08	159	8	08
008	SPC	16-11	043	+	-55	082	RCLI	36 46	121	0	00	160	8	08
009	R/S	51	044	STOI	35 46	083	STO4	35 04	122	RCL6	36 06	161	x	-35
010	*LBLC	21 13	045	RCL6	36 06	084	GTO8	22 08	123	RCL2	36 02	162	STO7	35 07
011	STO2	35 02	046	-	-45	085	*LBL3	21 03	124	+	-55	163	RTN	24
012	PRTX	-14	047	ABS	16 31	086	RCL6	36 06	125	2	02	164	RCL4	36 04
013	R/S	51	048	RCL8	36 08	087	PRTX	-14	126	÷	-24	165	*LBLb	21 16 12
014	*LBLD	21 14	049	X>Y?	16-34	088	RCL4	36 04	127	÷	-24	166	RCL3	36 03
015	STO3	35 03	050	GTO2	22 02	089	PRTX	-14	128	.	-62	167	-	-45
016	PRTX	-14	051	RCLI	36 46	090	SPC	16-11	129	0	00	168	ABS	16 31
017	SPC	16-11	052	STO6	35 06	091	RCL2	36 02	130	0	00	169	.	-62
018	RCL2	36 02	053	GTO0	22 00	092	RCL6	36 06	131	4	04	170	0	00
019	+	-55	054	*LBL2	21 02	093	+	-55	132	2	02	171	0	00
020	2	02	055	GSBd	23 16 14	094	2	02	133	5	05	172	5	05
021	÷	-24	056	RCL2	36 02	095	÷	-24	134	RCL6	36 06	173	8	08
022	STO6	35 06	057	RCL3	36 03	096	PRTX	-14	135	RCL2	36 02	174	2	02
023	RCL3	36 03	058	-	-45	097	RCL6	36 06	136	+	-55	175	x	-35
024	+	-55	059	RCL0	36 00	098	RCL4	36 04	137	2	02	176	1	01
025	2	02	060	RCL9	36 09	099	+	-55	138	÷	-24	177	.	-62
026	÷	-24	061	÷	-24	100	2	02	139	x	-35	178	5	05
027	STO4	35 04	062	RCL1	36 01	101	÷	-24	140	+	-55	179	3	03
028	GSBa	23 16 11	063	RCL7	36 07	102	PRTX	-14	141	STO9	35 09	180	5	05
029	*LBL8	21 08	064	÷	-24	103	RCL5	36 05	142	RTN	24	181	+	-55
030	GSBb	23 16 12	065	+	-55	104	PRTX	-14	143	*LBLc	21 16 13	182	STO5	35 05
031	*LBL0	21 00	066	RCL5	36 05	105	RCL9	36 09	144	RCL6	36 06	183	RTN	24
032	GSBc	23 16 13	067	1/X	52	106	PRTX	-14	145	RCL4	36 04	184	*LBLa	21 16 11
033	RCL4	36 04	068	+	-55	107	RCL7	36 07	146	+	-55	185	.	-62
034	RCL3	36 03	069	÷	-24	108	PRTX	-14	147	2	02	186	5	05
035	-	-45	070	STOA	35 11	109	SPC	16-11	148	÷	-24	187	STO8	35 08
			071	RCL5	36 05	110	RCLA	36 11	149	.	-62	188	RTN	24
			072	÷	-24	111	PRTX	-14	150	0	00	189	R/S	51
			073	RCL3	36 03	112	SPC	16-11	151	0	00			
			074	+	-55	113	RCL8	36 08	152	0	00			

The author

Calvin R. Brunner is a principal engineer with Malcolm Pirnie, Inc. (2 Corporate Park Dr., White Plains, NY 10602). His experience includes the design of thermal reduction equipment for solid waste and sludges, and also operator assistance with incineration. He has presented seminars on incineration and on programming the TI-59, and has developed programs for heat transfer, steam flow, and the design of waste-disposal equipment. Holder of a B.M.E. from City College of New York and an M. Eng. from Pennsylvania State University, he is a licensed engineer in four states and a member of ASME.

Section IV
Heat Recovery

Analyses for the energy efficiency of a plant, a unit operation or a piece of process equipment are readily performed by using a single concept, "exergy." Here is the theoretical background for this concept and practical energy-optimization examples both for low-temperature separations in ethylene plants and for turboexpanders and compressors.

Victor Kaiser, Technip

Energy Optimization

☐ Common to all development in the chemical process industries (CPI) is the best utilization of the energy input to the process. To deal with such a requirement, the engineer has to select the methods that will yield a systematic approach to the problem.

Directly related to energy conservation are all of the following:
- Yields of main products.
- Recovery of main products.
- Specific energy consumption.
- Investment.
- Operability.
- Maintenance.

Resolving these six factors is our objective when treating an energy optimization problem. A new process or machine is a true improvement if we can show benefits in all of the above factors or, at least, in the sum of them.

The weakest link is the investment. We generally integrate this factor by conversion to monetary values. However, it is not certain that these values are representative of their energy and raw-material contents such as would be the case in a stable economy. Studies are conducted to evaluate such contents [1].

To be stressed in this article are operability and maintenance, which are equivalent to plant availability. These are of prime interest when a complete process study is performed. An improvement can easily prove illusory if these factors are not carefully evaluated.

Tools for the study

Exergy is the best single concept on which to base our energy-efficiency analysis, and is defined as *the maximum technical work that can be derived from a fluid or a system.* This concept has the advantage of being quite universal because it can be applied to chemical processes, combustion processes, biochemical or photochemical processes, and so on, as well as to physical systems. The exergy function is defined as:

$$E = (H - H_o) - T_o(S - S_o) \qquad (1)$$

It is important to notice the presence of a reference state, which is one advantage of the exergy function. It is evident that our conclusion about a specific process must depend on its environment—specifically on the cooling-medium temperature, T_o.

All CPI processes are open to the surroundings from which they receive feeds and energy and to which they discharge products, energy and effluents. Hence, the quality of these surroundings is of utmost importance to the process with respect to technical performance.

As long as we deal with physical transformations that

Originally published February 23, 1981

leave the molecular structure unchanged, we need only three parameters to define the surroundings:

1. Reference temperature, T_o. We assume this to be constant, no matter how much heat is exchanged with the surroundings. No heat can be exchanged with the surroundings at another temperature.

2. Reference pressure, P_o. This pressure is also assumed to be constant and independent of the amount of material exchanged with the surroundings.

3. Reference gravity, g_o. For processes where gravity forces are important, the acceleration due to gravity has to be set as a reference. For CPI plants, however, this is generally of no importance.

The three parameters are sufficient as long as we do not intend to study the dilution of chemical substances in the surroundings or extraction of these substances from natural resources. For example, in an air-separation plant, we will have to set a reference air composition. This leads to the necessity of defining all reference compositions in the surroundings of substances we deal with. The reference state has a large impact on the process evaluation.

We might fear that some fundamental difficulties would prevent us from finding unbiased reference states, leading to a lack of generality and an excessive sensitivity of the results toward the reference parameters. Fortunately, this is not true. Looking at the exergy function, Eq. (1), we notice that the reference values H_o and S_o will always cancel if we study exergy differences rather than absolute exergy values—the differences being calculated for steady-state systems. For such "flow systems" (i.e., open systems), this means that the material balance of each chemical species has to be maintained at all times without formation or loss or accumulation. This is true for industrial processes as long as chemical reactions are excluded. Any datum value can be selected for H_o and S_o for each pure compound separately, and the choice will have no influence on the exergy differences calculated.

If chemical reactions are involved, it is necessary to work with absolute entropies. Also, the datum, S_o, for each pure compound has to be selected in a manner so as to reflect the industrial reality. For example, in a combustion process where carbon dioxide is one of the products, the datum could be based on the ambient temperature and partial pressure typically encountered in combustion processes. This choice is not important if we compare such processes among themselves. In this situation, we can use the standard values for H_o and S_o (ideal state at standard pressure and temperature), and obtain precise conclusions about the value of each process relative to the others.

If the dilution of carbon dioxide in the surrounding air would be the process of interest, then another datum point should be selected, such as the condition of the carbon dioxide in nature.

From the preceding discussion, we can comprehend that the exergy concept always leads us to very carefully consider the technical reality related to the process.

Exergy analysis cannot be fully applied in the absence of computer programs that calculate the thermodynamic properties of mixtures and pure compounds over wide pressure and temperature ranges. For specific unit operations, the exergy fluxes can often be calculated from the defining parameters of the unit operation, with a minimum of calculations of physical properties of the material flows to and from the system. A few examples extensively used in the analysis of cryogenic separations will be given in this article.

The exergy balance

No exergy-conservation law applies for real systems. The exergy balance for a steady-state open system can be written as:

$$\Sigma E_{\alpha i} + \Sigma W_j = \Sigma E_{\omega k} + L \qquad (2)$$

Eq. (2) expresses that: The sum of all incoming exergy due to material streams plus the sum of all work exchanged with the surroundings (+ sign for incoming work, − sign for outgoing work) is equal to the sum of all outgoing exergy due to material leaving the system, plus the lost work or exergy gap, L. L is also called the irreversibility, and represents the irretrievably lost work. It is no physically existing form of energy, but rather a measure of the amount of potential work destroyed by the system due to irreversible processes.

There are a few important rules that apply to the signs of the various functions. These rules are all deduced from the definition of exergy:

$$\text{If } P \geqslant P_o, \text{ then } E \geqslant 0 \qquad (3)$$

Lost work, L, is always positive, so:

$$L > 0 \qquad (4)$$

The actual exchanged work can be positive, negative or zero:

$$W_j \gtrless 0 \qquad (5)$$

By using Eq. (4), we can introduce a concept (generally called effectiveness or reversibility) that measures how far we are from an ideally reversible system.

For each process or unit operation, the total supplied work and/or exergy must be calculated. Then, this total work can be compared to the exergy effectively received

Factors relevant to energy consumption   Table I

(Basis: 200,000-metric-ton/yr ethylene plant, naphtha cracking)

Yield of olefins and aromatics	65-68	% by weight
Recovery of main products	99-99.5	%
Ethylene yield	30-35	% by weight
Total specific-energy input	7-8	kWh/kg of ethylene
Total compressor power	0.8-1	kWh/kg of ethylene
Heat of reaction, overall	1.5	kWh/kg of ethylene
Investment/kg of annual capacity	2-2.5	FF/kg of ethylene
Operability, major turnovers	2-3	yr
Availability, long term	90-95	%
Maintenance	3	%/yr of investment

by the processed stream and/or the work produced. This will become clear in the examples to be given. In general terms, it is possible to define effectiveness.

Looking at Eq. (2), we can define the effectiveness for an overall system if we focus the analysis on the technical work exchanged:

For $\Sigma(E_{\alpha i} - E_{\omega k}) > 0$,

$$\eta = -\Sigma W_j/\Sigma(E_{\alpha i} - E_{\omega k}) = 1/[1 - (L/\Sigma W_j)] \quad (6a)$$

For $\Sigma(E_{\alpha i} - E_{\omega k}) < 0$,

$$\eta = -\Sigma(E_{\alpha i} - E_{\omega k})/\Sigma W_j = 1 - (L/\Sigma W_j) \quad (6b)$$

Eq. (6a) and (6b) state that the effectiveness is the useful work or the exergy produced, divided by the work or exergy supplied to the system.

For example, if the purpose of the process is to upgrade the exergy of stream n (to the exclusion of any other stream), then:

$$E_{\alpha n} - E_{\omega n} < 0$$

Eq. (6b) applies (provided that i, $k \neq n$). Hence, the effectiveness, η, becomes:

$$\eta = -(E_{\alpha n} - E_{\omega n})/[\Sigma W_j + \Sigma(E_{\alpha i} - E_{\omega k})] \quad (7a)$$

$$\eta = 1 - L/[\Sigma W_j + \Sigma(E_{\alpha i} - E_{\omega k})] \quad (7b)$$

Eq. (6) and (7) are quite similar in mathematical form, but Eq. (6) applies to an entire system, whereas Eq. (7) puts stream n in particular focus.

Effectiveness, η, is always less than unity, but can be zero or negative. This simply means that the lost work can be larger than or equal to the quantity of useful work produced.

Let us now look at certain unit operations to find generally applicable relationships for calculating both the lost work and the effectiveness without calculating the exergy fluxes, E, of the various material streams.

Heat exchange

Here, the following relationships are important:

$$\Delta E_H = \Delta Q - T_o \int_{out}^{in} dQ/T = \Delta Q[1 - (T_o/T_{mH})] \quad (8)$$

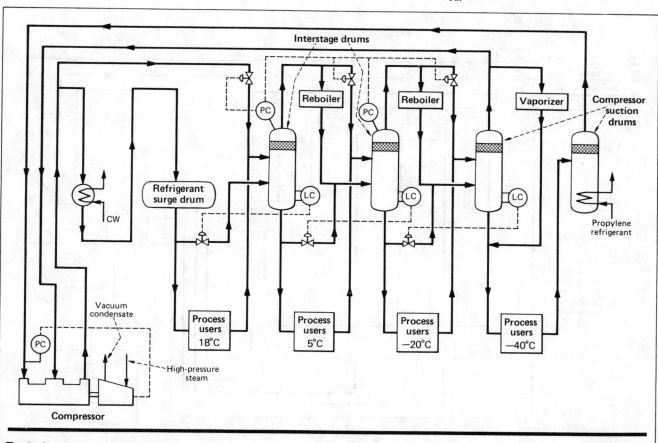

Typical propylene-refrigerant cycle   Fig. 1

For the hot side of the heat exchanger, differences are calculated as inlet minus outlet conditions, so:

$$\Delta Q > 0 \qquad (9)$$

If $T_{mH} < T_o$, then $\Delta E_H < 0$, $\Delta E_C > 0$, and the cold-side relationship is:

$$\Delta E_C = -\Delta Q[1 - (T_o/T_{mC})] \qquad (10)$$

Combining these relationships, we obtain the overall equation for the lost work as:

$$L = \Delta E_H + \Delta E_C = \Delta Q T_o \Delta T_m/(T_{mC})(T_{mH}) > 0 \qquad (11)$$

Based on Eq. (7b), we determine the effectiveness:

$$\eta = 1 - (L/\Delta E_C) = -\Delta E_H/\Delta E_C \qquad (12)$$

Closed-loop refrigeration cycle

For a closed-loop refrigeration cycle, effectiveness can be defined as the ratio between all outgoing exergy fluxes through the refrigerant users and the compressor power. By analogy to Eq. (6b), this can be written as:

$$\eta = -\Sigma\Delta E_{Ri}/W \qquad (13a)$$
$$L = \Sigma\Delta E_{Ri} + W \qquad (13b)$$

where:

$$\Delta E_{Ri} = +\Delta Q_i[1 - (T_o/T_{Ri})], \text{ and } \Delta Q_i > 0 \qquad (14)$$

ΔE_{Ri} is the exergy flux of the i-*th* cold producer at the refrigeration temperature, T_{Ri}. If the refrigerant is not

Nomenclature

E	Exergy flow, W	
H	Enthalpy flux, W	
L	Lost work, W	
M	Mass flow, kg/s	
P	Pressure, Pa	
Q	Heat flux, W	
S	Entropy flux, W/°K	
T	Temperature, °K	
V	Specific volume, m³/kg	
W	Power, technical, W	
δ	Recovery factor	
ε, η	Effectiveness	

Subscripts

C	Cold side of exchanger
H	Hot side of exchanger
i, j, k	Sequence numbers
m	Integral average
o	Reference state
R	Refrigeration
s	Isentropic
α	Inlet conditions
ω	Outlet conditions

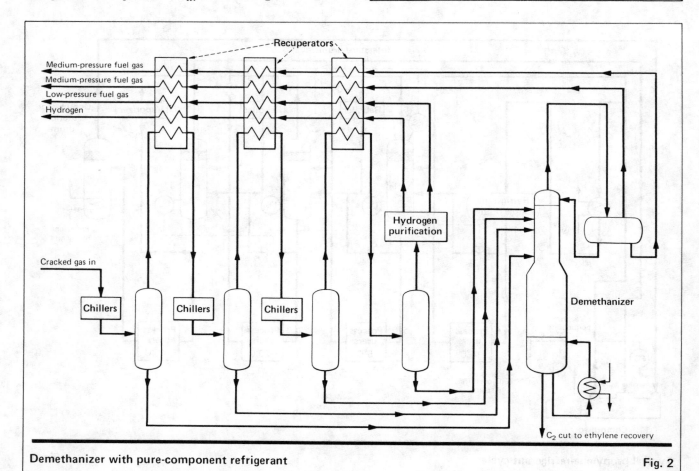

Demethanizer with pure-component refrigerant   Fig. 2

boiling at constant temperature, then T_{Ri} is an integral average temperature.

This relationship is very useful because η is independent of T_R for all practical purposes, and depends only on the type of refrigeration cycle, number of flash temperatures, pressure drops, and the compressor's polytropic efficiency.

Pressure drop in piping

It is useful to know the relationship for the lost work due to the pressure drop during isothermal flow in piping or equipment:

$$L = MV(T_o/T)\Delta P \qquad (15)$$

This lost work can be added to other items if a detailed study is made for an overall system. For heat exchangers, this term should be added to the lost work due to temperature approaches. In this way, it is possible to pinpoint the relative value of these various losses. Hence, decisions whether to lower the pressure drop and raise the exchanger area (and consequently lower the temperature difference) are put into a more realistic light.

Turboexpander

A turboexpander delivers shaft work and refrigeration, obtained from the cold-material stream leaving the machine. The effectiveness can be given separately for the two contributions. The results are:

Lost work overall:

$$L = (W - W_s)T_o/T_m \qquad (16)$$

Isentropic efficiency of the expander:

$$\eta_w = W/W_s \qquad (17)$$

Characteristics for a mixed refrigerant	Table II

Component	Composition, mole fraction
Methane	0.15
Ethylene	0.50
Propylene	0.35
Total	1.00

Heat of vaporization (−35/−40°C) 389 kJ/kg
Heat of subcooling (−35/−105°C) 170 kJ/kg
Molecular mass 31.5 kg/k-mole

Specific compression power for 1 MW 789.7 kW/MW
of process duty and no recuperators
($\delta = 0$)

For a flow of 100 k-mole/h, the vaporization curve of this refrigerant can be represented by:

$Q \quad = \quad aT^2 - bT + c$
$a \quad = \quad 0.00009505$, $(MW)(K^{-2})$
$b \quad = \quad 0.02692$, $(MW)(K^{-1})$
$c \quad = \quad 1.8626$, MW

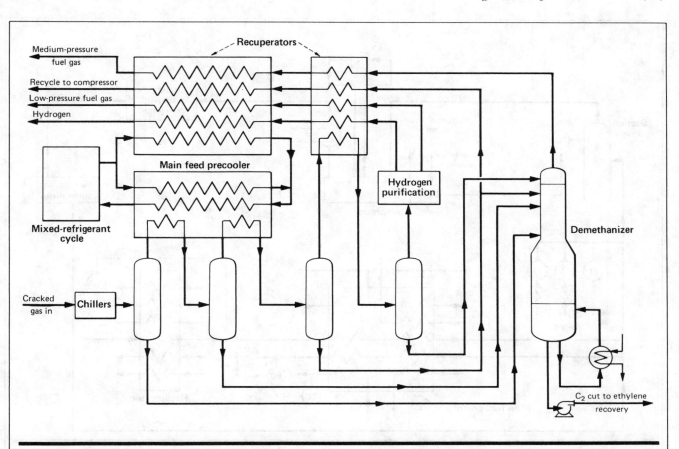

Demethanizer with mixed refrigerant   Fig. 3

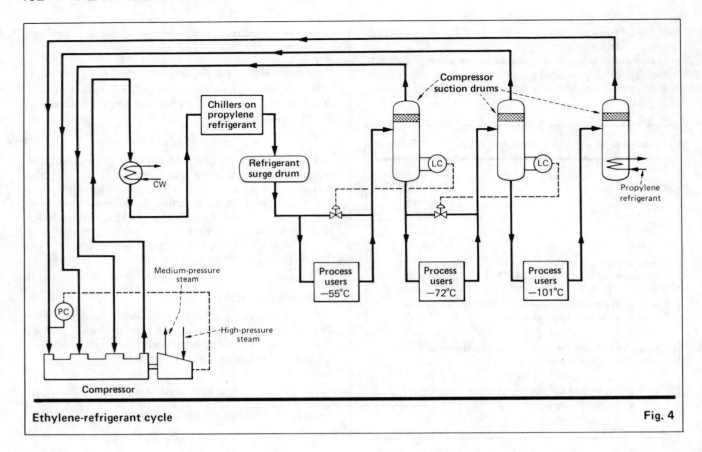

Ethylene-refrigerant cycle

Fig. 4

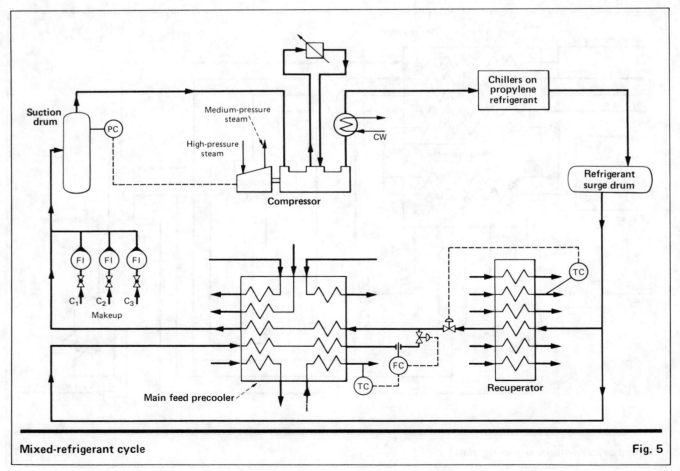

Mixed-refrigerant cycle

Fig. 5

Comparison of ethylene and mixed-refrigerant cycles at same duties and temperatures Table III				
Temperature, T		Heat flux,	Ethylene refrigerant	Mixed refrigerant
°C	K	ΔQ, MW	energy, ΔE, MW	energy, ΔE, MW
−101	172	0.2474	0.1999	0.1626
− 72	201	0.2194	0.1201	0.1060
− 55	218	−	−	−
Totals		0.4668	0.3200	0.2686
$\Delta E/\Delta Q$			0.6855	0.5754
$\dfrac{\Delta E \text{ (M)}}{\Delta E \text{ (E)}}$			0.84	

ΔE (M) represents energy difference for mixed refrigerant.
ΔE (E) represents energy difference for ethylene refrigerant.
Reference temperature, T_o, = 311K

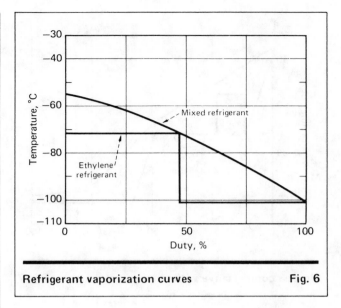

Refrigerant vaporization curves   Fig. 6

Refrigeration-power effectiveness:

$$\eta_R = 1 + (1 - \eta_w)[1 - (T_o/T_m)] \qquad (18)$$

Overall effectiveness, η, in accordance with Eq. (6a) is the shaft power versus input exergy, or:

$$\eta = \eta_w/(2 - \eta_R) \qquad (19)$$

Expander efficiency, η_w, is identical to the isentropic efficiency of the machine. Eq. (17), (18) and (19) clearly show that at low temperatures even a good value for η_w (such as 0.75) results in a modestly low figure for η_R. Once again, this reveals the difficulty of reaching good effectiveness at low operating temperature.

Compressor

Let us use the same symbols for the compressor as for the turboexpander. Here, s denotes exit conditions after an isentropic and adiabatic compression to the same pressure. We find the following relationships. (For the lost work, we will use Eq. (16).)
Isentropic efficiency:

$$\varepsilon_w = W_s/W \qquad (20)$$

Heating effectiveness:

$$\varepsilon_H = 1 + (1 - \varepsilon_w)[1 - (T_o/T_m)] \qquad (21)$$

Overall effectiveness:

$$\varepsilon = \varepsilon_w + \varepsilon_H - 1 \qquad (22)$$

If $T_m > T_o, \varepsilon_H > 1$. This means that the real machine has downgraded work to heat, which is not available from the ideal isentropic reference machine. Potentially, some work can be recovered from this heat. Of course, ε is always less than unity.

Heat leaks

Heat leaks, ΔQ, from a stream or equipment can be shown to produce the following lost work:

$$L = -\Delta Q[1 - (T_o/T_m)] \qquad (23)$$

where T_m is the integral average temperature of the

stream or fluid inside the equipment. We can convince ourselves that L is positive:

If $T_m < T_o$, then $\Delta Q > 0$, $L > 0$
If $T_m > T_o$, then $\Delta Q < 0$, $L > 0$

In such a way, it is possible to analyze many unit operations. However, not all can be reduced to general parameters. A rectifying column has to be defined by the incoming and outgoing material streams, so its "exergy balance" must use the enthalpy and entropy values of these streams. This is true whenever concentration gradients or steps are encountered.

Summary of exergy-balance analysis

For each processing step, it is possible to calculate the lost work by applying an "exergy balance."

Presenting a process analysis that shows minimum required work, lost work and actually exchanged work gives a very clear picture of the performance of each section. Examples of such analysis are presented in Ref. [4], [5] and [6].

The effectiveness can be used to check the performance of a processing step with the yardstick familiar to rotating-machine specialists. For very complex processes, effectiveness is not a versatile tool because overall effectiveness does not relate in a simple manner to the effectiveness of each part. On the other hand, the lost work in this case is very useful because the contribution of each part can simply be added up to yield the overall lost work, in quite the same way as with the real shaft-work.

Application to an ethylene unit

The ethylene production unit, as a key element in petrochemical production, is worth investigating in the terms discussed here. Table I summarizes typical values for the basic factors to be considered for energy optimization. In order to analyze such global figures, we can follow the method of Ref. [5], where application of the exergy analysis to the complete unit is discussed, to find areas of low effectiveness. Such a work tends only to highlight evident causes of low effectiveness, and is not

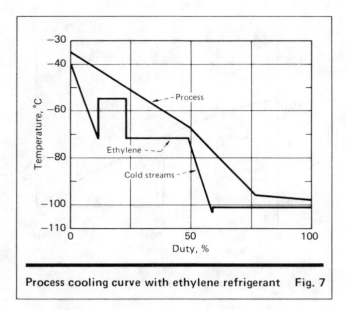

Process cooling curve with ethylene refrigerant   Fig. 7

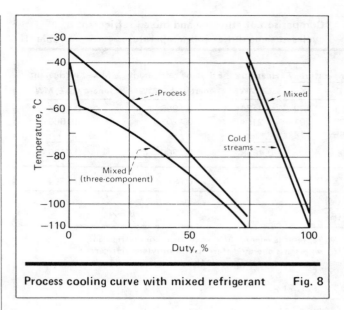

Process cooling curve with mixed refrigerant   Fig. 8

very rewarding. On the contrary, for comparing alternative processing schemes, exergy is the best tool to use.

Table I shows that the values of the several factors are not equally satisfactory—recovery being excellent, and yields and selectivity less favorable. For yields, we should, however, consider the kinetic optimum in relation to the technological constraints.

The investment value is in itself not very meaningful. When studying process modifications, we can, however, integrate the investment increments into optimization formulas.

Regarding the specific energy consumption, we notice that the useful energy (the heat of reaction) represents only one-fifth of the total energy input. One-tenth represents compressor power used for cryogenic gas separation. However, considering the conversion from thermal to mechanical energy, close to two-fifths of the energy inputs are really for compressor drivers. The remaining two-fifths of the energy input covers various process heating needs and all thermal losses.

From this general balance, we notice that the cryogenic fractionation, through the required compression power, represents about half of all thermal energy put through the plant. Therefore, improvements in gas separation are very desirable.

Optimization of refrigerant cycles

The refrigerant cycles and the cracked-gas compressor are the support for the cold fractionation throughout the unit. Once a scheme is selected, the problem is to fit the most efficient refrigerant cycle into the process. Ref. [2] provides an example of a study applied to the ethylene refrigerant cycles.

For refrigeration duties above $-35°C$, the largest part are constant-temperature heat exchangers. Hence, a pure-component refrigeration cycle is suitable. Fig. 1 shows a typical propylene-refrigerant cycle in an ethylene plant. By comparison, the gas-cooling section shown in Fig. 2 and 3 requires variable-temperature heat exchanges below $-35°C$, and variable-tempera-

ture cold recoveries from off-gases. In this case, a multi-component refrigerant is better suited to the process than a pure-component one.

Traditionally, ethylene has been used as a refrigerant so as to reach the low temperature of $-101°C$. It is interesting to study in detail the merits of each type of cycle. Ref. [3] and [4] discuss the process scheme when a mixed refrigerant is used. For this article, we have selected a refrigerant having the characteristics listed in Table II. Fig. 4 and 5 show typical arrangements for an ethylene and a mixed-component refrigerant cycle, respectively.

We begin by first comparing the theoretical refrigeration power that is required for either cycle for exactly the same duty. This corresponds to the curves shown in Fig. 6. Table III summarizes the data obtained. For this service, the mixed cycle has a base power requirement that is 16% lower than that of the ethylene cycle.

Fig. 7 and 8 show the cooling curves for the process, together with the vaporization curves of the refrigerant. We can calculate the effectiveness of these exchangers when operating against one or the other refrigerant, by using the methods described earlier in this article.

We will use the following relationships and data to perform the several computations for this example:

■ Calculation of exergy values for refrigeration.
Ethylene refrigerant:

$$\Delta E = \Delta Q - T_o(\Delta Q / T_R) \qquad (24)$$

Mixed refrigerant:

$$\Delta E = \Delta Q - T_o[2a\Delta T - b\ln(T_2/T_1)] \qquad (25)$$

Process: $\Delta E = \Delta Q - T_o k \ln(T_2/T_1)$ (26)

■ Heating and cooling curves.
Mixed refrigerant (see Table II for constants):

$$\Delta Q = aT^2 - bT + c \qquad (27)$$

Process:

$$\Delta Q = kT + Q_o \qquad (28)$$

Ethylene cycle:

For $-35° > T > -67°C$, $k = -0.1427$
For $-67° > T > -98°C$, $k = -0.08407$

Mixed cycle:

For $-35° > T > -70°C$, $k = -0.007726$
For $-70° > T > -105°C$, $k = -0.005815$

■ Recovery factor.

As shown in Ref. [2], the ratio, δ, of recuperator duty to total process duty is an important factor in cycle op-

---

Ethylene cycle, 1 MW of total refrigeration duty (without recuperators)   **Table IV**

Temperature, T,°C	Heat flux, ΔQ,MW	Enthalpy flux, ΔE_H,MW	Temperature, T,°C	Enthalpy flux, ΔE_C
-35/-67	-0.6367	-0.2567	-55	0.1186
-67/-98	-0.3633	-0.2310	-72	0.1955
			-101	0.2936
Total	-1.0000	-0.4877		0.6077
		A		B

Exchange effectiveness: $-A/B = 0.80$

Reference temperature, $T_{Q'} = 311$ K

Mixed-refrigerant cycle for 1 MW of total refrigeration duty   **Table V**

Temperature, T,°C	Heat flux, ΔQ, MW	Enthalpy flux, ΔE_H, MW	Temperature, T,°C	Enthalpy flux, ΔE_C, MW
-35/-70	-0.5700	-0.2356	-58/-110	0.5937
-70/-105	-0.4300	-0.2923		
Total	-1.0000	-0.5279		0.5937
		A		B

Exchanger effectiveness: $-A/B = 0.89$

Reference temperature, $T_{Q'} = 311$ K

Performance of chilling train   **Table VI**

	Line	Ethylene refrigerant	Mixed refrigerant
Total cooling duty, MW		7.173	10.004
Recuperator duty, MW		2.184	2.605
Net cooling duty, MW		4.989	7.399
Total exergy, process, MW	A	-3.499	-5.281
Total exergy, cycle, MW	B	4.360	5.939
Exchanger effectiveness (without recuperators)	-A/B	0.80	0.89
Exergy process (with recuperators), MW	C	-2.445	-3.906
Exergy cycle, MW	D	3.228	4.393
Net exchanger effectiveness	-C/D	0.76	0.89

Reference temperature, $T_{Q'} = 311$ K

---

timization. This ratio is directly related to the mixed-cycle power as follows:

$$\text{Compression power} = 789.7\Delta Q(1 - \delta), \text{ kW} \quad (29)$$

After performing the several calculations, we obtain the results shown in Tables IV and V.

From this example, we have a means of comparing the merits of various exchanger arrangements. From the cooling curves, we also see that the relatively good effectiveness of the ethylene cycle has been obtained at the expense of very close temperature approaches, down to 3°C at the coldest end.

The advantage of a mixed fluid is quite clear when we consider that the mixed refrigerant has a temperature difference with the process fluid of not less than 5 deg, and that this temperature approach can be adjusted in operation by changing the refrigerant composition. The same applies to the cold-gas recuperators. When the cold gases are exchanged against cracked gas, their effectiveness is quite low due to the large temperature difference. By using the mixed refrigerant to recover the cold, a perfectly constant temperature difference of 5 deg can be maintained throughout the exchanger—greatly improving the effectiveness of the recovery.

As another example, we have calculated the performance of the various sections of a 500,000-metric-ton/yr ethylene unit, cracking naphtha. Table VI summarizes the results.

Comparing exchanger effectiveness with and without recuperators, we see that the ethylene cycle is less efficient because the net effectiveness (taking into account the recuperators) is less than the one based on gas cooling (0.76 vs. 0.80, from Table IV). Table VII summarizes the results concerning the refrigerant cycles themselves for the whole system, including the demethanizer.

Analyzing the results of Table VII:

■ Starting from nearly equal total process duty, we first deduct all recuperators in the -35 to -110°C temperature interval. This gives the net refrigeration duty to be supplied by the refrigerants.

■ The recover factor, δ, is a valuable index that is directly related to the compressor power for mixed refrigerants. In relative and absolute values, it is more favorable in the mixed-refrigerant application.

■ Next, we consider the process-side received exergy (line C) in relation to the refrigerant-cycle supplied exergy (line D). The ratio, $-C/D$, indicates the overall exchanger efficiency. Comparing with Table VI, this value is equal for the mixed cycle in both tables, and higher for ethylene in Table VII because the duty for the demethanizer condenser is included. This unit has a very good efficiency, being a nearly constant-temperature heat exchange. Overall, the mixed cycle is still about 13% more efficient.

■ The ethylene and mixed-refrigerant cycles do not supply the total exergy by themselves. Part of it is supplied by the propylene compressor up to the reference temperature of 311°K (38°C), as shown on line E. We see that the mixed-refrigerant cycle demands only half the power of the ethylene cycle. This is one outstanding feature of its application, and results in a better load balance between the compressors. This is especially favorable for large units where the propylene compressor intake volume is sometimes substantial.

It is possible to show the exergetic efficiency of each cycle (see Table VIII). The low efficiency of the ethylene cycle is due to several factors such as suction pressure drop at low absolute pressure, low intake temperature, and condensing duty concentrated at the $-40°C$ level.

Going back to Table VII, we can attribute a real power to the propylene compressor if we take its overall efficiency to be 0.55. We obtain the values on line F in Table VII. Adding to it the real compressor power for the mixed-cycle or ethylene-cycle compressor, we get the total shaft power, line (F + G), required to supply the exergy (line D). Calculating the overall average efficiency, D/(F + G), we see that both systems are quite similar. The poor efficiency of the ethylene cycle is covered by its smaller contribution overall. However, looking at the overall process efficiency, C/(F + G), the bonus of the mixed cycle, about 16%, is apparent. This advantage can also be seen when we compare the total compressor power, (F + G), showing the mixed-cycle system to represent 80% of the ethylene-cycle system.

Looking at the plant overall, the total compressor-power difference between the systems is smaller, because a recycle stream is treated by the charge-gas compressor in the mixed-cycle scheme. This adds 1.0 MW to the total power. So the ratio for total power between the mixed- and ethylene-refrigerant cycles becomes:

$$(9.04 + 1.0)/11.276 = 0.89$$

This ratio is the overall direct comparison between the two schemes. But a better cold recovery from the demethanizer reboiler, which favors the mixed-cycle scheme, should be taken into account for a global comparison (see Ref. [7]). We see that the mixed-cycle scheme introduces better exergy efficiency at some key points. Due to its better adaptation to the process requirements, it has better potential for improvement.

The example treated here shows how a complicated process system can be decomposed into a sequence of unit operations. The exergetic effectiveness analysis can be applied to each subsection, and even to each equipment item. Due to the definition of lost work, as a balance difference, the contribution of each section to the lost work can be added up to finally arrive at the total for the whole system.

Application to a turboexpander

A turboexpander receives a stream with properties identified by index α. The stream leaving the expander has properties identified by index ω. An ideal isentropic and adiabatic expansion would produce a stream leaving the expander with properties identified by index s. The following relationships hold:

First law of thermodynamics:

$$H_\alpha + W = H_\omega \qquad (30)$$

And, in accordance with Eq. (2):

$$E_\alpha + W = E_\omega + L \qquad (31)$$

For isentropic expansion:

$$S_\alpha = S_s \qquad (32)$$

Performance of refrigerant and power cycles — Table VII

	Line	Ethylene refrigerant	Mixed refrigerant
Process temperature, °C		−35/−98	−35/−105
Total cycle duty, MW	A	10.280	10.004
Recuperators, MW	B	2.184	2.605
Net cycle duty, MW		8.096	7.399
Recovery factor, δ	B/A	0.212	0.260
Energy of process, MW	C	− 4.221	− 3.906
Energy of cycles, MW	D	5.368	4.393
Energy of propylene cycle, MW	E	3.627	1.760
Propylene cycle, MW (real power)	F=$\frac{E}{0.55}$	6.594	3.200
Real cycle power, MW	G	4.682	5.840
Total compressor power MW	F + G	11.276	9.040
Overall exchanger effectiveness	−C/D	0.79	0.89
Overall cycle efficiency	D/(F+G)	0.48	0.49
Overall process efficiency	C/(F+G)	0.37	0.43

Relative process effectiveness = 0.43/0.37 = 1.16

Reference temperature, T_o = 311K

Exergetic-cycle efficiencies — Table VIII

Exergy	Line	Ethylene refrigerant	Mixed refrigerant
Total cycle, MW	D	5.368	4.393
Propylene cycle, MW	E	3.627	1.760
Net ethylene/mixed, MW	D−E	1.741	2.633
Real ethylene/mixed compressor power, MW	G	4.682	5.840
Cycle efficiency	(D−E)/G	0.37	0.45

Example for performances of turboexpander/compressor — Table IX

	Turboexpander		Compressor	
Inlet temperature, °C	−100		38	
Inlet temperature, K	173		311	
Outlet mean-temperature, T_m, K	100		384	
Reference temperature, T_o, K	311		311	
T_o/T_m	3.11		0.81	
Machine efficiency, η_w or ϵ_w	0.75	0.70	0.75	0.70
Refrigeration effectiveness, η_R	0.47	0.37	....	....
Heating effectiveness, ϵ_H	....	....	1.05	1.06
Overall effectiveness, η or ϵ	0.49	0.43	0.80	0.76

But, the technical or isentropic power is given by:

$$-W = H_\alpha - H_\omega \qquad (33)$$

$$-W_s = H_\alpha - H_s \qquad (34)$$

For the ideal machine, Eq. (31) is:

$$E_\alpha + W_s = E_s \qquad (35)$$

Using Eq. (35), Eq. (31) can then be written as:

$$W - W_s + E_s - E_\omega = L \qquad (36)$$

Effectiveness for shaft power, η_w:

$$\eta_w = 1 - [(W - W_s)/(-W_s)] = W/W_s \qquad (37)$$

Effectiveness for refrigeration, η_R:

$$\eta_R = 1 - [(E_s - E_\omega)/(-W_s)] \qquad (38)$$

But: $\qquad E_s - E_\omega = H_s - H_\omega - T_o(S_s - S_\omega) \qquad (39)$

Eq. (33) and (34) yield:

$$H_s - H_\omega = W_s - W \qquad (40)$$

Because the exhaust pressures for the real and ideal machines are equal, we know from thermodynamics that:

$$H_s - H_\omega = \int_{S_\omega}^{S_s} T dS = T_m(S_s - S_\omega) \qquad (41)$$

If the specific heat is constant, the integral average temperature, T_m, is calculated as:

$$T_m = (T_s - T_\omega)/[(\ln (T_s/T_\omega)] \qquad (42)$$

Finally: $\qquad E_s - E_\omega = -(W - W_s)\left(1 - \dfrac{T_o}{T_m}\right) \qquad (43)$

From Eq. (36):

$$L = (W - W_s)(T_o/T_m) \qquad (44)$$

Now, we can rewrite Eq. (38) as:

$$\eta_R = 1 - \left[(W_s - W)\left(1 - \dfrac{T_o}{T_m}\right)\right]/(-W_s) \qquad (45)$$

$$\eta_R = 1 + (1 - \eta_w)[1 - (T_o/T_m)] \qquad (46)$$

The term $(E_s - E_\omega)$ is the refrigeration power equivalent to the duty between T_s and T_ω. This refrigeration duty is lost because the real-machine exhaust temperature is T_ω rather than T_s. So, η_R represents the refrigeration-power effectiveness measured against the theoretical shaft power, W_s.

Using the effectiveness definition of Eq. (6a), we find the following relationship from Eq. (44):

$$\eta = 1/\left[1 - (T_o/T_m)\left(1 - \dfrac{1}{\eta_w}\right)\right] \qquad (47)$$

Finally, η, η_w and η_R are related as follows:

$$\eta_w = \eta(2 - \eta_R) \qquad (48)$$

For the compressor, we note that all equations, including Eq. (44), hold without any modification. However, for the shaft effectiveness, ε_w, by analogy to Eq. (6b), we will use the definition:

$$\varepsilon_w = 1 - [(W - W_s)/W] = W_s/W \qquad (49)$$

The lost work remains unchanged as per Eq. (36).

Hence, the overall exergy effectiveness, ε, according to Eq. (6b) is:

$$\varepsilon = 1 - [(1 - \varepsilon_w)(T_o/T_m)] \qquad (50)$$

Following an exact analogy with the turboexpander analysis, we can define a "heating" effectiveness, ε_H, as:

$$\varepsilon_H = 1 - [(E_s - E_\omega)/W] =$$
$$1 + (1 - \varepsilon_w)[1 - (T_o/T_m)] \qquad (51)$$

Finally: $\qquad \varepsilon = \varepsilon_w + \varepsilon_H - 1 \qquad (52)$

If no value is attached to the heat to be recovered from the compressor discharge, then only ε_w is technically meaningful.

A numerical example comparing the turboexpander and the compressor shows the various effects. The final results are tabulated in Table IX. From the table, we see that the amount of cold not recovered from the turboexpander represents a sizable loss. Hence, refrigeration and overall effectiveness are low. For the compressor, the heating to be recovered from the actual machine relative to the ideal machine is low. Here, ε_H is larger than unity because the real machine downgrades work to heat that can be recovered to the extent of 5% or 6% for compressor efficiencies of 0.75 or 0.70, respectively. But this is marginal, so the overall efficiency remains good.

Acknowledgement

This work has been performed as part of the Technip/TechniPetrol development program concerning olefins production. Thanks are due to M. Watrin and D. Gilbourne of Technip for their active support, and to C. Pocini and M. Picciotti of TechniPetrol for their help and advice.

References

1. "Economie d'Energie en Raffinage et Petrochimie," Technip, Paris, 1976.
2. Picciotti, M., Optimize Ethylene Plant Refrigeration, *Hydrocarbon Process.*, May 1979, pp. 157–166.
3. Kaiser, V., Becdelièvre, C., and Gilbourne, D., Mixed Refrigerant for Ethylene, *Hydrocarbon Process.*, Oct. 1976, pp. 129–131.
4. Kaiser, V., Salhi, O., and Pocini, C., Analyze Mixed Refrigerant Cycles, *Hydrocarbon Process.*, July 1978, pp. 163–167.
5. Maloney, D. P., U.S. Dept. of Energy Workshop, Aug. 14–16, 1979.
6. Kenney, W. F., Improving Energy Use: Thermodynamic Analysis for Research Guidance, Paper No. 48a, AIChE Meeting, San Francisco, Nov. 25–29, 1979.
7. Kaiser, V., Heck, G., and Mestrallet, J., Optimize Demethanizer Pressure for Maximum Ethylene Recovery, *Hydrocarbon Process.*, June 1979, pp. 115–121.

The author

Victor Kaiser is process supervisor for Technip, Cedex 23, 92 090 Paris La Defense, France. He heads a group of engineers engaged in process engineering and design of ethylene plants. Previously, he was plant process engineer with Lonza (Switzerland), and then joined Lummus (Paris) as a process engineer. He has a Diplom Ingenieur Chemiker and a Ph.D. in chemical engineering from the Federal Polytechnic Institute (Zurich). He is also a member of the Assoc. Française des Techniciens du Pétrole, and of the Schweizerischer Ingenieur- und Architektenverein.

Energy conservation in process plants

Here is a rundown of the many ways in which a company can reduce usage of increasingly expensive energy.

Robert Aegerter, Northern Petrochemical Co.

☐ One cannot effectively minimize energy use in a plant without first identifying and quantifying major losses. Hence, an energy audit is a necessary first step.

Energy reduction in a petrochemical plant such as ours (as in most chemical process industry plants), is more easily accomplished by reducing fuel and steam use than by cutting electrical use. Therefore, the first plantwide energy audit should be a steam balance.

For this, all operating units must conduct the balance over the same time period. All steam users, and all flows between operating units, should be correctly identified. An initial steam balance may have errors arising from poor metering or incorrect assumptions that will appear when it is mathematically impossible to close the balance. Any errors should be identified as "unaccounted for" flows. Where errors are apparent, additional metering should be placed on large steam users, to minimize "unaccounted for" flows on subsequent steam balances.

The first plant steam balance may have some large inaccuracies. Flow of steam through turbines, vent and letdown valves, and most major steam users may not have been measured. These flows may be determined from manufacturers' design data for turbines, from process conditions and valve manufacturers' data for steam letdown and vent flows, and from energy balances for major steam users. Although inaccuracies may exist in an initial steam balance, it still will indicate areas of large potential energy reduction.

However, flowmeters should be installed on all turbines, and steam letdown and vent installations. Good instrumentation can identify problems and quickly pay for itself. Note that a steam balance need only be accurate enough to predict the result of system changes.

One must correctly evaluate the potential savings identified in energy audits. Any savings calculations should reflect actual plant operating costs such as maintenance, electricity, purchased fuel, and boiler-feedwater treatment chemicals. Special care should be taken not to count the same saving twice. For example, one project may significantly reduce low-pressure steam use, and thereby result in an excess of low-pressure steam. Another project may then eliminate the excess steam created

Horizontal split-case pump with tandem drivers— an electric motor and a high-efficiency steam turbine

by the first. Instead of counting the saving twice, the estimated saving should be the net from the two.

Optimum steam balance

Many people believe that an optimum steam balance occurs when:

1. No steam is being vented to the atmosphere or condensed in an excess-steam condenser.

2. Steam letdown between pressure levels is being minimized by using the pressure to operate various combinations of motor- and turbine-driven equipment.

This is untrue; an optimum balance occurs when:

1. Steam is being generated from as many waste-heat sources as economically feasible.

2. High-pressure steam is generated only to supplement the low-pressure steam demand that is not met by waste-heat recovery.

3. The steam demand at every lower pressure level is minimized by waste-heat utilization.

4. All continuous-operation steam turbines are of the highest efficiency that is economically feasible.

5. The most efficient turbines that can be operated intermittently are run, when required, to minimize steam letdown between steam pressure levels.

6. Enough flexibility is designed into the system to minimize venting or inefficient condensing of steam.

7. Reliability of plant operation is not jeopardized by changes in the steam system.

Energy-reduction projects

The energy-reduction measures that a petrochemical plant undertakes fall into three categories:

Originally published September 3, 1984

1. Projects unaffected by the steam balance.
2. Projects that reduce excess steam.
3. Projects that reduce steam demands.

The following are projects that can be carried out independently of a steam system:

As steam production decreases, careful consideration should be given to determining when it is economically feasible and reliable to shut down a boiler. Boilers normally operate more efficiently when loaded over 50% of design rating. When steam demand is reduced below 50% of design load on those then operating, managers should consider shutting one down, and increasing the load on the others.

Increased attention to the demineralized water system and boiler-feedwater chemical treatment program can improve water quality and increase boiler cycles. Also, reduced boiler blowdown can decrease boiler fuel costs and boiler-feedwater chemical costs.

Additional boiler feedwater and feed-preheat tubes can be installed in stacks of boilers and furnaces, to reduce their fuel requirements. However, stack temperatures must be maintained above the acid dewpoint temperature of the flue gas, unless tubes are specially coated to prevent corrosion.

Process-header liquid pressures should be examined, to see if excessive pressure drop is being taken across control valves. Pressures can be reduced by trimming impeller diameters, slowing the driver speed, or both. Motors can be slowed by installing a variable speed controller or by using a two-speed motor. Turbines can be slowed by changing the governor setting.

If worn-out exchangers are replaced, larger piping and lower-pressure-drop exchangers should be considered, to see if larger equipment is economically feasible. Lower-pressure-drop exchangers and piping can reduce the horsepower requirements of compressors and pumps.

High-efficiency drivers should be substituted for motors or turbines that need replacement. The incremental cost of the higher-efficiency driver can usually be easily justified because of its lower operating expense.

Low-level heat sources can be used to preheat furnace feeds. The higher furnace-feed temperature reduces the furnace fuel-gas requirements and adds additional heat to the stack, which can be further recovered by using additional economizer surface area.

Excess-oxygen analyzers and controllers should be installed on all furnace stacks to control the furnaces' excess-oxygen levels.

If the utility unit's boilers have internal superheaters, the superheat temperature should be operated as close as possible to the safe upper temperature limit of the boiler and all downstream equipment. The higher superheat temperatures allow the boilers to operate at lower excess-air levels. This mode of operation both reduces boiler fuel requirements and decreases turbine steam demand, owing to the higher superheat temperature.

Weekly backflushing of interstage coolers can reduce the horsepower requirements of a large ethylene-unit compressor. Such backflushing can reduce interstage temperatures during summer months.

A clear-plastic strip door can be installed on warehouse loading doors to reduce air infiltration.

Color-coded outside lighting switches will allow personnel to shut off unneeded lights during daylight hours.

An economic-insulation-thickness program can be run and compared with the plant's insulation standards. It may mean updating the present standards, and perhaps reinsulating over existing insulation.

Reducing use of low-pressure steam

The following projects can reduce excess low-pressure steam (note that whenever steam demand is reduced, another change must be made on the system to reduce the supply of steam, to conform to the reduced demand):

An automatic upstream pressure controller can be installed on the deaerator to take up pressure swings in the low-pressure-steam header. The deaerator pressure control can be integrated into an existing low-pressure-steam control system. When the low-pressure system begins to overpressure, more steam is added to the deaerator by the pressure controller. The potential excess steam condition is eliminated and more preheat is added to the boiler feedwater. When the low-pressure header needs additional steam, the deaerator pressure is automatically reduced, thus eliminating the demand.

To further increase the steam system's flexibility, the pressure rating of the deaerator should be investigated to find if a higher rating is feasible.

Nuclear-industry-grade positive-shutoff valves can be installed on steam letdown control stations. Such valves provide tighter shutoff and better wear characteristics than conventional ones. Higher-quality valves are justifiable with today's high energy costs. Valves should be checked quarterly to verify positive shutoff.

Along with the deaerator pressure control, selecting which motors or turbines should be operating can minimize the steam venting, or steam letdown, between steam pressure levels. When a selection of several turbines is available to minimize steam letdown, the most efficient turbine with a steam rate that most closely matches the steam letdown flow is operated.

Consider replacing all continuous-operation back-pressure turbines over 100 hp with the most efficient turbines available. (New well-designed turbines can operate with over 50% efficiency).

Larger steam headers or dual headers between operating units can be installed to allow effective steam transfer. These headers not only can reduce steam venting, but also can reduce steam letdown flow.

If a turbine is designed to exhaust steam into the low-pressure steam header, it can be replaced with a highly efficient turbine that exhausts into an intermediate-pressure system. Changing the operating conditions of steam turbines can provide a quick remedy for some unbalanced conditions.

Increased steam-system flexibility and plant reliability can be achieved with motor/turbine tandem drivers (see photo on p. 93), and automatic startup on turbines, where motors are the primary drivers. The reliability of these systems is ensured by:

a) Maintaining both the motor and turbine simultaneously on line with the tandem driver option. The selection of which driver is carrying the equipment load is determined by the turbine governor. If the turbine is the selected primary driver, its speed is set slightly above the motor's synchronous speed; conversely, if the motor is to be the selected driver, the turbine's speed is set just below the motor's synchronous speed. The turbine governors are stroked weekly to ensure smooth operation.

b) Maintaining a turbine ready to start up instantaneously on low line pressure or another process condition. The turbine is maintained in a ready condition by:

■ Keeping the turbine case hot, by leaving the exhaust valve open.

■ Keeping the inlet line and turbine case free of wet steam, by trapping the line and case into the next lower level of steam pressure.

■ Controlling flow to the turbine inlet by using a solenoid switch connected to the process variable. If the process condition is met, the switch sends a signal to the inlet steam valve, which opens at a controlled rate to start the turbine.

Excess-steam condensers can be removed from the plant. These finned fan condensers, which are maintenance problems, only mask large energy waste.

Many times, engineers immediately think of installing steam users to eliminate excess steam. New steam users should be installed only if:

■ A feasible method does not exist to reduce steam entering the overpressured system.

■ The steam is more valuable used in the process than for its heating value.

Reducing low-pressure steam demand

See that steam-leak repairs are made promptly. Any large leak, especially of high-pressure steam, deserves immediate repair by either plant maintenance personnel or a professional online leak-repair service.

Incorporate welded-bonnet valves into piping specifications. Frequently, valves in medium- or high-pressure service leak at the bolted bonnet. Welded-bonnet valves, which cost no more, eliminate bonnet leaks.

Keep tower reflux to a minimum to reduce energy requirements of reboiler and reflux pumps.

Use low-level heat sources whenever possible to supplement steam requirements. If a process stream must be heated from 60 to 250°F, instead of using strictly steam for the purpose, use a low-level heat source to preheat it, and then use steam only to meet the remaining heating requirements. One excellent heat sink for low-level heat recovery is boiler makeup water. The water can be heated to within 20°F of the operating temperature of most deaerators without causing problems from dissolved oxygen. Stainless-steel exchangers and piping must be installed where non-deaerated water is heated above 160°F.

Recover additional steam generation from waste heat. At plant locations where steam is being generated from waste-heat sources, projects can be installed that further generate steam at the same or a lower pressure level.

Return condensate to the boilers, wherever economically feasible. Advantage should be taken of the various steam pressure levels in the plant to recover condensate. For example:

■ Always trap high- or medium-pressure steam into the next available lower-pressure steam header.

■ Any large flow of condensate that is discharged to the ground should be flashed at the lowest available pressure level. For instance, boiler blowdown should be flashed into a tank that vents into the deaerator steam supply line. The remaining unflashed condensate can be used as an additional heat source before discharging.

The flash steam from condensate flash tanks should be recovered by:

■ Flashing the steam into the low-pressure system.

■ Installing a steam condenser on top of the flash tank that recovers heat to a process stream.

■ Raising the flash tank pressure and utilizing the flash steam for process use. For example, to recover the heat that is lost to the atmosphere through the condenser, the pressure of the condensate flash tank can be raised and the flash steam can be used in the process instead of low-pressure steam.

If condensate is being discharged to the sewer because of an impurity that restricts it from being used as boiler feedwater, it can sometimes be used in other locations. For example, blowdown from boilers can be flashed from a flash drum into a low-pressure steam header and the remaining condensate can be used as a heat source.

It is important to install positive-shutoff valves on steam vents. A normal valve that is operated for several years will leak several hundred to several thousand pounds/hour depending on its size and service. Positive-shutoff valves quickly pay for themselves.

A stethoscope, or a contact ultrasonic instrument, is the best device to use in testing steam traps. The "sound testing" method is superior to others when visual observation is not possible.

Along with projects, better process control can reduce steam system operating costs. Place meters on all major steam users and on all flows entering and leaving operating units. A computer can be installed in the utility unit to continuously monitor the plant's steam balance. Utility-unit operators can monitor the plant's steam balance and make changes in their unit or recommend changes to other units to optimize the steam system.

Adding more metering and a computer takes much of the guesswork out of steam system optimization.

It is important to use a plantwide approach to monitoring energy use. Both operating-unit and plant energy costs and use should be closely followed to determine progress in energy reduction. The operating units' and the plant's energy efficiency (the number of Btu consumed/pound of production) and the energy costs and use trends should be routinely plotted on charts. Such charts will enable superintendents and managers to quickly assess the plant's present energy status.

The author

Robert Aegerter is Energy Management Engineer for Northern Petrochemical Co. (NPC), P. O. Box 450, Morris, IL 60450; tel: (815) 942-7266. His former positon at NPC was Area Engineer in the utility and ethylene plants. He holds a B.S.M.E. from Iowa State University, and is a registered professional engineer in the state of Illinois.

How to evaluate heat recovery via high-temperature fluid media

Cost data and a procedure are presented for making a preliminary estimate of the economic viability of using a heat-transfer fluid in a waste-heat-recovery system.

*Walter F. Seifert, John Beyrau, Gallie Bogel and **Louis E. Wuelpern**, The Dow Chemical Co.*

☐ Heat that would otherwise be wasted is generally recovered by being transferred into combustion air (Fig. 1a), or into steam and subsequently into process or utility streams via exchangers (Fig. 1b). Instead of into water, the heat can be transferred into a heat-transfer fluid, and from it into another stream (Fig. 1c).

Because recovered heat can be transported more easily at higher temperatures to further distances by a heat-transfer fluid than by high-pressure steam, a system using a heat-transfer fluid often offers a more attractive investment return, particularly when stack-gas exit temperature exceeds 600°F and furnace duty surpasses 30 million Btu/h.

An approach to economic evaluation

In a study of the economic factors governing heat recovery via intermediate organic heat-transfer media, Dow Engineering Co. (a subsidiary of The Dow Chemical Co.) compiled and analyzed an extensive body of equipment and operating cost data. From this study was developed a simplified approach to the economic evaluation of different sizes of heat-recovery systems involving various operating conditions.

The approach provides a tool for estimating: (1) potential annual savings and offsetting annual expenses; (2) required capital investment; and (3) investment payback period. Payback period, in years, is defined as: initial investment divided by the difference between annual savings and expenses.

A broad-stroke, first-phase evaluation procedure, the approach primarily determines whether a more detailed and accurate second-phase analysis is warranted. The only data needed include: furnace duty (the energy output to a process stream), stack-gas exit temperature, fuel cost, and system operating factor.

Example calculations are presented for systems ranging from 30 to 250 million Btu/h furnace duties, to illustrate the approach. In the examples, the following assumptions are made: stack-gas exit temperature = 750°F; fuel cost = $7.50/million Btu (No. 2 fuel oil); and

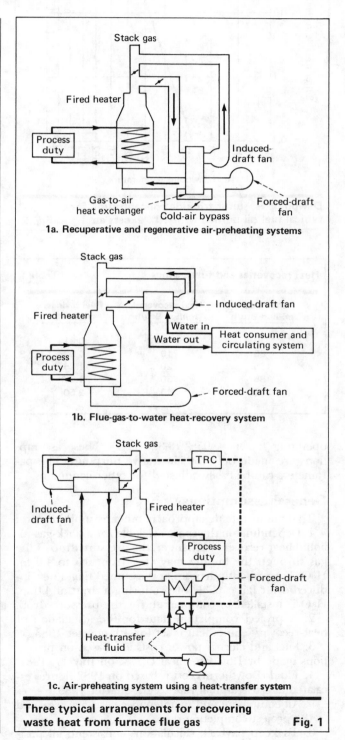

1a. Recuperative and regenerative air-preheating systems

1b. Flue-gas-to-water heat-recovery system

1c. Air-preheating system using a heat-transfer system

Three typical arrangements for recovering waste heat from furnace flue gas   **Fig. 1**

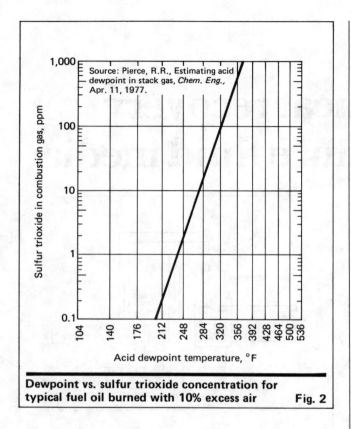

Dewpoint vs. sulfur trioxide concentration for typical fuel oil burned with 10% excess air   **Fig. 2**

Heat recoveries and fuel savings		Table I
Furnace duty, million Btu/h	Heat recovered, million Btu/h	Fuel savings, million $
30	5	0.31
65	10	0.62
125	20	1.25
185	30	1.87
250	40	2.50

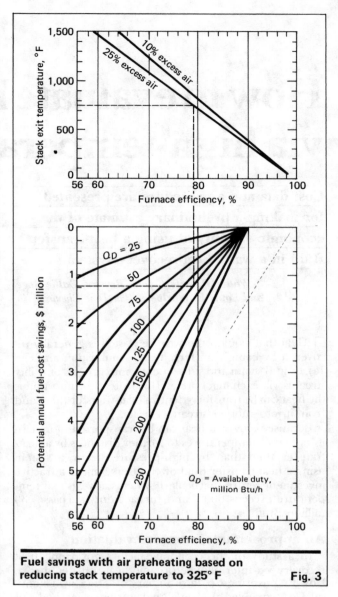

Fuel savings with air preheating based on reducing stack temperature to 325°F   **Fig. 3**

operating factor = 95% (8,320 h/yr). These assumptions are made only for illustrative purposes. The parameters can be easily adjusted for other conditions.

General assumptions

To demonstrate the approach, other assumptions are:

1. Depending on the space available for a stack-gas-to-liquid heat-recovery exchanger, the pressure drop of the gas through the bundle may vary from 0.5 to 4.0 in. H_2O. For present purposes, it is assumed that a new induced-draft fan capable of producing a draft of 4.0 in. H_2O is installed to compensate for the pressure drop.

2. A project completion date for the installation of heat-recovery equipment is set for 4th-quarter 1983.

3. Fuel and electric-power costs are based on projections made by Dow Chemical U.S.A. on July 17, 1981.

4. Fluid (Dowtherm) cost is based on 1982 figures escalated at a 6% annual rate to 4th-quarter 1983.

5. All equipment and maintenance costs are escalated to the project completion date.

6. To keep payback calculations consistent, all post-startup cash flows are adjusted to constant dollars. Other interest and inflation rates can be applied.

It should be noted that payback period is defined on a before-tax basis, because tax rates vary with time and company. A rate based on a company's forecast should be used to determine an after-tax payback period.

Estimating potential savings

Energy savings depend on furnace-duty rating, and on total drop in stack-gas exit temperature achieved through heat recovery. As a practical matter, exit temperatures seldom drop below 280–325°F, the acid dewpoint range for most sulfur-containing fuels (Fig. 2).

The exterior tubewall (and the base of fins) temperature of a gas-to-liquid exchanger should be higher than the dewpoint temperature of the stack gas. The extra cost for highly-corrosion-resistant exchangers is seldom justified by the value of additional heat recovered below the dewpoint temperature.

Fig. 3 provides a simple means of estimating total available savings based on reducing the stack-gas exit

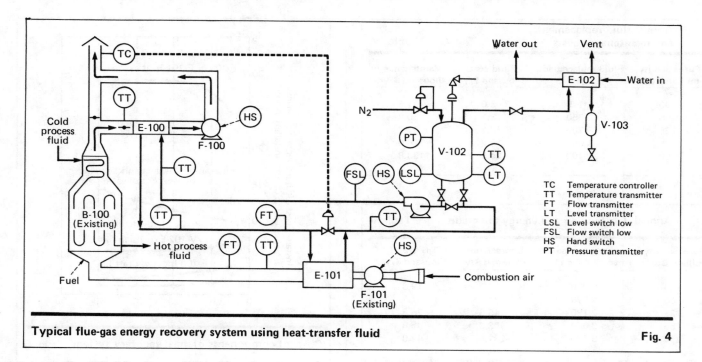

Typical flue-gas energy recovery system using heat-transfer fluid   **Fig. 4**

temperature to 325°F. Enter the upper graph at your current stack-gas exit temperature and move horizontally to the right, to the correct excess-air plotline for your system. Next, drop to the curve of the lower graph that most closely corresponds to your furnace duty (interpolation may be necessary). Lastly, move horizontally to the left to determine the fuel savings.

For an exit temperature of 750°F, 25% excess air and a 125-million-Btu/h furnace, the dashed line in Fig. 3 shows an annual fuel saving of about $1.25 million.

The curves in Fig. 3 are based on a fuel cost of $7.50/

million Btu and a 95% operating factor (8,320 h/yr). Estimated fuel saving can be adjusted via simple proportioning of different fuel costs and operating factors. For example, if the operating factor were 90% and the fuel cost were $6.00/million Btu for the 125 million Btu/h furnace, the adjusted fuel saving would be: [(90%/95%) ($6.00/$7.50)] $1.25 million, or $0.95 million.

Examples: costs, savings and paybacks

Heat recoveries and fuel savings corresponding to a range of furnace duties from 30 to 250 million Btu/h are

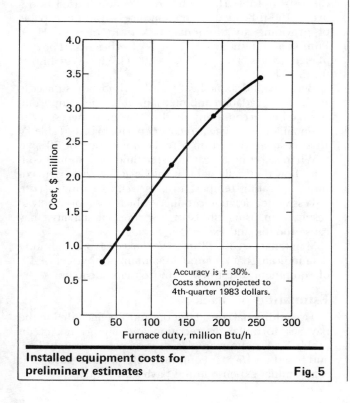

Installed equipment costs for preliminary estimates   **Fig. 5**

Accuracy is ± 30%.
Costs shown projected to
4th-quarter 1983 dollars.

Costs of installed equipment and heat-transfer fluid			Table II
Furnace duty, million Btu/h	Installed equipment, million $	Initial filling, gal	Fluid cost, thousand $
30	0.790	1,750	30.3
65	1.270	3,500	60.7
125	2.180	7,000	121.4
185	2.900	10,500	182.1
250	3.470	14,000	242.8

Expenses of running heat-transfer-fluid pump and draft fan				Table III
Furnace duty, million Btu/h	Pump bhp*	Induced-draft fan bhp*	Total bhp*	Electrical cost, million $/yr
30	4.1	16.4	20.5	5.8
65	8.2	32.8	41.0	11.8
125	16.4	65.6	82.0	23.0
185	25.6	98.6	123.2	34.6
250	32.8	131.4	164.2	46.1

*kW = bhp (brake horsepower) x 0.7457/0.85

Annual fluid-replacement and maintenance costs			Table IV
Furnace duty, million Btu/h	Fluid replacement, gal	Fluid cost, thousand $/yr	Maintenance, thousand $/yr
30	175	3.0	31.6
65	350	6.1	50.8
125	700	12.1	87.2
185	1,050	18.2	116.0
250	1,400	24.3	138.8

Annual depreciation and energy tax credit			Table V
Furnace duty, million Btu/h	Installed equipment cost, million $	Depreciation, thousand $/yr	Tax credit, thousand $/yr
30	0.790	52.7	19.8
65	1.270	84.7	1.7
125	2.180	145.3	26.8
185	2.900	193.3	35.8
250	3.470	231.3	42.9

Payback estimates for the five examples			Table VI
Furnace duty, million Btu/h	Total initial investment, thousand $	Annual profit (savings less expense), thousand $	Payback period, yr
30	820.3	228.7	3.59
65	1,330.7	486.8	2.73
125	2,301.4	1,007.3	2.28
185	3,082.1	1,545.7	1.99
250	3,712.8	2,102.3	1.76

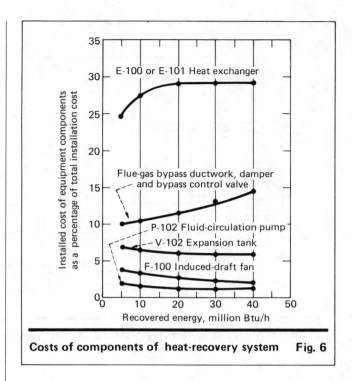

Costs of components of heat-recovery system   Fig. 6

presented in Table I—for a system in which the stack-gas exit temperature of 750°F is dropped to 325°F and the furnace burns 25% excess air, and for which the operating factor is 95% and fuel costs $7.50/million Btu.

The savings are based on reduced fuel costs resulting from preheating the furnace air with recovered heat, as in the typical system shown in Fig. 4. The savings, however, represent general values of recovered heat; as such, they would also be meaningful in the analysis of systems that supply heat to other process services.

Equipment costs—First-phase estimates of installed equipment costs for the five examples (Table II) reflect the equipment and instrumentation in Fig. 4. The estimates, based on case histories at Dow's cost locations, are taken as a national baseline. This technique is used by Dow Engineering Co. for making first-phase estimates. Table II costs have been escalated to 4th-quarter 1983.

The relationship between equipment costs and furnace duty is shown graphically in Fig. 5. Although this curve represents a baseline for first-phase estimates, it should be pointed out that second-phase estimates will, for the most part, fall within ±30% of the baseline.

Heat-transfer fluid costs—Also included in each first-phase investment estimate is the cost of an initial filling of heat-transfer fluid (Table II). These costs are based on using 350 gallons (about 3,200 lb) of fluid for each million Btu/h of recovered heat, and a price of $1.75/lb.

Operating and maintenance costs—Operating expenses include the electrical cost of running the heat-transfer-fluid pump and the induced-draft fan, and the cost of partially replacing fluid (required because of mechanical losses and thermal degradation). The estimated electrical costs in Table III are based on a pump sized to deliver a 150-ft head, and on an induced-draft fan capable of providing an additional stack draft of 4 in. H_2O. Pump and fan motors average 85% efficiency. The cost of electricity is taken as $38.50/1,000 kWh (kW = bhp × 0.7457/0.85).

Because all organic heat-transfer fluids are subject to thermal degradation and mechanical loss, it is important to take into account any fluid replacement necessary to maintain optimum heat-recovery performance. Table IV fluid costs are based on a replacement rate of 10%/yr.

When selecting a heat-transfer fluid, one should consider how thermally stable it will be at the system's maximum operating temperature, as well as its initial cost. Excessive degradation products, which are more likely to result from using fluids of low thermal stability, may cause fouling and exchanger damage.

Maintenance costs (Table IV) derived from operating data indicate that an annual maintenance budget of 4% of equipment investment is more than adequate.

Estimating payback

Two additional considerations must be included in the payback formula: depreciation and energy tax credit (Table V). For the examples, depreciation is straight-line, and based on 15-year equipment life, with no salvage. It is an annual expense in the payback formula.

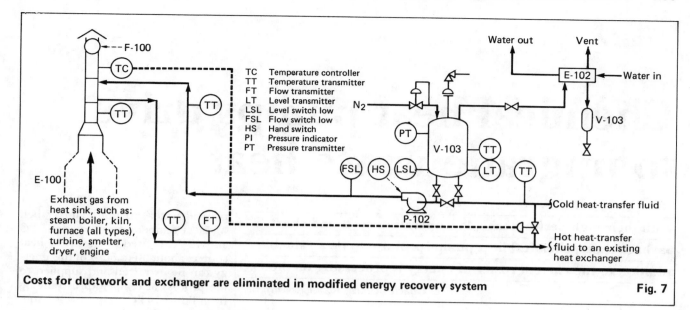

TC	Temperature controller
TT	Temperature transmitter
FT	Flow transmitter
LT	Level transmitter
LSL	Level switch low
FSL	Flow switch low
HS	Hand switch
PI	Pressure indicator
PT	Pressure transmitter

Costs for ductwork and exchanger are eliminated in modified energy recovery system   Fig. 7

To determine the gain from the 10% investment tax credit, a marginal tax rate must be applied. For estimating purposes, the anticipated tax credit must be converted into an equivalent before-tax amount, which should then be spread over the 15-year amortization period: Equivalent tax credit = [0.1 (initial equipment investment)]/[15 (1.00 − marginal tax rate)].

Assumed in the examples is a maximum corporate tax rate of 46%, which yields the equivalent annualized before-tax credit deductions shown in Table V.

The payback formula can now be refined to reflect the impact of the investment tax credit: Payback in years = (initial equipment investment + cost of initial fluid filling)/(fuel saving + equivalent tax credit − operating and maintenance expenses − depreciation).

Applying the values calculated for the terms in the refined payback formula yields the first-phase payback estimates in Table VI for the examples. The estimates indicate that an intermediate-fluid system can yield a relatively quick payback. (Because the assumed parameters in the five examples were set conservatively, the estimated payback periods may reflect understated results.) This payback analysis yields valuable decision-making information.

Effects of design modifications

In many instances, one or more of the equipment items in Fig. 4 can be excluded from the equipment-cost estimate, because existing equipment is available. Obviously, any reduction in equipment purchase cost will shorten the payback period.

For first-phase estimating, Fig. 6 (which is based on cost studies of Dow Engineering Co.) can be used to derive reasonably accurate figures for adjusting downward equipment-cost estimates. (Note that the equipment-cost percentages are related to heat recovered.)

To illustrate, assume the initial conditions of the previous examples—i.e., a 250-million-Btu/h furnace and an expected heat-recovery rate of 40 million Btu/h. Now, however, assume that recovered heat can be directed to an existing exchanger and that bypass duct-

work is not required (Fig. 7). The estimated cost for the exchanger and ductwork can, therefore, be eliminated.

Based on Fig. 6, eliminating the exchanger and ductwork reduces equipment cost 29.2% and 14.3%, respectively. Thus, the total equipment-cost estimate in the previous example ($3.470 million) would be lowered by 43.5%, or $1.512 million. The adjusted equipment cost of $1.958 million shortens the payback to one year.

Adjustments for economic factors, such as inflation and the cost of capital, should have only minimal effect on projects having such short payback periods. In most cases, their influence would be negligible compared with possible variations in estimated capital costs. Economic adjustments become more important in second-phase estimates.

Freedom from patents of Dow Chemical Co., or others, is not to be inferred.

The authors

Walter F. Seifert, research associate, has been with Dow Chemical U.S.A. for 23 years. For the past 17 years, he has been responsible for the development and technical support of Dow's high-temperature heat-transfer fluids. Coauthor of four papers on high-temperature fluids, two on secondary oil recovery and one on fluid mechanics, he has been awarded four patents involving new heat-transfer fluids and two dealing with friction- reduction agents. He holds a B.S.Ch.E. from the University of Colorado, and is a member of AIChE.

John Beyrau, senior research engineer with Dow Chemical U.S.A., joined the company in 1977, upon receiving a M.S.Ch.E. from the University of Virginia. Now with the company's Functional Fluids Technical Services and Development Group, he had previously worked with a number of product groups, providing technical support for plant startups, implementation of computer control systems and coordination of the design phases of capital projects. He holds a B.S. in chemistry from Belmont Abbey College (Belmont, N.C.).

Gallie Bogel is an engineering specialist with the Engineering and Construction Services Group of Dow Chemical Co. He has had 37 years of diversified experience in plant design and operation, including specialization in heat transfer and energy recovery. Among recent assignments are projects for recovering energy from geothermal brine and from the exhaust gas of gas-fired turbines in a cogeneration power plant. He holds a B.S.Ch.E. from Texas A&M University.

Louis E. Wuelpern is a heat-transfer specialist in the Functional Fluids Technical Services and Development Group. For the past 12 years, he has concentrated on high-temperature heat-transfer fluids, including test methodology, thermal degradation and engineering design. His previous experience was in research, pilot-plant development and operations. He holds a B.S.Ch.E. from Michigan Technological University (Houghton, Mich.).

Chemical heat pumps drive to upgrade waste heat

With their high performance, low operating costs and short payback times, chemical heat pumps promise to be a valuable addition to the current roster of energy reprocessing methods.

E. Charles Clark, *Rocket Research Co.*

☐ Recently, several companies have been investigating chemical heat pumps as a cost-effective means of capturing low-grade waste heat and reusing the heat at increased temperatures. Application areas include industrial processing, air or gas moisture-removal, and energy storage.

These chemical heat pumps operate without mechanical compressors or turbines, as in conventional units.

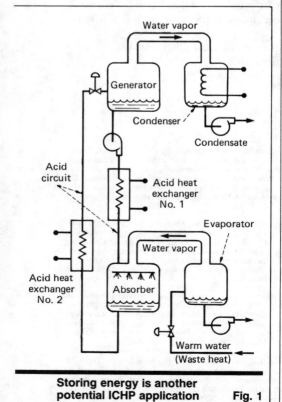

Storing energy is another potential ICHP application   Fig. 1

Instead, waste heat—e.g., from industrial wastewater such as condensate, cooking water or cooling water—is used to drive an exothermic chemical reaction, most often between water and a soluble chemical such as lithium bromide or sulfuric acid. Higher-quality heat is then recovered from the reaction products, which are at higher temperatures—because of the exothermic reaction—than the waste heat introduced to the system.

Such heat pumps, besides operating at higher temperatures, require only 5 to 25% of the electrical power needs of mechanical-compression heat pumps. Indeed, in the last two years, chemical heat pumps capable of boosting waste heat to 165°C have been commercialized in the U.S., Europe, Japan, and the U.S.S.R.

Systems capable of upgrading process heat to higher temperatures (up to 240°C) are also verging on commercialization. Last year, under funding by the Department of Energy, Rocket Research Co. (RRC, Redmond, Wash.), a division of ROCK-COR, Inc., successfully completed development of a 44-kW sulfuric-acid-based heat pump capable of supplying process heat at such higher temperatures. RRC expects—by summer's end—to finalize a joint venture with at least one other participant for commercializing the design.

OPERATION—Heat pumps operate by inducing a flow of heat from a colder body to a warmer one—in RRC's Industrial Chemical Heat Pump (ICHP), this is done via a sulfu-

ric acid/water reaction. Thus, when water at temperatures between 54°C and 121°C is used as the waste-heat source, about 30 to 50% of this energy can be extracted and supplied to the process at temperatures ranging from 66 to 193°C, respectively. And, with an additional heat source—say stack gases at 175°C—the pump's output can be further boosted to 240°C.

The ICHP system (Fig.1), centering around the sulfuric acid/water circuit, consists of six primary components: the generator, condenser, absorber, evaporator, and two heat exchangers.

■ *Acid circuit*—The sulfuric acid, contained in a closed-loop circuit, flows continuously between the generator and absorber. Dilute acid, introduced into the generator, is flash-distilled into concentrated acid and water-vapor streams by maintaining the pressure in the generator at a low level (0.5 to 1.5 psia). The generator pressure is controlled by the water temperature in the condenser. The concentrated acid is removed from the bottom of the generator and pumped into the heat exchanger (#1) where it is heated by approximately 30 to 50% of the total waste-heat input into the system.

Next, the concentrated acid enters the absorber with its vapor pressure at this stage considerably lower than that in the absorber. The absorber pressure, which is between 2.2 and 30 psia, is controlled either by the pressure of the water vapor generated in the evaporator (with the remaining 50 to 70% of the waste-heat input), or by the direct injection of low-pressure high-quality steam.

The water vapor from the evaporator condenses into the acid, diluting it. The resulting absorption releases both the heat of dilution of the acid and the heat of vaporization of water, raising the temperature of the dilute

Originally published February 20, 1984

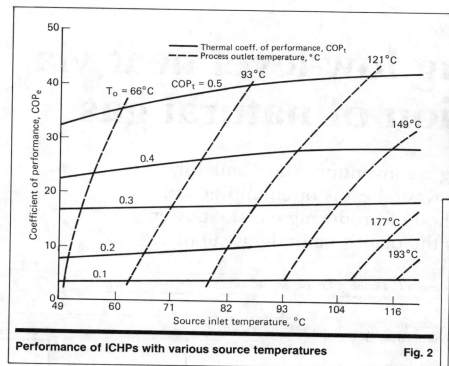

Performance of ICHPs with various source temperatures   **Fig. 2**

acid by about 20 to 80°C. The dilution process continues until the acid solution reaches an equilibrium corresponding to the absorber pressure.

Completing the cycle, the hot, dilute acid then flows through the heat exchanger (#2) and on to the generator because of the pressure differential between the two. Sensible heat, the product of the ICHP, is removed from the acid stream in the heat exchanger and delivered to the process at temperatures (between 66 and 240°C) significantly higher than that of the waste-heat source.

Acid concentration depends on the desired output temperature, waste-heat temperature, and condenser temperature. The dilute acid concentration ranges from 60% for moderate-temperature applications to 90% for high-temperature ones; the range for concentrated acid is 3 to 4% higher. An added benefit of using sulfuric acid is that no crystallization occurs, allowing the system to tolerate greater fluctuations in waste-heat and cooling-water temperature cycles.

■ *Water circuit*—In the ICHP scheme, both the generator and condenser as well as the absorber and evaporator are connected as pairs with little or no pressure drop. Since the waste heat introduced into the evaporator is about 40 to 70°C hotter than that withdrawn from the condenser, the absorber/evaporator pressure is 2 to 29 psia higher than that of the generator/condenser.

Water vapor produced in the generator is cooled—by industrial water or groundwater—in the condenser. The cooling-water temperature, usually between 5 and 50°C, should be as low as possible to maximize temperature and energy output. The condensate is then either discarded or routed to the evaporator.

PERFORMANCE—The performance of industrial heat pumps is typically defined as an electrical coefficient of performance (COP_e), which equals the heat rate to the hot process (in Btu/h) divided by the electrical power (converted to Btu/h) required to pump water and acid through the system. (Essentially, the heat extracted from the waste-energy source is not taken into account for performance evaluation since it would normally be rejected.)

Also, a thermal coefficient of performance (COP_t)—also called the output-to-input heat ratio—determines if sufficient waste heat is available, at reasonable cost, to supply the desired process heat.

As an example, calculated COP_e as a function of the waste-heat temperature source is presented in Fig. 2 for various values of COP_t and output temperature. Assumptions for these calculations are a condenser inlet temperature of 10°C, and 5°C heat-exchanger approach temperatures.

As can be seen, both COPs increase as waste-heat temperature is increased, or as the output temperature is lowered. Typical COP_t values vary from 0.5 to 0.1—i.e., the waste-heat supplies should be 2 to 10 times the process needs; however, system improvements, now under study, could double or triple the lower COP_t figure for high-temperature applications. Further, by using a second heat source at a higher temperature, the output temperatures in Fig. 2 can be raised by about the same amount as the difference in temperature between the two waste sources.

EQUIPMENT—The 44-kW test unit, shown in the photograph, is packaged in a modular configuration with ease of maintenance in mind. Materials of construction include glass, Teflon, cast silicon iron, ceramics, and special alloys. The two acid heat exchangers were specially designed to overcome the pressure and temperature limitations of commercial units operating in such environments.

A preliminary design for a 586-kW unit using two waste-heat sources has been completed. Based on production quantities of 100 units per year, this ICHP is estimated to cost $183,000. Equipment payback (including installation), based on energy savings, is approximately two to three years at 1984 energy prices.

The author

E. C. Clark, a project manager at Rocket Research Co., is responsible for chemical heat pump development. Previously, he worked as a technical consultant in Italy and as a project engineer in various areas of the Voyager and Viking space programs. He has a B.S. in Engineering from California State University.

Recovering low-level heat via expansion of natural gas

Finding a conventional heat sink for low-level waste heat is often impractical. This method of producing electric power employs the plant's natural-gas supply.

Richard C. Doane, S.I.P. Engineering, Inc.

☐ In most large chemical-processing plants and petroleum refineries, fuel-gas pressure is controlled by the natural-gas makeup, as shown in Fig. 1. The supply pressure is reduced across control valves while enroute to the fuel-gas blend tank.

Using the conditions in the illustration as an example, a pressure drop of 220 psi is available for performing mechanical work. The first law of thermodynamics, Q(heat added) $-$ W(work done) $=$ ΔH(enthalpy change), requires, however, that heat be added to the gas to keep the enthalpy constant. Without this heat, fuel users would consume more gas, due to the lower energy content per unit mass.

(Note that as the gas passes through a control valve, no work is done and no enthalpy change occurs; in the turboexpander, work is done and the enthalpy decreases.)

The low temperature of the natural-gas line makes it an ideal sink for a low-level waste-heat stream. However, heating the gas is not of itself a practical means of energy conservation. Fuel-gas distribution systems are normally complex, and insulation is impractical.

So, any heat that would be put into the fuel gas would be lost to the ambient air. Salvaging energy from a waste-heat stream thus requires the addition of heat and mechanical work.

Modified fuel-gas systems

Fig. 2 shows recovery of waste heat, and conversion of the recovered energy to work. A conventional shell-and-tube heat exchanger transfers heat from a hot gasoline stream to the natural gas. Previously, this heat had been lost to the atmosphere in the air cooler shown in the figure.

An expander recovers mechanical work from the hot natural gas, and drops the gas pressure to that of the blend tank. With expanders controlled by variable inlet guide nozzles, overall efficiencies greater than 80% are possible [1]. Efficiency is measured as the ratio of the electrical power to the work done by the gas.

Although the expander could be directly coupled to a

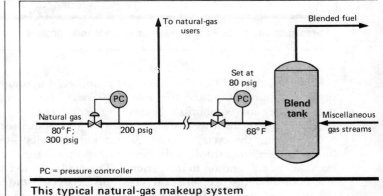

This typical natural-gas makeup system can be modified to recover low-level heat   Fig. 1

pump or compressor, maximum flexibility is achieved by generating electric power. An induction generator, which receives field excitation from the plant power grid, is preferred to a synchronous generator.

(An induction generator uses the plant's power supply to excite its electromagnet—this device is said to be "externally excited." The synchronous type is "self-excited" and will generate electricity even during a plant power failure.)

This preference is due to its low cost, safety and reliability. The power generated is tied to the plant through the nearest substation.

Let us now discuss the details of the modified system to recover low-level heat.

Natural-gas supply pressure

Fig. 2 shows a 10-in. line that delivers nominally 300-psig gas to the expander. A decision was made to build the new line and expand the gas at maximum pressure, rather than using the existing one. This decision depends primarily on the length of the new line and the maximum operating pressure of the existing line.

The old line pressure could have been raised to 275 psig. However, the new line is not long and it will be paid

Originally published April 2, 1984

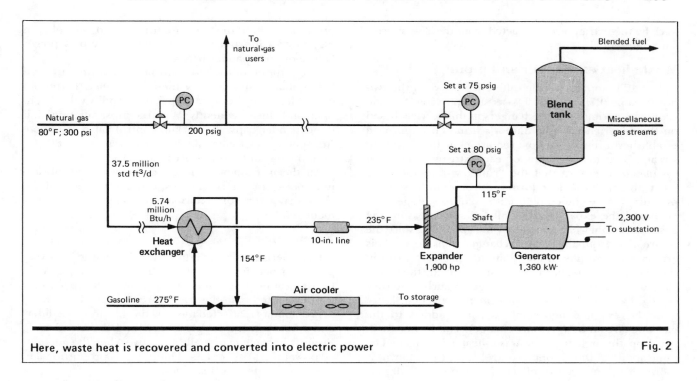

Here, waste heat is recovered and converted into electric power   Fig. 2

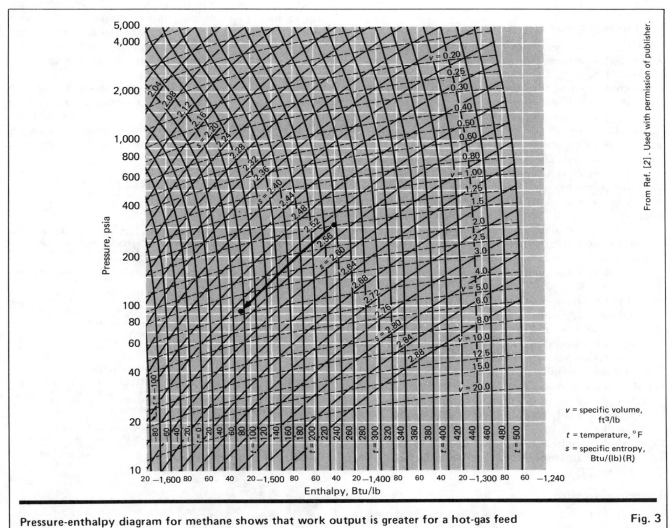

Pressure-enthalpy diagram for methane shows that work output is greater for a hot-gas feed   Fig. 3

back by the extra power extracted from the 25-psi pressure differential.

Waste-heat exchanger and piping

The 40°F approach—the difference between the two hot streams (275 less 235°F)—as shown in Fig. 2 is the economic optimum for exchangers in this service, based on the experience of the author's firm. This approach establishes a minimum flow requirement for the waste stream. The heat content of each stream should be calculated to ensure that the "rundown" temperature meets this criterion. Rundown is the amount by which a product is to be cooled before it is sent to storage.

Should the expander shut off for some reason, the gasoline stream will be cooled by the air cooler, since the return line from the heat exchanger runs to it. This arrangement results in an additional pressure drop in the waste-heat stream. The stream selected, therefore, should not be too viscous or require such extensive piping that a costly pump replacement is needed.

In the example here, only 6 psi is added to the gasoline-stream ΔP (pressure change). This includes a 2-psi drop through the new waste-heat exchanger. The rangeability of the gasoline control valve (not shown) is more than enough to offset the imposed hydraulic resistance.

Expander/generator

The expander/generator can be purchased as a pre-assembled, skid-mounted unit complete with lubrication system. The speed of this induction-type power generator is controlled by the plant's grid frequency (60 Hz), since the generator's electromagnet uses plant power. Modern expanders have outstanding turndown and turnup. Gas flow can vary from 50—125% of the design value without significant efficiency loss. The induction generator will normally be about 96% efficient.

The expander outlet temperature of 115°F is higher than that of the entering gas in the blend tank in Fig. 1 (68°F). As mentioned previously, this increase in sensible heat will eventually be lost to the surroundings, and cannot be included in the energy saving. Why, then, should the expander be heated above 68°F?

The reason is that more power results from the expansion of a hotter gas, assuming a constant ΔP (pressure). Fig. 3, from Ref. [2], shows the thermodynamics involved here. The arrow traces the isentropic expansion for methane under the conditions in Fig. 2. The slope of the isentropic line in Fig. 3 increases as the inlet gas temperature is lowered. The enthalpy change is greater for a hot gas feed than it is for a cold one. Since the first law of thermodynamics requires that $W = -\Delta H (Q = 0)$, the shaft work is also greater.

Although the pressure-enthalpy diagram for methane provides reasonably accurate estimates of power generated (within ± 5%, depending on natural-gas composition), a more rigorous method is needed to determine the expander discharge temperature. This method is found in Ref. [3].

Instrumentation

In Fig. 2, the pressure is controlled by regulation of the expander inlet guide nozzles. The controller that previously served this purpose (the second controller in Fig. 1) now functions as a backup in case of a power failure or an expander trip.

The setpoint of the old controller is adjusted to 5 psi below that of the primary (new) controller. If the old controller were also set at 80 psi, the standby valve might open and close frequently. With the 5-psi differential, if the expander trips, the backup control valve will open. An operator can manually raise its pressure to 80 psi, until the expander is put back into service.

All the instrumentation associated with the lubrication, power monitoring, overspeed control, etc., of the expander/generator is normally supplied by the equipment vendor.

Economics

Considering rising energy costs, this energy-recovery project is becoming attractive. Installed at a U.S. Gulf Coast location, the system in Fig. 2 has an earning power (or discounted-cash-flow rate-of-return) of 41%.

Profitability is affected most by the gas pressure differential between the supply header and the blend tank. Projects become marginally attractive when the generated power is 100 kW or less. This power is proportional to the pressure drop × mass flow.

On the other hand, there is no practical economic upper limit to recovering power. Expander/generator units can be staged or arranged in parallel. Each stage may have up to a 12:1 pressure ratio. Single-stage machines have been built to generate as much as 4,650 kW of power [4].

To conserve fuel, conversion of waste-heat energy to electric power is economically superior to exchanging heat: In the Gulf Coast area, the cost of purchased electricity is about three times that of fuel, as measured in the same energy units. Profitability will vary with location.

References

1. Holm, J., and Swearingen, J. S. (of Rotoflow Corp., Los Angeles), The Application of Turboexpanders for Energy Conservation, paper presented at Amer. Soc. of Mechanical Engineers' Energy Technology Conference and Exhibit, Houston, Sept. 18, 1977.
2. "Engineering Data Book," 9th ed., pub. by Gas Processors Suppliers Assn., Tulsa, Okla., p. 17-7, 1972.
3. Perry, Robert H., and Chilton, Cecil H., "Chemical Engineers' Handbook," 5th ed., pp. 24-34 — 24-36, McGraw-Hill Book Co., New York, 1973.
4. Personal correspondance with Barry Wood, senior sales engineer, Rotoflow Corp., Houston, Feb. 14, 1983.

The author

Richard C. Doane is a senior process engineer with S.I.P. Engineering, Inc., P.O. Box 34311, Houston, TX 77034. Tel: (713) 946-9040. He is involved with energy conservation for the chemical-process and oil-refining industries. Previously, he was process supervisor for Hercules, Inc., at Bayport, Tex. He holds B.S. and M.S. degrees in chemical engineering from Northeastern University, and is a professional engineer in the state of Texas.

Use hydraulic turbines to recover energy

Hydraulic turbines, or centrifugal pumps running in reverse, can be used to recover energy often lost across reducing valves and orifices. Here are some of the things you need to know in order to use this technique.

Navneet Chadha, *Davy McKee Corp.* *

☐ Owing to continually rising energy costs, it has become increasingly attractive to use a hydraulic turbine to recover energy from high-pressure liquid streams and use it to drive a pump or other piece of rotating equipment. There are only a few makes of conventional hydraulic turbines marketed; a convenient accepted alternative is to use centrifugal pumps running in reverse.

Pumps are readily available in many sizes and configurations, are easier to operate, and are cheaper than turbines. Most of the literature published so far [1,2] has dealt with the selection of a pump to be used as a hydraulic turbine and has compared the performance of a hydraulic turbine with that of the same unit functioning as a pump. This article discusses some process design considerations for a typical hydraulic-turbine installation, including the instrumentation required for its safe and efficient operation.

A hydraulic turbine can be used in any continuous process in which a high-pressure liquid is let down to a lower pressure across an orifice or control valve. By replacing the pressure-reducing device with a hydraulic turbine, a large portion of the energy that would otherwise be lost as heat is recovered as shaft work.

A typical application is in the hydrocracking process where reactor effluent is released from a high-pressure separator drum to a low-pressure one; another is in gas-treating processes, where a high-pressure, rich liquor stream is let down from an absorber to a low-pressure stripper. In many applications, dissolved gases will be present in the process stream; these will evolve as pressure is lowered across the hydraulic turbine.

The associated problems of erosion and reduced capacity are usually not significant, unless the proportion of gas is more than 50% of the liquid at the turbine outlet conditions [3].

Economic justification

Flowrates for hydraulic-turbine installations generally vary from about 250 gpm to over 5,000 gpm, whereas differential pressures are usually in the 100-psi to 2,500-psi range. However, there is no rule of thumb as to the individual levels of flow and differential pressure required to justify such an installation. Rather, it is the product of these two terms, and hence the horsepower recoverable, that governs the economics.

Most multistage hydraulic-turbine applications are in the 1,000-hp to 100-hp range. The 100-hp lower limit is the economically feasible value suggested by pump manufacturers, on the basis of present power costs. Based on a 5¢/kWh power cost, even such a small turbine accrues an energy saving of:

$$100 \text{ hp} \times 0.746 \text{ kW/hp} \times \$0.05/\text{kWh} \times 8,000 \text{ h/yr} = \$29,840/\text{yr}$$

It is obvious that the key variable in determining the feasibility of a turbine installation is the energy saving, which, for a particular application, depends on the power

---

*The author is now working for Sherex Chemical Co. See "The author" on the last page of this article.

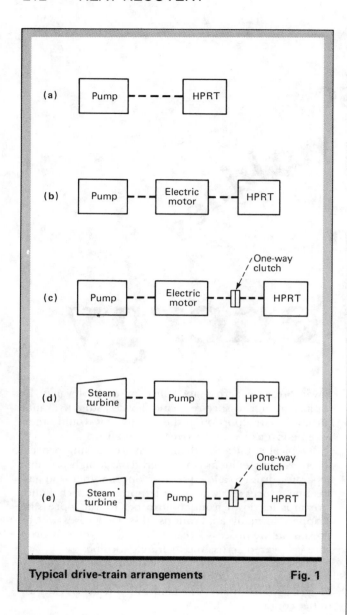

(a) Pump ---- HPRT

(b) Pump ---- Electric motor ---- HPRT

(c) Pump ---- Electric motor ---- One-way clutch ---- HPRT

(d) Steam turbine ---- Pump ---- HPRT

(e) Steam turbine ---- Pump ---- One-way clutch ---- HPRT

Typical drive-train arrangements   **Fig. 1**

payout calculations with those for discounted cash flow (DCF) or some other acceptable profitability criterion.

Typical drive-train arrangements

Fig. 1 shows some typical arrangements in which the hydraulic turbine can be used as a process driver. The simplest arrangement, in which an HPRT provides all the power to drive a pump, is shown in Fig. 1a. If a spare pump were required, it would be driven by an electric motor or steam turbine. Arrangements (b) and (c), involving a pump, electric motor and HPRT, are the most common. An electric motor is a good speed governor and is readily available with a double extended shaft; therefore it is placed between the pump and the HPRT.

The arrangement shown in (c) is similar to (b) except for the one-way clutch that is used to connect the HPRT with the electric motor. The principle of operation of the one-way clutch and why it is used are discussed under the next heading. In (d) and (e), a steam turbine is used as the auxiliary driver in place of an electric motor. A steam turbine also serves as a good speed governor, provided the HPRT power rating is not significantly greater than that required by the pump or other driven equipment. Note that in this arrangement the pump is placed in the center because a steam turbine is not commonly available with shaft extensions from both ends.

The one-way clutch

In most cases, the horsepower recovered by the HPRT is not sufficient to independently drive a pump. It is necessary, therefore, to use an electric motor or steam turbine in tandem with the HPRT, to make up the difference in horsepower. The HPRT's performance characteristics are such that when the flow through it is less than about 40% of its best efficiency point, it will begin to absorb power, and thus impose a drag on the auxiliary driver. In addition, an overload condition will occur if the auxiliary driver is sized only to make up the horsepower difference and not for the full brake horsepower required by the driven equipment.

This problem of drag imposed on the auxiliary driver at low flows can be solved by using a one-way clutch on the HPRT output shaft, as shown in Fig. 1c and 1e. As liquid flow through the HPRT increases, it accelerates until it reaches the auxiliary driver speed, at which point the one-way clutch engages, reducing the auxiliary driver's power consumption. The clutch will also automatically disengage if the HPRT speed falls, due to a reduction of flow through it. This allows the pump and its auxiliary driver to operate independently of the HPRT during startup and other low-flow situations, and lets the HPRT be isolated for maintenance during normal operation.

The one-way clutch is usually supplied by the hydraulic-turbine vendor, with the clutch sized on the basis of the required clutch torque and speed of the application. The shaft sizes of the turbine and the driven equipment do not figure in the selection of the clutch size, since the clutch is connected in place through suitable sizes of flexible gear-type couplings.

Hydraulic-turbine specification

The general requirements of API Standard 610 for centrifugal pumps can be applied to hydraulic turbines,

cost. The power cost, in turn, could depend on whether electricity were purchased from a utility company or produced inplant as a part of a cogeneration scheme.

To judge the profitability of this project, the payout time can be calculated as follows:

Payout (yr) = Installed capital cost/Energy saving/yr

When a hydraulic power-recovery turbine (HPRT) is installed in an existing plant it repays its capital cost in 1½ to 2½ years in most cases. The capital cost for this project is essentially the cost of the turbine plus that of associated instrumentation, valves and piping. Rather than estimating the turbine cost from that of an equivalent pump, it is preferable to get a quote from the pump manufacturer since the price will also normally include the cost of turbine accessories such as the one-way clutch and the overspeed trip. As the project becomes more defined, economic parameters such as the timing of the initial investment, depreciation schedule and expected maintenance costs for the HPRT are likely to be firmed up. At that point, it is a good idea to complement the

with all references in the standard to pump suction and discharge interpreted as applying to the turbine outlet and inlet, respectively. The most important consideration for turbine applications is the choice of the casing type. Horizontally-split-case multistage construction is not allowed if any of the following operating conditions is specified:

- An operating temperature of 401°F or higher.
- A flammable or toxic liquid with a specific gravity of less than 0.7 at the operating temperature.
- A flammable or toxic liquid with a pressure greater than 1,000 psig at the turbine inlet.

A centrifugal-pump data sheet can be used for hydraulic turbine specification with the following provisos:

- No excess margin should be included in the HPRT design flow, as it would then recover less than the maximum power when passing normal flow.
- The specified design-head should be as high as the system allows, to maximize power recovery. Therefore, the HPRT flow-control valve and inlet/outlet lines should be sized to have a nominal pressure drop.
- The speed at which the driven equipment is to be operated should be indicated. In case the auxiliary driver is an electric motor, this value is approximately the synchronous speed of the motor, e.g., 3,600 or 1,800 rpm.
- If the liquid is likely to flash as its pressure is let down across the HPRT, the degree of flashing should be indicated on the data sheet. This information will be useful to the manufacturer in selecting a turbine that can handle the increased volume due to gas formation, without a reduction in normal capacity.
- To prevent cavitation, a certain amount of pressure is required at the HPRT outlet. The higher the specific speed, the higher the required exhaust head, and the more susceptible the HPRT is to cavitation. In most applications, however, the HPRT speed cannot be reduced. Also, the backpressure available at the site is fixed by the process and cannot be increased. It is, therefore, the manufacturer's responsibility to supply a turbine that has a minimum of cavitation under the specified conditions.
- If the flow through the HPRT will vary, the power recovery calculations should be based on average flowrate rather than at the best efficiency point. This is especially important in justifying the cost of an HPRT installation. Brake horsepower developed by the hydraulic turbine can be predicted from:

$$\text{bhp} = \text{hhp} \times \eta_T = (Q\Delta P/1,714)\eta_T \qquad (1)$$

where: hhp=hydraulic horsepower recovered; η_T=turbine efficiency; Q=flow through the HPRT, gpm; and ΔP=corresponding total differential head at flowing conditions, psi.

To determine turbine efficiency, the pump manufacturer's assistance is required. This is because it is necessary to select the pump to be used as a hydraulic turbine, by a procedure similar to that described in Ref. [1].

The first step is to convert the turbine-design conditions to pump-design conditions, using conversion factors for capacity and head provided by the pump manufacturer. With the capacity, head and rpm known, a pump can be selected that has its best efficiency point at

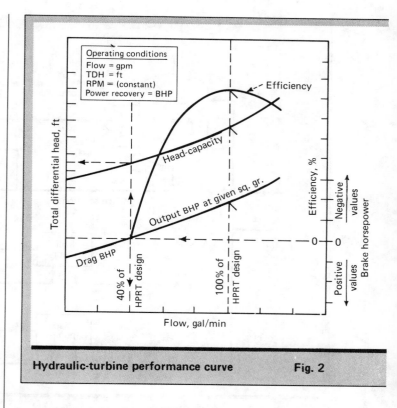

Hydraulic-turbine performance curve   Fig. 2

these conditions. The selected pump's efficiency is then changed to turbine efficiency, again using a conversion factor.

Conversion factors for capacity and head vary from 1.1 to 2.2, and those for efficiency from 0.92 to 0.99, depending on the speed of the application. In actual practice, however, these conversion factors may not be readily available, and the pump manufacturer may supply only a turbine-efficiency value.

Performance curve

Based on the operating conditions specified, the turbine vendor will select a "pump" and supply a head-capacity curve for design rpm (similar to Fig. 2). Note that since the turbine is essentially like a restriction orifice in a relatively fixed differential-head system, its differential head increases with flow. The head capacity curve can be used to size the HPRT flow-control valve, which must consume any difference between the head available from the process and the head that the HPRT is capable of absorbing at any given flowrate. The output bhp curve gives the flowrate at which zero bhp (zero efficiency) occurs. As mentioned before, this usually occurs around 40% of turbine design flow. At flowrates lower than this, the hydraulic turbine consumes power, and, therefore, it should always be operated to the right of this point.

Seal system

Tandem mechanical seals are recommended for hydraulic-turbine installations because of the high pressures usually involved. In a tandem seal arrangement, if the inner seal fails, the secondary seal serves as a backup, since both the seals are designed for the same pressure and temperature conditions. A small (2 to 3 gpm) circu-

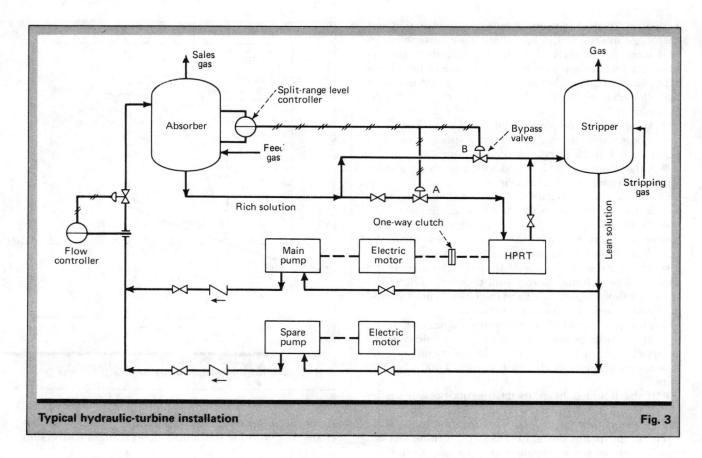

Typical hydraulic-turbine installation   Fig. 3

lation of a secondary low-pressure fluid is maintained through the buffer zone between the two seals. Inner-seal failure is detected by a change of pressure in the buffer zone. An external flow of clean and cool liquid (at constant pressure) is provided to flush the turbine seals; orifices are used to control this flow.

During normal operation, with the outlet isolation valve(s) open, the turbine stuffing boxes are exposed to the same pressure as the outlet (or low-pressure end) of the turbine. However, during startup it is possible that the operator may inadvertently open the inlet isolation valve(s) prior to opening the outlet valve(s), thus exposing the mechanical seals to the higher inlet pressure. Therefore, the mechanical seals (and all the turbine stages) should be designed to withstand the highest inlet static pressure they can be exposed to during this contingency.

Lubrication system

Lubrication of the turbine bearings is usually accomplished using a pressure-lubrication system consisting of a lube-oil reservoir, a shaft-driven main lube-oil pump, an auxiliary lube-oil pump, an oil cooler and a filter. The turbine may have its own lubrication system, or it may be lubricated from a common header supplying oil to the hydraulic turbine, auxiliary driver and pump.

Controls and instrumentation

The important parameters in hydraulic-turbine operation are flow, head and speed. The head available to the turbine is fixed by the process, whereas flow and rpm can vary, and therefore need to be controlled.

Flow control

A simplified schematic of a typical hydraulic-turbine installation is shown in Fig. 3. In this arrangement, a pump is used to charge a vessel, and the HPRT recovers energy, from the high-pressure-liquid stream leaving the same vessel, to drive the charge pump. The electric motor in tandem with the HPRT provides the horsepower requirement not covered by the HPRT. Flow to the HPRT is usually on split-range level control; normal flow is through Control Valve A in the inlet line. Control Valve B opens on high level in the inlet vessel, allowing liquid to be bypassed around the HPRT.

Flow-Control Valve A is preferably located upstream of the HPRT rather than downstream. This allows excess pressure due to lower-than-normal flow through the HPRT to be consumed upstream. As a result, the HPRT outlet pressure will be constant for all flowrates, and the mechanical seals will operate at a constant pressure well below their rated maximum. This reduces seal wear and increases seal operating life.

Throttling the flow upstream of the turbine may cause some gas or vapor release in the turbine inlet line. However, this is not likely to be a problem since the flow-control valve is normally selected to produce a low pressure-drop.

Speed control

The method of speed control depends on the drive-train arrangement:

HPRT connected directly to driven unit—In this case, the turbine speed is controlled by liquid bypassing the turbine at excess flows, and inlet-pressure throttling at flows less than design, so that at the design point the turbine head/capacity—rpm relationship is satisfied. The balance point is always determined by the horsepower and rpm requirements of the driven unit, as sensed by the turbine at its coupling.

HPRT and auxiliary driver connected to driven unit—When an electric motor or steam turbine is used as an auxiliary driver, no external means of *HPRT* speed control is necessary. The auxiliary driver will hold the rpm constant and make up horsepower as required, to allow the *HPRT* to satisfy its head—capacity—rpm curve at almost any flowrate.

Instrumentation is provided for manual startup of the turbine; for ensuring a smooth switchover from manual to automatic operation; and for protection of the seal system. The "smooth switchover" can be accomplished by supplying an independent, manually controllable instrument-air supply to Flow-Control Valve A in the turbine inlet line.

The air pressure of the manual signal to the control valve is matched with that of the level-controller signal to the same valve before switching operation from manual to automatic and vice versa. The hydraulic turbine can have its own local control panel with indicators for flow, inlet pressure, outlet pressure, and inlet-vessel level; controller for seal-flush pressure; alarm for high pressure in the seal buffer zone; auto/manual switch for turbine startup and shutdown.

Shutdowns

The turbine should shut down on high speed, low speed, low lube-oil pressure and low seal-flush pressure (or flow). All the shutdowns essentially close Control Valve A in the turbine inlet line and open Bypass Valve B (Fig. 3).

Overspeed occurs when there is zero load on the shaft (if the clutch breaks, for instance) and can be predicted using a method suggested by pump manufacturers. It is based on the pump affinity law, which states that the total differential head (*TDH*) varies with the ratio of speeds squared, or:

$$N_2 = N_1(H_2/H_1)^{1/2} \qquad (2)$$

Where for overspeed operation: $H_1 = TDH$ at zero bhp and design speed; $H_2 =$ design *TDH*; $N_1 =$ design speed, rpm; and $N_2 =$ maximum overspeed, rpm. The overspeed trip should be set at a value lower than the calculated value of N_2.

For most applications, the maximum speed is limited to about 140% to 155% of design speed, provided that the design head and flow are always available to the *HPRT*. In actuality, at overspeed operation the turbine's inefficiencies become so great that the flow through it is choked.

The low-speed trip is usually set around 40% of design flow, i.e., the point at which the turbine is not generating any power. However, in arrangements using a one-way clutch, it may be sufficient only to sound a low-flow alarm, if the auxiliary driver has been sized for the full load.

In any case, the turbine should be shut down when flow through it falls to about 20% of design, at which point it may become noisy and overheat.

Installation

The hydraulic turbine is usually located at the point of power consumption (provided physical space is available), and the liquid is piped to and from that point. However, dissolved flammable gases are often present in the process stream—gases that could leak to atmosphere in the event of bearing or seal failure. In such cases, it is imperative to check the adequacy of spacing of the turbine from other equipment that could be a source of ignition.

In most applications, the turbine provides power to a pump that also has an electric-motor- or steam-turbine-driven spare; therefore, it can be installed in existing plants during normal operation. To accommodate the hydraulic turbine and the one-way clutch, it is preferable to extend the driven equipment's baseplate (and foundation) rather than provide a separate one.

When the turbine is used in flashing service, vibration and erosion of the downstream piping may occur. Excessive vibration of the pipework can affect turbine alignment—therefore, the downstream piping must be securely held in place.

In conclusion, one must be careful to purchase the hydraulic turbine from a manufacturer that has run its pumps backwards and can guarantee the hydraulic performance of the machine. Also, because of the high pressures and power outputs usually involved, a mechanical-design review of the turbine installation is necessary, with emphasis on checking of casing pressure and shaft stress limits, and on the design of the seal and bearing systems. A procedure for startup and shutdown of the hydraulic turbine should be written up, giving proper attention to startup precautions, commissioning of the seal and lube-oil systems, and operation of manual valves. The operators can then successfully bring the turbine onstream in an organized manner.

References

1. Buse, F., *Chem. Eng.*, Jan. 26, 1981, pp. 113-117.
2. Shafer, L., and Agostinelli, A., *Power*, Dec. 1981, pp. 87-88.
3. Franzke, A., *Hydrocarbon Process.*, Mar. 1975, pp. 107-110.
4. Purcell, J. M., and Beard, M. W., *Oil Gas J.*, Nov. 20, 1967, pp. 202-207.
5. Jenett, E., *Chem. Eng.*, Apr. 8, 1968, pp. 159-164.
6. "Design Characteristics of Hydraulic Power Recovery Turbines," Bulletin No. 4090, Pacific Pumps Div., Dresser Industries Inc., Huntington Park, Calif.
7. "API Standard 610, Centrifugal Pumps for General Refinery Services," 6th ed., American Petroleum Institute, Washington, 1981.

The author

Navneet Chadha is a Process Engineer with Sherex Chemical Co., P.O. Box 9, Mapleton, IL 61547, where he is responsible for providing design and engineering support for Sherex's fatty-chemicals facility. While preparing this article, he worked in process engineering for Davy McKee Corp. He has a B. Tech. from the Indian Institute of Technology (Kanpur, India), and an M.S. from the University of Kentucky, both in chemical engineering. He is a member of AIChE, and a licensed professional engineer in Ohio.

Saving heat energy in refractory-lined equipment

High-temperature equipment operating under severe conditions must use the least amount of energy in order to economically process many materials.

James E. Neal and *Roger S. Clark,* *Johns-Manville Corp.*

☐ Refractories are vital to processes using great heat, such as those found in the chemical, hydrocarbon, metallurgical and ceramic industries. Originally, refractories were seen merely as a means for containing heat within a given space, but they are now considered critical to the success of an efficient energy-conservation program and a cost-efficient manufacturing process. The proper use of refractories also promotes safety and health for the workers and creates a more comfortable environment overall.

In our discussion, we will provide information on the types and properties of refractories, refractory selection, installation and/or construction, heat-flow theory, and economics.

A basic approach

The traditional approach to insulation is to only consider the amount of heat escaping through the insulation when we should instead be thinking of the amount of heat contained or restricted from flowing.

If we look at insulation in a resistive sense, we cannot help but be impressed by its performance. For example, let us examine:

1. The radiative heat transfer between two radiantly black surfaces, each 1 ft². One surface is at 100°F, and the other at the temperature, T, as indicated by the abscissa in Fig. 1. This heat transfer does not include convective and conductive heat flow of the air, both of which are small by comparison.

2. The heat transfer resulting from placing 1 in. of insulation in the form of a refractory fiber having a density of 10 lb/ft³ between the two plates.

There will be quite a difference in heat transfer for each condition. For instance, suppose we have a hot-face temperature of 2,000°F. In this situation, and with no insulation, we find a radiant heat flow of over 60,000 Btu/(h)(ft²) from Fig. 1. With insulation, the total heat flow from the hot face to the cold face is about 1,500 Btu/(h)(ft²). Over 58,000 Btu/(h)(ft²) has been retained. This is a ratio of 40 to 1.

Obviously, it would be economically as well as practically impossible to operate this system without the insulation. The capacity of the power source that would be necessary to maintain the temperature of 2,000°F would have to be increased 40 to 50 times if insulation were not used.

Insulation-selection criteria

An insulation allowing no heat flow would be ideal, of course. Several insulation systems approach this goal, but they must operate in a vacuum. In most applications, this condition is impractical to realize. Insulation selection is not limited to resistance to heat flow. Other criteria that may be involved are: temperature limitations, thermal-shock resistance, coefficient of expansion, strength, hardness, compressibility, specific heat, erosion resistance, chemical resistance, environmental resistance, space limitations, ease of application, and economics for the various refractory materials.

The choice is generally a compromise governed to a great extent by the cost of insulation and its installation. However, the increasing cost of energy introduces a new parameter, life-cycle cost. This is based on the initial cost plus maintenance and operating costs (in-

Originally published May 4, 1981

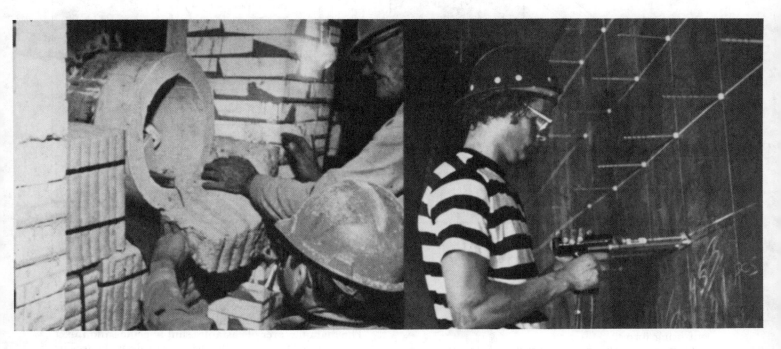

cluding energy cost) for a reasonable life expectancy of the system, rather than the lowest initial cost.

Compromise begins with the initial conception of the refractory or insulating material. The components of the refractory/insulating system are critical parts in this selection.

Types of refractories

The essential properties of the available refractories are shown in Table I. The basic advantages and disadvantages for each material are:

Dense castables, for use in temperatures to 3,300°F, are modern monolithic versions of the age-old fired-clay brick. Their insulating qualities are marginal. However, they have lower thermal conductivity than fired brick, and provide greater strength and lower first cost than true insulating firebrick (IFB).

True insulating firebrick is available to 3,300°F as are insulating castables. In general, these materials weigh substantially less than dense refractories, have much better insulating qualities (being two to four times more resistant to heat transmission), but are not as strong. In most applications, the thicknesses called for by nonstructural considerations provide ample strength.

Refractory fiber products such as felts and blankets are available up to 3,000°F. They have three major advantages: excellent insulating values, ease of application, and low heat storage. Low heat storage is of great advantage in cyclical heat applications, such as for a periodic kiln. Heat absorbed by the walls of such a kiln, and then lost when the kiln cools, can exceed the amount of heat actually used for production.

Calcium silicate, mineral-wool, fiberglass boards and blankets, all at the lower fringe of refractory-temperature ranges, are often used because of their combination of qualities. For example, calcium silicate usually offers structural-strength and high thermal-insulating values, and is not affected by moisture or humidity. Mineral wool and fiberglass generally combine the properties of light weight, heat resistance, low conductivity and high sound absorption.

Castables

Castables, also called refractory concretes, are available in dense and insulating versions. Castables are supplied dry, and are mixed with water before installation. They are installed by pouring, trowelling, and pneumatic gunning, or occasionally, by ramming. As a result, castables give a smooth, practically jointless construction to monolithic hearths, walls and roofs of furnaces and kilns.

Applications of castables in industries or processes using heat vary from aluminum furnaces or reformers for ammonia plants, to zinc furnaces. The growing use of castables stems directly from their advantages. Castables save money during installation because they go into place faster and easier. Mixing, conveying and placement of the unfired castable mixture are often mechanized. Furthermore, castables save time by eliminating the cutting of brick to fit a furnace or kiln, and eliminate the need for inventories of special brick shapes.

Gunning castables into place requires more skill than pouring them, and uses more material, but gunning

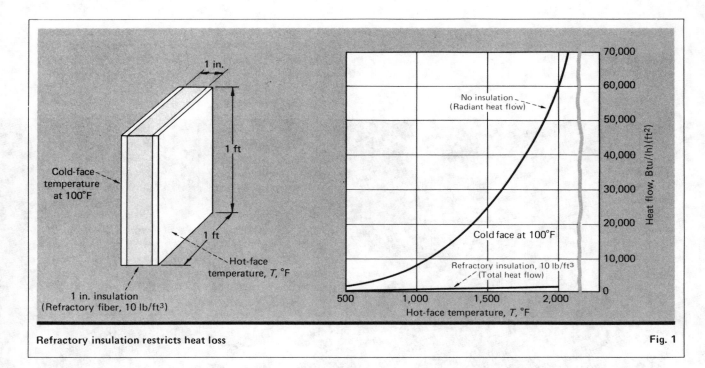

Refractory insulation restricts heat loss   Fig. 1

applies more material in less time than any other method. It also makes possible the installation of castables to horizontal, vertical or overhead structures without using forms.

Castables also have a number of performance advantages. First, dense castables have low permeability, partly because of the nature of the material and partly because the structure has few joints. As a result, castables are usually the best lining for vessels or chambers operating at other than atmospheric pressures. Second, most castable refractories have good volume stability within the specified temperature range, and many have good resistance to impact and mechanical abuse.

Installation practices for castables may cause prema-

ture failure and uneconomic life of the refractory. The reason is that the user or installer of such castables performs most of the ceramic manufacturing processes. The manufacturer of castables simply grinds and mixes raw materials together, then bags and ships them. The user adds water (whose amount and quality are critical to complete the mixing operation), and fires the emplaced castable to the necessary refractory quality.

The amount of water required for casting is probably the most commonly neglected variable associated with castable installation. Some castables are more sensitive to water requirements than others. Too much or too little water can produce poor physical characteristics, such as poor strength or poor abrasion resistance.

Refractories for use in fired heaters   Table I

	Mineral-wool block	Calcium-silicate block	Haydite, vermiculite	Insulating castables	Dense castables	Insulating firebrick	Dense firebrick	Refractory-fiber boards and shapes	Refractory-fiber blankets	Refractory-fiber modules
Maxium operating temperature, °F	1,500	2,000	1,800	2,800	3,300	3,200	3,300	2,600	2,600	2,600
Thermal insulation	1	2	3	3	4	3	4	2	1	1
Heat storage	1	1	2	3	4	2	4	1	1	1
Thermal-shock resistance	4	4	2	2	1	2	1	1	1	1
Chemical resistance	4	4	4	2	1	2	1	3	3	3
Erosion resistance	4	3	3	2	1	2	1	3	4	3
Strength	4	3	2	2	1	2	1	3	4	2
Resiliency	2	3	4	4	4	4	4	2	1	1
Acoustical insulation	2	3	3	3	4	3	4	2	1	2
Installed cost	Low	Medium	Low	Medium	Medium	Medium	Medium	High	Medium	High

Code: 1 - Excellent, 2 - Good, 3 - Fair, 4 - Poor.

Unfortunately, there are more failures of dense-castable installations than there should be. In practically all cases, failure can be pinpointed to improper preparation and installation of the material.

Physical properties of castables

To choose the right castable refractory for any application, we must balance the requirements of the process against the cost and capability of the castable. Capabilities and other characteristics of castables, such as maximum service temperature and modulus of rupture, are obtainable from manufacturers' data sheets. Test procedures for these physical properties are established by the American Soc. for Testing and Materials (ASTM), and are conducted by the refractory-manufacturers' laboratories. Other characteristics, such as the amount of water to be added to the castable, are determined by actual tests.

Chemical properties of castables

In refractory compositions, the alumina-silica ratio roughly indicates the refractoriness of the composition: the higher the alumina (Al_2O_3) content, the higher the refractoriness. Lime (CaO) in the composition appears as a result of the binder, and also contributes to the ultimate refractoriness.

Iron oxide (Fe_2O_3) becomes important in refractory compositions when the process being carried out in the refractory-lined vessel involves a reducing atmosphere. For example, carbon monoxide in the vessel's atmosphere at certain temperatures can react with the iron oxide to deposit carbon within the lining. If the reaction continues, the carbon will cause the lining to crack. Likewise, a hydrogen atmosphere will reduce Fe_2O_3.

Thermal expansion of castables

Unlike brick, castables are not fired before installation. Only after prolonged exposure to heat do castables acquire the reversible expansion and contraction characteristics found in fixed ceramics. Because of this, castables show the results of complex forces working on them during the initial heatup. The fired aggregate expands according to its chemical composition, the cement phase shrinks as it loses the water of hydration, and the castable itself expands or contracts as it sinters and mineralogical reactions occur.

Most of what we know about the thermal expansion of refractories is based on brick. This knowledge is difficult to transfer to castables. From an application standpoint, two factors are important:
- Fireclay castables do not normally require an expansion allowance.
- Certain dense, highly refractory castables have substantial reversible thermal expansion, up to 2,000°F or more. Castables of this type may require an expansion allowance in massive installations.

Insulating firebrick

The objective of any insulation manufacturing process is to create a product of great porosity. After all, only dead-air spaces insulate. The surrounding material is useless in the insulating sense, and it is there only to provide structural strength.

Insulating firebrick (IFB), made to this objective, reduces heat loss by two thermal mechanisms. As compared with dense refractories, IFB has high resistance to heat flow and, being lightweight, stores very little heat. Initially, IFB was applied to cyclically operated equipment, allowing energy conservation from reduced heat flow and low heat storage. Later, IFB was used in continuous heat-processing equipment such as tunnel kilns. In such equipment, IFB's high insulating ability conserves fuel and lowers the cost of construction by reducing the thickness of the refractory system.

Let us compare the capabilities and limitations of insulating firebrick with dense firebrick. We will find that IFB is:
- Highly resistant to heat flow.
- Very light in weight, and stores much less heat. The light weight expedites installation and bricklaying. Moreover, the supporting structure and foundations can be lighter and less expensive. This weight reduction can be as much as 80%.
- Structurally self-supporting at elevated temperatures because it has the best compressive strength of any insulating refractory. IFB is compatible with dense firebrick to firm up the whole construction.
- Machined to finished shape. It has tighter tolerance than dense fireclay brick. This provides a tighter refractory construction, less heat loss through the joints, and faster installation.
- Low in impurities that can adversely affect refractory performance—for example, a furnace atmosphere having hydrogen as one of its constituents.
- Available in a large variety of shapes, both as supplied by the manufacturer and as made from slabs without mortared joints.
- Readily cut or sawed on the job.

On the other hand, we find IFB to be:
- Typically lower in strength than dense fireclay brick. This will influence the design of large structures.
- Prone to hot-load deformation failure from heavy loads at high temperatures.
- Generally not good for abrasion resistance. For example, it should not be used in a flue that would have high velocities and/or abrasive matter present.
- More prone to thermal and mechanical spalling than most dense refractories. This prohibits its use in some very abusive applications. For example, the slab-and-billet reheat furnace found in rolling mills.
- Incompatible with a chemical environment that will flux an alumina-silica refractory body.

Shapes and sizes

IFB is available in almost all the standard refractory shapes in addition to special shapes such as arch, key-wedge, and circle brick, and in nonstandard thicknesses (obtained by mortaring two or more pieces).

The classification by ASTM (C 155-70), as shown in Table II, groups the brick in accordance with bulk density (a good indication of thermal conductivity) and the behavior of the reheat change test.

The reheat change test for each group is conducted at a temperature 50°F below the normal maximum recommended service temperature. For example, Group 23 brick (ordinarily referred to as 2,300°F brick, and recommended for this temperature) is tested at 2,250°F. A maximum of 2% linear shrinkage is allowed.

If precise operating conditions are known, fairly exact factors of safety can be determined. This is seldom the case. Hence, a rule of thumb has been developed by refractory users that a 200°F safety factor will be allowed between the furnace design temperature and published temperature-use limits.

For example, a Group 23 brick would not be used for a furnace having a design temperature above 2,100°F. This rule has evolved through long experience. Its rationale is the difficulty of operating heating equipment at a precise design temperature for long periods of time. During that period, we can expect temperature variations within the furnace, and it is almost inevitable that an upset condition will occur where the design temperature will be exceeded.

Excellent information on construction methods is published by refractory manufacturers for their products, and is equally applicable to IFB. These handbooks include design shortcuts, tables of brick shapes, calculation methods, and many details for building and supporting refractory structures.

Brick equivalents

Because refractory brick differs in size from products such as building brick, the refractories industry has developed a term to describe what is meant by a "brick." This term is a "brick equivalent" (BE), and the brick having the dimensions of 9 in. by $4\frac{1}{2}$ in. by $2\frac{1}{2}$ in. is defined as one brick equivalent. Thus, a brick having dimensions of 9 in. $\times$ $4\frac{1}{2}$ in. $\times$ 3 in. is 1.20 BEs.

To illustrate the concept of BE, let us consider a furnace wall that might require 1,200 BEs. The wall could be built from twelve hundred $9 \times 4\frac{1}{2} \times 2\frac{1}{2}$ bricks, or one thousand $9 \times 4\frac{1}{2} \times 3$ bricks. Or, it could be constructed from four hundred $11\frac{1}{4} \times 9 \times 3$ bricks because this brick contains 3.00 BEs.

The idea of brick equivalents is useful in estimating the brick requirements. Calculations can be done first in brick equivalents, and then converted into the exact number of pieces of each size desired. The information in Table III is helpful in estimating quantities. To calculate the number of BEs needed for walls of other thicknesses, use multiples of the values given in Table

III. These same values can be applied to curved walls by using the outside wall dimensions.

Mortar and mortaring methods

An incorrect selection of mortar not only slows down bricklaying, but can compromise the strength and life of the resultant structure. The mortar must have good water-retention properties, and be a strong air-setting one. Good air-setting properties give the refractory wall proper strength and stability. Brick not strongly bonded will loosen and bulge, making it necessary to rebuild the lining. Furthermore, IFB is most frequently used in furnaces where the temperatures are not high enough to develop a good bond in a heat-setting mortar. Even in furnaces operated as high as 2,600°F, the temperature drop through the IFB is so great that sufficient temperature for a ceramic bond may exist only in the first $\frac{1}{2}$ to 1 in. of the hot face of the brick. The remainder of the wall will not be bonded.

The most common and best mortaring practice is to use "dipped joints." The mortar is thinned to a creamy consistency, just pourable, and the brick dipped into this slurry before being placed in the wall. Mortar supplied wet to trowel consistency will commonly require additional water to reach this creamy consistency. One precaution in dipping is to minimize the amount of mortar deposited on the surface of the brick that will become the hot face of the furnace lining. This coated face is prone to spalling.

Troweled joints are often used but should be discouraged because they require a degree of skill mastered by only a few brickmasons. However, troweling is often used on crowns.

Insulating aggregates

Insulating aggregates are a comparatively unfinished form of refractory. There are two common types: expanded and burn-out:

Expanded aggregate is made by expanding a clay in such a manner that aggregate particles are formed having internal porosity and light weight. Such aggregates have very fine insulating properties. However, expanded insulating aggregates are not readily available. Most producers use their production in-house. An exception is haydite, a clay that bloats upon being heated. It is a low-temperature refractory of limited use. A common application involves mixing it with calcium aluminate cement and vermiculite, and gunning the resulting castable into low-temperature petrochemical heaters.

Burn-out aggregate is made by the burn-out IFB processes. Both reject-brick and dobies [molded blocks of ground clay or refractory materials, crudely formed, and fired] are crushed into aggregate. Various temperature grades are available. The crushed aggregate will often have excessive fines that can detract from its insulating value.

Insulating aggregates are used as fills in places such as tunnel-kiln roofs, flat suspended arches, and car-bottom furnace floors. The major use of the insulating aggregates is in the form of castables in which the aggregate is combined with calcium aluminate cements. Refractory manufacturers formulate these castables

Classifications for insulating firebrick		Table II
Group	Reheat* change, °F	Bulk density,† lb/ft³
16	1,550	34
20	1,950	40
23	2,250	48
26	2,550	54
28	2,750	60
30	2,950	68
32	3,150	95
33	3,250	95

Source: ASTM C 155-70
*Not more than 2% when tested at indicated temperature.
†Not greater than indicated density.

into numerous combinations of aggregate sizing, aggregate temperature grades, and types of calcium aluminate cement.

Refractory fibers

Many insulating fibers will block radiation, stop convective currents, and still have a minimum amount of thermal conduction. For high-temperature applications (1,000 to 3,000°F), fibers classified as ceramic or refractory are used. In general, these refractory fibers are considered in four broad categories:
- High-silica-leached and fired-glass fibers.
- Flame-attenuated pure silica fibers.
- Alumina-silica fibers.
- Pure metal-oxide fibers.

The silica fibers have their major application in the aerospace industry. They are very expensive and currently have limited use in the industrial field. Chemically produced metal-oxide fibers are also quite expensive, but their high-temperature capabilities are meeting some very critical needs. For example, pure alumina fibers can be used for insulations up to 3,000°F.

The bulk of the refractory fibers in use are the alumina-silica ones because of their relatively low cost and good thermal properties. Our discussion will primarily deal with alumina-silica fibers, but the principles generally apply to the other fiber types. The easiest way to understand the differences between refractory fibers is to examine their temperature limits (see Table IV).

Characteristics of refractory fibers

The features most important for standard applications are fiber diameter and thermal stability. Fiber diameter is influenced by complex manufacturing considerations. The alumina-silica fibers have average diameters ranging from 2.2 to 3.5 microns. The finer fiber produces a lower thermal conductivity at light density and high mean temperature, but has a marginal effect at higher densities.

The thermal stability of refractory fibers is, in most cases, the more important characteristic. The maximum temperature limit of any fiber is set by the manufacturer to indicate the point at which the fiber becomes thermally unstable. A phenomenon known as devitrification begins to set in at approximately 1,800°F. This crystal-forming process is responsible for shrinkage, loss of strength, and general physical degradation that occurs in such fibers at high temperature. It is a time/temperature phenomenon that rapidly accelerates at 2,300 to 2,400°F.

Let us consider the amount of shrinkage at 2,300°F. Products from different manufacturers, all rated for 2,300°F, have linear shrinkages ranging from 3.5 to 8.8%. Obviously, the thermal stability of the product depends on the amount of shrinkage. Hence, a user of refractory fibers should be aware of the difference in shrinkage of the various refractory fibers and should determine the appropriate design criteria for acceptability. The most efficient insulating product is of no value if it does not have structural integrity.

No matter which fiber is chosen, there are several characteristics inherent to all. Thermal conductivity of refractory fiber is very low compared with brick and castable refractories at the same temperature range. Low density allows a lighter steel-support structure in furnace design. Low heat storage means faster heating and cooling time in periodic kilns. Resiliency of the fiber allows it to be compressed and be packed in various areas, and also makes it resistant to mechanical shock and vibration.

Refractory fibers are completely resistant to thermal shock. They are used in severe applications such as the furnace doors of heat-processing equipment. The fibers are chemically stable except for certain acids and strong alkalies that attack them. After wetting with oil or water, the properties will return when the product is cleaned and dried.

Most products made from refractory fibers use a high percentage of fiber. Hence, the characteristics for the fibers generally apply to finished products.

Health and safety

The airborne particulates generated by handling a refractory fiber are classified as a nuisance dust that does not produce a significant toxic effect. The fiber can cause throat irritation if inhaled, and mechanical skin irritation on contact. After working with the fibers for a period of time, most people no longer react to the skin-irritating effects. It is recommended that workers exposed to the dust wear long sleeves, gloves and safety goggles. Respirators are also recommended where airborne concentrations exceed the limit for a nuisance dust.

Refractory-fiber products

Refractory fibers are processed into products such as felts, blankets, vacuum-formed shapes, sprays and paper. Almost all of these products are available for use from 1,600 to 2,600°F.

The basic material is the refractory fiber in bulk form. It is inorganic and can operate continuously at

Brick equivalents for different wall thicknesses	Table III
Wall thicknesses, in.	Brick equivalents, BE/ft^2
2½	3.6
4½	6.4
9	12.8

1 ft^3 = 17.07 BE
BE = 101.25 in.^{3}

Temperature limits for refractory fibers	Table IV
Refractory fiber	Temperature range, °F
Fluxed alumina-silica	1,600 to 1,800
Alumina-silica	2,300 to 2,400
Modified alumina-silica	2,600 to 2,700
Pure metal oxides	3,000 to 3,100

the maximum recommended service temperature. In single-use applications, the fiber can be taken well above this temperature limit.

There are three varieties of bulk fiber: cleaned or uncleaned, lubricated or unlubricated, and long or short length, or a combination of these.

Uncleaned fiber still has the "shot" or unfiberized material in it, and normally has the longest fiber length. It is not reprocessed before being used as packing or loose fill in expansion joints, furnace-crown cavities, or wall cavities. The long fiber length tends to tie the bulk material together. Lubrication added to bulk fiber makes handling easier. This uncleaned fiber is the most inexpensive type.

Cleaned fiber has been processed in order to separate the shot from it. Certain applications such as aerospace and paper making cannot tolerate a large amount of unfiberized material. The process of shot separation also tends to break the fiber into shorter lengths.

Lubricated fibers contain an organic lubricant that is added in the fiberizing process to give the fibers a thin coating. This allows them to be more easily worked when further handling is required, such as in packing joints. The lubrication allows the fibers to slip on one another, and also significantly reduces dusting. However, a lubricant must not be used on fiber intended for wet mixing since such fiber will not properly disperse in water.

All bulk forms of refractory fiber are highly resilient and are often used to reduce brick cutting in the field, to make emergency repairs, and to pack void spaces in refractory construction. Bulk-fiber density will range from a loose fill of 6 lb/ft^3 to a maximum for packed material of about 12 lb/ft^3. Packing to higher densities will cause the fibers to break up and lose resiliency.

Refractory-fiber felts and blankets

Refractory fibers in the form of felts and blankets are used as linings for furnaces, kilns, and other high-temperature equipment.

Felts are organically bonded mats in roll or sheet form. The product has good cold strength and is used when a semirigid insulating sheet is required. Upon firing, however, the phenolic binder burns out. This causes two problems. First, the binder emits an objectionable odor and irritating fumes. Second, the fired product without the binder is quite weak, which can be serious if strength is required for the material to stay in place.

High-density felts (10 lb/ft^3 or higher) are formed in the same manner as regular-density felts, but the binder is cured in a pressure press to achieve felt densities of up to 24 lb/ft^3. Such high-density felts are one answer where a lower-thermal-conductivity felt of exceptional strength is needed or where space (i.e., thickness) for a felt is limited.

Blankets are completely inorganic (i.e., no binder is used with the refractory fiber). Originally, the blankets were not "needled." In one process, the refractory-fiber mat was sent directly from the collection chamber through a set of heated compression rolls in order to heat-treat the mat into a blanket. In another process, the fiber was water-washed to remove the shot, and then refelted into a blanket. With either process, the resulting product was weak. The product from the first process was subject to laminar failure when flexed, while the weakness in the product from the second process was due to the short fibers.

The un-needled blanket is being replaced by an inorganic needled product. Here, the mat is run from the collection chamber through a needling machine. The needler has thousands of barbed needles that move up and down through the blanket. As the needles penetrate, the barbs hook some of the fibers and pull them through the blanket, binding the blanket together. Needling laces the blanket together with its own fibers. As these fibers become interlocked, the product becomes an integral unit having excellent flexibility. After needling, the blanket is heat-treated to relax the fibers and bring them into intimate contact with each other.

After firing, the strength of the blanket is as great as before because the interlocking mechanism is not affected by heat. This is very important when lining furnaces and kilns where the anchor spacing for the blanket may be as wide as 18 in.

Needled blankets have many applications such as (a) removable insulating blankets for turbines where the blankets must be flexible and resistant to physical abuse, (b) insulation wraps for investment-casting molds and for stress-relieving field welds, and (c) high-temperature sound-absorption systems. Since blanket properties such as strength vary with the manufacturer, detailed product performance data should be studied.

Another material made predominantly of refractory fiber is a board for expansion joints in brick constructions. This board is a refractory felt having a high phenolic-binder content. As brick structures are laid up, a space must be provided for the brick to expand, but this space should also be packed so heat will not escape.

The standard practice is to carefully gauge the joint spacing, and build a free-standing wall away from the first wall. This is tedious and time-consuming, and requires the stuffing of the space at a later time if insulation is desired. An expansion-joint board provides the necessary spacing in this type of construction, as well as the fiber to fill the joint. It is rigid enough to build against, yet resilient enough to spring back and refill the joint when the structure cools. Much time and installation cost can be saved by using such board.

A thin and flexible paper is made from refractory fibers in thicknesses ranging from $\frac{1}{32}$ to $\frac{1}{8}$ in. The fiber is held in a latex-binder matrix. This paper is expensive but its qualities are desired in high-temperature gaskets, linings for combustion chambers, and expansion-joint materials for molten-metal troughs. It can also be used for thermal insulation wherever small clearances are encountered.

Applications for refractory fibers

Perhaps the largest single application of refractory fibers is for the lining of furnaces and kilns. Table V lists the several categories of furnaces that use these fibers as the primary insulation in the roof and walls. Since service conditions for each application vary, different types of fibers will be required. For lining heating equipment with refractory-fiber materials, there are two basic

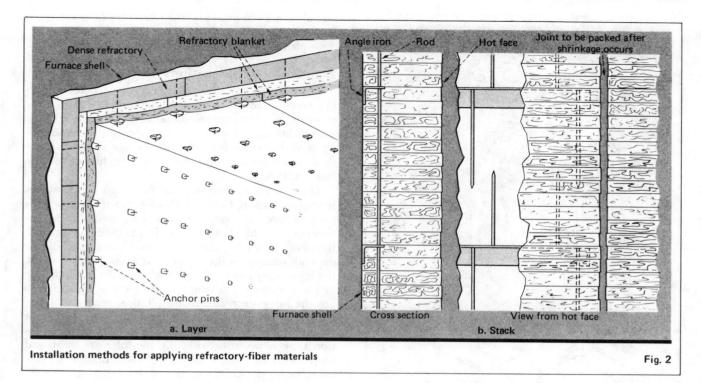

Installation methods for applying refractory-fiber materials   Fig. 2

methods—layer or stack construction. Typical construction features for each method are shown in Fig. 2.

Layer construction involves impaling one thickness of insulation on top of another on anchors attached to the furnace shell, until the required thickness is achieved. However, major differences exist in layer construction for furnaces operating up to 2,250°F and those up to 2,600°F in the anchor system and in the type of hot-face fibers. The layer method also enables the use of various types of refractory felts or blankets in combination. For example, a needled blanket (useful to 1,600°F) becomes the backup insulation for the hot-face refractory-fiber material. The needled blanket is less expensive than the hot-face alumina-silica lining, so cost savings are possible.

Stack construction uses refractory fibers or blankets cut into strips that extend from the hot surface to the cold surface of the furnace. These pieces are stacked upon each other in the field or formed into a module that is subsequently mounted as a single unit.

Anchoring systems

In layer construction, a variety of anchors are used to meet varying temperature and operating conditions.

Applications for refractory fibers	Table V

Type of furnace	Temperature, °F
Brick firing	1,800 to 2,200
Ceramic firing	1,800 to 3,250
Forging	2,250 to 2,450
Heat treating (metal)	1,500 to 2,100
Petrochemical heaters	1,600 to 2,100

All forms of alumina-silica fiber require anchors that will match, or exceed, these conditions.

Metallic anchors are available in a variety of lengths and alloys for temperatures to 2,250°F. These anchors are welded to the furnace shell, and refractory-fiber insulations and blankets are impaled on them. An anchor washer is then installed to hold the insulation in place.

Ceramic anchors are used where temperature limits exceed 2,250°F. One system uses a Type 304 or 310 alloy stud, welded to the furnace shell. A ceramic pin is threaded over the metal stud. A ceramic washer locks onto the head of the ceramic pin. A major advantage for this system is that a less expensive alloy can be used for the stud as long as the ceramic pin extends into the refractory lining far enough.

Another method uses a ceramic locking cup. The cup extends into the insulation 2 in., and locks onto the standard metal anchor. The cup is then filled with either bulk fiber or insulating cement. The ceramic cup does not penetrate the lining very far, and requires the use of a high-alloy stud. Even then, the temperature on this stud could easily exceed 2,250°F in high-temperature furnaces.

Ceramic anchors are required in carbon-rich reducing atmospheres. The nickel in alloy-metal anchors catalyzes the reduction of carbon monoxide, and carbon accumulates around stud locations where the temperature is 1,000 to 1,200°F, to cause fiber deterioration.

Modular construction

The modular system is a method of prefabricating the refractory blanket into 12-in.-square modules that can be rapidly installed. The modules are attached to the furnace steel and eliminate the layer-by-layer buildup.

In one patented modular system, the strips of blanket

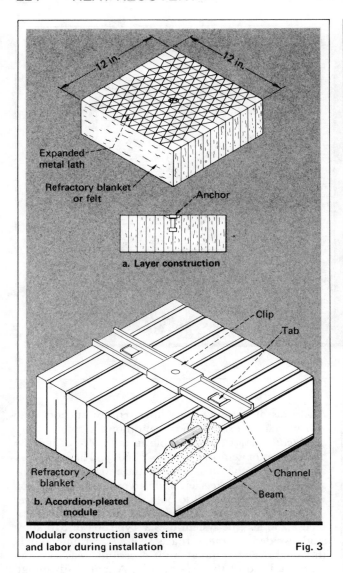

Expanded metal lath

Refractory blanket or felt

Anchor

a. Layer construction

Clip

Tab

Refractory blanket

Channel

Beam

b. Accordion-pleated module

Modular construction saves time and labor during installation   Fig. 3

or felt are stack-constructed. The stacks are attached to a 12-in.-square expanded-metal lath by using a refractory cement. The module is then fastened to the furnace shell with a one-step blind-welding and threading operation. In another version, the lath is welded directly to the furnace shell. The strips are compressed when mounted to the metal lath. The module tends to expand to fill the shrinkage joints that occur at high temperatures. However, the joints in this system do open, and must be stuffed with bulk fiber or blanket strips during the first cooldown cycle.

Another patented module is made with a single piece of inorganic, needled, refractory-fiber blanket 12 in. wide. This blanket is accordion-pleated so each section is as deep as the final thickness of the lining (between 5 and 12 in.). Stainless-steel support beams are buried in the fiber pleats next to the cold face of the block. A stainless-steel suspension channel attached to the beams fits into a stainless-steel clip that is fastened to the furnace wall.

Banding and kraft-covers keep the pleated blanket in compression prior to and during application. After the module is attached to the furnace shell, the compression restraints are removed and the refractory fiber expands. Since shrinkage occurs at high temperature, the accordion pleats expand and offset any shrinkage.

Economic considerations

There are several tradeoffs to consider when comparing layer methods to stack or module ones. In layer construction, less-expensive backup insulation is easily used. Stack and module construction employs one material that extends from the hot surface to the furnace shell. Material costs, of course, will be higher for the stack method.

The opening of shrinkage joints at high temperatures is another factor. Layer methods offset the joints from layer to layer, and provide a small amount of compression between the blanket edges. In the modular method, the joints fill during fiber expansion. The success of this depends on how well the job is installed. In some cases, the joints may need to be packed after the initial firing.

The undisputed advantages of modular construction are the large decreases in installation time and labor costs. The buildup of layers and the intricate pin-layout required in layer construction are eliminated. Installation time is saved because the modules are light and more easily handled than large rolls of blanket or felt. Essentially, two-thirds of the onsite labor time is replaced by shop prefabrication. Thus, installation costs and furnace downtime are reduced.

Installation advantages also come from the hidden anchoring system of the modules. Since each module has the hardware located close to the insulation's cold face, lower-cost alloys may be safely used at furnace temperatures that would normally require ceramic anchors. Atmospheric deterioration of the anchors is eliminated because the hardware is located in a relatively cool zone. When repair or replacement is necessary, only the damaged modules are removed. With blankets or felt rolls, such spot repairs are difficult, and large areas must often be replaced.

Layer construction has a definite advantage over modular in relation to thermal efficiency. Fiber orientation in layer construction is perpendicular to the direction of heat flow; whereas in the stack module (Fig. 3a), the fibers are parallel to the flow. This orientation can increase thermal conductivity 20 to 40% as indicated by manufacturers' tests. Hence, more insulation is needed in stack-module construction than in layer construction to achieve the same thermal resistance. The accordion-pleated module (Fig. 3b) will have a thermal efficiency that is greater than the stack module but not as good as the layered construction, due to the orientation of the blanket. Here, the fiber orientation is both parallel and perpendicular to the heat flow.

Retrofit applications

Rising fuel costs are causing a closer look at kiln and furnace operations in relation to heat losses. If the unit is inefficient but the refractory lining is still serviceable, its efficiency can be increased without replacing the entire refractory lining. This can be done by installing a refractory-fiber blanket directly to the hot face of the existing surface. Since refractory fibers do not soak up a lot of heat, they do not require large amounts of heat to

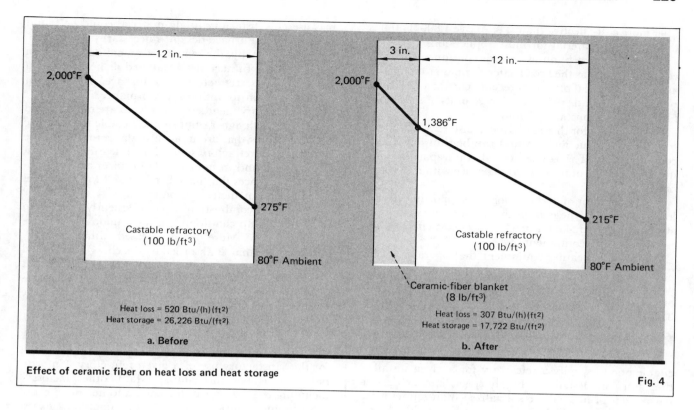

Heat loss = 520 Btu/(h)(ft²)
Heat storage = 26,226 Btu/(ft²)

a. Before

Heat loss = 307 Btu/(h)(ft²)
Heat storage = 17,722 Btu/(ft²)

b. After

Effect of ceramic fiber on heat loss and heat storage

Fig. 4

reach equilibrium conditions. In this application, the overall heat loss will be reduced because of additional insulation and the elimination of cracks on the hot face that allow heat to escape. Generally, fiber insulation can be attached to the existing refractory surface by using the wallpaper or modular-veneering technique.

Wallpaper construction—Major cracks or spalled areas should be repaired. A desired stud pattern is laid out. Layers of blanket are then impaled on the studs to the desired thickness. The reduction in overall heat loss for a typical installation is shown in Fig. 4.

Veneer construction—A module is made from strips of refractory fiber 12 in. long, turned edge-grain up and held together with an open-mesh organic cloth. Using an air-setting refractory mortar, the modules are installed over refractories such as dense brick, insulating firebrick, bubble alumina, semisilica brick, castables of any density, and plastic and ramming mixes. The module's thickness can vary from 1½ to 5 in. The organic cloth burns off when the furnace is first fired.

Performance

Insulating firebrick (IFB) and refractory-fiber materials are used in heating equipment for many process operations. Such heating equipment does not normally undergo aggravated chemical attack such as that found in a basic-oxygen steel furnace, or the aggravated physical abuse found in a rotary cement kiln.

Insulating firebrick and refractory fibers have very long service lives. It is not uncommon to find IFB linings in steel-annealing furnaces last more than 20 years. Refractory fibers are completely resistant to thermal shocking—so cyclic heating equipment can be elevated and reduced in temperature as quickly as the burner system will allow. On the other hand, an IFB-lined fur-

nace must be heated and cooled with reasonable caution to prevent thermal spalling.

Neither material resists mechanical abuse such as a direct blow by a steel ingot. IFB will withstand higher air velocities and will better withstand minor abuse such as gouging and small blows.

Compensating for the thermal movement of IFB makes the design and construction of an IFB-lined furnace more difficult. Refractory fibers experience none of this thermal movement.

Refractory fibers are not considered a good backup insulation to dense refractories, because they are resilient and compressible. On the other hand, IFB is thoroughly compatible with dense-refractory construction.

Refractory fibers are the optimum material for periodic heating equipment up to 2,250°F. IFB will be favored above 2,250°F. For cyclic heating equipment operating at any temperature, it is doubtful that dense refractories would be specified unless abuse of the lining or chemical resistance were a major factor.

In a very strong thermal-shocking environment (to 2,600°F), refractory fiber would be selected. Its low heat storage favors periodically operating heating equipment, and the lining will not deteriorate from thermal spalling. In such an environment, the choice is not between IFB and refractory fibers, but between refractory fibers and dense refractories. IFB is too prone to thermal spalling in viscous situations.

Until 1973, dense refractories were used almost exclusively for continuous-operating heating equipment such as tunnel kilns. Now, IFB in these applications saves energy, and reduces foundation requirements and plant space. Refractory fibers are seldom used because low heat storage is not sufficient in such equipment. Only at low mean temperatures (below 1,800°F) is a refractory fiber more cost-effective than IFB. Above 1,800°F, IFB is the optimum material for most applications.

There is an exception as regards continuous-heating equipment such as the fired heater common to refineries and petrochemical plants. In recent years, some furnace builders have made portions of these units of refractory fibers. The reasons are twofold:

■ Downtime for these units is expensive. With refractory-fiber construction, the unit can be shut down and cooled rapidly if repair is needed; the repair is done quickly, and the unit returned to service with minimum downtime.

■ The amount of and cost for the supporting steel is lower. Typically, these units are tall, and the savings in structural steel make refractory fibers cost-effective for portions of the equipment.

Almost any heating equipment that operates above 2,500 to 2,600°F is not practical for refractory fibers. The higher the temperature, the less cost-effective they are. This applies to fibers that are useful to 2,800 to 3,000°F.

Refractories in furnace design

A basic knowledge of refractories is not enough to enable one to specify a refractory for a given installation. Furnace design is a highly specialized field, and few aspects of it are more specialized than refractories. Hence, refractory manufacturers can be invaluable for selecting the optimum material. They do not design heat-processing equipment but do furnish technical data and suggestions on applications.

Specific information is required regarding a kiln or furnace and its operation. Before an intelligent recommendation can be made for a refractory, one should have a complete understanding of the unit, including operating data such as function, cycle time and temperature range; and type of fuel, including its characteristics and impurities. After analyzing the need for a refractory, choosing the optimum product, and engineering and building the refractory system, the next step is performance evaluation.

Refractory application skills are largely developed through on-the-job experience. Moreover, many decisions are based on judgment—carefully made after evaluating a great deal of information. Although innumerable factors influence refractory performance, most come under one of the following subjects: heating-equipment design, refractory-system design and type of refractory, operating practice, refractory quality, and workmanship.

Heating-equipment design

Furnaces are designed to accomplish consistently and economically certain results, such as production of high-pressure steam, reduction of an ore, refining of a metal, or heat treating. The efficiency of heat-processing equipment is normally reported as the ratio of production to fuel consumption.

The effect of furnace design on refractory life too often receives only secondary consideration until after the furnace is in operation, and high refractory costs force a redesign. There are cases where an otherwise efficient and successful furnace design was abandoned

because no satisfactory lining material could be found or developed. Furnace design is necessarily the first requirement. Ideally, the refractory-system design is then done, and the furnace design altered as needed and if possible. Thus, refractory and furnace people must work as a team to optimize performance.

The refractory members of the team are not expected to be furnace designers, but should be able to recognize design features that are inherently dangerous from the standpoint of refractory service. Such features are most frequently found in home-made furnaces and in well-designed furnaces that have been altered to reduce fuel consumption or increase production.

Insufficient combustion space (frequently referred to as "a bottled-up condition") is a common cause of refractory failure. More heat is released within the combustion space than is absorbed by the charge, dissipated through the furnace walls by heat flow, or carried out with the flue gases. As a result, refractory-wall temperatures can approach flame temperatures (above 3,000°F). Coupled with impurities, such temperatures will quickly destroy even high-quality refractories.

It would be convenient to say that a heat release of X Btu/(h)(ft^3) of combustion space were the upper limit for the satisfactory use of a given refractory. Unfortunately, it is not this simple. Many other factors—including refractory-wall thickness and height, effect of the amount of insulation, fuel type, amount of excess air, and especially the amount of heat absorbed by the charge or the water tubes—influence the amount of heat that can be released without disastrous effects. Judgment based on experience must be the guide, rather than a formula.

Theoretically, a single combustion space operates at a uniform temperature. This rarely occurs in practice. Only one thermocouple can control temperature in a firing zone, and it is usually placed to control the temperature of the ware or load.

Furnace drawings should be carefully studied to determine the location of control and shutdown thermocouples, and of furnace areas that could be expected to exceed the thermocouple settings. One must study the intended thermocouple settings to determine the maximum controlled-upset condition in the furnace. From this, factors of safety for refractory selection can be established, including possible zones where different refractories can be used.

When portions of the furnace lining (division walls, bridge walls, piers, door jambs, or the nose of an arch) are exposed to high temperatures on more than one side, they frequently fail before the rest of the lining. Failure may be due to load, shrinkage, excessive vitrification and spalling, slagging, or a combination of these. Sometimes, a change in the design of the refractory lining or in the refractory material will eliminate, or at least improve, conditions in the vulnerable spots.

Many burner designs are in use, and some can affect refractory selection. For example, a long-flame front burner can cause flames to lick the opposite wall. A bag wall or muffle in front of a burner can reradiate heat onto the refractories around the burner. A top-mounted flat-flame burner can direct excessive heat to a roof or onto a skew.

Refractory-system design

The study of a refractory system should involve three specific areas, whether one is designing a system or evaluating a refractory. These are: (1) thermal—the refractory system must resist the attack of the heat that is expected; (2) structural—the refractory structure must be mechanically sound; and (3) chemical—the system must resist the attack of whatever chemical materials are present.

Thermal considerations

Temperature—Obviously, the refractory system must withstand the expected temperature of the heat-processing equipment in which it is to be used. However, this is much too simple. The hot-load strength of the refractory, while it might appear to be a mechanical consideration, must be studied as a thermal consideration because it is the thermal design of the system that will determine the refractories to be used.

Load-bearing strength—The refractory's load-bearing strength, measured as hot-load deformation, is not usually of major importance where the refractory is exposed to heat on only one face. In most instances, there is a fairly steep temperature gradient through the lining, and the load is largely carried by that portion of the lining cool enough to be below the temperature at which any softening may occur.

Load deformation—In other refractory constructions, hot-load deformation is the primary factor for refractory selection. Potential trouble spots would arise when the refractory was exposed to high temperature on more than one face, or where the lining would be heavily insulated, thus decreasing the temperature drop through the wall. The ceramic process of why deformation takes place is covered by Norton,* and others.

There is no ironclad rule or formula to predict the amount of deformation that occurs in actual service. A fireclay refractory may experience hot-load deformation when the refractory is exposed to soaking heat, or when the temperature gradient through the refractory is very slight. Although it is risky to make a general rule, it is safe to say that hot-load deformation may occur in a fireclay brick, including IFB, when the refractory will experience anywhere within it a temperature of 2,250°F, or above, with a load of more than 10 psi. At higher temperatures, say above 2,400°F, even lower temperatures within the refractory, and lower loads, may cause trouble via hot-load deformation over an extended period of time.

Backup insulation—Since a load-bearing refractory system might use dense refractory, it is necessary to consider the effect of heavy backup insulation. As an example, let us consider a wall built of 13½ in. of firebrick, and backed with 1 in. of insulating block. With a hot-face temperature of 2,600°F, about 4 in. of the inner brick wall will be exposed to a temperature above 2,250°F. With a concentrated heavy load on the inner portion, this part of the wall could deform. However, with a uniformly distributed load over the entire 13½ in. of brick, there would be little chance of trouble because the relatively cool 9½ in. of brick in the back

would carry the entire load, and the wall would remain structurally stable.

This example illustrates the occasional necessity for trial-and-error approaches in refractory-system design. Hot-load deformation properties are used to compare one manufacturer's refractory brick to another's. Experience must be added to hot-load properties to determine which refractories tend to work in given situations, and which do not.

Expansion—Expansion can cause a wall to bow, cause pinch spalling, and create other problems. Many failures have been identified as due to not providing for expansion. The correct procedures, allowing for expansion spaces, are available in refractory manufacturers' installation and construction manuals.

Thermal spalling—IFB has been used in some thermal-spalling situations but is not considered as resistant to spalling as dense castables, fireclay brick, or fireclay-ramming mixes.

Structural considerations

Roof construction—Heat-processing equipment generally requires a roof. The options include a sprung arch or dome, a flat suspended arch, and variations of these. A sprung arch is the least expensive approach, but is limited to small spans. Otherwise, one must use a flat suspended arch, or division walls so that several arches can cover the required roof area. Arches and domes are more successful with low-rise (down to 2 in./ft or even 1½ in./ft of span) roofs, particularly at higher temperatures. This helps to eliminate pinch spalling.

Most insulating refractories require some sort of anchoring. This is one of the most critical aspects of castable and IFB constructions. Therefore, each application requires careful study in order to get the optimum anchoring system.

Abrasion/erosion resistance—Usually, this is a straightforward problem because the materials' limitations are known. What is not appreciated, however, is that erosion occurs more rapidly in turbulent-flow zones such as a corner than in nonturbulent zones such as a straight flue. Thus, the design must recognize that if erosion can occur, it will certainly occur in the turbulent areas. These areas might benefit from a more abrasion/erosion-resistant construction.

Permeability—An important factor with IFB is its very high permeability. Insulating-refractory constructions commonly use lower-temperature grades of refractory as the cold face is approached. If a flow of hot gases can take place through the lining, then damage to the backup insulation can be expected. Hence, we should be very cautious in furnaces where a high positive pressure exists within the furnace, and flow to the outside is possible.

Mechanical spalling—This occurs from high differential stresses placed on the refractories. Refractory fibers, being resilient, do not encounter this problem.

Chemical considerations

Refractories are affected by the action of the atmosphere in the heating equipment, and by chemical attack on permeable materials.

In heat-treating equipment, nonoxidizing atmo-

*Norton, F. H., "Refractories," 4th ed., McGraw-Hill, New York, 1968.

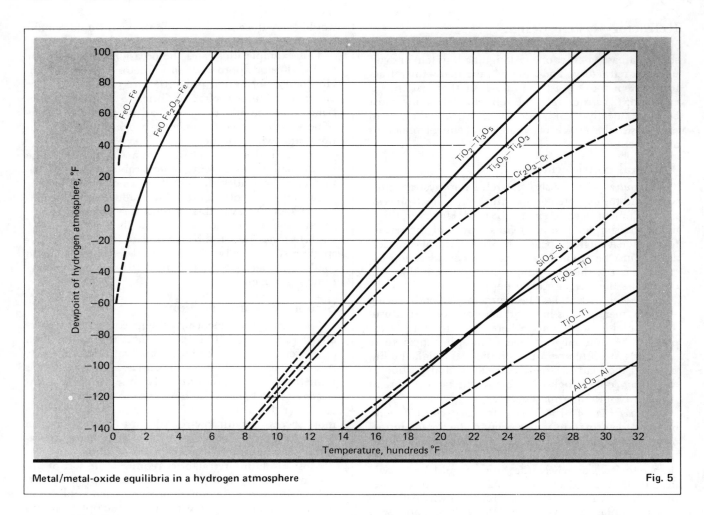

Metal/metal-oxide equilibria in a hydrogen atmosphere   Fig. 5

spheres create some problems for insulating refractories. Low oxygen pressure (i.e., concentration) reduces Fe_2O_3 to FeO. Since FeO is less refractory than Fe_2O_3, this condition promotes hot-load deformation of the refractory at temperatures above 1,800°F. This effect will also reduce SiO_2 to SiO (a gas), which also causes refractory failure. Upon cooling, the SiO tends to form deposits on surfaces in heat exchangers, boilers and reformers.

Equilibria conditions for a number of metal/metal-oxide combinations in a hydrogen atmosphere are shown in Fig. 5 for the relationship between dewpoint and temperature. As long as operating conditions are maintained to the left and above a particular curve, the metal's oxide will be stable. If the conditions are to the right and below a curve, the oxide will be reduced. This chart is true only for IFB and high-density brick. Refractory fibers should not be used above 900°F, or at a dewpoint lower than −20°F.

As the dewpoint is lowered, the service temperature must also be lowered to prevent a particular oxide from being reduced. For example, if the dewpoint for SiO_2 is −60°F, the temperature must be maintained below 2,400°F to prevent reduction of SiO_2.

A disintegration triggered by the catalytic decomposition of carbon monoxide or hydrocarbons such as methane also occurs in prepared atmospheres. Let us review the mechanism for failure due to this reaction. Ferric oxide (Fe_2O_3) is present in the refractory in local-ized concentrations. This is converted to iron carbide (Fe_3C) that catalyzes the decomposition of carbon monoxide at 750 to 1,300°F to carbon dioxide and carbon. The carbon is deposited on the Fe_3C. Carbon builds up on the catalytic surfaces that are under stress, and ultimately causes disintegration of the refractory. In many cases, such stresses are severe enough to burst the steel shell of the furnace.

A similar condition exists in hydrocarbon atmospheres, which persists up to 1,700°F.

The risk of carbon disintegration can be minimized by using refractories above the carbon-deposition temperature, and by using products having low iron content. This effect is more noticeable in dense-brick or castable refractories.

Thermal conductivity

Furnace atmospheres also have an effect on the thermal conductivity of insulating refractories. Let us review how the thermal conductivity of the furnace gases affects these refractories. Fig. 6 shows the thermal conductivities for air and hydrogen. The component of the thermal conductivity affected by changing the gas constituents in the furnace atmosphere is the gas conduction. It is quite apparent that the gas (normally air) in the pores of an insulating refractory can be readily replaced by other gas constituents found in the furnace atmosphere.

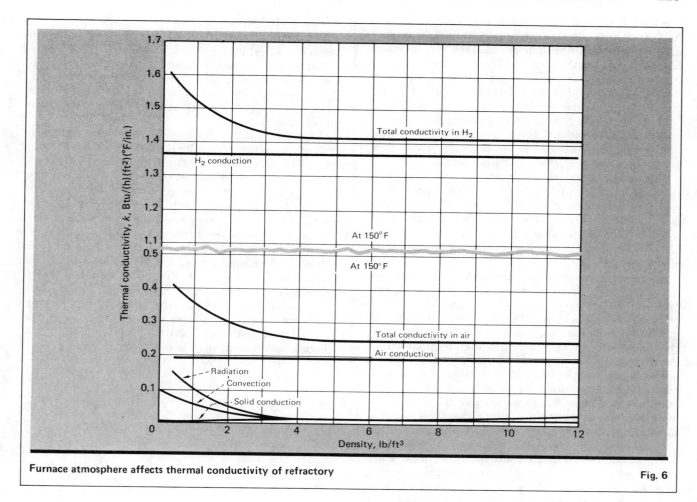

Furnace atmosphere affects thermal conductivity of refractory   **Fig. 6**

Replacement or dilution of this air constituent by other gases will change the insulation's k value. The amount of change is dependent on the k value of the replacement gas, and the porosity of the insulation. Most gases involved in a heating atmosphere have essentially the same k value as air (Fig. 7). However, hydrogen has a very high k value, causing a significant change in insulation effectiveness.

Other compounds can be found in small quantities in many furnace atmospheres. They can originate in the fuel, the refractories, or even the charge in the kiln or furnace. One of these is vanadium pentoxide (V_2O_5) from low-grade fuel oils such as Bunker C. Another vanadium compound, sodium vanadate, may be found in oil flames as droplets. It appears to decompose and cause alkali attack at about 1,240°F.

Sulfur occurs in fuels and in some clays. Depending on its content and chemical form, it can become part of a furnace atmosphere as sulfur oxides.

Alumina-silica (45 to 54% Al_2O_3 range) maintains the highest hot strength of the several refractory materials cycled at 1,400 to 1,800°F in the presence of an SO_3 atmosphere.

Test data show that disintegration of the refractory may occur if Na_2SO_4, $MgSO_4$, $Al_2(SO_4)_3$, or $CaSO_4$ are formed in a sulfur dioxide atmosphere.

Insulating refractories are seldom used in installations where chemical attack is expected. Unfortunately,

it often arises unexpectedly. Primarily, such attack is due to the permeability of insulating refractories. Many times, dense refractories resist chemical attack, not because of chemical resistance but because their very high density and low permeability prevent damaging materials from entering.

Slagging is defined by the American Soc. of Thermal Manufacturers as "the destructive chemical reaction between refractories and external agencies at high temperatures, resulting in the formation of a liquid with the refractory."

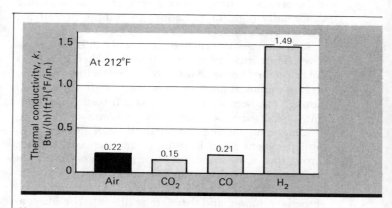

Hydrogen conductivity affects insulation effectiveness Fig. 7

Attack from mill scale occasionally causes slagging. As ferrous-metal objects are heated, their surfaces oxidize, and the resulting iron oxides flake. These oxides fall onto the hearth, and will readily attack an alumina-silica refractory because the iron oxides act as a fluxing agent. Should the oxides become airborne, they can contaminate refractories on the furnace wall.

Other materials heated in a furnace can throw off other oxides, considered to be fluxes for an alumina-silica refractory. Incinerators are the worst of all; they burn or oxidize everything put into them.

Refractory fibers are troublesome in fluxing situations. Experience dictates extreme caution when applying such fibers above 1,800°F. However, these fibers have worked well in incinerator afterburners above 1,800°F, perhaps because the fluxes by then have become so oxidized that they are no longer able to attack the refractory.

Chemical attack of refractories has become much easier to diagnose. Chemical analysis of refractory specimens can be made quickly and inexpensively by most refractory manufacturers. The methods used include spectrographic analysis, x-ray diffraction, and wet chemistry. The best approach is to analyze specimens of the damaged refractory, and of the original refractory before firing. On the basis of these data, a decision is made. One comparison may be between the damaged refractory and its original composition. If a fluxing element is found that should not be present, its source is searched for. If the source cannot be eliminated, an alternative refractory must be used.

Direct reactions such as a flux from a ceramic glaze being fired and an alumina-silica IFB are relatively easy to evaluate. However, there are some secondary reactions in which a material in the furnace atmosphere will act as a catalyst, reacting with a second element in the atmosphere and a component in the refractory. These are considerably more difficult to evaluate.

Operating practices

Premature refractory failures may arise because of unintentional changes in furnace conditions. These changes, while unobserved by the operator, may be of long enough duration to destroy a refractory lining but not long enough to be considered a change in practice. The changes may arise from: burners being out of adjustment, furnace atmospheres that are highly reducing for a period of time, furnace temperatures becoming abnormally high, or variations in composition of the raw materials or fuels.

Such changes in practice are usually difficult to discover. They leave no record other than their effects on the refractory lining. Failure of the refractory may not become apparent until weeks or months after the actual damage has been done. This makes investigation difficult, and often raises credibility problems.

Sometimes, the premeditated abuse of heat-processing equipment arises in order to get increased furnace throughput and/or greater return on assets. In almost every case, operating a furnace and refractory system beyond its capacity will reduce refractory life. This abuse is common, and there are really only two alternatives. One, stop doing it; or two, expect a shorter life from the refractories.

Workmanship

It is easy to blame brickmasons for every failure of refractory brick. By the time failure occurs, the evidence is practically destroyed. While good brickwork is extremely important, there are other factors involved. These probably occur more frequently than either poor brickmasonry, or refractory products that are not up to standard quality.

These other factors include the proper size and location of expansion joints, the rise of a furnace arch, and the thickness of a wall. These are not factors of masonry workmanship, and should be taken care of in the design of the refractory system. In practice, many design details are left to the brickmason doing the work. In some cases, this is the best way to handle such details.

Sometimes, however, brickmasons called on to do refractory work have only had experience in the building trades. Such masons make mortared joints up to $\frac{1}{2}$ in. thick. This practice is not compatible with IFB where commonly the mortar may not even be visible. The problem seldom happens in large companies employing their own masons. Smaller companies should keep this in mind, and make sure that the person in charge of the masons understands high-temperature-furnace work.

The authors

James E. Neal is manager of refractory engineering and technical services for Johns-Manville Corp., Ken-Caryl Ranch, Denver, CO 80217. He has had more than 35 years of experience with refractories. A graduate of Rutgers University, he has a B.S. in ceramic engineering. Mr. Neal is a member of the American Ceramic Soc., Tau Beta Pi and Keramos. He is chairman of the refractory-fiber committee for the American Soc. for Testing and Materials.

Roger S. Clark is merchandising manager of refractories for Johns-Manville Corp., Ken-Caryl Ranch, Denver, CO 80217. He has been associated with the refractory industry for many years. Prior to joining Johns-Manville in 1972, he was sales manager of Babcock & Wilcox's ceramic fiber division. He has a B.S. in mechanical engineering from Texas A&M University, and is a member of ASME and the American Ceramic Soc.

Better integration of process equipment and operations pays off twofold, through simplicity in structures, and the simultaneous best use of both energy and capital.

Bodo Linnhoff and *John A. Turner,*
Imperial Chemical Industries Ltd.

Heat-recovery networks: new insights yield big savings

☐ Process design usually involves the optimization of individual unit operations, with little effort spent on the optimal integration of overall processes as systems.

In this article, we will demonstrate that overall integration can be neglected only at one's peril. Here, we will concentrate on heat-recovery networks as one aspect of total process integration. We will present a new approach with solved examples, and describe industrial studies carried out at ICI.

In doing so, we will attack the conventional belief of an inevitable tradeoff between heat recovery and capital cost. Reasons will be given as to why tradeoffs need not always exist in the context of networks. The argument is borne out both by a solved example and by the results achieved at ICI. In most cases, design changes have been found that are less expensive in both energy and capital.

Specifying the problem

The heat-recovery network (or heat integration) problem is defined by a series of hot and cold process streams requiring either the cooling of hot streams or the heating of cold streams. A typical specification for such a problem is shown in Table I.

The required temperature changes may be achieved by using interchangers between process streams and/or heaters and coolers supplied with utilities. The design task is that of identifying the optimum network of interchangers, heaters and coolers with respect to annual operating and capital costs. The maximum heat load handled by any particular heat-transfer unit may be constrained by the minimum allowable temperature approach between the hot and cold streams, ΔT_{min}.

Two possible networks for the data given in Table I are shown in Fig. 1. The first (Fig. 1a) is the "no-fuss, no-complication design" and is usually the one regarded as having the lowest capital cost. The second (Fig. 1b) is a network having the lowest utility usage. Most design engineers assume that the optimal solution must lie between these two extremes. Later in this article, we will challenge this assumption.

The methods used at ICI for tackling the problem are somewhat unconventional. They involve two phases: data analysis and network design.

Data analysis yields targets that correspond to the performance characteristics of the economic optimum network. These targets are invaluable when dealing with complex integration problems, and give confidence to an engineer's current design or stimulate the engineer to search for a better solution. Furthermore, the targeting procedures give a new insight into a very simple but fundamentally important decomposition of the original problem into several parts. With this insight, even complex industrial problems can be easily solved by hand.

Data analysis

The cost of a heat-recovery network, like the cost of any process plant, is expressed in terms of annualized capital and operating costs. Capital costs depend pri-

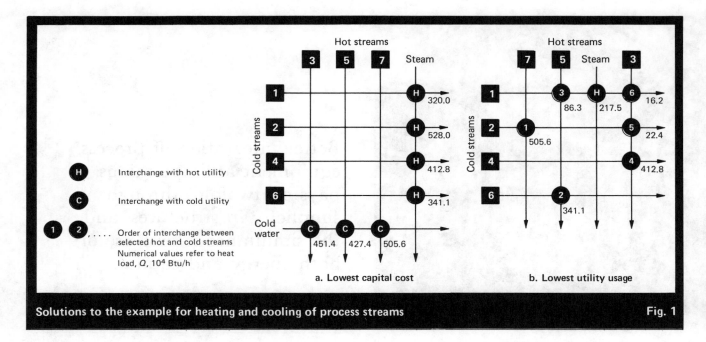

Hot streams

Steam

H 320.0
H 528.0
H 412.8
H 341.1

Cold streams

H Interchange with hot utility
C Interchange with cold utility
① ②.... Order of interchange between selected hot and cold streams. Numerical values refer to heat load, Q, 10⁴ Btu/h

Cold water

C C C
451.4 427.4 505.6

a. Lowest capital cost

Hot streams

Steam

86.3 217.5
505.6
341.1

16.2
22.4
412.8

b. Lowest utility usage

Solutions to the example for heating and cooling of process streams   **Fig. 1**

marily on the number and sizes of the heat-transfer units (interchangers, heaters, coolers), while operating costs are dominated by total utility usage (both hot and cold). Targets are therefore set for the economic optimization of heat-recovery networks, based on the minimization of utility usage, number of transfer units, and total heat-transfer area. Simultaneously, the targeting procedures define the problem decomposition into a number of design principles.

In Fig. 2, we will represent the heat-recovery problem via a box through which the hot and cold streams flow—giving up and receiving heat, respectively. Heat taken from the utility source is labelled Q_h, while heat rejected to the utility sinks is labeled Q_c. For heat balance, the difference between hot and cold process-stream enthalpies (i.e., ΔH) must equal the difference between utility heat flows:

$$Q_c - Q_h = \Delta H$$

ΔH is a constant, depending on the process stream data only. Hence, any increase in Q_h must lead to a corresponding increase in Q_c. And, any saving in source utilities must lead to a saving in sink utilities—giving a double incentive for minimizing utility usage. The crucial question is: What is the minimum possible value for Q_h in any given problem?

Minimum utility requirements

An algorithm to predict the minimum utility requirements has been previously published [3] and will now be briefly described with reference to Example 2 (Table II). A step-by-step description of the algorithm is given in an accompanying box on p. 59.

Any heat-recovery problem covers a temperature range that is bounded by the hottest and coldest process-stream temperatures. This temperature range is divided into a number of temperature intervals (see box for definition), as shown in Fig. 3. Hot utility heat, Q_h, is supplied to the first interval. The heat to be rejected

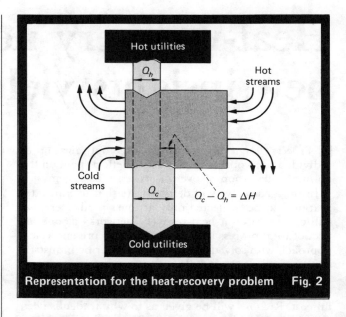

Representation for the heat-recovery problem   **Fig. 2**

Hot utilities

Q_h

Hot streams

Cold streams

Q_c

$Q_c - Q_h = \Delta H$

Cold utilities

Data for Example 1   **Table I**

Process stream no. type		Temperature		Heat capacity flowrate, c_p 10^4/(Btu)/(h)(°F)	Heat load, Q 10^4 Btu/h
		Supply T_S °F	Target T_T °F		
1	Cold	200	400	1.6	320.0
2	Cold	100	430	1.6	528.0
3	Hot	590	400	2.376	451.4
4	Cold	300	400	4.128	412.8
5	Hot	471	200	1.577	427.4
6	Cold	150	280	2.624	341.1
7	Hot	533	150	1.32	505.6

$\Delta T_{min.} = 20°F$

Data are for problem 7SP2 in Ref. 5.

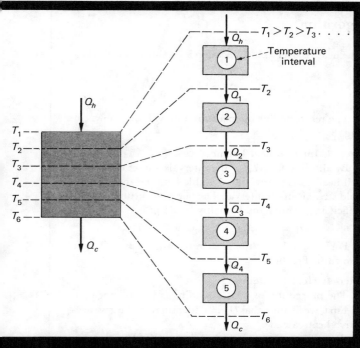

Define temperature intervals for cascading   **Fig. 3**

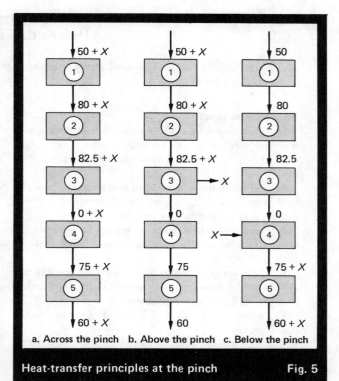

a. Across the pinch   b. Above the pinch   c. Below the pinch

Heat-transfer principles at the pinch   **Fig. 5**

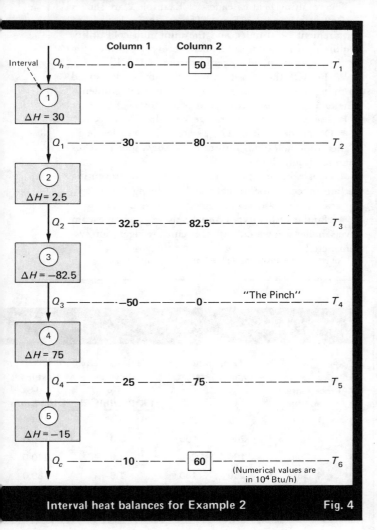

Interval heat balances for Example 2   **Fig. 4**

to the sink utility, Q_c, leaves the last interval. Definition of the temperature intervals ensures that a minimum driving force ($\Delta T_{min.}$) is maintained at all points.

Just as Q_h and Q_c are related by the overall process-stream enthalpy balance, the heat flows passed on between intervals, Q_i, are related by the interval heat balances, ΔH_i:

$$Q_i - Q_{i-1} = \Delta H_i$$

Given the temperature intervals and the interval heat balances, it is possible to calculate all intermediate heat flows (and ultimately Q_c) for any given value of Q_h. To demonstrate this, let us again consider Example 2, with the interval heat balances now specified, as given in Fig. 4.

Column 1 in Fig. 4 shows the heat flows resulting from a value of $Q_h = 0$. Between Intervals 3 and 4, the heat flow is negative (-50 units), i.e., an infeasible condition. (Negativity implies that heat flows against the natural temperature gradient.) To remedy this situation, an extra 50 units of heat (i.e., the largest infeasible flow) are supplied from the external source. The resulting heat flows are all non-negative, and are shown in Column 2 of Fig. 4.

Column 2 yields three important items of information. First, the minimum heat supply from external heat sources, necessary for feasible heat flows, is 50 units. Second, the corresponding minimum heat load on external sinks is 60 units. Third, there is a point in the temperature range having zero heat flow. From now on, we will call this point "the pinch."

Problem decomposition

Three important constraints regarding the pinch:
■ Do not transfer heat across the pinch (Fig. 5a).

Description of algorithm

The algorithm to predict minimum utility requirements and the heat-recovery pinch location is established here via a four-step procedure:

Step 1: Data input

In its simplest form, the algorithm is restricted to solving problems having the following characteristics:

■ The streams have a constant heat-capacity flowrate.

■ The minimum allowable temperature approach for heat transfer, $\Delta T_{min.}$, applies to all potential process/process and process/utility matches.

■ There are no processing, safety or plant-layout reasons to prevent certain pairs of streams being matched with each other.

■ There is only one hot utility level (at the extreme hot end of the problem) and one cold utility level (at the extreme cold end of the problem).

Examples 1 and 2 (Tables I and II) are problems of this type.

Step 2: Temperature intervals

The temperature intervals are defined by "interval boundary temperatures." In turn, these are defined by stream supply temperatures, T_S, and stream target temperatures, T_T, adjusted for $\Delta T_{min.}$. For example, the interval boundary temperatures can be defined by unadjusted cold-stream temperatures, and by hot-stream temperatures minus $\Delta T_{min.}$. Any duplications of interval boundary temperatures are ignored. Let us consider the data for Example 2 in Table II.

Stream no.	Stream type	T_S, °F	T_T, °F	Adjusted temperatures, °F	°F	Order*
1	Cold	120		120		T_6
			235		235	T_2
2	Hot	260		250		T_1
			160		150	T_5
3	Cold	180		180		T_4
			240		240	T_3
4	Hot	250		240		Duplicate
			130		120	Duplicate

*Temperature order ranges from hottest to coldest.

The resulting five temperature intervals are shown in Fig. 3.

Step 3: Interval heat balances

By allowing for $\Delta T_{min.}$, the intervals are set up so that full heat-transfer is always feasible between the hot and cold streams in an interval. The net heat surplus or heat deficit is given by the enthalpy balance:

$$\Delta H_i = (\Sigma_{Hot}C_P - \Sigma_{Cold}C_P)(T_i - T_{i+1})$$

For Example 2, this generates the heat balances shown in Fig. 4.

Step 4: Heat cascading

The procedure for heat cascading is in two stages.

First, we assume that $Q_h = 0$ identifies the largest negative heat cascade.

Second, we eliminate the largest negative heat cascade by adding heat from an external hot-utility source into the first interval.

The resulting heat cascades for Example 2 are shown in Columns 1 and 2 of Fig. 4. The minimum hot-utility requirement is 50 units, and the minimum cold-utility requirement is 60 units. The heat-recovery pinch occurs at the interval-boundary temperature of 180°F

Within ICI, the algorithm has been further developed to manipulate the complexities of industrial problems. The extra facilities include provisions to:

■ Use nonlinear temperature/enthalpy data.

■ Define dependent $\Delta T_{min.}$s for heat transfer (e.g., higher approach temperatures for vapor/vapor matches than for liquid/liquid ones).

■ Constrain "forbidden matches" [11] by preventing certain process/process matches from being made (e.g., for reasons of plant layout, safety, corrosion, etc.).

■ Handle utility types and levels (e.g., plant steam levels, boiler feedwater, hot-oil circuits, refrigeration systems, etc.).

Source: "The Problem Table" algorithm in Ref. 3.

Any heat flow across the pinch must result in the same amount of heat being added to every other heat flow in the problem in order to maintain the interval heat balances. Thus, a transfer of X units of heat across the pinch must result in increased utility requirements (Q_h and Q_c) by X units each.

■ Do not use cold utility above the pinch (Fig. 5b). Above the pinch, systems having a minimized heat flow do not reject any heat. Despite this, if we choose to reject heat into a utility sink, we must provide the same amount of heat through utility heating. Thus, X units of cold-utility usage above the pinch result in X additional units of hot-utility usage.

■ Do not use utility heating below the pinch (Fig. 5c). Below the pinch, systems having a minimized

Data for Example 2				Table II

Process stream no. type	Temperature supply, T_S °F	target, T_T °F	Heat-capacity flowrate, C_P 10^4 Btu/(h)(°F)	Heat load, Q 10^4 Btu/h
1 Cold	120	235	2.0	230.0
2 Hot	260	160	3.0	300.0
3 Cold	180	240	4.0	240.0
4 Hot	250	130	1.5	180.0

$\Delta T_{min.}$ = 10°F

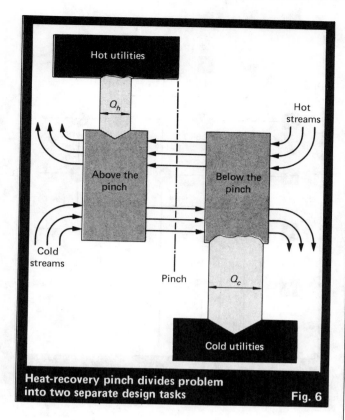

Hot utilities

Q_h

Hot streams

Above the pinch

Below the pinch

Cold streams

Pinch

Q_c

Cold utilities

Heat-recovery pinch divides problem into two separate design tasks   Fig. 6

heat flow do not absorb any heat. Any heat supply through utility heating, X, must therefore lead to a cold-utility requirement of X, over and above the minimum necessary.

These three principles suggest a general approach for overall process-heat integration. We divide the problem at the pinch, and design the parts "above the pinch" and "below the pinch" separately (Fig. 6).

Minimum area and number of units

Having established the minimum utility requirements, there is then an opportunity to minimize capital cost by minimizing (a) the number of units of heat-transfer equipment and (b) the total heat-transfer area. Minimization of heat-transfer area is achieved through even distribution of available driving forces. Algorithms for calculating minimum total surface area are given elsewhere [2,6]. We find that this target is not normally important because the total surface area is fixed to within a few percent by one's specifying the utility usage for the network. Therefore, we will concentrate on the target for the minimum number of units, which is particularly important due to the associated costs for piping, foundations, maintenance, etc.

Euler's theorem from graph theory can be applied to the heat-recovery network problem to predict the minimum number of heat-transfer units [7]. The targeting equation developed from this theorem depends only on the number of process streams and utility levels, and not on the process or utility data (temperature, heat capacities, etc.). In its simplest form, the equation may be written as:

$$U_{min.} = N - 1$$

where $U_{min.}$ is the minimum number of process and utility transfer units, and N is the total number of process stream and utility heat sources and sinks.

One of the corollaries of this equation is that if each match in a network brings at least one stream to its target temperature (or exhausts a utility source or sink), the network must be one with minimum units. Thereby, it is irrelevant whether the entire stream (or utility) heat load is transferred or only a residual load left from previous matches. This leads to the design rule of "ticking off" streams.

Let us consider again the network in Fig. 1b for Example 1. Match 1 between Streams 2 and 7 ticks off Stream 7 by taking its full load (505.6 units). Similarly, Match 2 ticks off Stream 6. However, Match 3 ticks off Stream 5 by taking its residual load (86.3 units). Matches 4, 5 and 6 and the heater continue this ticking off procedure. Applying the targeting equation, with $N = 8$ (i.e., seven process streams + one utility heat source), confirms that the design has the minimum number of units.

Network design

The network design procedure used at ICI has been developed for hand application so that engineering judgment can influence the development of designs. Our experience indicates that it would be misguided to code design rules into a computer program. In chemical processes, every practical design problem is different, due to different safety constraints, control objectives, materials limitations, etc. Thus, design decisions should not be made by a computer, on the basis of temperature and heat-load information only, but by an engineer having a knowledge of the process.

The hand design procedure involves four important steps: (1) problem decomposition, (2) heat interchange options and design constraints, (3) ticking off streams, and (4) utility placement. Let us examine each:

■ *Problem decomposition*—The heat-recovery pinch (if one exists) divides the problem into two separate tasks (Fig. 6). Above the pinch, the design should consist only of interchangers and heaters; below the pinch, the design should consist only of interchangers and coolers. There should be no interchanger transferring heat across the pinch.

If we have completely separate design tasks, we must apply the targeting equation for the minimum number of units to each task individually. The target for the minimum number of units for the entire problem is then given by:

$$U_{min.} = U_{min.(a)} + U_{min.(b)}$$

where $U_{min.(a)}$ and $U_{min.(b)}$ are the minimum number of process and utility transfer units above and below the pinch, respectively.

■ *Heat-interchange options and design constraints*—Heat-interchange options are identified on the basis of temperature feasibility. The hot- and cold-stream temperatures must always be different by at least $\Delta T_{min.}$. In general, this constraint will be satisfied by several matches, but immediately above and below the pinch the problem is tight (the driving forces approach $\Delta T_{min.}$), and the matching options are limited.

Sometimes, there may be only one design option at the pinch—an "essential match." Unless this match is made, the resulting design will suffer in terms of heat transferred across the pinch and, therefore, create increased utility usage (Fig. 5a). For this reason, the design task should always be started at the pinch, and developed by moving away from the pinch in both directions. This is a new insight into heat-recovery network design and is discussed in detail elsewhere [8].

Occasionally, essential matches involve stream splitting [8]. Here, the engineer compares the costs and controllability problems of stream splitting against the cost of more than the minimum utility requirements. Then, the engineer chooses whether or not to stream split. In other situations, an essential match may be forbidden, e.g., for reasons of plant safety. Again, the engineer must make a decision either to accept the energy penalty or to overcome it—e.g., by using an intermediate heat-transfer medium such as hot oil.

In all such situations, the engineer's task is made easier by an elegant extension of the minimum utilities algorithm [11]. This predicts the energy penalty of any forbidden matches at the targeting stage prior to design. For example, forbidding a match between Streams 1 and 7 in Example 1 would result in a predicted energy penalty of zero because there is a minimum utility solution without this match (Fig. 1b).

Once away from the pinch, many design options exist. Then, the procedure allows decisions to be made based on the engineer's judgment and knowledge of the process. For example, a match may be undesirable for reasons of safety or controllability, or a particular match may be preferred for reasons of plant layout.

■ *Ticking off streams*—The load on each chosen interchanger, heater or cooler is fixed by following the ticking-off rule. If a match fails to tick off a stream, the resulting design will almost always have more than the minimum number of units. If splitting is needed to tick off a stream, the engineer has a tradeoff to consider—the cost of stream splitting vs. the cost of more than the minimum number of units.

■ *Utility placement*—The placement of utility heaters and coolers in a network is closely related to the placement of interchangers. A desired location for, say, a startup heater may be achieved by careful choice of the heat interchange options.

Representation

If we are to keep all options open during design, we need a representation of the problem that does not accidentally suppress possible schemes. At first sight, we may think that the representation shown in Fig. 1 is suitable. However, we find that the order of hot streams is changed in Fig. 1b compared with Fig. 1a. Unfortunately, it is impossible to represent the solution in Fig. 1b without either rearranging the order of hot streams, as shown, or rerouting Stream 3 (Fig. 7). It is quite likely that use of this representation could accidentally hide good designs, depending on what order for the streams has been initially chosen.

A better alternative, developed at Leeds University [4] and now used at ICI, is the grid representation (Fig. 8). Here, the hot streams run from left to right

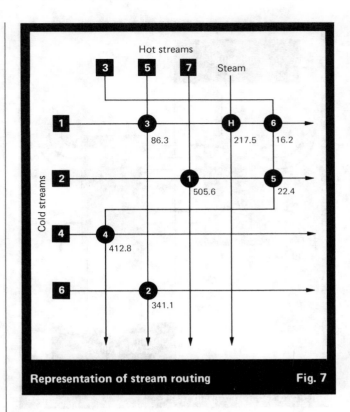

Representation of stream routing   **Fig. 7**

across the top of the diagram, and the cold streams from right to left at the bottom. A process interchanger is shown as two circles, connected by a vertical line. With this representation, any design can be developed using any sequence of hot and cold streams in the empty grid, and any design change is possible without reordering or rerouting streams.

Examples illustrate techniques.

To illustrate the use of the targeting concepts, the grid representation, and the design procedure, we will now develop the minimum-utility minimum-unit solutions for Examples 1 and 2.

Example 1—Application of the minimum utilities algorithm to the data for Example 1 (see Table I) shows that 217.5 units of utility heating, and no utility cooling, are required. As explained in Ref. 8, this implies that the problem has no pinch. However, the design is started from the most constrained part of the system. This is the cold end because cold-utility usage is not allowed. Therefore, all hot-stream target temperatures must be achieved by process interchange. At the hot end, we can use utility heating to achieve some or all of the target temperatures.

Fig. 8a shows the problem data represented in the grid, complete with heat-capacity flowrates (mass flowrate multiplied by specific heat capacity) and stream heat loads. Stream 7 must be cooled to 150°F by matching it with Stream 2 at 100°F. This is an essential match since none of the other cold streams are cold enough, and cold utility is not to be used. Choosing a heat load of 505.6 units ticks off Stream 7, leaving a residual load of 22.4 units and a corresponding temperature of 416°F in Stream 2. Comparing temperatures in Match 1 shows that heat transfer is feasible. Similarly,

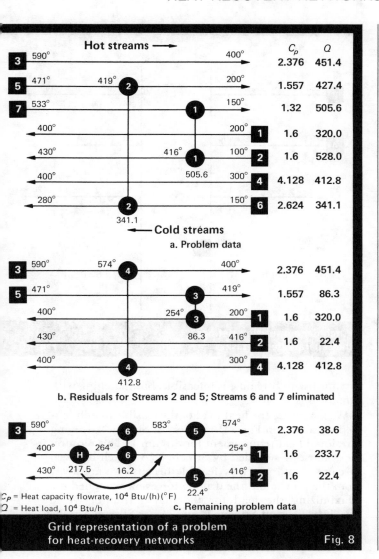

a. Problem data

b. Residuals for Streams 2 and 5; Streams 6 and 7 eliminated

C_P = Heat capacity flowrate, 10^4 Btu/(h)(°F)
Q = Heat load, 10^4 Btu/h

c. Remaining problem data

Grid representation of a problem
for heat-recovery networks   Fig. 8

the match between Streams 5 and 6 is essential (because there is no other stream capable of cooling Stream 5 to 200°F), and ticking off Stream 6 is feasible.

For clarity, the grid is now redrawn with the remaining data. Streams 6 and 7 are completely eliminated, and Streams 2 and 5 are represented only as residuals (Fig. 8b). Streams 3 and 5 can both be cooled by either Stream 1 or Stream 4. Therefore, there are two matching-options: match Stream 3 with 4 and Stream 5 with 1 (as shown); or match Stream 3 with 1 and Stream 5 with 4. For this example, the choice is arbitrary. In practice, the engineer would choose the option on the basis of safety, control, plant layout, etc. The heat loads are fixed by ticking off Streams 4 and 5, and the feasibility of the matches is confirmed by checking the corresponding temperatures.

Again, the remaining data are drawn in the grid of Fig. 8c. The remaining design can be solved by either of two options. One uses two interchangers and a heater, as shown; the other uses one interchanger and two heaters. Both options are consistent with the targets for minimum utility and minimum number of units. Having chosen the first option, the matching sequence can be investigated. In this case, it is shown to be better in terms of overall cost to move the heater to a lower temperature duty (see dotted arrow in Fig. 8c). The temperature feasibility of Match 6 is maintained.

The complete minimum-utility minimum-unit design is drawn in the grid form in Fig. 9. Inspection of this design will show it to be equivalent to that presented in Fig. 1b.

Example 2—Column 2 of Fig. 4 summarizes the results of the minimum-utility analysis for Example 2. The pinch temperature, T_4, corresponds to hot streams at 190°F and cold streams at 180°F. The separate design tasks defined by the pinch decomposition are taken account of in the grid representation in Fig. 10. For

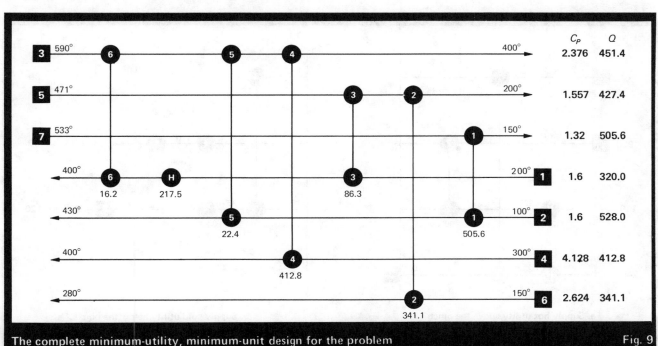

The complete minimum-utility, minimum-unit design for the problem   Fig. 9

each design task, we have noted the target utility usage and the target for the minimum number of units.

Let us consider the development of the separate designs to achieve these targets.

■ **Above the pinch**

Since utility cooling is not allowed above the pinch, Streams 2 and 4 must both be cooled to 190°F by process interchange. Streams 1 and 3 are both available at 180°F. Therefore, there are two design options for the engineer to consider:

1. Match Stream 2 with 1 and Stream 4 with 3. The relative heat-capacity flowrates (from Table II) for Stream 1 ($C_P = 2.0$) and Stream 2 ($C_P = 3.0$) are such that the temperature approach at the hot end of this match must always be closer than the temperature approach at the cold end. As the match is situated immediately above the pinch, the cold end is already at $\Delta T_{min.}$, and the match is infeasible for any load. We derive the general rule that for temperature feasibility in matches immediately above the pinch: $C_{P(cold)} \geqslant C_{P(hot)}$. Since one of the required matches is infeasible, this first option must be rejected.

2. Match Stream 2 with 3 and Stream 4 with 1. Both these matches are feasible based on the inequality of the C_P. Maximizing their loads, to tick off Streams 2 and 4, produces Matches 1 and 2 in Fig. 11a. The residual heating requirements for Streams 1 and 3 are supplied via heaters from the hot utility. The resulting "above the pinch" design complies with targets for utilities and minimum number of units.

■ **Below the pinch**

Stream 1 must be brought to 180°F by process interchange with either Stream 2 or Stream 4. The pinch is now at the hot end of any proposed interchanger, and the C_P inequality must be reversed. We obtain the general rule for feasibility immediately below the pinch: $C_{P(hot)} \geqslant C_{P(cold)}$. Based on this inequality, matching

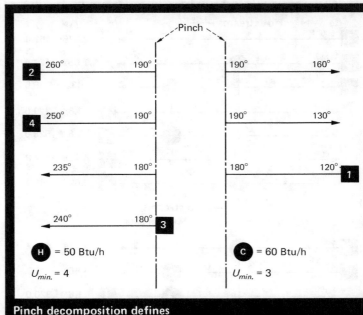

Pinch decomposition defines
the separate design tasks   **Fig. 10**

Stream 1 with Stream 4 is infeasible, but matching with Stream 2 is feasible.

We maximize the heat load of the feasible match (circled number 3 in Fig. 11b) to tick off Stream 2, leaving a residual heat requirement for Stream 1. This requirement must be supplied by Stream 4. A match between Streams 1 and 4 is now feasible because we are away from the pinch, and the driving forces have opened up. Maximizing the load for the match (number 4 in Fig. 11b) ticks off Stream 1, and leaves Stream 4 to be satisfied by utility cooling.

The resulting "below the pinch" design (Fig. 11b)

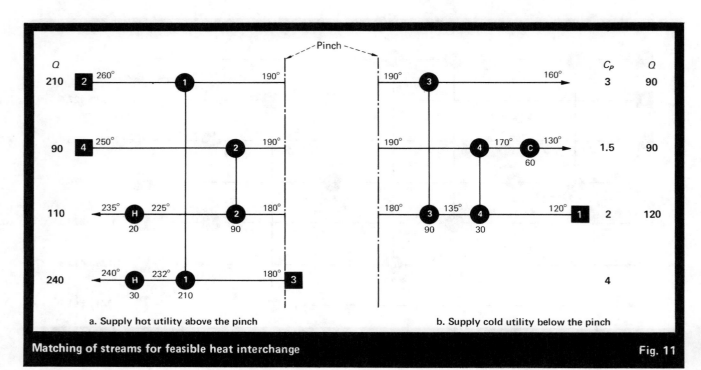

a. Supply hot utility above the pinch   b. Supply cold utility below the pinch

Matching of streams for feasible heat interchange   **Fig. 11**

features minimum utility usage and minimum number of units. Combining this design with that developed for "above the pinch" (Fig. 11a) produces the complete minimum utility network, shown in Fig. 12.

Clearly, these examples are fairly simple. In large industrial studies, the task of generating a network can be complex. However, experience has shown that even large petrochemical and refinery processes are not too difficult for hand procedures. There are two reasons for this:

First, as long as alternative designs all feature the same degree of heat recovery and the same number of units, cost differences are within the bounds of capital cost uncertainty in preliminary design, and a global optimization is pointless. In other words, the cost-optimality curve is flat, and the decision as to which solution is optimal is best left to the engineer who bases the final decision on the overall plant-layout and operability requirements.

Second, to complement the C_P inequalities introduced previously, we have several other design rules. For more-complex problems, a step-by-step procedure is available that uses all these rules. A full description of this material is given in Ref. 8.

Capital energy tradeoffs

Examples 1 and 2 have been chosen to illustrate two basic types of heat-recovery network problems: pinched and unpinched (or threshold) problems. Example 1 is unpinched and Example 2 is pinched. A detailed description of both types is given in Ref. 8.

Here, in a brief discussion, we will concentrate on their rather different capital/energy tradeoffs. So far, we have designed networks for minimum utility usage and minimum number of units, accepting a given value for $\Delta T_{min.}$, and respecting the pinch decomposition. What are the implications as to the overall network cost for each type if we now choose different $\Delta T_{min.}$s, or if we choose to violate the pinch by transferring heat across it?

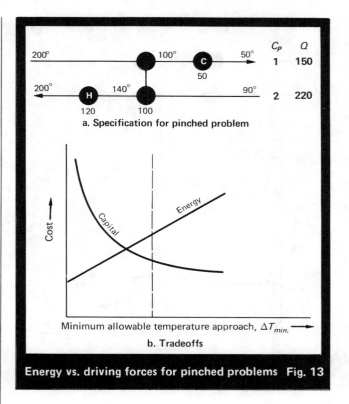

a. Specification for pinched problem

b. Tradeoffs

Energy vs. driving forces for pinched problems   Fig. 13

Energy vs. driving forces

Let us first consider pinched problems because these are most common. Resolving a pinched problem always requires both utility heating and utility cooling, whatever the specified value for $\Delta T_{min.}$. A simple two-stream pinched problem is illustrated in Fig. 13a. As shown, the interchanger is designed for $\Delta T_{min.} = 10°F$. Decreasing $\Delta T_{min.}$, say to 5°F, allows us to increase the interchanger load and reduce utility costs. The energy saving must, however, be traded off against the increase in capital cost due to lower driving forces and, therefore, increased surface area. This utility/surface area tradeoff applies equally well to networks of hot and cold streams. Fig. 13b shows how increasing $\Delta T_{min.}$ decreases capital costs but increases energy costs.

Now, let us consider the case of unpinched problems (Fig. 14). These are characterized by a threshold value, $\Delta T_{th.}$, for $\Delta T_{min.}$: At or below $\Delta T_{th.}$, the problems require only one utility (hot or cold). While above $\Delta T_{th.}$, both utilities are needed. The value of $\Delta T_{th.}$ is completely problem dependent. For the problem shown in Fig. 14a, specifying a value for $\Delta T_{min.} > 30°F$ incurs the need for utility heating and more utility cooling. However, a value for $\Delta T_{min.} < 30°F$ would not result in any saving in utility usage because we must always remove at least 50 units for overall enthalpy balance. Therefore, above $\Delta T_{th.}$, we have the normal, pinched-problem utility/surface area tradeoff; while below $\Delta T_{th.}$, we have no tradeoff (Fig. 14b).

Energy vs. number of heat-transfer units

For Example 2, we earlier described how the minimum-number-of-units target applies separately above and below the pinch (Fig. 15a). Clearly, we could target

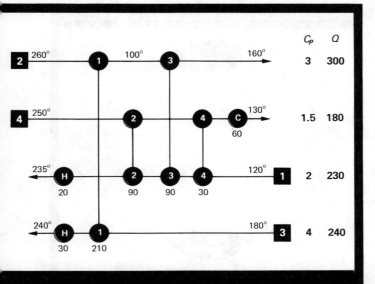

The complete minimum-utility network   Fig. 12

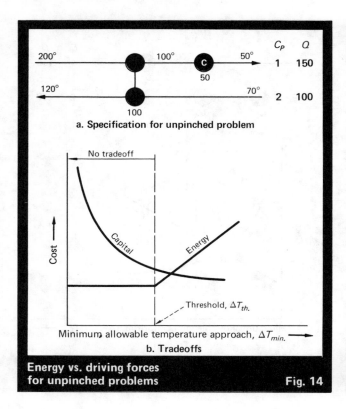

a. Specification for unpinched problem

b. Tradeoffs

Energy vs. driving forces for unpinched problems   Fig. 14

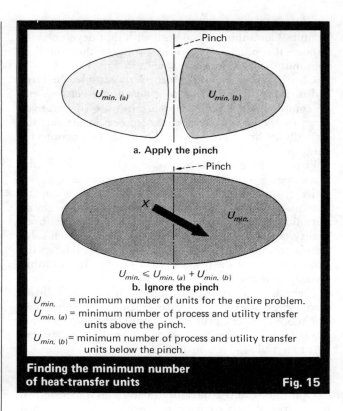

a. Apply the pinch

b. Ignore the pinch

$U_{min.} \leqslant U_{min.\ (a)} + U_{min.\ (b)}$

$U_{min.}$ = minimum number of units for the entire problem.
$U_{min.\ (a)}$ = minimum number of process and utility transfer units above the pinch.
$U_{min.\ (b)}$= minimum number of process and utility transfer units below the pinch.

Finding the minimum number of heat-transfer units   Fig. 15

for $U_{min.}$ for the entire problem, ignoring the pinch decomposition (Fig. 15b). The difference between the two targets represents the possibility of eliminating transfer units by heat transfer, X, across the pinch. The magnitude of this energy sacrifice will depend on the problem.

A procedure that evaluates X for any desired unit elimination is described elsewhere [8]. Applying this procedure to the minimum-utility solution for Example 2 (Fig. 12) produces the network shown in Fig. 16. This network achieves the target as described in Fig. 15b. It has two transfer-units fewer than the design in Fig. 12, at an energy sacrifice of 30 units. (The energy sacrifice is far less than the original load handled by the elimi-

nated units.) Clearly, Fig. 16 represents an attractive alternative to the minimum-utility solution.

For unpinched problems designed at or below $\Delta T_{th.}$, there is no tradeoff. With no pinch, there is no problem decomposition, and the $U_{min.}$ target always applies to the entire problem.

A new approach

The approach adopted at ICI to the design of heat-recovery networks is summarized in Fig. 17. Data analysis, prior to design, allows rapid identification of the extent of possible process improvement by establishing practical, cost-related performance targets. These capi-

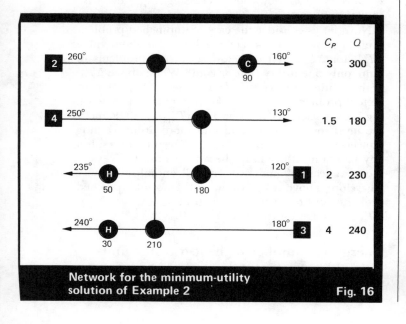

Network for the minimum-utility solution of Example 2   Fig. 16

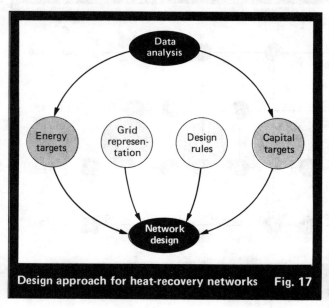

Design approach for heat-recovery networks   Fig. 17

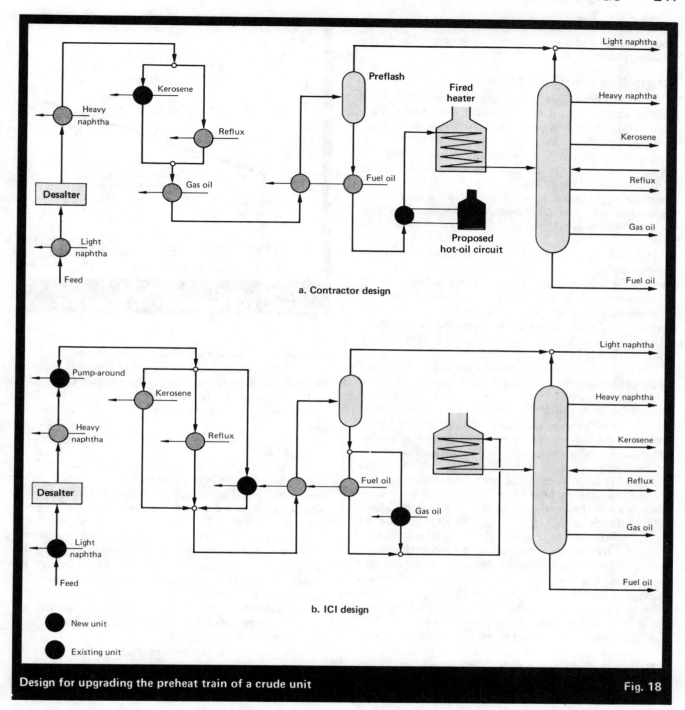

a. Contractor design

b. ICI design

Design for upgrading the preheat train of a crude unit   Fig. 18

tal and energy targets are almost always achievable if consideration is taken of practical constraints such as operating flexibility and startup. Any failure to design a near-optimal network is immediately recognized and the engineer is challenged to try again.

For the design task, the engineer is given the targets, grid representation, and design rules. The representation helps the engineer see the options, while the design rules are used to check the preferred options for operational and economic feasibility.

We can draw an analogy between network design and a game of chess. The grid represents the chess board and pieces, while the design rules represent the rules of the game. In any particular situation, the engineer sees the options available and decides on the best "move" to make, based on judgment. The role of the targets in this analogy is intriguing: It is easy in chess to overlook chances of reaching, say, mate in two. Instead, the game is played on for many moves, following a perfectly feasible and seemingly optimal path. On the other hand, network performance targets will always tell the engineer that a mate in two is possible.

Project applications

Many studies based on the approach discussed have been undertaken in the various divisions of ICI [1]. Two recent studies will now be described. One relates to the design of a new plant; the other, to the upgrading of an existing plant.

Upgrading a preheat train

A conventional crude unit was to be upgraded. This unit produced the primary fractions from raw crude: light naphtha, heavy naphtha, kerosene, gas oil, and fuel oil (Fig. 18a). To increase the capacity of the unit and enable it to handle a different feedstock, a feasibility study involving the redesign of the preheat train was commissioned. This study was subject to two design constraints: (1) maximize the use of existing heat-transfer equipment, and (2) only propose modifications that can be implemented during the routine annual shutdown.

The initial work was done by a contractor whose recommendation was to modify the preheat train by adding a hot-oil circuit (supplied by a separate fired heater) at the existing furnace inlet (Fig. 18a). This was thought necessary because the capacity of the existing furnace could not handle the increased throughput. The scheme was flowsheeted, and a preliminary costing study performed.

To check the contractor's recommendation, the minimum-utility targeting procedure was applied. Surprisingly, this indicated that it was possible to design a heat-recovery network for the increased throughput, with no need for supplementary heating, and not even requiring the full use of the existing furnace capacity.

Detailed analyses showed why the existing design was so far from optimal—the gas-oil interchanger was transferring large amounts of heat across the pinch. To "repair" the design, heat from below the pinch was recovered from the overheads of the main column via a column pump-around (Fig. 18b). This allowed the load on the gas-oil unit to be reduced. The load was further reduced and eventually eliminated by recovering more heat from the fuel oil in a third branch of the existing two-way split. Thus, gas oil (the second hottest product) became available to replace, above the pinch, the supplementary heating located before the furnace. The advantages of this new scheme turned out to be:

■ Elimination of the supplementary heating, yielding a fuel saving of approximately $1.2 million/yr.

■ Removal of the separate furnace and hot-oil circuit, providing a saving of capital.

■ Three-way splitting, initially thought to be an operability drawback. Later, more-detailed analysis revealed that three-way splitting led to increased operating flexibility for different feedstocks.

The importance of the stimulus given by target setting is illustrated by Fig. 19. The minimum fuel usage for the upgraded throughput (related to the existing furnace capacity) is plotted for different assumptions of $\Delta T_{min.}$ Designs that lie on the line use minimum fuel, while designs above the line use excess fuel and cooling capacity. The contractor's design required roughly twice the minimum fuel usage. This indicates grossly inelegant features.

Exact minimum fuel usage was found to involve an almost complete rebuilding of the existing network, and this was unacceptable. However, aiming for a fuel usage

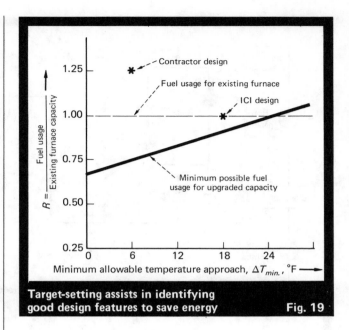

Target-setting assists in identifying good design features to save energy   **Fig. 19**

corresponding to the existing furnace capacity (horizontal line in Fig. 19) led to designs that reused much of the existing installation. The quickest way to identify such designs was by elimination of the "worst offender," i.e., the biggest match transferring heat across the pinch.

New design

The study for an inorganic-bulk-chemicals process aimed to improve the heat-recovery network for a newly developed flowsheet. A large part of the process was designed as a package unit by a contractor; while the remainder was ICI-based. The process was quite complicated, with several areas of batch operation, many unit operations, and many streams. Prior to the analysis, the problem was reduced in complexity by eliminating some of the streams from consideration for either of two reasons:

■ Availability—Being partially batch, the process included streams of intermittent availability, which were dealt with individually.

■ Physical state—Some streams were unsuitable for conventional countercurrent heat exchange because of their composition, e.g., slurries.

Furthermore, the remaining streams were divided into two distinct groups corresponding to the package design (Area A) and the ICI design (Area B). One of the issues to be examined was the benefits to be obtained, if any, from inter-area integration. The problem analysis then consisted of three comparisons of actual to predicted utility performance: (1) whole process, (2) Area A, and (3) Area B. The results for steam usage for this process:

	Original usage, %	Target usage, %	Scope, %
Whole process	100	88	12
Area A	93	92	1
Area B	7	0	7

We found that the sum of the scopes for Areas A and B was 8%; while the scope for the whole process was 12%. The difference of 4% was therefore the scope possible if we were prepared to allow inter-area integration. this 4% scope was not considered sufficient to prejudice independent operation of the two areas. Therefore, with a scope of only 1% in Area A, the analysis highlighted Area B as worthy of further design effort.

Based on the target for steam usage, knowledge of allowable driving forces and observance of important design constraints, several optimum or near-optimum networks were quickly identified. All the schemes were better than the original design in terms of operating and capital costs. The preferred alternative in respect to operability, startup, etc., was accepted. The scheme represented a saving of $300,000/yr in steam costs, together with savings in cooling-water usage and capital.

In the past, the overwhelming complexity of the total process, and the corresponding need to compartmentalize design responsibilities, had prevented the design team from finding what with hindsight was quite a simple solution.

Other project applications

Table III lists some of the project applications at ICI. The table indicates whether a study was for a new flowsheet, or a modification to an existing process, or both, and gives the energy savings and capital implications of the new schemes. In all cases, these were accepted as being operable, controllable and safe by the project teams. In most of the studies, the capital outlay for the energy-efficient schemes was either similar to, or less than that for, the original scheme.

Prior to the availability of these novel design procedures, process improvements through better heat integration had relied on experience from designing successive processes over periods of many years. The stimulus for these improvements often came from the so-called "learning curves." For example, Fig. 20 shows the learning curve for a specialty-chemicals process. Not knowing the lower limit on possible performance, the engineer was satisfied to approach it with only marginal improvements in successive designs.

Now, with the targeting procedures, the engineer can confidently predict the lower limit and is no longer satisfied with step-by-step improvements. The resulting breakaway from the learning curve of the specialty chemical is illustrated in Fig. 20. This break is equally true for the other studies cited in Table III.

Significance

The significance of the targeting and design procedures described here goes beyond providing a new technique for putting together networks of interchangers, heaters and coolers. We suddenly learn that in the past we have been seriously lacking in our understanding of optimal network structures.

First, there was the failure to recognize that there is a heat-recovery pinch.

Second, the project applications have highlighted three important and common misconceptions in process design. These are the energy-saving surprises (to be discussed shortly).

Identification of the heat-recovery pinch [7] has been a major step in our understanding of heat-recovery network design. Any heat transfer across the pinch necessarily incurs an extra requirement for heating and cooling. Thus, when designing new flowsheets, we divide the problem at the pinch, and design each area separately. For plant modifications, we identify the interchangers that transfer heat across the pinch and look for the simplest way to eliminate them. Without this insight, we have been previously "shooting in the dark" when placing process heat interchangers.

The pinch has also proved significant in other areas of process design. For example, it is used for integrating heat engines and distillation columns into process networks [9,10].

The energy saving surprises

■ *Design time*—We have found that once engineers are familiar with the new procedures, performance targets give them confidence in early basic-design decisions. Therefore, the engineers concentrate their effort where needed. As a result, much of the usual trial-and-error in design is eliminated and better designs are obtained in a shorter time span.

■ *Control*—None of the studies cited in Table III has uncovered unusually difficult control problems. Broadly, the reason is that the schemes do not propose integration in place of no integration, but simply propose more appropriate integration. This often follows from deliberate choices of design options to achieve good control and desirable process responses.

■ *Capital cost*—The accepted belief that one has to spend capital in order to save energy is frequently not true for networks. How can this be? Let us again consider the two solutions for Example 1 (Fig. 1):

The solution of Fig. 1b uses less of the hot utility than the solution of Fig. 1a, and no cold utility. Comparing total heat transferred within each network shows a considerable reduction for the solution of Fig. 1b, as shown in Fig. 21. A smaller heat load requires less heat-transfer surface, and this saving more than compensates for the

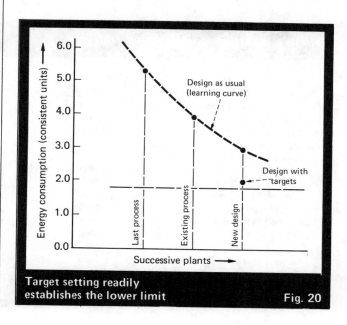

Target setting readily establishes the lower limit

Fig. 20

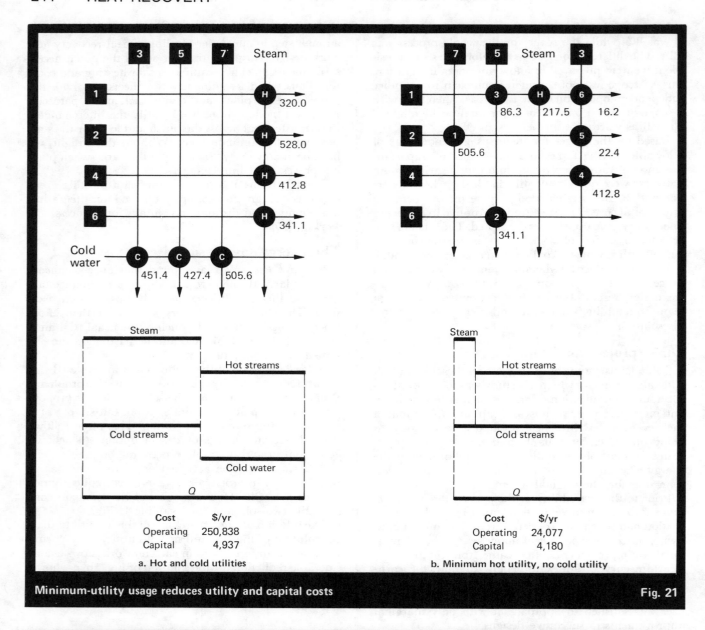

Cost $/yr
Operating   250,838
Capital   4,937

a. Hot and cold utilities

Cost $/yr
Operating   24,077
Capital   4,180

b. Minimum hot utility, no cold utility

Minimum-utility usage reduces utility and capital costs   Fig. 21

Results of applying network analysis to projects   Table III

Process	Facility*	Energy savings available $/yr	Capital cost expenditure or saving $
Organic bulk chemical	New	800,000	Same
Specialty chemical	New	1,600,000	Saving
Crude unit	Mod	1,200,000	Saving
Inorganic bulk chemical	New	320,000	Saving
Specialty chemical	Mod	200,000	160,000
	New	200,000	Saving
General bulk chemical	New	2,600,000	Unclear
Inorganic bulk chemical	New	200,000 to 360,000	Unclear
Future plant	New	30 to 40%	30% saving
Specialty chemical	New	100,000	150,000
Unspecified	Mod	300,000	1,000,000
	New	300,000	Saving
General chemical	New	360,000	Unclear
Petrochemical	Mod	Phase I 1,200,000	600,000
		Phase II 1,200,000	1,200,000

*New means new plant; Mod means plant modification.

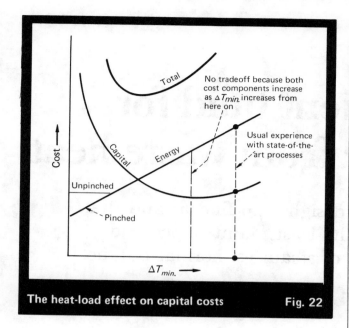

No tradeoff because both cost components increase as ΔT_{min} increases from here on

Usual experience with state-of-the-art processes

The heat-load effect on capital costs   **Fig. 22**

reduction in driving forces associated with tighter driving forces.

In other words, we observe a heat-load effect and a driving-force effect on surface area. Combined, both effects provide a net reduction in heat-transfer area. Therefore, we save capital by saving energy. The cost comparison in Fig. 21 is based on consistent costing data from Ref. 5. The network for minimum utility usage features a 90% reduction in utility costs and a 15% reduction in capital costs.

The rather surprising influence of the heat-load effect on capital cost has forced us to redraw the graphs first presented in Fig. 13b and 14b. Beyond a certain value for ΔT_{min} (as shown in Fig. 22), the heat-load effect begins to dominate, and capital costs rise even though driving forces are increasing. To the right of this minimum in capital, we can never have a tradeoff. Unfortunately, many of today's state-of-the-art processes lie in this region.

The energy-saving surprises suggest that the stimulus to save energy, following the escalation of fuel costs, may be somewhat double-edged. This stimulus tends to place emphasis on optimizing unit operations with the well known, and well understood, capital/energy tradeoffs.

We should become less preoccupied with unit-operation design and strive for better process integration—thereby conserving energy and capital. Philosophically, the ultimate stimulus to the engineer should be that of elegance in design, a stimulus independent of fuel costs and fashions. Simplicity in structures and the simultaneous best use of both energy and capital would have been desirable in the past and always will be. Without insights such as those presented here, there was no way of knowing just how inelegant some of our present-day processes really are.

Acknowledgements

The authors wish to express their gratitude to ICI Ltd. for permission to publish this article. They also express their thanks to all people in process design and project management, inside and outside ICI, who have contributed to the successful completion of the case studies.

References

1. Boland, D., and Linnhoff, B., *Chem. Eng. (London)*, Apr. 1979, p. 222.
2. Hohmann, E. C., "Optimum Networks for Heat Exchange," Ph.D. Thesis, University of Southern California, 1971.
3. Linnhoff, B., and Flower, J. R., *AIChE J.*, Vol. 24, p. 633 (1978).
4. Linnhoff, B., "Thermodynamic Analysis in the Design of Process Networks," Ph.D. Thesis, University of Leeds, U.K., 1979.
5. Masso, A. H., and Rudd, D. F., *AIChE J.*, Vol. 15, p. 10 (1969).
6. Nishida, N., Liu, Y. A., and Lapidus, L., *AIChE J.*, Vol. 23, p. 77 (1977).
7. Linnhoff, B., Mason, D. R., and Wardle, I., *Computers and Chemical Engineering*, Vol. 3, p. 295 (1979).
8. Linnhoff, B., and Hindmarsh, E. H., Submitted for publication in *Chem. Eng. Sci.*
9. Townsend, D. W., and Linnhoff, B., Designing Total Energy Systems by Systematic Methods, paper presented at Total Energy Design in Process Plants conference, I. Chem. Eng./Soc. for the Chemical Industry, London, Apr. 29, 1981.
10. Dunford, H. A., and Linnhoff, B., Energy Savings by Appropriate Integration of Distillation Columns into Overall Processes, paper no. 10, Cost Savings in Distillation symposium, Leeds, U.K., July 1981.
11. Mason, D. R., and Linnhoff, B., Predicting Minimum Utility Requirements in Constrained Heat Recovery Problems, Submitted to *AIChE J.*

The authors

B. Linnhoff   J. A. Turner

Bodo Linnhoff is responsible for process design methods at Imperial Chemical Industries Ltd., Corporate Laboratory, P.O. Box 11, The Heath, Runcorn, Cheshire WA7 4QE, U.K., As of January 1982, he will take up a chair in chemical engineering at the University of Manchester Institute for Science and Technology. He has worked for Holderbank AG and ICI Ltd, and has taught at the Swiss Federal Institute of Technology (ETH), in Zurich, and at Imperial College (London). He holds an M.S. in mechanical engineering from the ETH and a Ph.D. in chemical engineering from Leeds University (U.K.).

John A. Turner works on network design aids at ICI's Corporate Laboratory, which he joined in 1978. His work involves the development of procedures for the design of heat-exchanger networks, and the second law analysis for processes. He holds a B.S. in chemical engineering from Imperial College (London).

Selecting an efficient fluid for recovering power from waste heat

A basic problem in power-recovery design is to find a fluid that generates large net horsepower for the least maintenance and investment. Here are pros and cons of steam vs. normal butane.

Opal R. James and *Sea K. Fan*, B & C Associates, Inc.

☐ To begin, let's first review a power-recovery system using commercial butane (95% butane) as the power fluid (Fig. 1). The process represents a Rankine cycle.

In the system, the working fluid, condensed at about 80 psia and 130°F, is taken from the surge tank by means of a multistage centrifugal pump that boosts its pressure to about 1,040 psia. It flows through a shell-and-tube heat exchanger, in which it is heated to 288°F, then to a vaporizer, where it is further heated to 500°F. Afterwards, the fluid expands in a radial-inflow turbo-expander, producing work for compression or for generation of electrical power. It exhausts from the turbo-expander at about 95 psia and 362°F, is cooled in the preheater to 158°F, and is condensed in an air cooler. If cooling water is available, the air cooler may be replaced by a shell-and-tube or an evaporative cooler.

If water were the power fluid, the equipment would be the same except for the turboexpander, which would be replaced by a steam turbine, and the preheater, which would be eliminated. The pressure of the steam would vary from 150 to 600 psig.

Qualities of an ideal power fluid

The ideal power fluid is readily available at a reasonable price, and is not flammable, toxic or corrosive. Its vapor pressure at the condensing temperature is above atmospheric pressure, to prevent leakage of air into the system. Its saturated vapor line on the temperature-vs.-entropy chart should be vertical, to ensure that it will not expand into the two-phase region. Its specific vapor volume is small, minimizing the size of pipe and equipment. Lastly, the pump work it requires is much

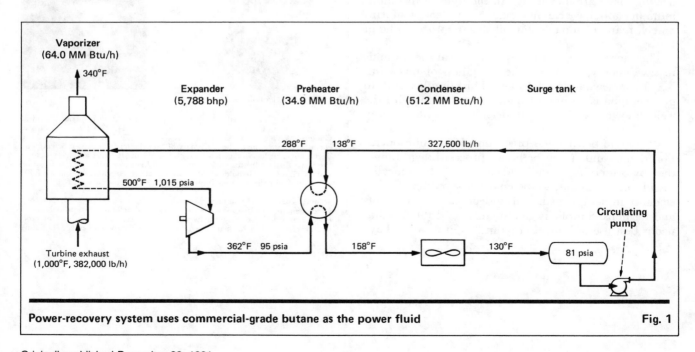

Power-recovery system uses commercial-grade butane as the power fluid

Fig. 1

Originally published December 28, 1981

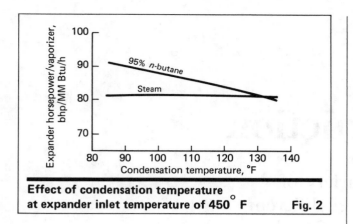

Effect of condensation temperature at expander inlet temperature of 450°F   **Fig. 2**

less than the power generated by the expander (turbo-expander or steam turbine).

Pros of a water system

Water, which is reasonably priced and is neither flammable nor toxic, fulfills most of the foregoing requirements. Furthermore, whereas the pump work in a steam system takes only about 1% of the horsepower generated by the turbine, it amounts to 10% of the shaft work produced by the turboexpander in a butane system. These considerations—together with the fact that, at condensation temperatures above 130°F, the horsepower output per Btu of vaporizer duty is higher in the steam system (Fig. 2)—would lead one to conclude steam to be the better power fluid choice.

Cons of a water system

However, steam has qualities that make it undesirable, even impractical, for many applications.

The steam cycle is not a closed one. It must be blown down periodically to remove precipitants.

Water is corrosive because it contains dissolved oxygen, which is expensive to remove. Moreover, its removal does not guarantee its absence, because it may leak in afterwards. Dissolved salts also accelerate corrosion and are costly to remove.

At a condensation temperature of 130°F, the vapor pressure of water is 2.22 psia (whereas that of butane is 80 psia). Because no system can be completely air-proof, air leakage into the surge tank under such a vacuum is unavoidable. When this occurs, the efficiency of the steam cycle drops, and corrosion speeds up.

Water also poses a freezing problem in winter. Although this can be designed for, it adds initial and operating costs that are not a part of a butane system.

Some states require that an operator attend a steam system at all times, which adds costs. A mandatory annual inspection lowers operation time. No similar requirements are imposed for butane systems.

In some cases, the heat sources may be low-temperature (between 300 and 350°F, for example). Under such conditions, the steam pressure at the turbine inlet must be lowered. This increases the size of the inlet nozzles and lessens the actual horsepower recovered. Moreover, so as to maintain some superheat, the pressure has to be dropped below 30 psia. This makes the whole cycle economically unfeasible. With butane, the pressure can be dropped to 515 psia, and a heat rate in the vaporizer of 20,618 Btu/(bhp)/(h) can still be achieved.

In cold weather, the condensation temperature can be dropped below 130°F, and a 40°F difference from ambient temperature can still be maintained. Under such circumstances, air coolers can still be used to achieve condensation. The effect on cycle performance of lowering the condensation temperature can be seen in Fig. 2, which shows that lowering the condensation temperature of butane helps boost the horsepower output per quantity of heat absorbed but that this does not happen with steam. It does not because steam's expansion ratio increases only slightly through the turbine as condensation temperature is lowered.

Cons of butane as a power fluid

The chief disadvantage of butane as a power fluid is that it is inflammable. However, careful design can contain this hazard.

Another drawback is that, compared to steam, more of it must be circulated to generate the same amount of horsepower. For example, 373,507 lb/h of butane (vs. 66,235 lb/h of steam) must be circulated to generate a gross output of 6,600 bhp. However, because of steam's much larger specific volume (50 ft³/lb vs. 1.5 ft³/lb for butane), the difference in size of piping and equipment is insignificant. This can be seen from a consideration of the size of piping at the expander outlet, where pipe diameter is the largest. Here, the steam system requires 20-in.-dia. Schedule 10 piping, and the butane system 18-in.-dia. Schedule 10 piping.

References

1. Baudat, N. P., and James, O. R., "Application of a Power Recovery System to Gas Turbine Exhaust Gases," ASME Gas Turbine Conference, Mar. 1978.
2. Baudat, N. P., and Darrow, P. A., Power Recovery in a Closed Cycle System, *Chem. Eng. Prog.*, Feb. 1980.
3. Trojan, P. K., and Flinn, R. A., "Engineering Materials and Their Applications," Houghton Mifflin Co., Boston, 1975.

The authors

Opal R. James is Vice-President of B & C Associates, Inc. (5400 Mitchelldale, Suite A-3, Houston, TX 77092; telephone 713-680-3251). Previously, she was a process engineer and manager of data processing with McDermott-Hudson Engineering. A graduate of Texas Tech University with B.S. degrees in chemical engineering and in industrial engineering, she is a registered engineer in the state of Texas.

Sea K. Fan is a process engineer with B & C Associates, Inc. He holds a B.S. degree in chemical engineering from Lowell University (Massachusetts) and an M.S. degree in chemical engineering from Pennsylvania State University. A member of AIChE and Soc. of Omega Chi Epsilon, he is currently pursuing an M.B.A. degree at the University of Houston.

Heat-pipe construction

These devices are available in a variety of
materials and configurations to meet different
uses. Here is a summary of what is offered.

K.S.N. Raju and *Virender Kumar Rattan*, Panjab University

☐ In recent years, heat pipes have found increasing use in industry. Unlike many conventional heat exchangers, these devices require no power supply and can operate on small temperature differentials between heat source and heat sink. Also, heat pipes transfer more heat per unit area than do solid-metal conductors. Because these devices have no moving parts, they are inherently reliable and trouble-free.

Basic operation

A heat pipe consists of a sealed tube that contains a wick and a working fluid (see Fig. 1). In the evaporator section of the pipe, heat is transferred from the source through the container wall to the wick, where the working fluid evaporates. The vapor moves along the pipe to the condenser section. In between, the pipe is adiabatic. At the condenser section, the vapor condenses and heat is transferred from the container. Condensate is returned to the evaporator section by capillary action—the wick contains capillary pores.

Container

The main function of the container is to isolate the working fluid from the outside environment. The container must be thin enough to transfer heat effectively, yet strong enough to withstand internal pressure caused by the heated vapor.

Desirable properties of the container material include a high strength-to-weight ratio, ease of fabrication, high thermal conductivity, low cost and high wettability.

The container must be compatible with the working fluid. Table I (taken from S.W. Chi's "Heat Pipe Theory and Practice," Hemisphere Pub. Co., Wash., D.C.) lists compatibilities for some common materials and working fluids.

Copper, aluminum and stainless steel are the most widely used materials for the container. Since copper tubes are available in a variety of sizes, these find the greatest use. Aluminum is used in aerospace applications due to its light weight; stainless steel is employed in high-temperature service when the working fluid is a liquid metal.

Originally published December 17, 1979

248

Working fluid

Working fluids are available in a variety of temperature ranges—for cryogenic to liquid-metal temperature applications.

Several properties are important in a working fluid. A high surface tension is needed to enable the heat pipe to operate against gravity, by generating a strong capillary force. A low viscosity ensures a minimum resistance to flow, while a high latent heat of vaporization maintains a low pressure drop. Other properties are: good wetting characteristics to aid heat transfer; high thermal conductivity to minimize the radial temperature gradient and to reduce the possibility of nucleate boiling at the surface; moderate vapor pressure, since low vapor pressure results in low vapor density and high pressure drop in vapor flow; chemical stability and compatibility with wick and container; high density; and low cost.

No fluid has been found truly suitable for temperatures from 250 to 350°C. Table II lists fluids recommended for temperatures from 20 to 1,800°C.

Wick

In selecting a wick, pore size is critical, since the maximum capillary head is obtained by using a wick with the smallest pores. When the pipe has to work against gravity, smaller pores are needed. The smaller the pore, the less permeable the wick is to gases and liquids. Hence, the effects of permeability and capillary action must be balanced in determining pore size.

The wick should have a large specific surface and a low density. Thickness of the wick should be optimized, since a thicker wick will increase the heat-transfer capability of the heat pipe.

Wicks can be made from several materials. Here, we will consider homogeneous wicks, which are made of one material; but wicks can also be composites of two or more materials.

Often, wicks are meshes or twills of materials such as stainless steel, nickel, copper or aluminum. Felts and metal foams are also used, as well as fibrous materials and sintered powders.

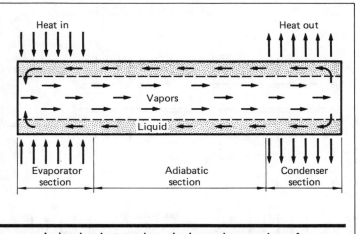

In its simplest version, the heat pipe consists of a wick and working fluid sealed in a container   Fig. 1

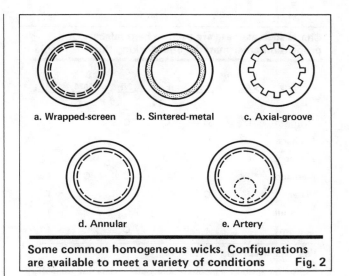

Some common homogeneous wicks. Configurations are available to meet a variety of conditions   Fig. 2

Fig. 2 shows five of the most popular homogeneous-wick constructions:

Wrapped-screen—Used in cryogenic and moderate-temperature heat pipes, such wicks consist of two or more fine-mesh screens. Altering the tightness of the wrap changes the resistance to flow.

Sintered-metal—Used if a small temperature drop across the wick is needed, since heat is transmitted by the sintered-metal structure. Due to the varying amounts of deformation that occur in sintering, it is not possible to predict the exact geometric configurations of such wicks.

Axial-groove—Used for zero-gravity, as well as cryogenic to liquid-metal, heat pipes.

Annular—Used when a small resistance to liquid flow is needed. However, resistance to heat flow is large if the thermal conductivity of the liquid phase of the working fluid is low. Capillary action results from a thin annulus that holds the working fluid.

Artery—Used generally in spacecraft, these high-performance wicks transport liquid with a minimum pressure drop. For best operation, the artery should be shut off from the vapor space.

Control methods

By making changes in the basic structure of the heat pipe, units can be constructed that will provide temperature control. Some of the basic types are discussed here and are shown in Fig. 3, taken from Chi.

Excess-liquid heat pipe—This device has a reservoir at its condenser end. When heat is transferred to this unit, excess liquid is swept by the vapor toward the liquid reservoir. When the temperature of the pipe's evaporator section is above that of the condenser section, the device functions as a typical heat pipe. Reversing the temperature gradient will sweep the liquid to the end of the pipe opposite from the reservoir, thereby inactivating that section for rejecting heat. Under such conditions, the heat pipe acts an insulator and is sometimes called a thermal diode.

Vapor-flow modulated heat pipe—This unit throttles vapor flow in the evaporator, using a throttling valve. A pressure difference—and a corresponding temperature

difference—is created. A bellows filled with a control fluid reduces the valve opening as the temperature increases. Thus, upping the heat load is offset by a lowered vapor temperature in the condenser section. Such a device, also called a variable-conductance heat pipe, keeps part of the heat pipe at a relatively constant temperature when heat-source or heat-sink temperatures vary.

Gas-loaded heat pipe—This device has a gas reservoir next to the condenser. A noncondensing gas is used; this is swept along, collecting in the condenser and partly blocking part of the condenser. In the configuration shown in Fig. 3, the noncondensing temperature is maintained close to that of the ambient.

Failure of heat pipes

Heat pipes can fail for many reasons: Corrosion can weaken the container and, if internal, can clog or deteriorate the wick; sometimes, the wick can dry out if capillary action becomes insufficient to transport enough liquid to a part of the heat pipe; or the working fluid can deteriorate due to high temperature.

Applications

Heat pipes are used over a wide range of working temperatures, including values to 3,000°C. Units are employed as single pipes or pipe bundles. A bundle of tubes provides low thermal resistance. Tubes are made in lengths to 100 ft.

Heat pipes are used in the chemical process industries on a limited basis. Here are a few examples of such uses:

High-temperature heat pipes are used for heat recovery from devices such as incinerators, boilers and process heaters. Sun Co. (Radnor, Pa.) uses heat pipes to recover heat from stack gases at about 700°F to preheat ambient air to around 100°F for a reformer feed in an oil refinery. The heat-pipe system consists of four rows of 22 stainless-steel pipes, 1 in. dia. and 60 in. long with carbon-steel spiral fins. The hot end of the pipe is 22 in. long; the cold end is 18 in.

In another application, Dow Chemical Co. (Midland, Mich.) uses a heat pipe to recover waste heat from a gas-fired furnace. The heat recovered from stack gases is

Charts such as these are used to help select proper container material and working fluid   Table I

Fluids	Al	Cu	Fe	Ni	SS[a] 304	Ti
Nitrogen	C[b]	C	C	C	C	
Methane	C	C			C	
Ammonia	C		C	C	C	
Methanol	I	C	C	C	C	
Water	I	C		C	C[c]	C
Potassium				C		I
Sodium				C	C	I

[a]SS = stainless steel.

[b]C = compatible; I = incompatible; blank = data not available.

[c]Possible hydrogen generation.

Recommended working fluids and their temperatures of operation   Table II

Temperature range, °C	Fluid
20-40	Ammonia
50-200	Water
250-650	Mercury
400-800	Potassium
500-1,000	Sodium
1,000-1,800	Lithium

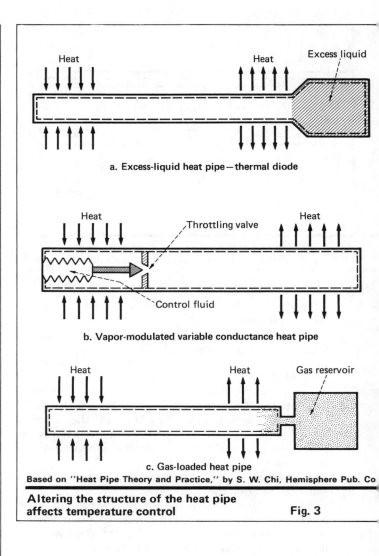

a. Excess-liquid heat pipe—thermal diode

b. Vapor-modulated variable conductance heat pipe

c. Gas-loaded heat pipe

Based on "Heat Pipe Theory and Practice," by S. W. Chi, Hemisphere Pub. Co

Altering the structure of the heat pipe affects temperature control   Fig. 3

used to heat the company's Dowtherm heat-transfer fluid, which serves as a heating medium in a distillation unit. Employing heat pipes requires a smaller heating surface than would be needed for a typical shell-and-tube exchanger.

Heat pipes have a potential application in controlling temperatures of exothermic or endothermic reactions in chemical reactors. Local hot spots in large reactors can be minimized by the use of heat pipes, mainly due to their compactness.

The biggest potential for heat pipes is as air-to-air exchangers, because of their compactness, moderate pressure drop, and good performance when operated countercurrently.

Low-temperature heat pipes are used for applications such as cooling electronic equipment, semiconductor devices, rotors and starters of motors.

Heat pipes are also finding use in controlling the temperature in plastic injection-molding and alloy die-casting processes, where removal of heat during solidification is important.

So far, most heat pipes have been used either in high-technology industries—e.g., as heat sinks for electronic and power-conversion devices—or in heating, air conditioning and ventilating systems.

The authors

K.S.N. Raju is professor of chemical engineering at Panjab University, Chandigarh, India 160014. His areas of interest include phase equilibrium thermodynamics, heat pipes and pollution control. He has taught at Panjab since 1960, and has been responsible for developing teaching programs and research activities. He also has done research at Shri Ram Institute for Industrial Research. Prof. Raju holds a Ph.D. degree in chemical engineering from Panjab and is a member of the Indian Institute of Chemical Engineers. He has published over 30 research papers and one book.

Virender Kumar Rattan is senior research fellow in chemical engineering at Panjab University. He is now working toward a Ph.D. degree, with specialization in heat pipes. He holds B.Sc. and M.Sc. degrees in chemical engineering from Panjab and is a member of the Indian Institute of Chemical Engineers.

MAXIMIZING ENERGY SAVINGS FOR HEAT ENGINES IN PROCESS PLANTS

Here is a novel method to integrate diesels, turbines, refrigeration machines and other such systems into the total process to effect large energy savings. Paybacks are typically short-term, and the method is equally effective for new and retrofit applications.

Eric Hindmarsh, David Boland and D. William Townsend
Imperial Chemical Industries PLC

Originally published February 4, 1985

New techniques have been developed to design plants for more efficient energy use. A previous *CE* article [1] that reported the results of ICI research showed that correct integration of heat-exchanger networks (HENs) into process plants results in large savings in operating costs. The article described a powerful procedure for combining HENs with the rest of the plant and demonstrated that, when it is correctly done, there is often a twofold advantage over incorrectly integrated designs. Correct designs are often structurally simpler (i.e., cheaper to build) and make best use of energy and capital.

Consider the total flowsheet

Design procedures for HENs are, however, only one element in considering an entire flowsheet in terms of an integrated thermodynamic analysis [2]. Methods for analyzing the total flowsheet encompass the appropriate integration of power systems, separation systems, reactors, etc., to yield the best practical design — one that employs the high-

251

est possible amount of energy recovery at a given capital cost. Further, such a design combines components on the bases of controllability, plant layout, safety, reliability and other relevant factors.

New techniques have been applied within ICI by a team of specialists [3]. When these are applied to the different systems in the plant, energy and capital savings often exceed those realized with HEN analysis alone — the effects are synergistic. Utility savings of over 50% of the flowsheet requirements are often realized, and with a short payback period [2].

Substantial energy savings can be realized through the correct blending of heat engines with chemical processes. Here, we will outline a novel method by which this can be accomplished.

The procedure will be illustrated by a case study that involves a refrigeration system. For the refrigeration compressors alone, a 15% reduction in the shaft-power requirements was achieved through appropriate integration. Table I

summarizes other cases in which the method has been applied to effect substantial energy savings.

Systems approach

In the structured approach to design, a process can be regarded as a set of interacting systems. Development of the best practical flowsheet involves understanding system design and system interactions. A flowsheet can be broken down into as many systems as are needed.

The HEN analysis is one system of the flowsheet; see Fig. 1. The design of this is well understood. Heat engines and their interaction with the remaining systems are covered here. Other systems involve utilities and distillation [4,5].

Heat engines

All machines involving heat and work interactions can be considered as either heat engines or reversed heat engines (refrigeration systems). We will now look at the basic thermodynamics of ideal heat engines and heat pumps. From this, we will develop the concepts needed to integrate ideal machines with the overall flowsheet. Then, such concepts will be extended to the integration of real machines.

The simplest ideal heat engine is the Carnot engine; see Fig. 2. Heat, Q_H, is absorbed at a constant hot temperature, T_H, work, W, is produced, and heat, Q_C, is rejected at a cooler temperature, T_C. The Carnot efficiency puts an upper limit on the maximum amount of work, W_{Max}, that can be produced:

$$W_{Max} = Q_H \left[1 - (T_C / T_H) \right] \qquad (1)$$

Cycle efficiency is a function only of the heat absorption and rejection temperatures, and heat rejection must be at a lower temperature than heat acceptance. The efficiency of ideal — and real — machines can be increased by raising the average heat-acceptance temperature or reducing the average heat-rejection temperature.

An ideal refrigeration system (or heat pump) is a reversed Carnot engine; see Fig. 2. A minimum amount of work, W_{Min}, is supplied to absorb heat at a lower temperature and

is referred to as the problem-table analysis. Details of this are found in [1,6].

Appropriate heat-engine integration

What then is the significance of the pinch in relation to the correct integration of heat engines with a process? To answer, consider a hypothetical process shown to the right of Fig. 4a. The process pinch, and minimum utility requirements, $Q_{H, MinProcess}$ and $Q_{C, MinProcess}$, are shown.

To the left of Fig. 4a is a Carnot engine; heat-acceptance and heat-rejection temperatures are hotter than the pinch temperature. The Carnot engine is assumed to have an efficiency of 33-1/3% (the ratio of power produced to heat absorbed). Then for each $3W$ units of heat absorbed, $2W$ are rejected at a lower temperature, and W units are converted into work.

The total hot-utility requirement for the engine and the process is $Q_{H, MinProcess} + 3W$. This requirement can be reduced by integrating the heat engine with the process, above the pinch; see Fig. 4b. In this integrated arrangement,

In these processes, proper integration of heat engines saved energy and money		Table I
Process	Saving, $ million/yr	Payback, yr
Food processing	0.2	< 1
Petrochemical	5.0	< 3
Evaporation system	0.8	1
Combined heat and power system	10.0	< 3

reject it at a higher one. The minimum work required is:

$$W_{Min} = Q_C \left[(T_H / T_C) - 1 \right] \qquad (2)$$

Once again, power requirements are a function of only the heat-source and heat-sink temperatures.

The pinch

The pinch concept and its application have been described elsewhere [1,9]. The same concept also applies to the design of heat-engine networks. In a correctly balanced HEN design, the pinch is a temperature across which there is no heat flow; i.e., no heat passes through the network at the pinch.

In Fig. 3, the pinch is shown partitioning a HEN design-problem into hot and cold regions. The problem is defined by a set of hot streams (that must be cooled) and cold streams (that must be heated). The hot end, which comprises all streams or parts of streams hotter than the pinch temperature, requires only process heat-exchange and utility heating. Utility cooling is not required. The cold end, which comprises all streams or parts cooler than the pinch, requires only process heat-exchange and utility cooling. Utility heating is not required. Thus, the region above the pinch is called a heat sink and the region below it, a heat source.

In all designs that minimize utility usage, heat must not be transferred across the pinch, from the sink to the source. Heat that is transferred across the pinch has to be supplied from a hot utility above minimum requirements and must be rejected to the cold utility as well.

Determination of the pinch and minimum utility requirements are made by using temperature-interval analysis. This

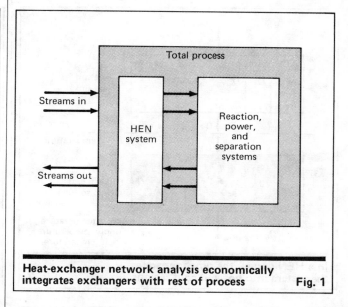

Heat-exchanger network analysis economically integrates exchangers with rest of process   Fig. 1

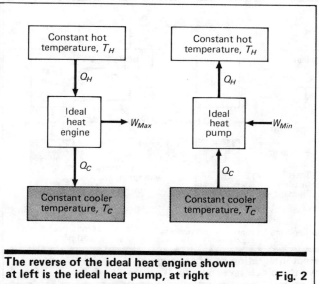

The reverse of the ideal heat engine shown at left is the ideal heat pump, at right   Fig. 2

the engine exhaust heat displaces the process hot utility, and the total utility requirement then falls to $Q_{H,\,Min_{Process}} + 3W$, a saving of $2W$ units of heat. Work W has been produced at an incremental efficiency of 100%, since the waste heat from the machine is used to displace process requirements.

Fig. 5a shows the same process, together with a Carnot engine for which both heat-acceptance and heat-rejection temperatures are colder than the pinch temperature. Again, an efficiency of 33-1/3% is assumed. The total hot-utility requirement is $Q_{H,\,Min_{Process}} + 3W$. When the heat engine is integrated with the process below the pinch, this requirement is reduced to $Q_{H,\,Min_{Process}}$, since the engine's heat requirements are supplied by heat rejected by the process; see Fig. 5b.

In Fig. 6, a Carnot engine is shown integrated with the process so that heat is absorbed above the pinch and rejected below it. The minimum utility requirement is not reduced from the total, nonintegrated requirements. The heat engine violates the process pinch by transferring heat across it. As a result, the total utility requirement of the integrated system is $Q_{H,\,Min_{Process}} + 3W$, the sum of the requirements for the two separate systems. This is expected, since the heat engine does violate the pinch.

Figs. 4 and 5 illustrate appropriate integration of a heat engine with a process. For a more detailed discussion, see [7]. Appropriate integration means that the engine's heat-acceptance and heat-rejection temperatures are both above or below the pinch. Such integration leads to substantial reductions in utility requirements over inappropriately integrated cases.

Integration of heat pumps

The above principles apply equally to heat pumps and refrigeration systems; see Fig. 7. Fig. 7a shows a heat pump integrated with the process sink, above the pinch. The pump requires work W to absorb heat from the sink and reject it to the sink at a higher temperature.

An enthalpy balance around the sink shows that the process hot-utility requirement is reduced by an amount equal to the shaft power supplied to the pump. Since fuel

---

Hot and cold stream data for the example listed in the box   **Table II**

Stream name	Supply temperature, °C	Target temperature, °C	Temperature, °C	Enthalpy, GJ/h
Separation-unit feed	30	−70	30	50
			−50	12
			−70	0
Mother liquor to remaining plant	−65	20	−65	0
			20	30
Solids melt	0	10	0	0
			10	12

$\Delta T_{min} = 10°\,C$

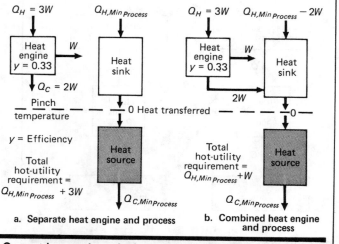

a. Separate heat engine and process

b. Combined heat engine and process

Correct integration of a heat engine above the pinch saves energy   **Fig. 4**

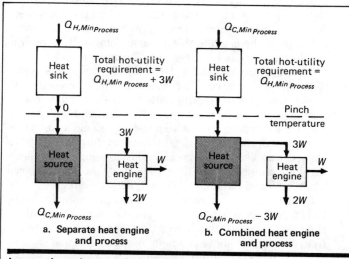

a. Separate heat engine and process

b. Combined heat engine and process

Integrating a heat engine with a process below the pinch saves energy   **Fig. 5**

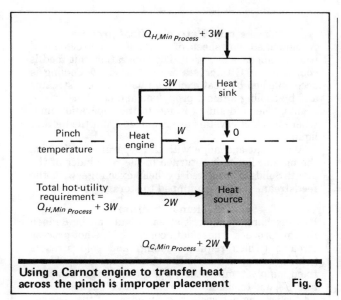

Using a Carnot engine to transfer heat across the pinch is improper placement   **Fig. 6**

costs will generally be much lower than work (electricity costs), the pump is not economical and is inappropriately placed.

Fig. 7b shows a heat pump integrated with the process source, below the pinch. Here, an enthalpy balance around the source shows that the cold-utility requirement is increased by an amount equal to the shaft power supplied by the pump. This is a ludicrous situation — work is used to heat the cold utility! Again the pump has been placed inappropriately.

Fig. 7c shows a heat pump integrated such that heat is upgraded across the pinch. The heat pump takes in work W to absorb heat below the pinch, where it is in surplus, and rejects it above the pinch, where it is needed.

An enthalpy balance around the entire system shows that the process hot utility is reduced by an amount equal to the shaft power plus the heat removed below the pinch. Cold-utility requirements are also reduced. This is appropriate integration of the heat pump with the process. For more information on devising such a design, see [7].

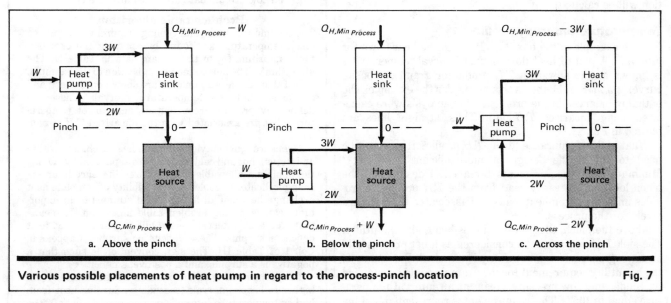

a. Above the pinch   b. Below the pinch   c. Across the pinch

Various possible placements of heat pump in regard to the process-pinch location   **Fig. 7**

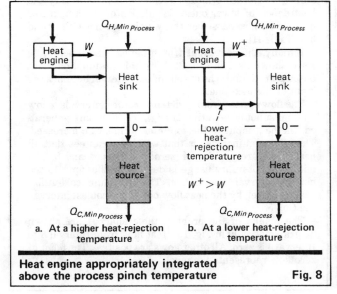

a. At a higher heat-rejection temperature   b. At a lower heat-rejection temperature

Heat engine appropriately integrated above the process pinch temperature   **Fig. 8**

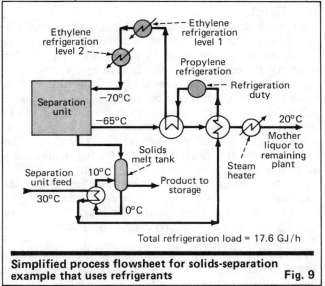

Total refrigeration load = 17.6 GJ/h

Simplified process flowsheet for solids-separation example that uses refrigerants   **Fig. 9**

Load vs. level

Consider Fig. 8a, which shows an appropriately integrated heat engine, above the process pinch. The work produced is a function of the heat-acceptance and heat-rejection temperatures — see Eq. (1) — and can be increased by either raising the acceptance temperature or lowering the rejection temperature. If one assumes that the engine's heat requirement is provided by the plant's highest-level utility, then the only option for improving efficiency is to lower the rejection temperature; see Fig. 8b.

Two obvious questions arise: To what extent can the rejection temperature be lowered? What heat load can be placed on the chosen rejection temperature? The problem-table analysis gives heat loads and levels, from which a T,H profile gives thermodynamic information about the process, and allows for optimization of load and level interactions between the process and heat engine.

A case study (see box) will be used to illustrate how load/level interactions between a heat engine and a process can be exploited to significantly improve the energy efficiency of processes. Minimum utility requirements and the pinch location will be calculated by an algorithm.

Temperature-enthalpy profile

For the example in the box, Fig. 10 gives the heat flows for intervals. A plot of heat flow against interval temperatures is known as the flowsheet T,H profile (or grand composite curve) [4,5,7]. Fig. 11 shows this profile for the example. The enthalpy balances in the problem-table analysis were calculated using linearized T,H data. Thus, straight lines are plotted in Fig. 11.

What is the significance of the T,H profile? How can it be used to aid in the design of more-efficient flowsheets? Minimum utility requirements for a HEN design and the pinch location are easily read from the T,H profile. It provides information on heat loads and associated temperature levels for the process.

Above the pinch, the process sink is heated by the utility. The sink T,H profile can be considered as a net process cold stream against which a hot utility is matched. For example, the hot-utility requirement for the example could be supplied by condensing low-pressure steam at an interval temperature equal to 105°C. This requirement is represented by Line AB in Fig. 11.

Alternatively, cooling water at 25°C could be used to supply the *heating* requirements. The supply of heat by the cooling water is represented by Line CD ion Fig. 11. For simplicity, the cooling-water line is drawn at constant temperature. In practice, there would be a temperature drop, which would be a function of the cooling-water flowrate.

The T,H profile of the heat source below the pinch can be considered as a net process hot steam, against which the cold utility (e.g., cooling water or refrigeration) is matched. Line EF represents the minimum utility cooling-requirements supplied by an extreme level of refrigeration.

In Fig. 12, comparison is made between the cascade diagram and the T,H profile. The cascade is divided by the pinch into a heat sink and a heat source. So is the temperature-enthalpy profile. An imaginary boundary is drawn around the sink and source.

The T,H profile clearly shows the load surpluses and deficits for the total flowsheet for all interval temperatures over which the process operates. Do not confuse this dia-

Case study: Petrochemical process

A simplified flowsheet of part of a petrochemical process appears in Fig. 9. The separation-unit feed is cooled to –70°C to crystallize out product. Cooling is accomplished by heat exchange with process streams and by cooling against propylene and ethylene refrigerants. The cold slurry is fed to the separation unit, where product crystals are removed from the mother liquor.

After heat exchange with the separation-unit feed, the mother liquor is returned to the remainder of the plant. Solids are melted by heat exchange with the feedstream, and then pumped to storage.

Problem definition

Heat-exchanger-network stream data can be defined for this process. These data consist of the hot process streams (which require cooling) and cold streams (which require heating). The streams are defined by a fixed supply temperature, T_S, a fixed target temperature, T_T, and temperature/enthalpy data that adequately set the heating and cooling requirements of the streams. Hot- and cold-stream data are listed in Table II.

Problem-table algorithm

The minimum heating and cooling requirements and the pinch temperature are found by applying the problem-table algorithm [6] to the stream data in Table II. The algorithm will be illustrated by application to the example.

In Table III, the stream data are shown on the left, and are divided into six temperature intervals. These are defined by process-stream supply and target temperatures that are associated with the linearized T,H stream data.

To ensure feasibility of complete heat exchange within an interval, hot and cold streams are separated by ΔT_{Min}, the minimum allowable temperature difference in an exchanger, heater or cooler. The feasibility of complete heat exchange between all hot and cold streams is an important feature of the problem-table algorithm. It means that, for each interval, there will be either a net heat deficit or surplus. These deficits or surpluses appear in Col. 1 of Table III. The sign convention is such that a negative is a surplus, and a positive, a deficit.

The problem-table algorithm allows heat transfer from higher to lower intervals (cascading). Surplus heat from higher temperature intervals can be used to satisfy the heat deficit of lower ones. Calculation of the amount of heat that can be passed on this way is shown in Cols. 2 and 3 of Table III.

Initially, the heat input from utilities is assumed to be zero (shown as 0 in Col. 2). With this assumption, heat outputs are found. The output of one interval becomes the input of the next one.

The flow of heat from high temperature intervals to low ones must not be negative. If negative values are generated, the heat input to the top interval must be increased. The minimum increase is that which guarantees that all heat flows are positive or zero; see Cols. 4 and 5. The minimum hot-utility usage is defined by the input to the hottest interval; see Col. 4. The minimum cold-utility usage is given by the heat flow out of the coldest interval; see Col. 5.

The results of the problem table analysis are shown diagrammatically in Fig. 10. Heat flow from Interval 3 to Interval 4 is zero. All other flows are positive. The point of zero flow represents the pinch.

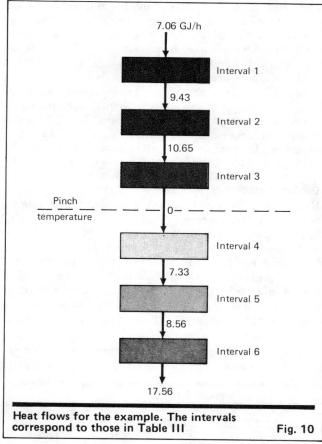

Heat flows for the example. The intervals correspond to those in Table III   Fig. 10

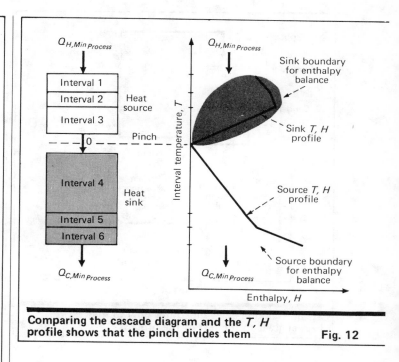

Comparing the cascade diagram and the T, H profile shows that the pinch divides them   Fig. 12

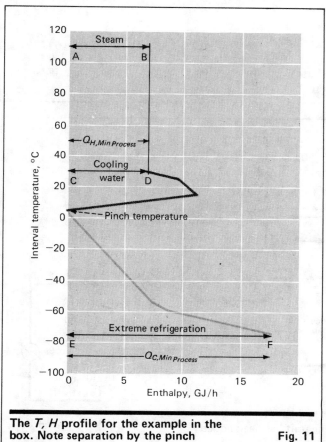

The T, H profile for the example in the box. Note separation by the pinch   Fig. 11

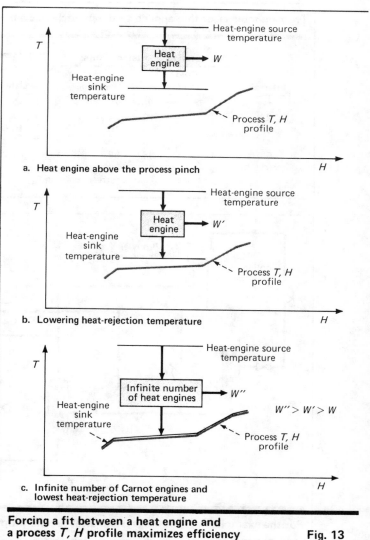

Forcing a fit between a heat engine and a process T, H profile maximizes efficiency   Fig. 13

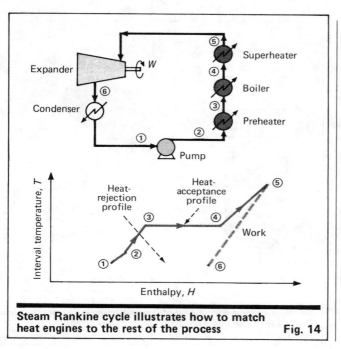

Steam Rankine cycle illustrates how to match heat engines to the rest of the process   **Fig. 14**

gram with simple composites of only hot or cold streams that have been used in design for years. The T,H profile has been formed from an analysis of the optimum balance between all hot and cold duties in the plant. Heat loads are those associated with any *minimum* utility design, and temperatures are interval temperatures.

This means that actual hot-stream temperatures are $\Delta T_{Min}/2$ hotter than the interval temperature, and cold-stream temperatures are $\Delta T_{Min}/2$ colder than the interval temperature. For example, given a ΔT_{Min} of 10°C, the actual condensing temperature of the steam represented by line AB in Fig. 11 is 105°C + 10°/2 = 110°C.

Profile matching

Here we will illustrate how the heat load and level information given by the process T,H profile can be used to exploit heat interactions between a process and a power system, to maximize cogenerated power.

Consider Fig. 13a, which shows an appropriately integrated heat engine above the process pinch. The T,H profile is chosen arbitrarily. If the engine's heat requirement is assumed to be provided by the highest-level utility, machine efficiency can be improved by lowering the heat-rejection

	Streams and temperatures			Column, GJ/h				
				1	2	3	4	5
	Cold		Hot		Accumulated heat		Heat flows	
Interval	streams	°C	streams	Deficit	Input	Output	Input	Output
1	20		30	−2.37	0	2.37	7.06	9.43
2	10		20	−1.22		3.59		10.65
3	0		10	10.65		−7.06		0
4	−60		−50	−7.33		0.27		7.33
5	−65		−55	−1.23		1.50		8.56
6	−80		−70	−9.0		10.50		17.56
	Stream heating		Stream cooling					

Problem-table algorithm applied to the petrochemical process listed in the box   **Table III**

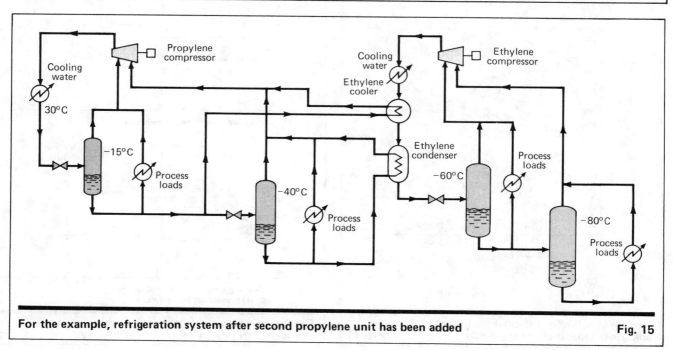

For the example, refrigeration system after second propylene unit has been added   **Fig. 15**

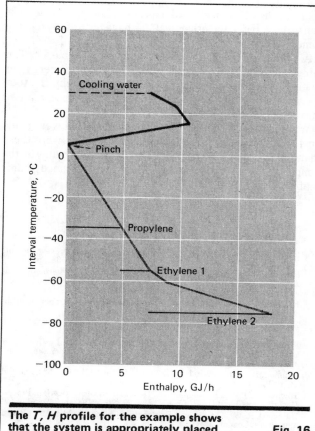

The *T, H* profile for the example shows that the system is appropriately placed   **Fig. 16**

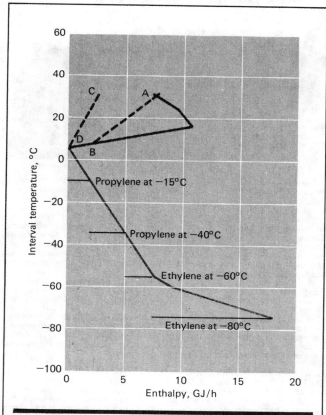

The *T, H* profile shows that a new refrigeration level and a heat pump can be added   **Fig. 17**

temperature; see Fig. 13b. The work output will increase from *W* in Fig. 13a to *W'* in Fig. 13b.

Further improvements are made by adding more Carnot engines and minimizing the heat-rejection temperatures of all engines. The maximum work obtainable is found by assuming an infinite number of engines whose combined exhaust-heat profile matches the process-sink profile; see Fig. 13c. The work output will increase further to *W"*.

Forcing a good fit between a heat engine and a process *T,H* profile extends the idea of appropriate placement to yield better overall integration schemes. Concepts of best fit follow from recent developments in applying profiles [4,5,7].

Choice of a real machine

There are many types of real heat engines, including turbines, diesel engines and combined cycles. Often, these engines are actually complex processes, consisting of flash vessels, heat exchangers, compressors, furnaces, etc.

Consider the steam Rankine cycle shown in Fig. 14. How do concepts of appropriate integration and load/level effects relate to such cycles? How can the process *T,H* profile be used to realize full and appropriate heat integration of the power-producing system with the process?

Appropriate integration involves blending the engine's heat-absorption and heat-rejection profiles with the process profile. Those for the steam Rankine cycle appear in Fig. 14. Linearized data were used, and all straight lines were drawn. Knowing these and the heat load/level characteristics of the process, the correct machine can be selected. Further, the design variables of the machine *or the process* can be optimized to ensure the best load/level interactions. For a full discussion, see [7]. These concepts will be demonstrated using the following example:

The procedures presented here will be applied to the case study described in the box. Its objective was to identify cost-effective retrofit schemes to improve overall performance of the large refrigeration system, integrated with the process. Fig. 15 shows the system after improvements were made.

The system is a dual-fluid cycle (propylene and ethylene), operating in cascade. In the original layout, the propylene circuit had only a –40°C flash; the –15°C flash is a retrofit. In the new design, propylene is condensed against cooling water and flashed twice. Vapors from each flash are returned separately to the compressor.

In the ethylene cycle, gas was compressed, condensed against cooling water, and flashed twice, first to –15°C, then to –40°C. Vapors from each were returned separately to the compressor. All of the process refrigeration duties were on the –40°C level.

HEN analysis—Table III, the problem table analysis, shows the minimum cooling requirement to be 17.6 gigajoules (GJ)/h. The actual plant refrigeration load in Fig. 9 is 17.6 GJ/h. Thus, there is no chance to improve the process HEN design. It is considered optimal.

Appropriate placement—The process *T,H* profile is redrawn in Fig. 16. The existing acceptance and rejection levels for the refrigeration system included propylene at –40°C, and ethylene at –70° and –80°C. Since refrigerant streams are cold streams (to be heated), these temperatures correspond to interval temperatures of –35°, –65° and –75°C, respectively. The heat-rejection level for the cooling water at 35°C corresponds to an interval temperature of 30°C.

Fig. 16 shows that the refrigeration system is absorbing

Two case studies show the synergistic effects of using systems analysis		Table IV
Plant	Synthesis technique	Energy-saving potential, GJ/h
Inorganic chemical plant	HENS	23
	Power systems	47
	Distillation systems	27
	Total flowsheet	56
Petrochemical plant	HENS	10
	Power systems	34
	Distillation systems	20
	Total flowsheet	40

heat below the pinch and rejecting it above the pinch. Therefore, the system is appropriately placed.

Exploiting load/level interactions—New refrigeration levels can be introduced to satisfy the process cooling requirements. Refrigeration shaft-power requirements can be minimized.

For retrofitting, the choice of new refrigeration levels is constrained by the existing system. It is desirable to avoid expensive modifications and to maximize use of existing hardware. The flash vessels allow heat to be absorbed or rejected with little change in the hardware. The flash temperatures have been placed on the T,H profile; see Fig. 17.

Fig. 17 indicates that a new refrigeration level at $-15°C$ can be introduced and will displace the load on the $-40°C$ level. The figure shows that the maximum load that can be placed on the $-15°C$ level (represented by a $-10°C$ interval temperature) is 1.75 GJ/h. This displaces an equal load from the $-40°C$ level. The net saving realized by this shift is a 5% reduction in shaft-power requirement.

Fig. 17 shows that it is possible to reject heat from the refrigeration cycle at temperatures lower than the cooling-water temperature. Since a flash vessel is not operating at the appropriate temperature, other streams in the refrigeration system must be found that will reject heat to the process. Subcooling of liquid propylene and ethylene can be used for this purpose. Heat can be taken out of streams below the pinch, upgraded with a heat pump, and rejected above the pinch.

Subcooling of propylene is represented by Line AB in Fig. 17; subcooling of ethylene, by Line CD. Net saving realized by subcooling is about a 10% reduction in refrigeration-system shaft power. Together with savings gained from use of more-appropriate heat-acceptance temperatures, the total shaft power saved is about 15%. Such a saving is realized by short-payback retrofits identified by applying the pinch method developed for HENs [9].

The shaft-power saving is due to a more appropriate thermodynamic match between the heat absorption/rejection profiles of the refrigeration system and heating/cooling requirements of the process. No attempt has been made to optimize the refrigeration machine itself, since the purpose here is to explore load/level interactions. In practice, correct optimization is essential.

Other uses

ICI has completed many projects, both internally and for other firms, using the approach described here. In most cases, in the analysis of a plant, a total thermodynamic approach has been used. All aspects of the process distillation, utility, HEN and other systems have been analyzed. For some processes, energy has been cut in excess of 50%.

A sample of these applicatons appears in Table IV. This table indicates the potential scope for saving energy via systems analysis and synergistic effects, whereby the total energy savings are often far greater than those realized through HEN design alone.

With hindsight, some of the modifications that were made for the examples in Table IV were simple. Why were these changes not identified earlier? This is probably due to the complexity of most process designs. Correct thermodynamic analysis cuts through this.

References

1. Linnhoff, B., and Turner, J. A., Heat Recovery Networks: New Insights Yield Big Savings, *Chem. Eng.*, Vol. 88, No. 22, Nov. 2, 1981, p. 56.
2. Boland, D., and Hindmarsh, E., Beyond HENS: Total Thermodynamic Approach to Process Design, paper presented at AIChE Diamond Jubilee Meeting, Washington, D.C., Oct. 31, 1983.
3. Boland, D., and Hindmarsh, E., Energy Management: Emphasis in the 80s, *The Chemical Engineer*, March 1983, p. 24.
4. Institution of Chem. Engineers, "User Guide on Process Integration for the Efficient Use of Energy," I.Chem.E. (Rugby, U.K.), 1982.
5. Hindmarsh, E., and Townsend, D. W., Heat Integration of a Distillation System Into Total Flowsheet—A Complete Approach, paper 88b, AIChE annual meeting, San Franciso, Nov. 1984.
6. Linnhoff, B., and Flower, J. R., *A.I.Ch.E.J.*, Vol. 24, 1978, p. 633, and Vol. 24, 1978, p. 642.
7. Townsend, D. W., and Linnhoff, B., Heat and Power Network in Process Design, Parts I and II, *A.I.Ch.E.J.*, Vol, 29, No. 5, 1983, p. 742 and p. 748.
8. Haywood, R. W., "Analysis of Engineering Cycles," 3rd ed., 1980, Pergamon Press, London, U.K.
9. Linnhoff, B., and Hindmarsh, E., The Pinch Design Method for Heat Exchanger Networks, *Chem. Eng. Sci.*, 1983, p. 745.

The authors

E. Hindmarsh   D. Boland   D.W. Townsend

Eric Hindmarsh is a member of ICI's Tensa Technology Consultancy Group, Petrochemicals and Plastics Div., P.O. Box 90, Wilton, Middlesbrough, Cleveland, TS6 8JE, U.K. He applies total energy synthesis techniques, generally on a consulting basis. In 1976, he graduated with a B.Sc. degree in chemical engineering from Loughborough University of Technology and joined ICI as an investigation manager. He then became a plant manager, and a commissioning manager for many small projects, before joining ICI's New Science Group as a senior research scientist in process synthesis techniques. He has been in his present position since Jan. 1985. The author of numerous papers on process synthesis, he is a member of the Institution of Chemical Engineers and is a chartered engineer.

David Boland leads a team of experts in process synthesis as a consultancy service for ICI at the above address, and has been active in process synthesis since 1979. He joined ICI in 1970, after receiving bachelor's and Ph.D. degrees in chemical engineering from Bradford University. Initially, he did trouble-shooting in an ICI operating plant. Later, he was in R&D and process engineering, and then a plant manager and commissioning manager. Before his present assignment, he was part of the company's engineering function, as a senior chemical engineer.

D. William Townsend is in R&D at ICI's New Science Group, P.O. Box 11, The Heath, Runcorn, Cheshire, U.K. He graduated from Cambridge University in 1974, and joined Courtaulds, Ltd., Coventry. Since 1978, he has been with ICI. Currently, he is involved with process synthesis methods, especially integrated heat and power systems. He is author of several papers on energy integration and coauthor of I.Chem.E.'s book, "User Guide on Process Integration for the Efficient Use of Energy." He is studying part-time for a Ph.D. degree at the University of Manchester, Institute of Science & Technology.

QUICK DESIGN AND EVALUATION OF HEAT REGENERATORS

Milorad P. Duduković and **P. A. Ramachandran**, Washington University

Such calculations can be complex and time-consuming. Here, the authors present a much simpler approach that can be solved using a calculator or referring to some graphs.

Heat regenerators are often employed when the use of heat recuperators is either uneconomical, because of the enormous heat-transfer areas required, or impractical, due to the likelihood of surface fouling by particulate-laden gases. The use of regenerators abounds in the metallurgical industry, glass manufacturing, air-separation plants, Fischer-Tropsch synthesis, storage of solar energy, etc. The above areas involve mainly large units processing enormous flowrates of gases. However, in Europe, the use of small, rotary (Ljungstrom) regenerators has been widespread for recovering heat from exhaust gases and preheating inlet air in commercial and residential oil- and coal-fired furnaces. With current consciousness about energy costs, an increased number of chemical engineers will likely be working with heat regenerators.

Unfortunately, most heat-transfer books offer only a passing reference on this subject, with the exception of Jakob [1] and Schack [2], which present a sound, but somewhat outdated, treatment. There are two monographs on this subject, one by Hausen [3] the other by Schmidt and Willmott [4].

Although the treatment in the above four texts is thorough, the jargon used in this field is somewhat alien to outsiders, especially chemical engineers. Moreover, if the user is searching for a simple and fast, yet accurate evaluation method of regenerator performance, the literature falls short. Design charts are presented [5,6], but they usually suffer from two major drawbacks: they are model-dependent and their notation requires considerable deciphering.

Here, we will familiarize chemical engineers with heat regenerator concepts, briefly review what these devices are, and discuss which types are frequently used and how they are modeled. Then, based on results of the authors' recent studies, we will present simple formulas for evaluation of regenerator performance. These do not rely on elaborate computer algorithms and can be used with the help of tabulated functions, pocket calculators, etc.

Principles of operation

In a heat regenerator, heat is not exchanged directly between hot and cold fluids across a separating solid wall. Instead, hot and cold gases flow alternately over solids that periodically absorb and release heat. Advantage is taken of high solids heat-capacity per unit volume, compared with that of gases, to make the solids an efficient medium to transfer heat between hot and cold gases. A regenerator typically operates cyclically. When hot gas flows past the solids, it heats them while being cooled. This heating period is followed by flow of cold gas over the hot solids—the cooling period.

Heat regenerators can be used either continuously to heat a cold gas and recover heat from a hot one, or to store thermal energy for use later on. In the former case, regenerators are usually employed in plant operations; in the latter, they are used for storage of solar energy. For continuous heating and cooling, the main task is to ensure maximum energy recovery (the best thermal efficiency); for the other use, the job is to find the fraction of energy that has been stored (heat storage factor) and predict heat leaks during storage. Here, we will mainly consider continuous regenerator operation, but will also address single-pass operation as part of the storage problem.

For continuous operation, two or more regenerators are used in parallel.

Originally published June 10, 1985

While hot gas passes into one, cold gas flows through the other, and gas flows are switched at appropriate times. This is called swing operation. The alternative is to rotate the solids between the hot and cold gas streams, as is done in a rotary (Ljungstrom) regenerator. Since the analysis of the rotary regenerator is analogous to that of swing units, we will concentrate our attention on swing regenerators only. Cocurrent operation takes place when hot and cold gases are introduced successively at the same end of the unit; countercurrent operation when the gases are introduced at opposite ends. Schematics are in Fig. 1.

Thermal efficiency

The key performance index for a regenerator is its thermal efficiency, E, which is defined as [1]:

$$E = \frac{\text{Heat actually transferred during time } \theta}{\text{Maximum possible heat transfer during time } \theta} \quad (1)$$

Assume, in continuous operation, that heat losses to surroundings are negligible and that mean physical properties, independent of temperature, can be used. Also, assume constant mass flowrate and gas inlet temperature. Then:

Nomenclature

A	Cross-sectional area of one channel, m²	Pe_{go}	$G'_g C_{pg} L/k_{ge}$, gas-phase Peclet number (Model B1)
a_s	External particle surface area per unit volume of the bed, m⁻¹	$Pe_{h,eff}$	$G_g C_{pg} L/k_{g,eff}$, gas-phase Peclet number (Model A2)
Bi	Rh/k_s, Biot number for hollow-cylinder model (Model B2)	Pe_p	$d_p G_g C_{pg}/2k_s(1-\epsilon_R)$, particle Peclet number (Model A1)
Bi_p	$d_p h_p/2k_s$, Biot number for particle	Pe_R	$RG'_g C_{pg}/k_s$, solid-phase Peclet number
Bi'	$R_p h/k_s$, Biot number for parallel-plate model (Model B1)	Pe_{sx}	$R_p G'_g C_{pg}/k_s$, solid-phase radial Peclet number (Model B1)
Bo	$d_p \bar{u}_g/\epsilon_R k_{m,eff}$, Bodenstein number for mass-transfer (Model B2)	Pr	$C_{pg}\mu_g/k_g$, Prandtl number
C_{pg}, C_{ps}	Mean specific heat for gas (g) or solids (s), kJ/(kg)(°C)	q	Heat storage factor
D	Temperature variance	R	Total radius of cylinder, m
D_h	Hydraulic dia., m	R_p	Half thickness of plates around rectangular or cylindrical flow channel, m
d_p	Particle dia., m	Re	$D_h G'_g/\mu_g$, Reynolds number
E	Thermal efficiency	Re_p	$d_p G_g/\mu_g$, particle Reynolds number
E_o	Overall thermal efficiency	r_c	Half height of flow channel, m
G_g	Superficial gas mass-velocity based on total cross-sectional area of regenerator, kg/(m²)(s)	Sc	Schmidt number
		St_p	$h_p a_s L/C_{pg}G_g$, Stanton number (Model A1)
G'_g	$G_g/\epsilon_R N$, Interstitial gas mass-velocity, kg/(m²)(s)	St'	$h/G'_g C_{pg}$, Stanton number (Models B1, B2)
$H(x)$,	Heaviside unit step-function	T	$R_p + r_c$, half height of channel and plate, m
h	Gas-solid heat-transfer coefficient, W/(m²)(°C)	T	Temperature—symbols have one or two subscripts, °C. (See listing of subscripts)
h_p	Gas-particle heat-transfer coefficient, W/(m²)(°C)	t	Dimensionless temperature—symbols have one or two subscripts. (See listing of subscripts)
J_h	Heat-transfer factor	u	Unit-step response (normalized breakthrough curve)
k_g	Gas conductivity, W/(m)(°C)	$\bar{u}_g$	Superficial gas velocity, m/s
k_{ge}	Axial heat effective transport coefficient in channel flow (Models B1, B2), W/(m)(°C)	V_R	Regenerator volume, m³
$k_{g,eff}$	Axial heat effective transport coefficient (Model A2), W/(m)(°C)		
k_{hz}	Axial heat effective transport coefficient in packed beds (Model A1), W/(m)(°C)		
$k_{m,eff}$	Mass-transfer dispersion coefficient, m²/s		
k_s	Solid conductivity, W/(m)(°C)		
$k°_{s,eff}$	Conductivity of packed bed with no gas flow, W/(m)(°C)		
L	Regenerator length, m		
l	$2L/d_p$		
$\bar{l}$	$L/(r_c + R_p)$		
M	$1/\sigma_D^2$		
M_s	Total mass of solids in the regenerator, kg		
$\dot{m}_g$	Gas mass-flowrate, kg/s		
N	Number of flow channels for parallel-plate or hollow-cylinder arrangement, also, number of heating periods		
$P(\chi^2;\nu)$	Chi-square probability distribution		
Pe_{gh}	$G_g C_{pg} L/k_{hz}$, gas-phase Peclet number (Model A1)		
Pe_{gm}	$\bar{u}L/k_{m,eff}$, Peclet number for mass transfer		

Greek letters

$\Gamma(x)$	Gamma function
ϵ_R	Regenerator voidage (porosity)
θ	Time, s
μ	$M_s C_{ps}/\dot{m}_g C_{pg}$, thermal mean residence time, s
ρ	r_c/T or r_c/R, dimensionless channel height (radius)
ρ_g	Gas density, kg/m³
ρ_s	Solids density, kg/m³
σ_D^2	Dimensionless variance of the impulse response
τ	θ/μ, dimensionless time
τ_s	Dimensionless switching time (for single pass)

Subscripts

c	Cooling period	opt	Optimum
e	Exit	p	Packed bed
g	Gas	sp	Single pass
h	Heating period	s	At switching or solids
i	Inlet		

Superscripts

°	Initial
⁻	Mean

Total heat actually transferred during time $\theta =$

$$\sum_0^\theta \left[\left(\binom{\text{Enthalpy of}}{\text{inlet gas}} - \binom{\text{Enthalpy of}}{\text{outlet gas}} \right) \binom{\text{Time}}{\text{interval } d\theta} \right]$$

$$\theta = (\dot{m}_g C_{pg})_h \int_0^\theta [T_{hi} - T_{he}(\theta)] \, d\theta \quad (2)$$

Maximum possible heat transfer would be accomplished if the temperature of the hot gas for the duration θ were reduced to the initial solids temperature, T_s^o, which, if everything operated ideally, would be equal to the inlet cold gas temperature, T_{ci}, and is taken as such.

Maximum possible heat transferred during time

$$\theta = (\dot{m}_g C_{pg})_h (T_{hi} - T_{ci}) \, \theta \quad (3)$$

Substituting (2) and (3) into Eq. (1), we get the following expression for thermal efficiency during the heating period:

$$E_h = \frac{\dfrac{1}{\theta} \int_0^\theta (T_{hi} - T_{he}(\theta)) \, d\theta}{T_{hi} - T_{ci}} = \frac{1}{\theta} \int_0^\theta [1 - t_{he}(\theta)] \, d\theta \quad (4)$$

We define dimensionless exit temperature by:

$$t_{he}(\theta) = (T_{he}(\theta) - T_{ci})/(T_{hi} - T_{ci}) \quad (5)$$

The higher $t_{he}(\theta)$, the hotter the exiting gas is during the heating period. This implies that a smaller amount of heat has been exchanged and, hence, efficiency is lower. Analogously, we define thermal efficiency for the cooling period, based on cold-gas exit temperature:

$$E_c = \frac{1}{\theta} \int_0^\theta t_{ce}(\theta) \, d\theta \quad (6)$$

where:

$$t_{ce}(\theta) = (T_{ce}(\theta) - T_{ci})/(T_{hi} - T_{ci}) \quad (7)$$

Often, the heat storage factor is also used [4]:

$$q = \frac{\text{Thermal energy actually stored in the solids}}{\text{Maximum thermal energy storage in the solids}} \quad (8)$$

$$q = \frac{(\dot{m}_g C_{pg})_h \int_0^\theta (T_{hi} - T_{he}) \, d\theta}{M_s C_{ps}(T_{hi} - T_{ci})} \quad (9)$$

Here, $M_s = \rho_s (1 - \epsilon_R) V_R$. Using the formulas for thermal efficiency and dimensionless gas exit-temperature (Eqs. (4) and (5)), we can reduce the expression for q to:

$$q = E\theta/\mu \quad (10)$$

where:

$$\mu = \frac{M_s C_{ps}}{\dot{m}_g C_{pg}} = \frac{\rho_s (1 - \epsilon_R) C_{ps} L}{G_g C_{pg}} \quad (11)$$

In some applications, a measure of exit-gas temperature-deviation around the mean may be desired:

$$D = \frac{1}{\theta} \int_0^\theta (t_{he} - \bar{t}_{he})^2 \, d\theta = \frac{1}{\theta} \int_0^\theta t_{he}^2 \, d\theta - (1 - E)^2 \quad (12)$$

where:

$$\bar{t}_{he} = \frac{1}{\theta} \int_0^\theta t_{he} \, d\theta = 1 - E \quad (13)$$

To relate the thermal efficiency, E, heat storage factor, q,

or temperature variance, D, to regenerator design parameters and operating conditions, we must evaluate the dimensionless exit-temperatures, $t_{he}(\theta)$ and $t_{ce}(\theta)$. Normally, this is done by solving an appropriate mathematical model for the regenerators. The type and form of the model depend on the type of regenerator and on the assumed dominant heat-transfer mechanisms.

Regenerator types

There are two types—packed beds (pebble heaters) and checkerwork structures (plate regenerators). These are sche-

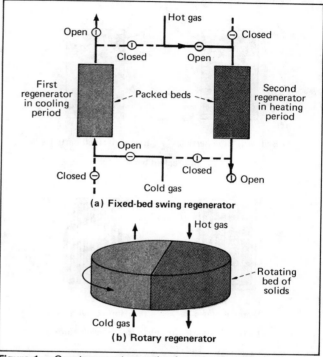

Figure 1 — Countercurrent operation for swing and rotary heat regenerators. Gas flows are in same direction for cocurrent flow

matically shown in Fig. 2. Packed beds have large or small particles (Fig. 2a). The particles are considered small when their conduction resistance is negligible compared with other resistances. Assume that all complex checkerwork structures can be represented by an equivalent setup having parallel plates or hollow cylinders (Fig. 2b).

Modeling of heat transfer in solid structures with gas flow is still based on the assumptions made by Singer and Wilhelm [7] in 1950 that several transport mechanisms are possible (see Ref. [7] for details).

Simplifying assumptions are made and only the most dominant heat-transfer mechanisms are preserved in arriving at a model for a particular regenerator type. The most commonly used models for packed beds and checkerwork structures are in the box.

Mathematical solutions for periodic regenerator operation for even the simplest of these models are quite involved and, in general, efficiency evaluations require extensive numerical computations [4]. Here, we will demonstrate how to use a much simpler, yet accurate procedure. The method is based on representing the unit-step thermal response of the regenerator (regenerator breakthrough curve) by a gamma distribution function of the system variance. The variance for each model is presented as a function of system parameters in the box.

With this information, we can show how to approach the two typical problems in heat regenerators:

1. Given hot-gas flowrate and properties, and inlet hot- and cold-gas temperatures, determine the regenerator size for the desired efficiency.

2. Estimate the performance (thermal efficiency, etc.) of an existing regenerator and suggest changes in operating procedures for improvements in efficiency.

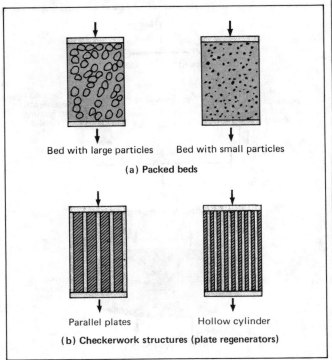

Figure 2 — Heat regenerators are modeled as either packed beds (large or small particles) or checkerworks (plates or cylinders)

Single-pass efficiency

The regenerator breakthrough curve can be represented with good accuracy by a gamma distribution function of the system variance as follows (see Fig. 3):

$$t_{he_{sp}} = u(\tau) = \frac{1}{\Gamma\left(\dfrac{1}{\sigma_D^2}\right)} \int_0^{\tau/\sigma_D^2} x^{(1/\sigma_D^2 - 1)}\, e^{-x} dx =$$

$$P\left(\frac{2\tau}{\sigma_D^2};\ \frac{2}{\sigma_D^2}\right) \quad (14)$$

σ_D^2 is tabulated in standard handbooks [8–10].

If system parameters are known, we can calculate μ and σ_D^2 for the appropriate model (see box) and, at a given dimensionless time, $\tau = \theta/\mu$, obtain the dimensionless exit temperature from Eq. (14). The appropriate value of $P(2\tau/\sigma_D^2;\ 2/\sigma_D^2)$ can be obtained either by interpolation of tabulated values [8–10] or by interpolation on Fig. 3. Both methods provide sufficient accuracy, considering the uncertainty in the values of heat-transfer parameters.

With M, single-pass efficiency up to switching time θ_s,

$\tau_s = \theta_s/\mu$ is obtained from Eq. (4); after some algebra it can be represented by:

$$E_{sp}(\tau_s) = 1 - \left(1 - \frac{1}{\tau_s}\right) P(2M\tau_s;\ 2M) -$$

$$\frac{(M\tau_s)^{M-1} e^{-M\tau_s}}{\Gamma(M)} \quad (15)$$

Values of single-pass efficiency are plotted as a function of M, with τ_s as parameter in Fig. 4. Once single-pass efficiency is known, q is found from Eq. (10).

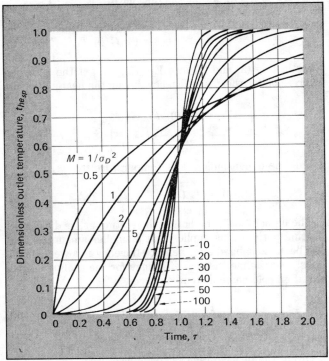

Figure 3 — For single-pass operation, the regenerator breakthrough curve is a gamma function of the system variance

$$q = E_{sp}\tau_s \quad (16)$$

If we select the thermal mean residence time for the switchoff time, $\theta_s = \mu$, and hence $\tau_s = 1$ (which we will later show leads to optimal performance for periodic, cocurrent, symmetric operation), the expression for single-pass efficiency can be considerably simplified for $M \geq 1.6$.

$$E_{sp}(1) \sim 1 - \frac{1}{\sqrt{2\pi M}} \quad (17)$$

Example 1

A thermal storage unit is composed of 24 parallel rectangular-flow channels 10 m long, 1 m wide and 2 cm high. The thickness of the Feolite storage material between channels is 10 cm, while the top and bottom channels have a 5-cm Feolite thickness between them and a material of negligible heat capacity that insulates the whole unit from the outside air. Hot air at a flowrate of 30,000 kg/h at an inlet temperature of 120°C is available periodically for 3 h. The storage unit is at 20°C initially.

(text continues on p. 266)

Models and equations

Note: The term in square brackets in Eqs. (a), (e), (h) and (l)≈1.

A1. Packed beds with large particles (spheres)
(Model of Sagara, et al. [14])

Accounts for: bulk transport in gas; eddy transport in gas; gas-solid convection; conduction in solids.

Neglects: gas conduction; solids point of contact transport; all radiation effects; accumulation term in gas phase; heat losses from walls or ends.

Dimensionless variance of the impulse response:

$$\sigma_D^2 = \underset{\substack{\text{Gas-solid film}\\\text{resistance}}}{\frac{2}{St_p}} + \underset{\substack{\text{Particle condition}\\\text{resistance}}}{\frac{2\,Bi_p}{5\,St_p}} +$$

$$\underset{\text{Eddy transport resistance}}{\frac{2}{Pe_{gh}}\left[1 - \frac{1}{Pe_{gh}}(1 - e^{-Pe_{gh}})\right]} \quad \text{(a)}$$

For gas-particle heat-transfer coefficient [15]:

$$\epsilon_R J_h = \frac{2.876}{Re_p} + \frac{0.3023}{Re_p^{0.35}} \quad \text{(b)}$$

$$J_h = \frac{h_p}{C_{pg}G_g}\left(\frac{\mu_g C_{pg}}{k_g}\right)^{2/3}$$

Suggested for "eddy" coefficient [12,16]:

$$Pe_{gh} \simeq Pe_{gm}$$

and

$$Pe_{gm} = Bo\frac{L}{d_p} \quad \text{(c)}$$

$$\frac{1}{Bo} = \frac{0.3}{Sc\,Re_p} + \frac{0.5}{1 + \dfrac{3.8}{Re_p Sc}} \quad \text{(d)}$$

For $Re_p > 10$, $Bo \simeq 2$

A2. Packed beds with small particles
(Pseudohomogeneous model [18])

Accounts for: bulk flow transport; gas-solid convective transport; conduction in solids (continuous phase); conduction at solids point of contact.

Neglects: conduction in gas; eddy transport in gas; conduction in solid particles; radiation effects; accumulation term in gas; heat losses from walls or ends.

Dimensionless variance of the impulse response:

$$\sigma_D^2 = \frac{2}{Pe_{h,eff}}\left[1 - \frac{1}{Pe_{h,eff}}(1 - e^{-Pe_{h,eff}})\right] \quad \text{(e)}$$

$$k_{g,eff} = k_{s,eff}^{\circ} + \frac{G_g^2 C_{pg}^2 d_p}{6h_p(1 - \epsilon_R)}$$

| Effective axial transport coefficient | Effective conductivity of the bed with no gas flow | Contribution to effective axial conduction by gas particle heat transfer |

$k_{s,eff}^{\circ}$ can be estimated from [18]:

$$\frac{k_{s,eff}^{\circ}}{k_g} = \left(\frac{k_s}{k_g}\right)^{[0.28-0.757\log \epsilon_R+0.057\log (k_s/k_g)]} \quad \text{(g)}$$

and h_p is found from Eq. (33). Other variables as defined in Model A1.

B1. Parallel-plate regenerators

Accounts for: bulk transport; eddy transport; conduction in solids perpendicular to flow; and gas-solid convective transport.

Neglects: conduction in gas; conduction in solids in direction of flow; radiation effects; accumulation term in gas phase; and heat losses from walls or ends.

Dimensionless variance of the impulse response:

$$\sigma_D^2 = \frac{2}{St'}\left(\frac{r_c}{L}\right) + \frac{2\,Bi'}{3\,St'}\left(\frac{r_c}{L}\right) +$$

$$\frac{2}{Pe_{go}}\left[1 - \frac{1}{Pe_{go}}(1 - e^{-Pe_{go}})\right] \quad \text{(h)}$$

Evaluation of parameters:

$$St' = 0.023\,Re^{-0.2}\,Pr^{-0.33} \text{ for } Re > 10^4 \text{ Ref. [4]} \quad \text{(i)}$$

$$Pe_{go} \simeq Pe_{mt}\frac{L}{D_h} \text{ (by analogy)} \quad \text{(j)}$$

$$Pe_{mt} = \frac{G_g D_h}{\rho_g k_{m,eff}}$$

$$\frac{1}{Pe_{mt}} = \frac{3 \times 10^7}{Re^{2.1}} + \frac{1.35}{Re^{1/8}},$$

$$\text{for } Re > 2{,}000 \text{ Ref. [16]} \quad \text{(k)}$$

B2. Hollow-cylinder regenerators (circular holes in solid storage material)

Accounts for: bulk transport; eddy transport; conduction in solids perpendicular to flow; and gas-solid convection transport.

Neglects: conduction in gas; conduction in solids in direction of flow; radiation effects; accumulation term in gas phase; and heat losses from walls and ends.

Dimensionless variance of the impulse response:

$$\sigma_D^2 =$$

$$\frac{2}{St'}\left(\frac{r_c}{2L}\right) + \frac{Pe_R\rho^2}{\bar{l}(1 - \rho^2)^2}\left[\frac{\rho^4 - 1}{4} - \frac{(1 - \rho^2)^2}{2} - \ln\rho\right]$$

$$+ \frac{2}{Pe_{go}}\left[1 - \frac{1}{Pe_{go}}\left(1 - e^{-Pe_{go}}\right)\right] \quad \text{(l)}$$

a. Determine thermal efficiency and the heat-storage factor for this single-pass operation of 3 h and the exit gas temperature at switchoff time.

b. Calculate the regenerator length to increase single-pass efficiency to at least 0.9 and evaluate heat-storage factor and exit gas temperature at switching time.

Data: ρ_s = 3,900 kg/m³; C_{ps} = 0.95 kJ/(kg)(°C); k_s = 2.1 W/(m)(°C); ρ_g = 1.1 kg/m³; C_{pg} = 1.0 kJ/(kg)(°C); $\dot{m}_g$ = 30,000 kg/h = 8.33 kg/s; μ_g = 2 × 10⁻⁵ kg/(m)(s); k_g = 0.028 W/(m)(°C).

Solution: Part a. Select model B1 from the box.

$$A = 1 \times 0.02 = 0.02 \text{ m}^2; \quad A_{TOT} = 24 \times 0.02 = 0.48 \text{ m}^2$$

$$G'_g = \dot{m}_g/A_{TOT} = 8.33/0.48 = 17.4 \text{ kg/(m}^2\text{)(s)}$$

$$D_h = \frac{4A}{2(\text{height} + \text{width})} = \frac{4 \times 0.02}{2(1 + 0.02)} = 0.039 \text{ m}$$

$$Re = \frac{G'_g D_h}{\mu_g} = \frac{17.4 \times 0.039}{2 \times 10^{-5}} = 33,900$$

$$Pr = \frac{C_{pg}\mu_g}{k_g} = \frac{1,000 \times 2 \times 10^{-5}}{0.028} = 0.71$$

$$St' = 0.023 \, Re^{-0.2} Pr^{-0.33}$$
$$= 0.023 \times (33,900)^{-0.2} \times (0.71)^{-0.33}$$
$$= 0.0032 \qquad \text{[Eq. (i)]}$$

$$h = St'G'_g C_{pg} = 0.0032 \times 17.4 \times 1,000$$
$$= 55.7 \text{ J/(m}^2\text{)(s)(°C)}$$

$$Bi' = \frac{R_p h}{k_s} = \frac{0.05 \times 55.7}{2.1} = 1.33$$

$$1/Pe_{mt} = \frac{3 \times 10^7}{33,900^{2.1}} + \frac{1.35}{33,900^{1/8}} = 0.0092 + 0.366 = 0.376$$

$$Pe_{mt} = 2.66 \qquad \text{[Eq. (k)]}$$

$$Pe_{go} \simeq Pe_{mt}\frac{L}{D_h} = 2.66 \times \frac{10}{0.039} = 683 \qquad \text{[Eq. (j)]}$$

$$\sigma_D^2 \simeq \frac{2}{St'}\left(\frac{r_c}{L}\right) + \frac{2Bi'}{3St'}\left(\frac{r_c}{L}\right) + \frac{2}{Pe_{go}}$$

$$= \frac{2}{0.0032}\left(\frac{0.01}{10}\right) + \frac{2 \times 1.33}{3 \times 0.0032}\left(\frac{0.01}{10}\right) + \frac{2}{603}$$

$$= 0.625 + 0.277 + 0.0029 = 0.9049 \qquad \text{[Eq. (h)]}$$

$$M = 1/\sigma_D^2 = 1/0.9049 = 1.105$$

$$\mu = \frac{M_s C_{ps}}{\dot{m}_g C_{pg}} = \frac{24 \times 10 \times 1 \times 0.1 \times 3,900 \times 950}{30,000 \times 1,000} = 2.96 \text{ h}$$

$$\tau_s = \theta_s/\mu = 3/2.96 = 1.01 \approx 1$$

$$E_{sp}(1) = 0.65 \qquad \text{[Eq. (15)]}$$

$$q = E_{sp}\tau_s = 0.65 \qquad \text{[Eq. (16)]}$$

$$t_{he_{sp}}(1) = u(1) = 0.626 \quad (\tau_s \sim 1) \quad \text{[Fig. 3 or Eq. (14)]}$$

$$T_{he} = T_{ci} + (T_{hi} - T_{ci})\, u = 20 + (120 - 20) \times 0.626 = 83°C$$

Part b. From the above, it is clear that the new L changes the thermal mean residence time and variance:

$$\mu_{new} = 0.296 \, L; \quad \sigma^2_{D_{new}} = 9.049/L$$

This means that: $\tau_{s_{new}} = 10.1/L; \quad M_{new} = L/9.049$

Trial and error is now required. From Fig. 4 we see that if we choose $M = 5$, then $L = 45.2$ m, $\tau_{s_{new}} = 0.223$ and $E_{sp} = 1$. Lower values of M are sufficient for $E_{sp} \approx 0.9$. We finally get $M_{new} = 2.15$, $L = 19.5$ m; $\tau_{s_{new}} = 0.52$, $E_{sp} \approx 0.9$ $q = 0.47$. From Fig. 3 for $M = 2.15$, $u\,(0.52) = 0.264$, hence T_{he} $(\theta = 3h) = 46°C$.

Example summary: Clearly, we can improve single-pass efficiency (i.e., the fraction of heat recovered from the hot-gas stream) by increasing regenerator length (capital cost) at the expense of a reduced heat-storage factor. The only true improvement would come by redesigning the unit.

Ideal regenerator

Now consider the operation of an ideal regenerator, i.e., one in which heat is transported by bulk flow of gas at a finite rate, while all other heat-transfer resistances perpendicular to flow direction are zero. The breakthrough curve of an ideal regenerator would be a step-up curve at $\tau = 1$, in Fig. 3. Clearly, this sharp rise is approached as $M \to \infty$ or $\sigma_D^2 \to 0$. This, in a regenerator with finite heat-transfer resistances, can be approached by increasing its length, since $\sigma_D^2 \propto 1/L$.

Since finite heat-transfer resistances define the length of the zone in which transport takes place (the height of the transfer unit), the longer the regenerator, the smaller the ratio of the height of the transfer unit to regenerator length and the higher the efficiency. For ideal regenerators:

$$[E_{sp}(\tau_s)]_{ideal} = 1 - \left(1 - \frac{1}{\tau_s}\right)H(\tau_s - 1) = \begin{cases} 1, & \tau_s \le 1 \\ 1/\tau_s, & \tau_s \ge 1 \end{cases} \quad (18)$$

where:
$$H(x - a) = \begin{cases} 0, & x < a \\ 1, & x > a \end{cases} \quad (19)$$

Continuous periodic operation

Swing regenerators are always onstream and can be operated cocurrently or countercurrently (Fig. 1). While countercurrent flow always yields higher efficiency, there may be other considerations that favor cocurrent operation. The efficiency for this case can be estimated rapidly by the method presented below.

We also must distinguish between balanced (symmetric) and unbalanced operation. Balanced operation is when the product of the gas mass flowrate and specific heat is the same for the hot and cold gas (the heating and cooling periods). Switching times are the same for both periods and so are efficiencies, provided that heat-transfer parameters are approximately the same.

Unbalanced operation is when the product of flowrate and specific heat are unmatched for heating and cooling, and efficiencies of the two periods are unequal, but real switching times are kept the same. Overall efficiency for both periods:

$$E_o = \frac{2E_h}{1 + (\tau_c/\tau_h)} = \frac{2E_h}{1 + (\mu_h/\mu_c)} \quad (20)$$

Cocurrent symmetric operation

Consider the operation of an ideal regenerator. As can be seen from Fig. 5, such a device has an efficiency = 1 if switching time is equal to the mean residence time ($\tau_s = 1$). For dimensionless switching times >1, efficiency is always <1. Dimensionless switching times <1 yield efficiencies ≤1.

However, even if the efficiency = 1, this implies that the same operation could have been achieved with a smaller regenerator of reduced thermal mean residence time so that the efficiency of unity is accomplished at $\tau_s = 1$. Thus, the consideration of ideal regenerators teaches us that the optimal switching time for cocurrent operation is the thermal mean residence time $\theta_s = \mu$ or $\tau_s = 1$. This finding holds for nonideal regenerators, too.

$$t_{ge}(\tau) = \sum_{n=0}^{\infty} [u(\tau - 2n\,\tau_s)\,H(\tau - 2n\,\tau_s) - u(\tau - (2n+1)\,\tau_s) \times$$
$$H(\tau - (2n+1)\,\tau_s)] \quad [\text{Ref. }[11]] \quad (21)$$

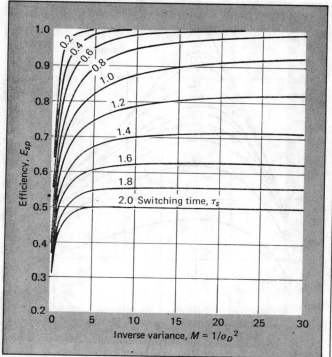

Figure 4 — Single-pass efficiency, found from switching-time and inverse variance, is used to calculate q

where the breakthrough curve $u(\tau)$ is given by Eq. (14) and the Heaviside unit step function $H(\tau)$ is defined by Eq. (19). The exit gas temperature can then be calculated during any heating period $2n\tau_s < \tau < (2n+1)\tau_s$ by superposition. After a sufficiently large number of cycles, steady-periodic-state is reached. The infinite series can be replaced by the finite sum and summation terminated when $n = N$ with:

$$N = 1 + \text{int}\left[\frac{1}{2}\left(1 + \frac{\chi_x^2}{2M\tau_s}\right)\right] \quad (22)$$

where χ_x^2 is such a value of the argument that $P(\chi_x^2; 2M) \geq 0.995$, which can readily be assessed from Fig. 3. A few examples of the effluent gas temperature in cocurrent operation are sketched in Fig. 6. Ideal regenerator operation is approached only at $\tau_s = 1$ and at high values of M.

The formula for thermal efficiency for cocurrent, periodic, symmetric operation is [11]:

$$E(\tau_s) = E_{sp}(\tau_s) + \sum_{n=1}^{N} [(2n-1)\,E_{sp}((2n-1)\tau_s) -$$
$$4n\,E_{sp}(2n\tau_s) + (2n+1)\,E_{sp}(2n+1)] \quad (23)$$

Eq. (23) can readily be evaluated using either Eq. (15) or Fig. 4 for single-pass efficiency, $E_{sp}(\tau)$. Fig. 7 shows the thermal efficiency of cocurrent, symmetric operation as a function of M, with τ_s as parameter. Clearly, optimal operation is achieved at $\tau_s = 1$, but the behavior with respect to τ_s is not monotonic. For optimal operation ($\tau_s = 1$), the efficiency can be calculated by an approximate formula, accurate within 5% for $M > 2.5$ [12]:

$$E(1) = 2E_{sp}(1) - 1 \quad (24)$$

Evaluation of thermal efficiency for cocurrent, symmetric

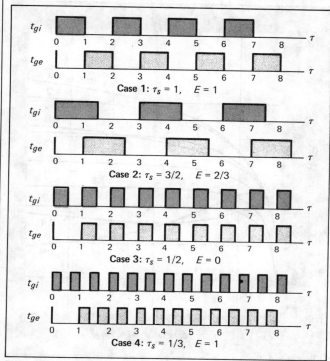

Figure 5 — Inlet and exit temperatures for ideal regenerators that are in cocurrent, symmetric operation

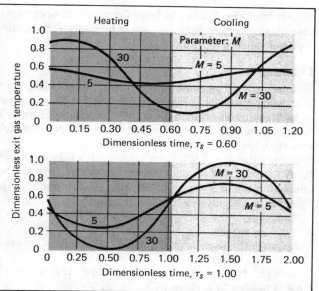

Figure 6 — Dimensionless exit-gas temperature profiles for periodic, cocurrent, symmetric operation of nonideal regenerators

operation now becomes simple. For a given system, we calculate all the physical properties over the temperature range and find their average values. Next, we select the appropriate regenerator and its model (see box) and calculate the heat-transfer parameter from appropriate correlations. Then we evaluate σ_D^2 and $M = 1/\sigma_D^2$. We should select optimal operation ($\tau_s = 1$) and calculate single-pass efficiency, E_{sp}, from Eq. (17) and optimal thermal efficiency for cocurrent operation from Eq. (24). If, for some reason, we are forced to use nonoptimal operation with $\tau_s \neq 1$, we can either calculate single-pass efficiency from Eq. (15) or esti-

symmetric and cocurrent. For a packed bed with large particles, an appropriate model is A1.

$$G_g = \frac{\dot{m}_g}{A} = \frac{4\dot{m}_g}{\pi \, d^2} = \frac{4 \times 72,000}{3600 \times \pi \times 4^2} = 1.59 \, \frac{\text{kg}}{(\text{m}^2)(\text{s})}$$

$$Re_p = \frac{d_p G_g}{\mu_g} = \frac{0.06 \times 1.59}{3 \times 10^{-5}} = 3,180$$

$$Pr = \frac{\mu_g C_{pg}}{k_g} = \frac{3 \times 10^{-5} \times 1020}{0.05} = 0.61$$

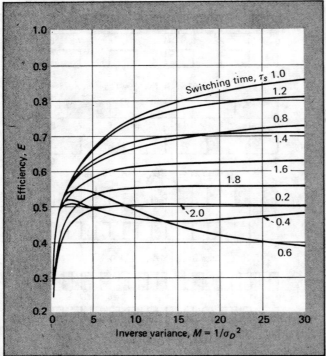

Figure 7 — Thermal efficiency of periodic, cocurrent, symmetric operation as a function of inverse variance

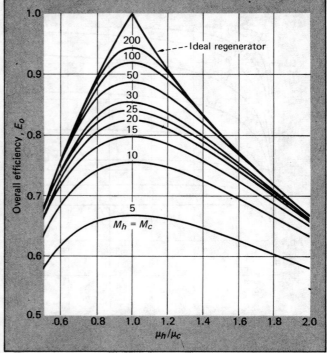

Figure 8 — For cocurrent, unbalanced, periodic operation, optimal overall thermal efficiency is bounded by efficiency of ideal unit

mate it from Fig. 4. The efficiency for cocurrent operation is then obtained either from Eq. (23) or by interpolation, using a proper value of τ_s, from Fig. 7.

Example 2

Two packed-bed regenerators are used in periodic, cocurrent swing-operation to recover heat from hot exhaust gases at 800°C and preheat inlet air at 100°C. Each regenerator is 30 m long and 4 m dia. Solid pebbles used are approximately spherical, with an average dia. of 6 cm.

Other data: $\rho_s = 3,900$ kg/m³; $C_{ps} = 1.0$ kJ/(kg)(°C), $k_s = 0.5$ W/(m)(°C); and $\epsilon_R = 0.4$. Mass flowrate of both the hot and cold streams is $\dot{m}_g = 72,000$ kg/h (55,600 Nm³/h). The following mean physical properties can be used for both streams: $\mu_g = 3 \times 10^{-5}$ kg/(m)(s); $k_g = 0.05$ W/(m)(°C); $\rho_g = 0.5$ kg/m³; and $C_{pg} = 1.02$ kJ/(kg)(°C).

a. Currently, a switching time of 7.2 h is used. What is the thermal efficiency?

b. What is the optimal efficiency, and what switching time is needed to achieve it?

Solution: First, we realize that, since $(\dot{m}_g C_{pg})_h = (\dot{m}_g C_{pg})_c$ and other physical properties are constant, operation is

$$\epsilon_R J_h = \frac{2.876}{Re_p} + \frac{0.3023}{Re_p^{0.35}} \qquad \text{[Eq. (b)]}$$

$$0.4 J_h = \frac{2.876}{3,180} + \frac{0.3023}{3,180^{0.35}} = 0.0009 + 0.01797 = 0.0189$$

$$J_h = 0.0472; \qquad \frac{h_p}{C_{pg} G_g} Pr^{2/3} = J_h$$

$$h_p = C_{pg} G_g J_h Pr^{-2/3}$$
$$= 1,020 \times 1.59 \times 0.0472 \times (0.61)^{-2/3}$$
$$= 106 \, \text{W/(m}^2)(°C)$$

$$St_p = \frac{6 \, h_p \, (1 - \epsilon_R) \, L}{d_p G_g C_{pg}} = \frac{6 \times 106 \times (1 - 0.4) \times 30}{0.06 \times 1.59 \times 1,020} = 118$$

$$Bi_p = \frac{h_p d_p}{2 \, k_s} = \frac{106 \times 0.06}{2 \times 0.5} = 6.36$$

Since $Re_p > 10 \qquad Bo = 2$

$$Pe_{gh} = Bo \, \frac{L}{d_p} = \frac{2 \times 30}{0.06} = 1,000 \qquad \text{[Eq. (c)]}$$

$$\sigma_D^2 \approx \frac{2}{St_p} + \frac{2\,Bi_p}{5\,St_p} + \frac{2}{Pe_{gh}} = \frac{2}{118} + \frac{2\times 6.36}{5\times 118} + \frac{2}{1{,}000}$$

$$= 0.0169 + 0.0216 + 0.002 = 0.0405 \quad [\text{Eq. (a)}]$$

$$M = 1/\sigma_D^2 = 1/0.0405 = 24.7 \approx 25$$

$$\mu = \frac{(1 - \epsilon_R)L\rho_s C_{ps}}{G_g C_{pg}} = \frac{(1 - 0.4)\times 30 \times 3{,}900 \times 1{,}000}{1.59 \times 1{,}020}$$

$$= 4.3285 \times 10^4 \text{s} = 12.0 \text{ h}$$

Part a. Current switching time is $\theta_s = 7.2$ h; Dimension-

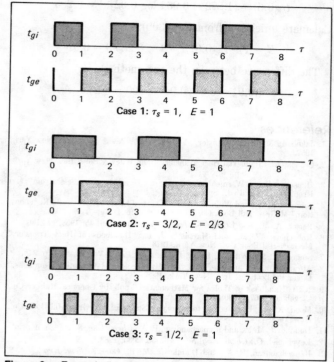

Figure 9 — For periodic, countercurrent, symmetric operation, inlet and exit temperatures for ideal regenerators

less switching time is: $\tau_s = \theta_s/\mu = 7.2/12 = 0.6$. From Fig. 7, for $\tau_s = 0.6$, $M = 25$, we read $E = 0.4$. Very low efficiency is obtained. The regenerators are poorly operated.

Part b. For symmetric cocurrent operation, optimal efficiency is obtained at $\tau_s = 1$. We need switching time $\theta_s = \mu = 12$ h. From Fig. 7 at $\tau_s = 1$, $M = 25$, we read $E = 0.84$. Without any redesign, just by changing the switching time to the optimal value we are able to improve the thermal efficiency by $100 \times (0.84 - 0.4)/0.4 = 110\%$!

If, however, other process requirements demand the switching time to be 7.2 h, we have two options to improve efficiency. We can either use countercurrent operation (to be discussed in Example 3) or reduce the length of the regenerators, so that the new thermal mean residence time equals the switching time $\mu_{new} = 7.2$ h. This requires:

$$L_{new} = L\frac{\mu_{new}}{\mu} = 30 \times \frac{7.2}{12} = 18 \text{ m}$$

$$\sigma_{D_{new}}^2 = \sigma_D^2 \frac{L}{L_{new}} = 0.0405 \times \frac{30}{18} = 0.0675$$

$$M_{new} = 1/\sigma_{D_{new}}^2 = 14.8 \approx 15$$

Fig. 7 gives at $M = 15$ and $\tau_s = 1$, $E = 0.80$. We double the efficiency by reducing the regenerator length from 30 m to 18 m. Clearly, in cocurrent operation, bigger does not necessarily mean better, and over-design may penalize process efficiency heavily.

Cocurrent unbalanced operation

Sometimes process constraints are such that we are unable to balance the product of mass gas flowrate and specific heat for the heating and cooling periods. Thus, $\mu_h \neq \mu_c$ and operation is unbalanced. If heat-transfer resistances are reason-

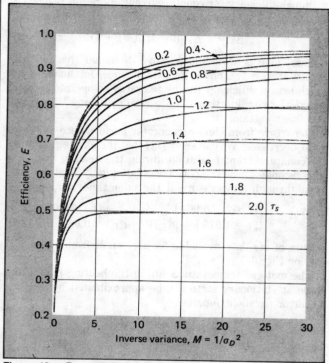

Figure 10 — For periodic, countercurrent, symmetric operation, thermal efficiency as a function of inverse variance

ably close for the heating and cooling periods, $\sigma_{Dh}^2 \approx \sigma_{Dc}^2$ and we need to operate at the same switching time for both periods. Optimal switching times are approximated by [12]:

$$\tau_{h_{opt}} = \frac{1}{2}\left(1 + \frac{\mu_c}{\mu_h}\right) \quad (25)$$

$$\tau_{c_{opt}} = \tau_{h_{opt}}\frac{\mu_h}{\mu_c} \quad (26)$$

If we operate at such optimal switching times, the maximum attainable overall efficiency is given in Fig. 8 as a function of μ_h/μ_c with M as the parameter. Optimal efficiency is bounded by the efficiency for an ideal regenerator, operated with the same μ_h/μ_c ratio.

If we are forced to nonoptimal, unbalanced operation, although this should be avoided if possible, the thermal efficiency can be calculated from formulas given in Ref. [11]. These are somewhat lengthy and are not reproduced here. They are simple enough to be programmed on a personal computer. Due to space limitations, an example will not be presented for this case.

Periodic countercurrent symmetric operation

This is a difficult case to calculate exactly [4]. Several simplified approaches have been suggested [11,12]. Consider the operation of an ideal regenerator (Fig. 9), from which it becomes clear that efficiency stays at 1 for $\tau_s \leq 1$, but diminishes for higher switching times, $\tau_s > 1$. Theoretical considerations of the nonideal regenerator [1] show that optimal operation is achieved at an infinitesimally short switching time, $\tau_s \rightarrow 0$, when the solids temperature-profile approaches the diagonal line $T_s = T_{hi} - (T_{hi} - T_{ci}) z/L$, where z is the axial distance measured from the hot-gas inlet. The maximal efficiency for countercurrent operation reached at $\tau_s = 0$ can be expressed as:

$$E_{h_{max}} = E_h(0) = M_h/(M_h + 1) \qquad (27)$$

This value is never more than 15% higher than the efficiency at $\tau_s = 1$. This is important since, fortunately, not much loss in efficiency can be tolerated to operate at more practical switching times, since zero switching time is impossible to attain.

Departure from ideal regenerator performance increases as M decreases. We can show [11] that the exit temperature for countercurrent operation during the heating period at dimensionless time (τ_s, $t_{he}(\tau_s)$) can be roughly related to the breakthrough temperature at the same time, $u(\tau_s)$, by:

$$t_{he}(\tau_s) = \begin{cases} [1.872 - 0.964 \log u(\tau_s)]^{-1}; u(\tau_s) < 0.06 & (28a) \\ [1 - 1.675 \log u(\tau_s)]^{-1}; u(\tau_s) \geq 0.06 & (28b) \end{cases}$$

Values of $u(\tau_s)$ are available for a given M from either Eq. (14) or Fig. 3.

The exit-gas temperature during the heating part of the cycle at stationary state can be approximated by either a linear or parabolic function:

$$t_{he} \approx \begin{cases} t_{he}(\tau_s)\dfrac{\tau}{\tau_s} & \text{for } 2 \leq M < 5 \text{ and all } \tau_s; \\ & \text{or for } \tau_s > 2, \text{ and } M > 5 \quad (29a) \\ t_{he}(\tau_s)\left(\dfrac{\tau}{\tau_s}\right)^2 & \text{for } M > 5, \text{ and } \tau_s < 2 \quad (29b) \end{cases}$$

Comparison of predictions by these formulas and the exact results computed numerically would indicate that the two are in reasonable agreement, especially for the mean temperature. Remember that mean temperature is of interest in efficiency calculations. Using Eq. (4) we get:

$$E_h = \begin{cases} 1 - \dfrac{t_{he}(\tau_s)}{2}; M \leq 5; \text{ or } \tau_s > 2, \text{ and } M > 5 & (30a) \\ 1 - \dfrac{t_{he}(\tau_s)}{3}; M > 5, \text{ and } \tau_s < 2 & (30b) \end{cases}$$

Efficiency for countercurrent symmetric operation is presented in Fig. 10.

Unbalanced countercurrent operation is more complex and the reader should consult the literature for complete treatment of it [2,4,11,13]. An approximate answer for efficiency can be obtained by using the above approach to calculate E_h and E_c each with its own μ, M and τ_s. The overall efficiency is then calculated from Eq. (20).

Example 3

For the conditions of Example 2, including the switching time of $\theta_s = 7.2$ h, find the efficiency if the system operates with countercurrent flow. Also calculate the maximum attainable efficiency. Take all the bed and gas data as given in Example 2.

Solution: Method 1—The following have been already calculated: $\sigma_D^2 = 0.0405$; $M = 25$; $\mu = 12$ h; and $\tau_s = 0.6$.

From Fig. 10, at $M = 25$, $\tau_s = 0.6$, we find $E = 0.94$. Maximum efficiency is at $\tau_s \rightarrow 0$ and is given by Eq. (27):

$$E_{max} = M_h/(M_h + 1) = 25/26 = 0.96$$

Method 2—Evaluate $u(0.6)$ at $M = 25$ from Fig. 3; $u(0.6) \approx 0.02$. Evaluate $t_{he}(0.6)$ from Eq. (28a):

$$t_{he}(0.6) = [1.872 - 0.964 \log 0.02]^{-1} = 0.285$$

Calculate efficiency from Eq. (30b):

$$E = 1 - t_{he}(0.6)/3 = 1 - 0.285/3 = 0.91$$

The difference between the two methods is:

$$((0.91 - 0.94)/0.94) \times 100 = 3.2\%.$$

References

1. Jakob, L. M., "Heat Transfer," Vol. II, John Wiley & Sons, Inc., New York, 1957.
2. Schack, A., "Industrial Heat Transfer," John Wiley & Sons, Inc., New York, 1965.
3. Hausen, H., "Wärmeübertragung in Gegenstrom, Gleichstrom and Kreutzstrom," 2nd ed., Springer-Verlag, Berlin, 1978.
4. Schmidt, F. W., and Willmott, A. J., "Thermal Energy Storage and Regeneration," McGraw-Hill Book Co. (Hemisphere Publ.), New York, 1981.
5. Peiser, A. M., and Lehner, J., *Ind. Eng. Chem.*, Vol. 45, 1953, p. 2166.
6. Rohsenow, W. M., and Hartnett, J. P., "Handbook of Heat Transfer," McGraw-Hill Book Co., New York, 1973.
7. Singer, E., and Wilhelm, R. M., *Chem. Eng. Prog.*, Vol. 46, 1950, p. 343.
8. Abramowitz, M., and Stegun, I. A., "Handbook of Mathematical Functions," Dover Publications, Inc., New York, 1965.
9. "CRC Handbook of Tables for Mathematics," pub. by Chemical Rubber Co., Cleveland, 1967, p. 908.
10. Mood, A. M. "Introduction to Theory of Statistics," McGraw-Hill Book Co., New York, p. 421, 1950.
11. Dudukovic, M. P., and Ramachandran, P. A., *Chem. Eng. Sci.*, 1985, in print.
12. Levenspiel, O., *Chem. Eng. Sci.*, Vol. 38, 2035, 1983.
13. Ramachandran, P. A., and Dudukovic, M. P., *Comp. Chem. Eng.*, Vol. 8, p. 377, 1984.
14. Sagara, M., Schneider, P., and Smith, J. M., *Chem. Eng. J.*, Vol. 1, 1970, p. 47.
15. Gupta, S. N., Chaube, R. B., and Upadhyay, S. N., *Chem. Eng. Sci.*, 29, 1974, p. 839.
16. Wen, C. Y., and Fan, L. T., "Models for Flow Systems and Chemical Reactors," Marcel Dekker, Inc., New York, 1975.
17. Vortmeyer, D., and Schaefer, R. J., *Chem. Eng. Sci.*, Vol. 45, 1979, p. 489.
18. Hlavacek, V., and Votruba, J., Steady State Operation of Fixed Bed Reactors and Monolithic Structures in "Chemical Reactor Theory—a Review" (Lapidus, L., and Amundson, N. R., eds.), Ch. 6, Prentice-Hall, Inc., Englewood Cliffs, 1977, p. 314.

The authors

Milorad P. Dudukovic is professor of chemical engineering and director for the Chemical Reaction Engineering Laboratory at Washington University, Campus Box 1198, St. Louis, MO 63130. Tel: (314) 889-6082. He is known for his research and publications in reaction engineering of multiphase systems. He is active in several professional organizations, including AIChE, ACS and the Amer. Soc. for Engineering Education, and consults to a number of companies. He holds a diploma (B.S. degree) in chemical engineering from the University of Belgrade, and M.S. and Ph.D. degrees from Illinois Inst. of Technology.

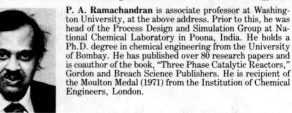

P. A. Ramachandran is associate professor at Washington University, at the above address. Prior to this, he was head of the Process Design and Simulation Group at National Chemical Laboratory in Poona, India. He holds a Ph.D. degree in chemical engineering from the University of Bombay. He has published over 80 research papers and is coauthor of the book, "Three Phase Catalytic Reactors," Gordon and Breach Science Publishers. He is recipient of the Moulton Medal (1971) from the Institution of Chemical Engineers, London.

Section V
Steam

How to optimize the design of steam systems

As energy and equipment costs increase, efficient steam systems become more important in the overall economics of process plants. Here is a method for modeling and optimizing an industrial steam/condensate system by using linear-programming techniques.

James K. Clark, Jr. and *Norman E. Helmick, Fluor Engineers and Constructors, Inc.*

☐ An iterative linear-programming (LP) algorithm can be used to minimize net costs in steam-distribution systems. These costs represent everyday operating expenses, or annualized costs including operating expenses and capital investment.

The LP technique will determine optimum values for the process-design variables, so as to achieve minimum cost. Typical variables include steam-header temperatures, turbine rates, letdown flowrates, makeup and desuperheating water flows, cooling/heating requirements and deaerator pressure.

When alternative designs are feasible, LP will also determine the optimum system configuration, based on selection between electric-motor or steam-turbine drivers, and selection and sizing of optional superheaters and condensate coolers.

The LP procedure is more flexible than methods based on solving a system of equations for heat and material balances. In addition, for a desired set of conditions with a specified objective, a final solution is obtained in a single computer run—thus eliminating the numerous solutions required by the case-study approach. Published work on the subject has emphasized methods based on the solution of heat and material balances with computer programs having little or no optimization capability [1,2,3,5]. A linear-programming

optimization approach has been proposed by Bouilloud [4], but it requires that all enthalpy values be known and that all thermodynamic calculations be done prior to the LP solution. It is up to the user to make repetitive trials based on different thermodynamic assumptions. Because of these requirements, the model is limited in what it can optimize.

When enthalpies are introduced as problem variables, direct LP solution becomes impossible due to the nonlinear nature of the energy-balance equations. The iterative technique described in the following sections gets around this difficulty by solving successive linear approximations of the nonlinear steam-balance equations. With each new solution, the linear approximations are improved. The final solution is achieved when all approximations are within a small tolerance of the actual nonlinear equations. Since each LP solution is optimal with respect to its linear approximations, the final solution must also be optimal.

Linear programming: a brief review

Linear programming is one of several mathematical techniques, known collectively as "mathematical programming," that attempt to solve problems by minimizing or maximizing a function of several independent variables. LP is the most widely used of these

Originally published March 10, 1980

methods, and is one of the best for analyzing complex industrial systems. Typical applications include determining the optimum allocation of resources (i.e., capital, raw materials, manpower and facilities) to obtain a particular objective such as minimum cost or maximum profit for the project.

In practice, optimal allocation of resources must be determined under conditions where there are alternative uses of resources and where physical, legal and managerial constraints must be met. Constraints take the form of upper or lower bounds that must be placed on resources or on operating conditions. For example, the availability of a raw material may be limited, or a process unit may have a maximum throughput. Other examples include the blending of petroleum products where sulfur contents, gasoline octanes, fuel-oil viscosities, etc., must meet upper or lower bounds to make salable products.

The LP procedure will maximize or minimize a linear objective function of the form:

$$\sum_{j=1}^{n} c_j x_j \qquad (1)$$

where the x_j are unknown problem variables, and the c_j are constant coefficients. The variables must be con-

strained by equality or inequality relationships, having the form:

$$\sum_{j=1}^{n} a_{ij} x_j \lesseqgtr b_i, \; i = 1,2 \cdots m \qquad (2)$$

where the x_j are the same problem variables as in Eq. (1), a_{ij} are constant coefficients, and b_i are right-hand-side constants. The m relationships indicated in Eq. (2) are called the constraint set. The objective function and the constraint set, together, form an LP model. Relationships indicated by Eq. (2) can be upper bounds (maximum constraints), lower bounds (minimum constraints), or equalities. The relationships are linear.

When developing a steam-system LP model, the engineer must derive the heat- and material-balance equations that relate the problem variables (i.e., process flows and enthalpies) to one another. In this article, most of the constraint set will consist of just such equations. However, it is also necessary to specify all of the upper and lower bounds that apply to the problem. Most of the inequalities will identify maximum and minimum temperature bounds. Finally, all LP solution codes require that all x_j be nonnegative (i.e., $x_j \geq 0$). This property helps to ensure feasible heat and material balances.

A number of computer codes have been developed to "solve" LP models. Many of these codes are available through the major computer manufacturers and service companies. Solution codes determine a set of values for the x_j that maximizes (or minimizes) the objective function while satisfying the relationships contained in the constraint set. Much has been written about the simplex algorithm that is used by all LP codes—such as the mathematics, options, and interpretation of results. These topics are beyond the scope of this article. For further information, see Refs. 6 and 7.

LP applied to steam systems

The constraint set that describes a typical steam system consists largely of the following four relationships:

1. Equations for the material and energy balances for each process unit and equipment item—this includes all headers, boilers, superheaters, turbines, and heat exchangers, flash drums, deaerators, etc.

2. An equation for each power demand (i.e., steam turbine), and each process-energy demand.

3. Upper and lower bounds for all independent variables.

4. Equalities and upper/lower bounds that are problem-dependent.

By way of example, let us consider the following steam header:

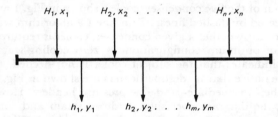

The x_j are input flows to the header. For each x_j, there is a corresponding enthalpy variable, H_j. (There is no loss of generality if some of the x_j or H_j are known, fixed values.) The y_i and h_i represent the header output flow and enthalpy variables, respectively. A material balance is expressed by Eq. (3); and the energy balance by Eq. (4).

$$\Sigma y_i - \Sigma x_j = 0. \tag{3}$$

$$\Sigma h_i y_i - \Sigma H_j x_j = 0. \tag{4}$$

Eq. (3) and (4) are not sufficient to describe the status of the header. The specific values of h_i depend upon the header pressure and state (liquid or vapor), as well as temperature. Thus, we must know the temperature, pressure and state in order to use a thermodynamic correlation to compute the enthalpy. Conversely, if the pressure, enthalpy and state are known, the temperature can be calculated. Throughout this article, header pressures are taken as known, constant values. Further, the output enthalpy variables, h_i, must all have the same value that we shall designate as H, e.g., $h_i = h_2 \cdots = h_m = H$.

Let k_1 be the enthalpy of saturated steam at the header pressure. It is apparent that if we wish to ensure a vapor state for all steam flows exiting the header, we must append the inequality:

$$H \geq k_1 \tag{5}$$

Process or equipment limitations normally require adding an upper bound on the header steam enthalpy. If k_2 is the maximum value, then the following inequality is also necessary:

$$H \leq k_2 \tag{6}$$

With the assumption of known header pressure, the relationships given by Eq. (3), (4), (5) and (6) are sufficient to describe the mass and energy balances.

Let us now consider a backpressure steam turbine, as shown:

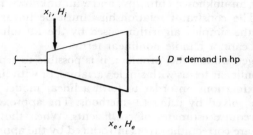

where H_i and x_i represent the input steam enthalpy and flow; H_e and x_e are the exhaust steam enthalpy and flow. The material balance is:

$$x_i = x_e \tag{7}$$

The energy balance is also the horsepower-demand equation:

$$H_i x_i - H_e x_e = D/C \tag{8}$$

In Eq. (8), C is a constant that converts the horsepower demand to Btu/h, and accounts for mechanical efficiency. Since the input steam is from a header with a known pressure, and the output enters a header with a known pressure, the exit enthalpy, H_e, can be expressed as a function of the input enthalpy, H_i:

$$H_e = \alpha + \beta H_i \tag{9}$$

The constants α and β in Eq. (9) will be derived later. Eq. (9) is problem-dependent and is only required if the header supplying the steam has a variable temperature (and hence, enthalpy). Further examples of problem-dependent constraints include bounds on fuel availability, boiler blowdown rate, and deaerator pressure.

Types of objective function

The objective function for a steam-system LP can range from very simple to quite complex. The simplest consists of a single variable. It represents either the steam-generation rate (in excess of that generated by process-heat recovery), or the total fuel consumed. In either case, the objective function is minimized.

An alternative objective function minimizes total operating expense. It includes costs for fuel, boiler-feedwater makeup, electric power, cooling-water makeup, and catalysts and chemicals for water treating. (The example that will be discussed in this article has as its objective the minimization of operating expense.) This type of objective function should be used by operating companies when the goal is minimum operating expense for an existing system.

A more comprehensive objective can be defined by including operating expense, and the cost of capital recovery plus return on investment for major equipment. For a new steam system in the design stage, this objective function represents the total variable cost of the system and, of course, is minimized.

The three types of objective functions described here are typical. Selection of a particular objective function depends on the purpose of the study.

The nonlinear problem

The relationships given by Eq. (4) through Eq. (8) may contain nonlinear terms such as Hx, where H represents an unknown enthalpy, and x an unknown flow-rate. The constraint relationships must be linear because the simplex algorithm (used by the LP solution codes) cannot handle nonlinear terms.

To get around this problem, it is possible to replace the nonlinear terms with single variables, or with linear approximations, in order to form a linear model that can be solved by using LP methods. The approximations require estimates of coefficients. When the estimates are correct, the errors introduced by the approximations become insignificant. The method to be described uses iterative estimation of coefficients, and subsequent problem solution, until a stable set of coefficients is found that satisfies maximum error requirements. The LP solution that utilizes this last set of coefficients is the solution that we are seeking.

In subsequent sections of this article, we will describe the linearization technique and the iteration strategy for obtaining the desired solution by means of an example of a typical steam system.

Modeling a steam system

A process flow diagram for a typical steam-distribution system is shown in Fig. 1. This system provides steam at three pressure levels: 850, 250 and 40 psig. Header temperatures can range from saturation to maximum superheat values. The optimum value for each header temperature will be determined by the LP procedure.

Process steam requirements and/or heating demands are indicated for each pressure level. Normally, such demands are given in constant Btu/h or lb/h. The diagram also shows that each pressure level has fixed sources of steam, which the system must accept.

The low- and intermediate-pressure levels each have one source where the volume and heat content are fixed. However, the temperatures of two other sources, going to the high- and intermediate-pressure headers, respectively, can be increased by superheaters. The latter units are shown by dashed lines to indicate optional equipment items. The LP procedure will determine the optimum temperature (fuel usage) for these superheaters. A zero fuel rate to a superheater simply means that the optimal solution does not include that superheater.

Power requirements

The four horsepower demands shown in Fig. 1 are designated by the symbol T/M. This indicates that the power demand can be satisfied by either a steam turbine (T) or a motor drive (M). The dashed lines for these units indicate that the proper selection will be made by the LP procedure. Three different types of tur-

bines have been included: backpressure, condensing, and condensing with extraction. In some cases, there may be a choice among turbine types for a given horsepower load, as well as a motor alternative. Here again, the LP procedure can choose the alternative required to minimize annual operating expense.

It is possible that a combination of turbine types, or both a turbine and a motor, will be selected to satisfy a horsepower demand. If the demand represents a single driver, such a solution is not physically possible, and it becomes necessary to make several case studies to select the optimum configuration. For each case study, all alternatives, except a single power source, are set to zero activity for each demand. Thus, if a power demand has been satisfied by both a motor and a turbine, two case studies are required. The solution with the lowest cost is optimum.

For the case where the horsepower requirement represents multiple drivers, combinations are acceptable. Individual driver types are selected by using the relative amounts of alternative energy sources. For example, if a power requirement represents ten drivers, and the LP solution indicates a 40%/60% split between turbines and motors, then four drives would be turbines, six would be motors.

Condensers, coolers and desuperheaters

Four of the five condenser/coolers shown in Fig. 1 are indicated by dashed lines; again, the LP algorithm will determine whether a given condenser/cooler is required for an optimum configuration. A zero cooling-water rate indicates that the subject cooler is not needed.

We notice that no desuperheaters are shown in Fig. 1 for the intermediate- and low-pressure headers. However, the diagram does show letdown steam and condensate going to each of these headers. This is all we need, since the LP procedure will determine the amounts of steam and condensate necessary to meet the mass and heat balances for a header. An LP solution that requires condensate to be added to a header indicates the need for a desuperheating station. If there is more than one steam flow that can be desuperheated, the design engineer has the option of selecting which streams to desuperheat. Again, this is an example of how the LP model helps to determine the optimum system configuration.

Deaerator pressure

A significant variable for the system is the deaerator pressure. It is a system variable because the enthalpies of the vent steam and the boiler feedwater reflect saturated values at the deaerator pressure, and these enthalpies are variables in the LP model. In practice, only one enthalpy, that of the boiler feed, is carried as a model variable. A linear correlation has been developed to represent the enthalpy of the vent steam as a function of the boiler-feed enthalpy over a practical pressure range.

Since boiler feedwater goes directly to all boilers and desuperheating stations, the enthalpy of the water has an important effect on the overall system design. The LP procedure chooses the optimum value for the deaerator pressure, as part of the final optimal solution.

The mathematical model of a steam system (such as

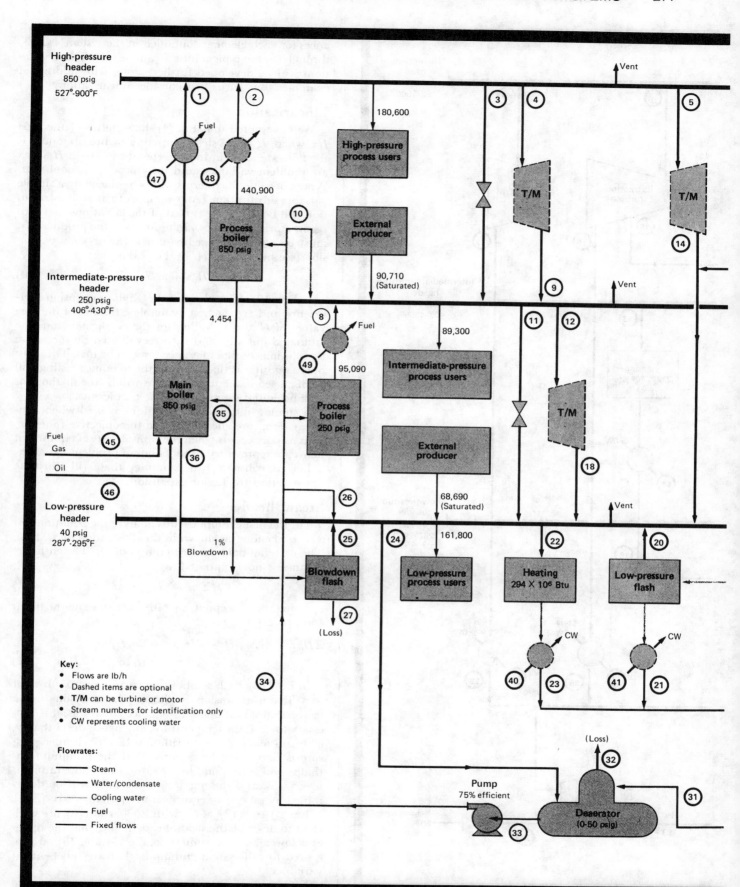

High-pressure
header
850 psig
527°-900°F

Vent

① ②

Fuel

47 48

440,900

⑩

180,600

High-pressure
process users

Process
boiler
850 psig

External
producer

③ ④ ⑤

T/M   T/M

⑭

90,710
(Saturated)

Intermediate-pressure
header
250 psig
406°-430°F

Vent

4,454

⑧

Fuel

49

95,090

⑨

89,300

Intermediate-pressure
process users

External
producer

⑪ ⑫

T/M

⑱

Main
boiler
850 psig

⑤35

Fuel
Gas 45

Oil

46

36

68,690
(Saturated)

Low-pressure
header
40 psig
287°-295°F

⑯26

Vent

1%
Blowdown

25

24

161,800

22

20

Blowdown
flash

Low-pressure
process users

Heating
294 X 10⁶ Btu

Low-pressure
flash

27

(Loss)

CW   CW

40 23   41 21

Key:
• Flows are lb/h
• Dashed items are optional
• T/M can be turbine or motor
• Stream numbers for identification only
• CW represents cooling water

34

Flowrates:

——— Steam
——— Water/condensate
- - - - Cooling water
—— Fuel
━━━ Fixed flows

(Loss)

32

Pump
75% efficient

33

Deaerator
(0-50 psig)

31

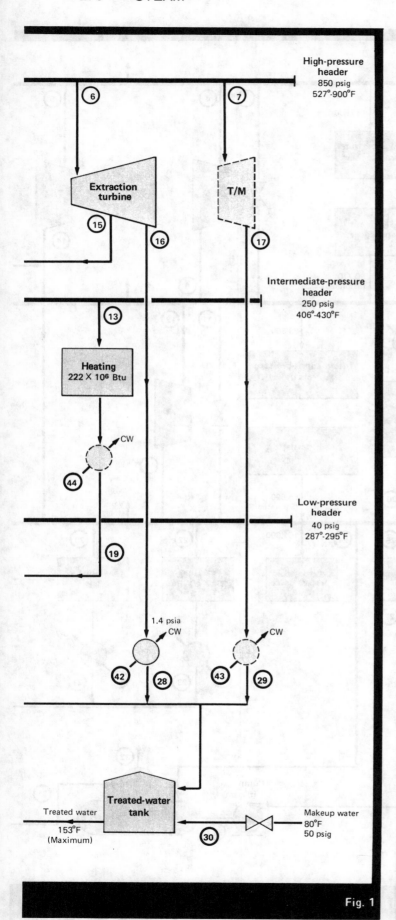

High-pressure
header
850 psig
527°-900°F

Extraction
turbine

T/M

Intermediate-pressure
header
250 psig
406°-430°F

Heating
222 × 10⁶ Btu

CW

Low-pressure
header
40 psig
287°-295°F

1.4 psia
CW

CW

Treated-water
tank

Treated water

153°F
(Maximum)

Makeup water
80°F
50 psig

Fig. 1

Fig. 1) will include equations for heat and material balances for each element contained in the system. Eq. (3) through (6) are typical for a steam header. These equations will be developed further, along with additional examples. But let us first consider linearization.

Linearization

We have noted that Eq. (4) shows terms of the type Hx, where H is the steam enthalpy in Btu/lb, and x is the flowrate in lb/h. In the general case, both H and x are problem variables, and their product is nonlinear. Where possible, such terms will be replaced by a simple variable substitution. (An example of variable substitution will be discussed as part of the equations that represent a primary boiler.) In most cases, the product, Hx, must be replaced by a first-order Taylor-series expansion (see p. 325 of Ref. 8). For example:

$$Hx \approx H_0 x + x_0 H - H_0 x_0 \qquad (10)$$

In Eq. (10), H_0 and x_0 are the Taylor-expansion coefficients, and are the best available estimates of the true values of H and x. Values for the coefficients must be estimated initially, and then reevaluated by successive LP solutions until a tolerance test can be met. The computer algorithms that perform the tolerance testing will be discussed in a later section, as will be the method for making initial estimates for the coefficients.

A similar linearization technique is used when nonlinear terms are encountered in the objective function. This usually occurs when the capital costs of equipment items are represented as functions of problem variables.

Let us consider two examples that will illustrate linearization by Taylor expansion.

Steam header

For the first example, let us analyze the equations for a steam header (as shown in the illustration on p. 118). The material balance is given by Eq. (3). The enthalpy balance is now expressed as:

$$H\Sigma y_i - \Sigma H_j x_j = 0 \qquad (11)$$

Using the concept of Eq. (10), we can write the linear form of Eq. (11) as:

$$\Sigma H_0 y_i + (\Sigma y_{i0})H - \Sigma H_{j0}x_j - \Sigma x_{j0}H_j = H_0 \Sigma y_{i0} - \Sigma H_{j0}x_{j0} \qquad (12)$$

In Eq. (12), each symbol that carries a zero subscript is a Taylor-expansion coefficient. In addition to the heat-balance and material-balance equations, it is necessary to limit the temperature, and hence the enthalpy, of steam leaving the distribution header (header pressure is fixed). We have noted that the minimum enthalpy will correspond to the saturation temperature as in Eq. (5), and the maximum enthalpy will be determined by process or equipment requirements [Eq. (6)].

Eq. (3) and (12), along with Eq. (5) and (6), are sufficient to describe the steam header. Note that the upper and lower bounds, similar to Eq. (5) and (6), are required for all system enthalpies that are independent variables.

Each steam header in Fig. 1 contains a flow labeled "vent." This flow represents steam or condensate that is thrown away to maintain valid header heat-and-material balances. Such a variable can be included in Eq. (3) and (12) as one of the y_i, i.e., one of the output flows. However, these variables deserve special consideration. Let v represent vent flow for the header shown on p. 118. Eq. (3) and (11) become:

$$v + \Sigma y_i - \Sigma x_j = 0 \qquad (13)$$
$$Hv + H\Sigma y_i - \Sigma H_j x_j = 0 \qquad (14)$$

If the system is well designed, the optimum value of v will be zero, in most cases. Therefore, when applying the Taylor expansion to the product Hv, we have:

$$Hv = H_0 v + v_0 H - H_0 V_0 \qquad (15)$$
$$v_0 = 0 \qquad (16)$$
$$Hv \approx H_0 v \qquad (17)$$

Substituting these relations into Eq. (14), and expanding the nonlinear terms, gives the final heat-balance equation:

$$H_0 v + \Sigma H_0 y_i + (\Sigma y_{i0})H - \Sigma H_{j0} x_j - \Sigma x_{j0} H_j = H_0 \Sigma y_{i0} - \Sigma H_{j0} x_{j0} \qquad (18)$$

Eq. (13) and (15) are the final set of heat- and material-balance equations for a typical header. A vent variable should be added to the equations for each steam header and condensate loop. For a well-designed system, all such variables will have a zero value. If more heat energy is available from process sources than can be used by the system demands, this will be indicated by nonzero vent values. For such cases, it may be wise to review the process-design and equipment specifications to determine whether the heat sources can be economically reduced or the system design modified.

Backpressure turbine

For the second example, let us consider the backpressure turbine (as shown in the illustration on p. 118). The LP module for this in matrix format is:

	x_i	x_e	H_i	H_e		RHS
(I)	-1	1			=	0
(II)	$-H_{i0}$	H_{e0}	$-x_{i0}$	x_{e0}	=	$x_{e0}H_{e0} - x_{i0}H_{i0} - D/C$
(III)			Z_1	-1	=	Z_2

where:

$$Z_1 = \left[1 - \rho_t\left(\frac{\partial \Delta h_s}{\partial H_i}\right)_p\right]$$

$$Z_2 = \rho_t\left[\Delta h_{s0} - \left(\frac{\partial \Delta h_s}{\partial H_i}\right)_p H_{i0}\right]$$

Row I in the matrix array is the material balance; Row II, the enthalpy balance; and Row III relates the output enthalpy, H_e, to the input enthalpy, H_i. C in the right-hand side (RHS) changes the horsepower demand to Btu/h. Row III expressed as an equation becomes:

$$\left[1 - \rho_t\left(\frac{\partial \Delta h_s}{\partial H_i}\right)_p\right]H_i - H_e =$$
$$\rho_t\left[\Delta h_{s0} - \left(\frac{\partial \Delta h_s}{\partial H_i}\right)_p H_{i0}\right] \qquad (19)$$

The development of Eq. (19) follows. Let Δh_s be the

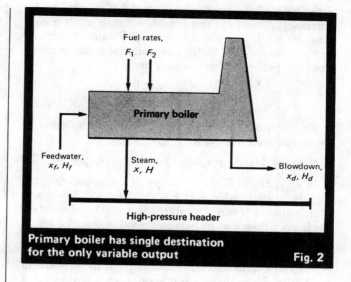

Primary boiler has single destination for the only variable output   **Fig. 2**

isentropic enthalpy change from inlet conditions for temperature and pressure to exhaust pressure. The actual work available is less than for an ideal reversible process. Hence, it is necessary to introduce a thermodynamic efficiency, ρ_t, defined as:

$$\rho_t = \frac{\Delta h_s - \int T ds}{\Delta h_s} = \frac{\Delta h}{\Delta h_s} \qquad (20)$$

The outlet enthalpy, H_e, is given by:

$$H_e = H_i - \Delta h = H_i - \rho_t \Delta h_s \qquad (21)$$

Since the exhaust pressure is fixed, the value of Δh_s depends only on the inlet conditions. Let Δh_{s0} be the value of Δh_s for the latest LP solution. Then Δh_s can be expressed:

$$\Delta h_s = \Delta h_{s0} + \left(\frac{\partial \Delta h_s}{\partial H_i}\right)_p (H_i - H_{i0}) \qquad (22)$$

$$H_e = H_i - \rho_t\left[\Delta h_{s0} + \left(\frac{\partial \Delta h_s}{\partial H_i}\right)_p (H_i - H_{i0})\right] \qquad (23)$$

Eq. (19) then follows from Eq. (23).

The partial derivative $(\partial \Delta h_s/\partial H_i)$ can be approximated numerically by using the values of H_{i0} and Δh_{s0} from the latest LP solution:

$$\left(\frac{\partial \Delta h_s}{\partial H_i}\right)_p \approx \frac{\Delta h_{s0} - \Delta h'_{s0}}{H_{i0} - H'_{i0}} \qquad (24)$$

The values of $\Delta h'_{s0}$ and H'_{i0} are computed by making a small temperature change (from the value of the last LP solution, which gave H_{i0} and Δh_{s0}), and calculating H'_{i0} and $\Delta h'_{s0}$ from the thermodynamic correlation for steam.

Substituting a variable

In this example, we will use substitution of a variable to remove a nonlinear term. A primary boiler whose output goes to a single destination, the high-pressure steam header, is shown in Fig. 2. The substitution is possible because the output has a single destination.

The diagram shows two fuels, F_1 and F_2, whose flowrates are expressed in lb/h. Associated with each fuel is a net heat of combustion, ΔH_1 and ΔH_2; and a boiler efficiency, e_1 and e_2. The blowdown rate, x_d, is a constant fraction, f, of the feedrate, x_f. Finally, F_1 has an upper bound of K, lb/h. The material-balance equations for boiler feed and steam are:

$$x_f - x_d - x = 0 \qquad (25)$$
$$fx_f - x_d = 0 \qquad (26)$$

The heat-balance equations must include heats of combustion for the fuels:

$$H_f x_f + e_1 \Delta H_1 F_1 + e_2 \Delta H_2 F_2 - Hx - H_d x_d = 0 \quad (27)$$

In Eq. (27), H is the enthalpy of steam entering the high-pressure header, and x is the flowrate. The minimum value of H corresponds to saturated conditions for the header pressure; its maximum value corresponds to the maximum amount of superheat that can be tolerated. H_d is the enthalpy of the blowdown, which is equal to the saturated liquid enthalpy at the pressure of the boiler. Since there is but a single destination for x, we can substitute $Hx = E$ into Eq. (27) to give:

$$H_f x_f + e_1 \Delta H_1 F_1 + e_2 \Delta H_2 F_2 - E - H_d x_d = 0 \quad (28)$$

where E represents the total energy delivered to the high-pressure header in Btu/h. Now, we note that E, x and H are related by:

$$E/x = H \qquad (29)$$

Further, let K_m be the minimum enthalpy value for H that corresponds to saturation conditions; and let K_h be the maximum value. Since H is an independent enthalpy variable, we must have the inequalities:

$$H = E/x \geq K_m \qquad (30)$$
$$H = E/x \leq K_h \qquad (31)$$

Finally, we have noted that the fuel rate, F_1, has an upper bound of K:

$$F_1 \leq K \qquad (32)$$

After expanding the term $H_f x_f$ in Eq. (28), and rearranging terms in Eq. (30) and (31), the final equations for the boiler model in matrix form are:

	x_f	H_f	x_d	x	F_1	F_2	E		RHS
(I)	1		-1	-1				$=$	0
(II)	f		-1					$=$	0
(III)	H_{f0}	x_{f0}	$-H_d$		$e_1 \Delta H_1$	$e_2 \Delta H_2$	-1	$=$	$H_{f0} x_{f0}$
(IV)					K_m		-1	$\leq$	0
(V)					$-K_h$		1	$\leq$	0
(VI)					1			$\leq$	K

Rows I and II are direct expressions of Eq. (25) and (26). Row III is Eq. (28) with the term $H_f x_f$ expanded, where H_{f0} and x_{f0} are the Taylor-expansion coefficients. Rows IV and V are Eq. (30) and (31) after rearranging, and Row VI corresponds to Eq. (32).

If steam from the boiler were supplying two, parallel, high-pressure systems, the substitution $Hx = E$ could not be made. This is simply because it is not possible to force the total heat, E, and mass, x, to be distributed in the same proportions at both destinations linearly.

The substitution can be, and should be, made when a

Modular LP models for a steam system   Table I

1. Steam boiler — variable flow and enthalpy
2. Process waste-heat boiler — variable Btu rate
3. Stand-alone superheater — variable Btu rate
4. Process user — fixed Btu demand
5. Process heater — fixed Btu demand
6. Steam header — variable temperature
7. Condensate flash — adiabatic, variable steam and Btu input
8. Deaerator — variable pressure
9. Backpressure turbine — fixed horsepower demand
10. Condensing turbine — fixed horsepower demand
11. Extraction turbine with condenser/cooler — fixed horsepower demand
12. Condenser/cooler — stand-alone
13. Pumps
14. Treated-water storage tank

boiler (or any process equipment item) has only one variable output and the output has only one destination. Upper and lower bounds, similar to Eq. (30) and (31), are usually required, as in the boiler example.

The complete model

The three examples that we have discussed are typical of what is required for obtaining linear equations to model process equipment. Space limitations preclude the development here of all the equations for the system shown in Fig. 1. A condensing turbine can be modeled by adding equations representing condensing and aftercooling to the model of the backpressure turbine. An extraction turbine can be modeled as two turbines in series with steam removal between them. A complete list of all mathematical models required for the system shown in Fig. 1 is given in Table I.

Input data and model definition

When solving a steam-balance problem, the first task is to define a model representing the desired configuration and operating requirements. This is accomplished by using input-data modules for each of the process units in the steam system. Table I shows the units for which data modules have been developed.

Each module is defined by completing a set of "fill in the blank" data sheets. Specific inputs for a given unit module include operating data required to define the unit's performance. This information is used to generate the constraint and heat- and material-balance equations for the unit. Examples of operating data would be a turbine's required horsepower, a process boiler's duty, or the exit temperature range for a superheater. Costs or revenues associated with the unit's operation or connecting process streams are also given.

The input-data sheets also identify the process streams connecting the units, along with the physical state of each stream. Data for these streams includes information defining their initial flow, pressure, enthalpy and stream type. Several stream types are possible: a stream may be either fixed or variable flow, and/or fixed or variable enthalpy, depending upon the process units connected to it.

Input data for process units

Table II

Boilers	Main	High-pressure process	Intermediate-pressure process
Rate, lb/h		440,900	95,090
Pressure, psig	850	850	250
Superheat range, °F	827-900	827-900	406-430
Blowdown rate, %	1.0	1.0	0.0
Heat-transfer efficiency, %	80.0	77.0	77.0
Fuel-gas heating value, Btu/lb	20,800		
Fuel-oil heating value, Btu/lb	17,300	17,300	17,300
Maximum fuel gas available, lb/h	20,000	0	0

Steam headers	High	Intermediate	Low
Pressure, psig	850	250	40
Maximum temperature, °F	900	430	295
Minimum temperature, °F	527	406	287

Turbines	H.P.-I.P.	H.P.-L.P.	I.P.-L.P.	Ex. #1	Ex. #2	H.P.-Vac.
Horsepower required, hp	11,170	13,555	2,217	19,080	7,241	2,633
Thermodynamic efficiency, %	75.9	68.55	65.51	86.67	78.9	67.8
Inlet pressure, psig	850	850	250	850	40	850
Exhaust pressure, psig	250	40	40	40	1.4 psia	1.4 psia
Condenser subcool, °F					0	0

Process heating	I.P.	L.P.	L.P. flash
Duty, million Btu/h	221.622	294.024	
Optional cooler:			
Minimum exit temp, °F	287	90	90
Maximum exit temp., °F	406	287	287
Maximum C.W. Δt, °F	70	70	70
C.W. inlet temp., °F	80	80	80

Process users/external sources	H.P.	I.P.	L.P.
Consumption, lb/h	180,600	89,300	161,800
Generation, lb/h		90,710	68,690

Treated-water tank	
Makeup temperature, °F	80
Makeup pressure, psig	50
Maximum exit temperature to deaerator, °F	153

Deaerator	
Maximum pressure, psig	50
Minimum pressure, psig	0
Minimum vent flow, lb/h	500
Feedwater pump efficiency, %	75

H.P. = High pressure, I.P. = Intermediate pressure
L.P. = Low pressure, Vac. = Vacuum, C.W. = Cooling water
Ex.#n = Extraction turbine, Stage n

In the example for the steam boiler, the flows designated as feedwater and steam have both variable flowrates and enthalpies. However, blowdown has only a variable flowrate, because the enthalpy is that of saturated water at the operating pressure of the boiler. When a stream is defined as having both variable flow and enthalpy, two Taylor-expansion terms and a constant are created to replace the single term that represents total energy in a heat-balance equation, according to Eq. (10).

Using the modular approach, any combination of units may be pieced together and solved. The user has great flexibility in modifying the configuration or operating data. Units may be added or deleted, and system designs may be altered by manipulating the input-data deck.

Solution strategy and technique

The system used to solve steam-balance problems consists of two computer programs. The first is an executive program that monitors and controls the solution strategy. The second can be any standard LP solution package. Problems are solved by repeated executions of, first, the executive program, and then the solution code, until specified convergence criteria are satisfied.

Our primary interest in this article is with the executive program because it makes all evaluations, decisions and model adjustments during the solution process. Its functions include: initial model construction from the input data, analyzing the LP solution output, checking for convergence, modifying the expansion coefficient values between iterations, and generating the LP-formatted input for each new iteration. The executive program includes correlations for generating thermo-dynamic properties of steam and water [9]. These are used in calculating the equation coefficients and right-hand-side values.

The purpose of the executive program strategy is to achieve a solution to the linearized model such that all Taylor-expansion coefficients are good approximations of their corresponding actual variable values. This is accomplished by using the recursive procedure as outlined in Fig. 3. As indicated, the solution strategy is divided into two phases: initialization and recursive-solution.

Initialization phase

The purpose of initialization is to obtain a consistent set of Taylor-expansion coefficients. These are estimated values of steam/condensate flowrates and enthalpies. To be consistent, they must satisfy the heat- and mate-rial-balance equations and the upper/lower bounds of the system model.

To begin initialization, all Taylor-expansion flow coefficients are set to zero, while the Taylor-expansion enthalpy coefficients are set to chosen "typical" values (as determined from input data). In addition, the en-thalpy variables are fixed at the same values chosen for the coefficients. All such values must be within the ranges defined by the upper and lower bounds that have been specified for the enthalpy variables. When completed, the above steps create a linear model having constant enthalpy values.

The linear model is then solved, and the values ob-tained for the steam and condensate flow variables be-come the initial Taylor-expansion flow coefficients for a second model. The initial Taylor-expansion enthalpy coefficients are simply the fixed values previously noted. Since all flow coefficients are initially zero, no conver-gence checking is done after the first LP solution.

To complete the initialization phase, the fixed status imposed on each enthalpy variable is removed, and the bounds are restored to their initial upper and lower lim-its specified by the input data. The system is now ready to obtain an optimum solution.

Recursive phase

After obtaining a consistent set of Taylor-expansion coefficients, the recursive phase begins. The second model is put in LP format by the executive program (using the Taylor-expansion coefficients just obtained) and is then solved by the solution code. The LP output is then interpreted, and the existing Taylor-expansion coefficients are compared with the corresponding flow and enthalpy solution values.

After each recursive pass, the executive program reads the LP solution output and checks for conver-gence of the Taylor coefficients. Fig. 4 outlines the con-vergence algorithm used. Both enthalpy and flow coef-ficients are checked against an allowable maximum fractional change, ε. In addition, flow coefficients may

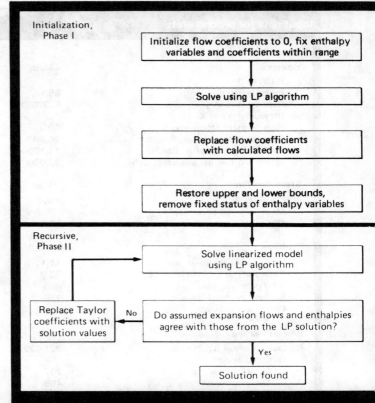

Recursive procedure for linearized steam model   Fig. 3

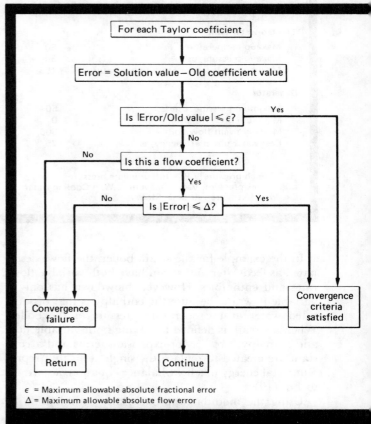

ε = Maximum allowable absolute fractional error
Δ = Maximum allowable absolute flow error

Convergence algorithm for Taylor coefficients   Fig. 4

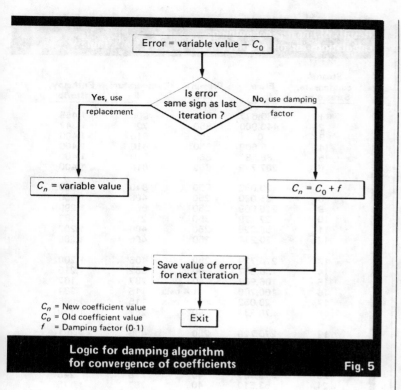

C_n = New coefficient value
C_o = Old coefficient value
f = Damping factor (0-1)

**Logic for damping algorithm
for convergence of coefficients**   **Fig. 5**

**Comparison of problem specifications for
different mathematical modeling systems   Table III**

Variable	H & MB	LP
Main-boiler superheat	Fixed	Upper and lower limit
Steam-level pressures	Fixed	Fixed
Steam-level temperatures	Fixed/calculated	Upper and lower limit
Process-generation superheat	Fixed	Upper and lower limit
Condensate subcool	Fixed	Upper and lower limit
Treated-water temperature	Fixed	Upper limit
Deaerator pressure	Fixed	Upper and lower limit

H & MB = heat-and-material balances
LP = linear programming

Utilities consumption and operating expense   Table IV

Utilities consumption*

Cooling water, thousand lb/h	5,138
Power, kW	1,034
Fuel gas, thousand lb/h	20.0
Fuel oil, thousand lb/h	20.6
Makeup water, thousand lb/h	279

Utilities costs

Cooling water	$ 10.00/million lb
Power	$ 0.03/kWh
Fuel gas	$ 1.21/million Btu
Fuel oil	$ 2.18/million Btu
Makeup water	$100.00/million lb
Annual operating costs	$ 9.96 × 10^6/yr

*Optimal solution values

satisfy convergence by meeting a maximum allowable error, Δ, in total flow. The total-flow tolerance is added to prevent streams having relatively small flowrates from hindering the convergence process. When changes in coefficients are all within the specified tolerance limits, a final solution has been found. If, however, any coefficient fails the test, all coefficients are recalculated, placed in the model, and another recursive pass is initiated through the program for the linearized steam model (see Fig. 3).

Stability of convergence

In certain cases, some variables may not follow a stable convergence pattern. They tend to fluctuate around the final solution value, delaying or even preventing convergence. This is especially true in the case of letdown flows between pressure levels, where small changes in header enthalpies can cause large changes in desuperheater-water or letdown-steam flows.

To solve this problem, damping logic (illustrated by Fig. 5) was included in the coefficient replacement procedure. As long as a coefficient is changing in a consistent direction, direct replacement of the coefficient with variable value is used. If the direction of the coefficient change reverses, the replacement is made by using the old value plus a specified fraction of the total change. This fraction, f, is specified as input data.

The sample problem that will be discussed converged in seven to eleven iterations, regardless of the problem's starting assumptions for header enthalpies, where ε was set to 0.0001, $\Delta = 5$ lb/h, and the damping factor, f, was 0.667.

Analysis of example problem

The steam-system design (outlined in Fig. 1) was optimized by using the recursive LP procedure. Variable flowrates for the steam system are color-coded as follows:

Color	Flow
Red	Steam
Blue	Water/condensate
Green	Cooling water
Orange	Fuel

All known (i.e., fixed) flowrates are colored black. Numbers beside each stream in Fig. 1 are for identification. A description of the problem is given on p. 119.

The numerical input data used for the LP solution are shown in Table II. These were obtained by first specifying a set of typical data that would be required to obtain a solution by using the traditional method of solving a set of simultaneous heat- and material-balance (H&MB) equations. The variables to be determined by the LP optimization were then removed from this list and replaced with a variable range. Each variable removed represents a degree-of-freedom that the LP procedure uses for obtaining the minimum-cost solution. Data in Table II are for illustrative purposes only, and do not represent operating data from an actual plant.

A comparison of the differences in problem specification between the LP and H&MB methods is shown in Table III. Note that ranges rather than fixed values are shown for most variables in the LP case. If desired, these variables can also be fixed, but since our purpose is that of minimizing costs, every effort should be made to give the LP algorithm the greatest possible flexibility. This means that variables should not be fixed unless required by equipment specifications or process design. As a result, less-specific input data are required for the LP procedure than for the H&MB method. Further, much of the required LP data is easily obtained from process and equipment requirements (loads, sources, and temperature limits), or from experience and engineering judgment (minimum steam temperatures).

The objective function of the example minimizes operating expenses. In this case, it includes the cost of fuel, makeup water, cooling water and power. The unit cost values are shown in Table IV.

The example LP-problem generated a linearized model containing 86 rows and 119 columns. Seven iterations (initialization + six recursive LP solutions) were required to obtain an optimum solution, using a coefficient-damping factor, f, of 0.667, and convergence tolerances, ε, of 0.0001, and, Δ, of 5 lb/h. The starting values, specified for the initialization phase, for this problem are those shown in Table II. Table V shows the results of the LP-problem solution for all steam, water and fuel flows. Steam-header vents are all zero as expected, and are not shown. The stream numbers in Table V correspond to those shown in Fig. 1. Utilities and total operating expense are given in Table IV.

Some points concerning the LP solution, which define important design requirements, include:

■ No electric motors were selected. All drivers are steam turbines.

■ An optional superheater with a duty of 6.3×10^6 Btu/h was added to the high-pressure process boiler (exit temperature 720°F). The intermediate-pressure superheater is not required.

■ Both the intermediate- and low-pressure headers indicated minimum enthalpies of saturated steam.

■ Treated-water tank temperature went to its maximum value of 153°F.

■ There is no letdown between the high-pressure (H.P.) and intermediate-pressure (I.P.) headers.

■ The desuperheating-water flow of 27,200 lb/h to the intermediate-pressure header indicates that a desuperheater is required on the H.P.–I.P. turbine exhaust. All other steam flows to the intermediate-pressure header are at saturated conditions.

■ The low-pressure (L.P.) header requires letdown steam and condensate. This indicates that the letdown steam plus the two turbine exhausts should be combined and introduced into a desuperheating station.

■ Deaerator pressure is at its minimum value of 0 psig.

■ Optional subcoolers were included on the low-pressure process condensate (3.5×10^6 Btu/h) and the low-pressure flash condensate (46.4×10^6 Btu/h). Since both coolers handle low-pressure condensate, the total duty of 49.9×10^6 Btu may be divided between the two streams (at the option of the design engineer).

Results of linear-programming calculations for the problem				Table V
Steam/ condensate, Stream no.	Flow, lb/h	Pressure, psig	Temperature, °F	Enthalpy, Btu/lb
1	430,677	850	900	1,455
2	440,900	850	720	1,347
3	0	850	810	1,400
4	276,609	850	810	1,400
5	186,587	850	810	1,400
6	207,729	850	810	1,400
7	20,052	850	810	1,400
8	95,090	250	406	1,200
9	276,609	250	564	1,298
10	27,179	850	213	183
11	59,325	250	406	1,200
12	70,234	250	406	1,200
13	270,729	250	406	1,200
14	186,587	40	362	1,216
15	106,969	40	287	1,167
16	100,760	1.4 psia	115	984
17	20,052	1.4 psia	115	1,066
18	70,234	40	287	1,120
19	270,729	250	406	381
20	36,520	40	287	1,175
21	234,209	40	90	58
22	319,933	40	287	1,175
23	319,933	40	276	245
24	53,510	40	287	1,175
25	2,535	40	287	1,175
26	4,383	850	213	183
27	6,270	40	287	256
28	100,760	1.4 psia	115	83
29	20,052	1.4 psia	115	83
30	279,070	50	80	48
31	954,024	0	153	121
32	500	0	212	1,150
33	1,007,034	0	212	180
34	1,007,034	850	213	183
35	435,028	850	213	183
36	4,351	850	527	521

Cooling water, Stream no.	Flow, gpm	Duty, million Btu/h	Fuel, Stream no.	Flow, lb/h	Duty, million Btu/h
40	100	3.5	45	20,000	332.80
41	1,325	46.4	46	7,600	105.18
42	7,265	90.77	47	8,026	111.01
43	1,578	19.72	48	4,974	66.30
44	0	0	49	0	0

Note: Stream numbers correspond to those on Fig. 1

■ The cooler on the intermediate-pressure condensate is not required.

To illustrate the advantages of the LP procedure, let us consider solving the minimization problem by using an H&MB program. In order to find the combination of conditions that minimizes operating expenses, many case studies would be required.

Each solution for a case study would represent the results based on an assumed set of fixed values for the problem variables given in Table III. If each of the nine variables [boiler superheat, header temperatures (3), process superheat, condensate subcool (2), treated-water temperature, and deaerator pressure] were investigated at only three levels, a possible 3^9 or 19,683 combinations would be required. If one assumes that many cases can be eliminated, based on past experience and engineering judgment, case-study optimization would still require a prohibitive amount of time and effort. In addition, the case studies use discrete values of the variables, which greatly increases the chance of missing the true optimum solution.

Even if a great many case studies are made, success in finding a near-optimum solution depends upon the sensitivity of the objective function (in this example, operating expenses) to changes in values of the operating variables. The greater the sensitivity, the greater the chances of not finding the optimum, or even a near-optimum, solution. By comparison, the LP algorithm uses continuous variables and will always find the true optimum solution for any problem where such a solution exists.

Applications and benefits

The LP method can be used as an operations-planning tool for existing steam systems to minimize operating expenses. This is especially true when one or more power demands can be satisfied by either electric motors or steam turbines. Maximum savings can be realized when there is significant flexibility in steam-header temperatures and deaerator pressure.

Incremental investments for existing steam systems can be studied by the LP method. Additions to an existing system can include new primary boilers, superheaters, desuperheaters, condensate coolers, turbine drives, etc. By including the total system in the LP model, the indicated savings due to an investment will not disappear because of unforeseen effects in parts of the system that are not directly associated with the new investment. Furthermore, we can be sure the LP procedure will adjust all aspects of system behavior to obtain maximum benefit from a new investment.

The greatest potential benefits of the LP method are in design engineering. For preliminary design, it is sufficient to minimize total steam or fuel consumption. For detailed design, the objective function should include all operating expenses and the daily capital-recovery plus return on investment of major equipment items. Principal benefits are derived from the ability of the method to choose the major design-parameter values that produce a truly optimized system. The design parameters include:

1. The ability to choose between turbine and motor drives, and/or between several turbine options. The incremental cost of primary-steam generation is included in such decisions because all steam boilers are part of the LP model, and because the total system cost is minimized.

2. Steam-header and condensate temperatures are problem variables and are part of the optimum solution. The design engineer need not estimate values for these variables to satisfy heat- and material-balance requirements. The procedure will select optimum temperatures so that all heat- and material-balance requirements are satisfied.

3. All inequality constraints, such as maximum or minimum temperatures, Btu demands, flowrates, etc., will be satisfied. This is a unique property of mathemat-

ical programming procedures. If the LP model is feasible, the LP-solution procedure will find a feasible and optimal solution. A lower-cost design is not possible, given the same problem definition.

4. Letdown-steam rates and desuperheating requirements will be determined by the LP solution. The design engineer need not "approximate" the requirements and recalculate all system heat-and-material balances to see whether the system is still feasible.

5. Cooling-water requirements and heat-exchange duties are determined as part of the LP solution.

6. Deaerator pressure is a system variable and becomes part of the optimum solution.

It is apparent that problem definition is more flexible for the LP method than for the traditional heat-and-material-balance simultaneous-equation approach, when the same level of economic sophistication is included. The reason is that estimates for heater temperatures, motor vs. turbine decisions, letdown flows, deaerator pressure, etc., are not required. The design engineer is able to concentrate on the demands of the overall system, its availabilities, and its upper and lower bounds.

References

1. Dodge, R. D., Gordon, E., Hashemi, M. H., and LaRosa, J., A Steam Balance Program Which Also Handles Combined Power Cycles, *Chem. Eng. Prog.*, July 1978.
2. Ruggerie, M. T. V., Automate Steam Balance, *Hydrocarbon Process.*, July 1977.
3. Slack, J. B., Steam Balance: A New Exact Method, *Hydrocarbon Process.*, Mar. 1969.
4. Bouilloud, P. H., Computer Steam Balance by LP, *Hydrocarbon Process.*, Aug. 1969.
5. Arnetin, R., and O'Connell, L., What's the Optimum Heat Cycle for Process Utilities?, *Hydrocarbon Process.*, June 1968.
6. Dantzig, G. B., "Linear Programming and Extensions," Princeton University Press, Princeton, N. J., 1963.
7. Gass, S. I., "Linear Programming Methods and Applications," McGraw-Hill, New York, 1958.
8. Zangwill, W. I., "Nonlinear Programming, A Unified Approach," Prentice-Hall, Englewood Cliffs, N. J., 1969.
9. Schmidt, E., "VDI Steam Tables," Springer-Verlag, New York, 1963.

The authors

James K. Clark, Jr., is a principal process engineer and systems analyst for Fluor Engineers and Constructors, Inc., 3333 Michelson Drive, Irvine, CA 92730. He joined Fluor in 1974 and has worked with and developed computer programs for various processes. Previously, he spent six years with Union Carbide Corp. as a senior process control engineer. He has a B.S. in chemical engineering and an M.B.A. from the University of Southern California.

Norman E. Helmick is chief of systems engineering for Fluor Engineers and Constructors, Inc., Irvine, CA 92730. He has over 29 years experience in chemical engineering, operations research and systems analysis, including extensive experience in developing linear programming models. He has a B.E. in chemical engineering from the University of Southern California and is licensed as a professional chemical engineer in California. He is a member of ACS, The Institute of Management Science, and Operations Research Soc. of America.

Steam accumulators provide uniform loads on boilers

Peak operating efficiency can be maintained when an accumulator meets the sudden demands of batch processes.

Norman Price, Consulting Engineer

☐ A steam accumulator is a pressure vessel that is partially filled with water. It permits the operation of boilers at a constant output equal to the average steam requirement.

During periods of low demand, surplus boiler steam is charged to the accumulator, below the liquid level in it. The steam condenses and the water is heated. As the water temperature increases, the pressure in the accumulator increases until it approaches that of the incoming steam.

During periods of high demand, the accumulator pressure is permitted to drop. Some of the water flashes to steam, and the temperature of the remaining water decreases to the saturation value for the reduced pressure. This process continues until the accumulator pressure approaches that required by the low-pressure steam users.

To use an accumulator, there must be some demand for steam at a pressure significantly lower than boiler pressure. As will be shown later, the fluctuating load need not be the low-pressure load.

To a limited extent, a boiler acts as an accumulator. An increase in steam demand causes a drop in pressure, and some of the boiler water is flashed to steam. This is more noticeable in a fire-tube boiler than in a water-tube boiler because the first holds more water per unit of steaming capacity.

However, a decrease in boiler pressure is undesirable. The combustion controls attempt to maintain boiler pressure by increasing the firing rate. When the load changes slowly, there is no problem. However, sudden increases in steam demand result in lower combustion efficiency, increased moisture in the steam going to the process or the superheater, decreased boiler life, and increased maintenance costs. And if a sudden demand exceeds the capacity of the boiler, or its ability to respond, the process is "short-changed." Plant production capacity becomes limited and, in some cases, product quality may be affected.

To correct this last situation, the immediate reaction is to call for more boiler capacity. Before proceeding in this direction, one should consider an accumulator. It

should reduce operating costs, and may even permit the shutting down of one boiler in facilities having a multiboiler installation.

Let us analyze the sizing procedures, consider typical hookups, and describe some industrial examples.

Accumulator sizing

For an existing plant, we begin by reviewing the boiler steam-flow recorder charts. For circular charts, we replot the data on rectangular coordinates. The area under the curve represents the total amount of steam (usually in pounds) supplied over a given time.

We determine the average flowrate by dividing the amount of steam by time, and plot this rate as a horizontal line on a rectangular graph. With an accumula-

Originally published November 15, 1982

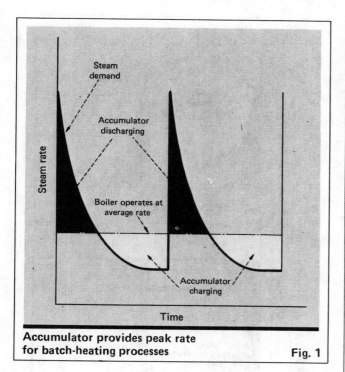

Steam
demand

Accumulator
discharging

Steam rate

Boiler operates at
average rate

Accumulator
charging

Time

**Accumulator provides peak rate
for batch-heating processes**

Fig. 1

tor, the boilers can be operated at this rate. The area above this line should, of course, equal the area below. The largest single area, above or below the average, represents the steam storage requirements.

If the recorder charts vary significantly from day to day, a suitable case (often the worst one) must be selected. Judgment is also required to determine a reasonable period for averaging. Do not average over 24 h if the fluctuating loads occur on only one shift. Allow for gradual changes in the boiler's firing rate (a departure from an average) to obtain a reasonable storage requirement for the unit.

Then, determine whether the low-pressure distribution system should be increased in size, i.e., diameter. This may be required so that the process will not be limited by steam flow.

For a new plant, the steam flowrates must be estimated and plotted from design data. Consideration must be given to the possibility of coinciding peaks. In the extreme case, where the steam demand changes rapidly from zero to a peak, an accumulator may be essential for efficient operation.

Fig. 1 illustrates a typical steam-demand curve for batch heaters whose cycles do not coincide or overlap. At the start of each cycle, the process fluid is cold and the steam demand is at a maximum. As the process temperature increases, the steam rate decreases because of decreasing temperature difference between the steam and the process. At the end of the cycle, steam is required only to compensate for heat losses.

If the boilers can supply steam at the average rate, the upper shaded areas represent steam supplied by the accumulator during periods of high demand, and the lower areas indicate steam supplied to the accumulator during periods of low demand. If the cycles are identical, these areas are equal and each represents the quantity of steam to be stored.

If two peaks coincide, the steam storage capacity must be doubled, and the steam lines to the process must be designed to handle double the previous flowrate. For the same daily production, the average steam flow is unchanged.

Accumulator pressures

The next step for either a new or existing plant is to establish the maximum and minimum accumulator operating pressures. The maximum pressure to be used in the calculation of storage volume is the boiler pressure less all line losses to the accumulator. The minimum pressure is the required process pressure plus all line losses (due to friction in the pipe and fittings) from the accumulator to the process.

From steam tables [2], we obtain the enthalpy of saturated liquid at the maximum accumulator pressure, h_{f1}, Btu/lb; enthalpy of saturated liquid at minimum accumulator pressure, h_{f2}, Btu/lb; average enthalpy of evaporation, h'_{fg}, Btu/lb. We can now calculate the change in enthalpy, Δh_f, as:

$$\Delta h_f = h_{f1} - h_{f2} \qquad (1)$$

In the accumulator, the sensible heat lost by the water between the maximum and minimum pressures equals the latent heat gained by the steam, or:

$$W_w(\Delta h_f) = W_s(h_{fg}) \qquad (2)$$
$$V_w = W_w v_f \qquad (3)$$
$$V_a = V_w/F \qquad (4)$$

where W_w = weight of water, lb; W_s = weight of steam, lb; V_w = volume of water, ft³; v_f = specific volume of water, ft³/lb; V_a = accumulator volume, ft³; and F = ratio of volume of water in tank, ft³, to volume of tank, ft³, dimensionless.

Accumulators are often designed to operate 90% full of water at the maximum pressure, i.e., $F = 0.90$. However, if entrainment problems are anticipated, a lower value of F is used. In any case, the vessel must be designed and constructed in accordance with the applicable unfired-pressure-vessel code.

Example illustrates procedure

An accumulator is required to store 1,000 lb of steam at a maximum pressure of 200 psig, and minimum pressure of 100 psig. The accumulator is to be filled with water to 90% of its capacity. What will be the required volume of the accumulator for these conditions?

From steam tables [2], we find that $h_{f1} = 361.9$ Btu/lb, $h_{fg(1)} = 837.4$ Btu/lb, and $v_f = 0.01847$ ft³/lb at the maximum pressure of 200 psig. Also from the steam tables, $h_{f2} = 309.1$ Btu/lb, and $h_{fg(2)} = 880.6$ Btu/lb at 100 psig. Substituting the appropriate values into Eq. (1) yields:

$$\Delta h_f = 361.9 - 309.1 = 52.8 \text{ Btu/lb}$$
$$h'_{fg} = (837.4 + 880.6)/2 = 859.0 \text{ Btu/lb}$$

From these data, we calculate the weight of water required, W_w, by substituting into Eq. (2) as:

$$W_w(52.8) = 1,000(859)$$
$$W_w = 16,269 \text{ lb}$$

The volume of water, V_w, from Eq. (3) will correspond to this weight:

$$V_w = 16,269(0.01847) = 300 \text{ ft}^3$$

Since $F = 0.90$, we find the vessel's volume from Eq. (4) as:

$$V_a = 300/0.90 = 333 \text{ ft}^3$$

Varying high-pressure demand

Fig. 2 depicts a system having a fluctuating high-pressure load, and a constant low-pressure load. The fluctuating load can be considered typical for batch heaters.

The boilers would be operated at the average total steam demand. At all times, the boilers must supply the requirements of the high-pressure users. When the total load is higher than average, the accumulator supplies part of the low-pressure load. When the total load is less than average, the boilers supply steam (difference between steam rate and demand) to the accumulator.

This system has one obvious limitation—the maximum high-pressure demand cannot exceed the boiler operating rate. Under the limiting condition, the entire low-pressure load is supplied by the accumulator, which is shown at the start of each cycle in Fig. 2.

Accumulator hookups

When only the low-pressure load fluctuates, the system can be operated as shown in Fig. 3. Steam flow to the accumulator and the low-pressure system is maintained at a constant rate equal to the average steam demand. When the actual demand exceeds this value, the accumulator supplies the difference. When the demand is less than the average, the surplus steam charges the accumulator.

Variations in the plant production rate will cause variations in the average steam demand, and the accumulator will tend to become either fully charged or fully discharged. Approach of either extreme condition is detected by sensing the accumulator pressure and resetting the flow controller. A somewhat oversized accumulator will help to prevent excessive resetting of the flow controller.

Pressure to the process is usually maintained constant by a pressure controller. This may not be required if the accumulator operating-pressure range is small and there are adequate controls on the process equipment.

A simplified diagram for a system in which the high-pressure load fluctuates is shown in Fig. 4. There is a preferential flow of steam to the high-pressure users. A backpressure controller maintains boiler pressure on the high-pressure system. The steam generated by the boilers and not required by the high-pressure users is the only portion permitted to overflow into the accumulator and low-pressure system. As the high-pressure load varies, a slight pressure change in the high-pressure header causes a compensating change in position of the backpressure control valve. If the accumulator approaches a fully charged or fully discharged condition, a signal can be sent to the boiler controls to reset the firing rate. Steam pressure to the low-pressure users is maintained by a pressure-reducing controller.

The instrumentation symbols in Fig. 3 and 4 are in-

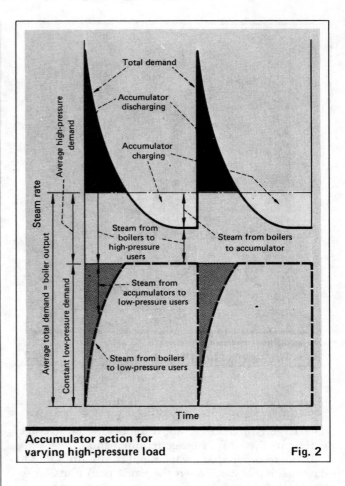

Accumulator action for varying high-pressure load   **Fig. 2**

tended to indicate functional requirements only. Specific hardware and degree of sophistication can be in accordance with plant preferences.

Applications

Fluctuating loads requiring the use of steam accumulators are usually found in batch processes. Several examples will be described:

Laminated-safety-glass heating—Glass is heated by direct contact with oil in an autoclave. The oil is circulated through an external tubular exchanger where it is heated by steam. The steam rate varies from maximum at the start of a batch to virtually zero during a holding period at the end of the cycle.

A packaged boiler supplies steam exclusively to the heaters. The boiler operates at the average rate, and an accumulator handles the peaks and valleys. To permit the use of an accumulator, the boiler pressure must be maintained about 100 psi higher than the required pressure at the exchangers.

Glass making—The boilers in this glass plant serve many types of users. Most of the loads require low-pressure steam; but the autoclave heaters (the fluctuating load) require steam at essentially boiler pressure. A backpressure control system, similar to Fig. 4, permits the boilers to operate at a substantially constant rate. Full steam pressure is maintained on the autoclave heaters, and the accumulator compensates for the load fluctuations.

Rubber vulcanization—Low-pressure steam is used to

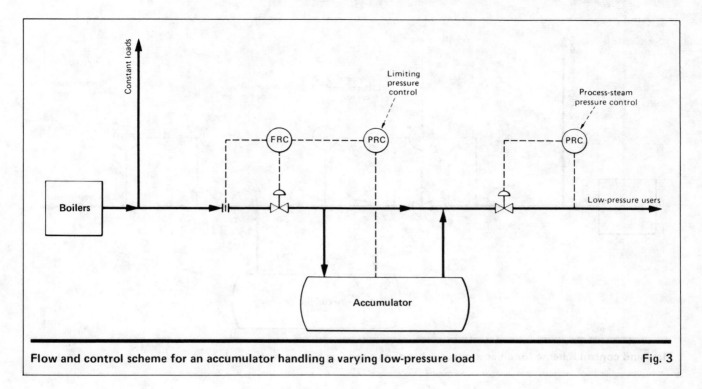

Flow and control scheme for an accumulator handling a varying low-pressure load   Fig. 3

heat the mold and its contents. Steam flow is a maximum at the start, then decreases during the cycle. Two or more vulcanizers may start their cycles at the same time. A flow controller on the steam supply to the accumulator, similar to that in Fig. 3, permits the boilers to operate at constant rate.

Footwear boarding—Nylon stockings are placed over leg-shaped forms and subjected to direct contact with steam for several seconds followed by about two minutes of zero demand. There is overlapping in the operation of several machines. Installation of an accumulator reduces boiler operating problems and increases production capacity. Flow control is used on the high-pressure steam, but there is no reducing station on the accumulator outlet.

Tobacco processing—Tobacco is moisturized by subjecting it to several cycles of vacuum, followed by low-pressure steam. A steam accumulator permits the boiler to operate at constant rate.

Pulp digesting—Digester heaters require large steam flows at the start of a cycle because the cooking liquid is cool. Steam rate decreases during the "cook." Accumulators have been used to smooth out the load.

Jet-engine testing—High-altitude conditions are simulated by the use of high-capacity steam-jet ejectors. The ejectors and low-pressure turbines require 700,000 lb/h of steam at 90 psig for 3 min. Hence, the accumulator stores 700,000 lb/h × 1/20 h, or 35,000 lb of steam.

Steam at 400 psig and 600°F is purchased from a utility at 35,000 lb/h. This allows one test every hour. Because of the superheat, more than one pound of water is evaporated in the accumulator for every pound of purchased steam condensed. In such cases, makeup feed is usually taken from the boiler feed pump. Since this facility has no boilers (and, hence, no feed pumps), a feed pump for the accumulator is required.

Electrical-equipment testing—Steam accumulators have also been used to simulate a loss-of-coolant accident (LOCA) in the testing of electrical equipment (e.g., solenoid valves, cables, motors, and motor-control centers) for use in nuclear-power generating stations.

A loss-of-coolant accident would be expected to involve a rupture in a steam line, and the sudden release of large quantities of steam, with rapid rise of temperature and pressure in the surrounding area. The performance of electrical equipment under such conditions must be determined in advance.

A LOCA test (IEEE Standard 323-1974) requires that the equipment be placed in a chamber and undergo large temperature and pressure increases by the direct injection of steam in periods as short as ten seconds. Performance of the electrical equipment is usually monitored before, during and after the injection of steam.

A steam accumulator may be charged at a slow rate over a convenient period, and discharged in 10 s without upsetting the boiler.

Additional considerations

If an entrainment problem is anticipated, a horizontal accumulator can be operated as low as 50% full. This level of fullness provides the maximum liquid surface for evaporation and the minimum vapor velocity leaving the surface. It also increases the vapor-liquid disengaging space in the accumulator. Of course, a larger vessel is required to store a given quantity of steam. An entrainment eliminator may also be installed either inside the vessel or in the accumulator discharge line.

Control of an accumulator involves some principles that are opposite those for boiler control. An accumulator must operate with a varying pressure and varying liquid level. At maximum pressure, the level is a maxi-

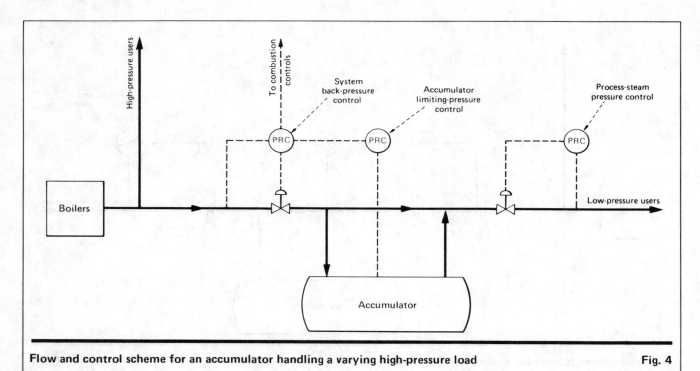

Flow and control scheme for an accumulator handling a varying high-pressure load   Fig. 4

mum. As the pressure decreases, some liquid is flashed into steam, and the equilibrium level drops. (The level may momentarily rise due to boiling.) There is a correct level for every pressure.

When the accumulator is charged with saturated steam, the level at a given pressure tends to stay constant from cycle to cycle. If over a period of time the level builds up because of heat loss (more incoming steam condensed than water evaporated), the excess water may be drained.

If the accumulator is charged with superheated steam, more than one pound of water is evaporated per pound of steam condensed. Treated makeup water must be added. Level control should correspond to a given pressure. To avoid decreasing the available steam-storage capacity, makeup should not be added when the accumulator is discharging.

When the boilers produce superheated steam and some low-pressure users require superheated steam while others use saturated steam, two low-pressure headers should be provided. The superheated-steam users should get steam directly from the boilers via a reducing valve. The saturated-steam users should be provided with saturated steam from the accumulator. A backpressure controller can be used to limit boiler steam flow to the accumulator.

Superheating the system

When some low-pressure users require superheated steam and the boiler produces saturated steam, an electrically heated superheater may be used. The heating elements may be sized for maximum demand, or the unit provided with sufficient mass and surface to permit some storage of heat during periods of low demand, followed by release during periods of high demand.

Occasionally, an electric boiler may be justified for a small average load. If the load is intermittent with sharp peaks, an accumulator may be required. Consideration should be given to combining the heating and storage functions, e.g., putting electric heating elements in the accumulator to provide for both the generation and storage of steam. The vessel must meet the requirements of the applicable boiler code.

An unfired steam generator is sometimes required for a process. This unit is a pressure vessel containing a tube bundle. Typically, water of a specified purity is vaporized in the vessel while boiler steam is condensed in the tubes. In other applications, heat is provided by available pressurized hot water or a hot process fluid. If the load fluctuates and if the steam generator can generate steam at a pressure significantly higher than required by the process, the unit can be designed to function as an accumulator.

References

1. Price, N., Accumulators Handle Fluctuating Steam Loads in Process Plants, *Industry & Power,* Sept. 1951.
2. Kennan, J. H., and Keyes, F. G., "Thermodynamic Properties of Steam," Wiley, New York, 1936.

The author

Norman Price is a consultant in process, environmental and energy engineering, 2117 Mather Way, Elkins Park, PA 19117. Telephone: (215) 886-1034. He was principal process engineer for Day & Zimmermann, Inc., when he wrote this article. His experience includes process and project engineering. Mr. Price has a B.Ch.E. from City College of New York, and an M.Ch.E. from New York University. He is a professional engineer in New York, New Jersey and Pennsylvania, a licensed marine engineer, and a member of AIChE and ACS.

Superheated-steam properties at a glance

V. Ganapathy*

☐ When engineers need to know the values of super-heated-steam properties, they generally go to the steam tables. However, frequent double-interpolation between temperatures and pressures can be cumbersome.

These two nomographs provide enthalpy and specific volume for superheated steam over the entire range of the steam tables. While the results are only approximations, they are accurate enough for many engineering applications.

*Struthers Thermoflood Corp., P.O. Box 753, Winfield, KS 67156

Examples

1. What is the specific volume of superheated steam at 650 psia and 800°F? *Solution:* On Fig. 1, connect $T = 800$ with $P = 650$, extend the line to the "V" scale and find the specific volume $V = 1.1$ ft³/lb.

2. What is the enthalpy of superheated steam at 650 psia and 800°F? If this steam is throttled to 300 psia, what is the final temperature? *Solution:* On Fig. 2, connect $P = 650$ with $T = 800$, extend the line to the "H" scale and read $H = 1,402$ Btu/lb. Connect this point with $P = 300$ to obtain $T' = 780°F$.

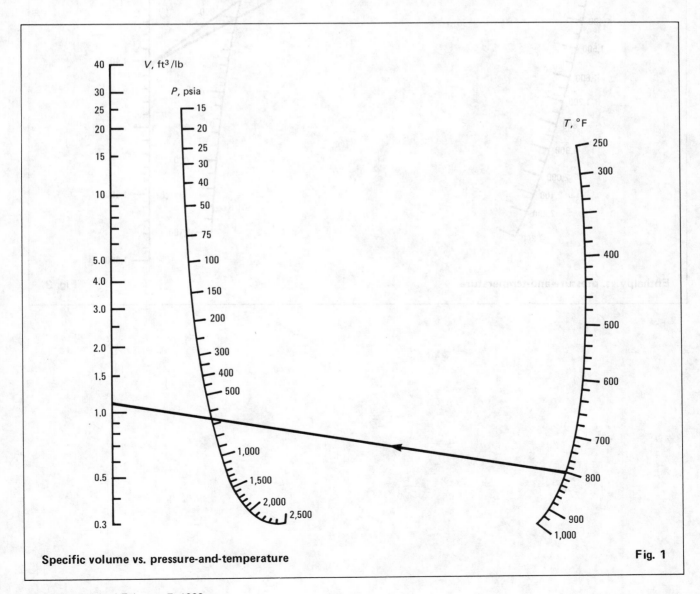

Specific volume vs. pressure-and-temperature

Fig. 1

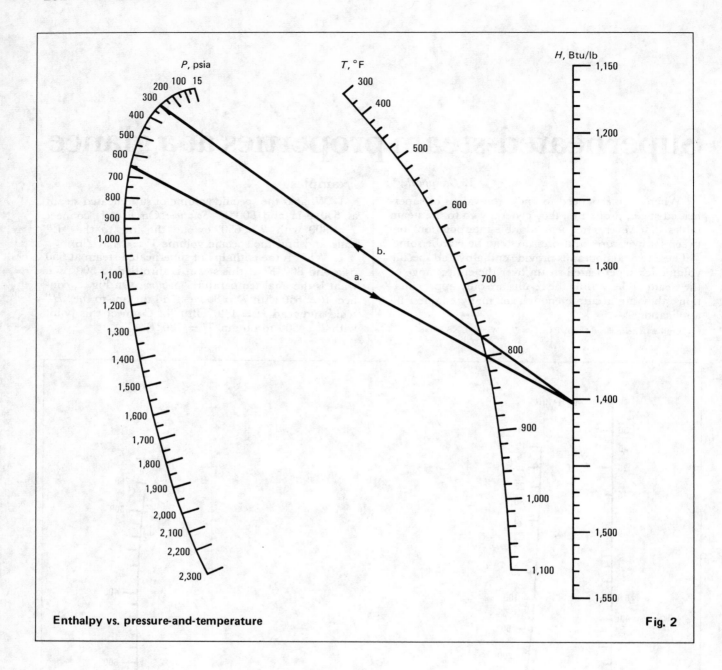

Enthalpy vs. pressure-and-temperature

Fig. 2

Steam traps— the quiet thief in our plants

In the first year of setting up a steam-trap program, one dollar spent on upgrading traps can return more than three dollars in energy savings.

☐ A malfunctioning steam trap is a great waster of energy. Almost any large plant in the chemical process industries has some (often many) energy-wasting traps.

The cost of energy is usually a large percentage of the cost of manufacturing. In most petrochemical plants, for example, it is the second most costly item (the greatest expenditure being for hydrocarbon raw materials). This energy portion of the cost of manufacturing ranges from 8 to 25%. Fuel used for producing steam is usually the major part of this. And most plants can save 10 to 20% of fuel costs simply by having a formal, active steam-trap program.

To carry this out requires someone on the plant's staff who is interested in energy, has a technical background (preferably an engineer), and can devote sufficient time to develop a simple, yet effective, program.

For field work, this engineer will need one mechanic or pipefitter for each 3,000 operating steam traps. For the first year, a return of $1 million in energy saving for each $300,000 spent in upgrading the system is the rule rather than the exception.

(For estimating purposes, you can assume that a modern petrochemical plant will have about one steam trap for each 100 lb/h of steam consumed.)

How to get started

The road to an effective steam-trap energy-saving program can be made much easier by following these six steps.

1. Select a steam-trap energy coordinator.
2. Develop a plant standard of four or fewer approved steam-trap types. (See the third article of this report, on p. 92.)
3. Monitor steam consumption for your different products (pounds of steam consumed per pound of product produced).
4. Go into selected plant areas, change out *all* steam traps and monitor steam savings.
5. Develop and implement a steam-trap checking procedure to detect bad traps (see the article on p. 275) and change them out using your plant trap standard.
6. Standardize on a set trap-checking frequency, say every six months, with plant-staff followup to make sure the program is working.

These six steps can be divided into two sections—Steps 1-4 are designed to solve *misapplication*, and Steps 5 and 6 are designed to eliminate *neglect*. Recognizing the possibility of trap misapplication and neglect goes a long way toward solving trap problems.

Energy coordinator

Selecting the right person as the steam-trap energy coordinator is the key to the success or failure of an effective steam-trap program. This person should be a senior (at least five years of plant service) employee who has a keen interest in energy saving. A desire to learn is more important than previous steam-trap experience.

Steam-trap manufacturers provide schools or seminars on steam traps—types, failures, applications, etc—and all are anxious to tell their stories. The local steam-trap representative can arrange for attendance at such classes. Also, various state, local and engineering society groups have knowledgeable people who will usually be glad to assist you by sharing their steam-trap experience. Such experience can be more valuable than vendor information because of the sometimes conflicting claims from the various trap manufacturers. An important thing to remember is that the person selected for this assignment should be free for at least a year to devote *full* time to steam traps and be willing and technically able to learn and understand steam-trap types, applications, failures, testing, and the like.

Establish a cost base

One of the best ways to show the effectiveness of your steam-trap program is to choose an area of 150 or more traps that does not now receive routine steam-trap maintenance. Instrument the steam flow to this area so that steam consumption can be totaled. Operate the area at "average" production or steam consumption for a reasonable time— say a month. Then inspect each trap and record the percentages of cold, properly working, and bad (blowing steam) traps.

Having done this, change out *all* traps, good as well as

bad ones, to the newly established plant steam-trap standard. Then measure the steam consumption for the same time, again say a month, and show the "average" saving per steam trap. This will normally convince the most skeptical manager of the effectiveness of a steam-trap program. You will also probably notice that the steam plumes from venting steam traps disappear in this area—a more visible sign of the program's effectiveness.

It is not uncommon for the average saving to be at least 10 lb/h per steam trap for systems whose pressure is under 500 psig. This is an approximate energy saving of $300 per year for each trap.

Total plant changeout

Armed with the success of the test area, you can now expand to the entire plant. Select a crew of plant pipefitters (maintenance mechanics) to be the steam-trap repair and replacement crew. These again should be energetic and knowledgeable field people. One field person for each 3,000 steam traps that are easily accessible will be sufficient. Your ratio may change to one field person for each 1,000 steam traps if the traps are scattered, not accessible, or in very poor condition initially. A crew of six workers can replace about twelve traps per day. So for a plant containing 3,000 traps, it will take about a year for a complete changeout. Newly installed steam traps should conform to your plant standard, which should address itself to easily accessible and maintainable steam-trap "banks."

Steam-trap checking procedure

During this changeout time, a comprehensive steam-trap checking program should be formulated by the plant energy coordinator. Each trap should be numbered and listed on both a steam-trap map and a steam-trap list. These will be the guides used to find each trap during the routine six-month test. The "trap map" should show a plot plan of each process area, with the traps numbered in letter groups of 50 or fewer traps. Example:

Area A	1A through 39A
Area B	1B through 54B
Area C	1C through 47C
Total traps in area	140

The "trap list" should have each letter trap-area on a single sheet (using both front and back). This makes expanding the listing of traps in each area easy, by either adding to an existing letter group or creating a new one. Information on the "trap list" should be:

- Trap number (1A, 5B, etc.).
- Service description (instrument tracing, column heater, etc.).
- Trap condition (cold, good, etc.).
- Operating pressure (10 psi, 50 psi, etc.).
- Type trap from the plant standard.

Each trap should then be permanently marked with a stainless steel tag on its condensate piping with the trap's respective number (1A, 5B, 47C, etc.). This allows easy trap identification when using the map and list. Note that the tag is attached to the piping—not the trap itself. Thus, if a defective trap is replaced, the number remains in place and does not have to be changed from the old trap to the new one.

Routine checking frequency

With all traps changed out to good ones and a plant standard established for specifying replacement traps, the system is ready for the first semiannual checkup. After the first year, a normal trap failure rate of less than 10% can be expected. For 600-psig service, you may have a 70% failure; but for a 60-psig service, a 2% failure is more likely with properly selected traps.

A normal six-month check frequency is recommended initially for all traps. This should be done in January (winter conditions) and June (summer conditions). The six-month routine check of every trap in the plant will not require large manpower expenditures. A skilled trap checker and a helper can check about 500 traps a day. So for a crew of two with a plant of 3,000 traps, only six working days (twelve man-days) are used. At an average of $100/man-day, the cost per check for each trap is only 40¢. One trap found and corrected from blowing only 40 lb/h will pay for all 3,000 traps being checked.

It is very important to use the same people to check the traps each time. Their knowledge from repeating checks will find and solve some of the difficult trap problems. It will also guarantee uniformity in checking. Most plants have found that a set trap-checking crew can check at least twice as many traps effectively in a given time as a crew whose permanent assignment is not traps.

Here is an example of how the checking would work. For a plant with 3,000 traps, all are checked in January by the permanent trap mechanic and a helper. This can be accomplished in less than two weeks. Then, during the next two months, the lone mechanic will replace all "bad" traps. For the remaining three and a half months, the trap assemblies that were changed out are disassembled, and new assemblies made up for use during the second six-month changeout.

A typical timetable for a 3,000-trap, single-mechanic system may look like this:

Task	Date
Winter steam-trap check	Jan. 8-Jan. 16
Replace bad 600-psig traps	Jan. 17-Feb. 14
Replace bad 425-psig traps	Feb. 15-Feb. 29
Replace bad 310-psig traps	Mar. 1-Mar. 7
Replace bad 190-psig traps	Mar. 8-Mar. 23
Replace bad 130-psig traps	Mar. 24-Apr. 6
Replace bad 60-psig traps	Apr. 7-Apr. 19
Replace bad 25-psig traps	Apr. 20-May 7
Repair trap assemblies	May 8-July 7
Summer steam-trap check	July 8-July 16

This completes the cycle and starts it over again.

Conclusion

There are four essentials to an effective trap program:
1. Select the right person to head the program.
2. Set up a plant trap standard.
3. Establish a routine trap-checking procedure.
4. Set up a permanent trap-maintenance crew.

These few people will save a minimum of $100,000/year/person in energy cost. Your plant will also be quieter, have fewer steam plumes, and have greater utility—all from a simple yet effective steam-trap program.

In-plant steam-trap testing

A planned check of all the plant's steam traps—at regular intervals—will pay off in energy savings.

☐ In many plants, most of the steam traps are actually steam wasters. The higher the price of energy, the more important it becomes to bring these losses under control. Following is the procedure used in our plant.

NO.	DESCRIPTION	CONDITION	STM. PRESS.	TRAP
AREA CFA	BLDG. 11	FLOOR 1st	DATE	
1A	Cond. tk. vent inst.		35	I
2A	Regeneration steam heater		500	II
3A	" " "		"	"
4A	H$_2$O recorder trap		35	I
5A	Process line tracers	.	35	I
6A	" " "		"	"
7A	" " "		"	"
8A	" " "		"	"
9A	" " "		"	"
10A	" " "		"	"

Fig. 1—In order to systematize steam-trap testing, you must first know what steam traps there are in your plant, and where each is to be found. Survey the plant and group the traps at each location into lots of 50 or fewer. Give each trap an identification number, and include it on a list. This list also shows trap service (tracing, drip leg, etc.), pressure (in psig) and a code for the type and size of trap.

Fig. 2—If your plant has standardized on a few types of traps, you can save space on the trap list by giving each type a code number. These are the codes for Du Pont's Victoria plant.

CODE	DESCRIPTION	SERVICE
I.	Bucket trap [Model___]	Light loads Low pressures
II.	Disk trap [Model___]	Light loads High pressures
III.	Bellows trap [Model___]	Moderate loads All pressures
IV.	Level chamber and control automatic	Heavy loads All pressures

Originally published February 9, 1981

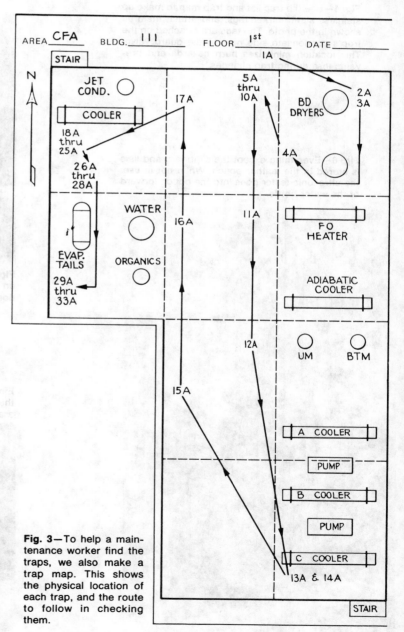

Fig. 3—To help a maintenance worker find the traps, we also make a trap map. This shows the physical location of each trap, and the route to follow in checking them.

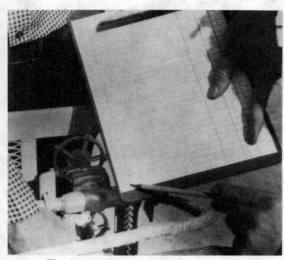

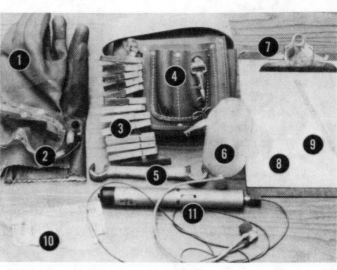

Fig. 5—Here are the tools and equipment used in checking the traps:
1. Gloves
2. Goggles
3. Spring-type clothespins (colored and plain)
4. Carrying pouch and belt
5. Valve wrench
6. Water-filled squeeze bottle
7. Clip board
8. Trap list and trap map
9. Pencil
10. "Maintenance required" tags
11. Ultrasonic sound detector.

Fig. 4—Use the trap list and trap map to make up stainless steel marker tags, and attach them as shown in the photo. The tags are attached on the trap's condensate side next to the guard valve. This location minimizes burn hazards and prevents loss during trap changes.

Fig. 6—Everything except the clipboard and lists is stored in the leather pouch. When not in use, the ultrasonic tester goes into the pouch forward of the water bottle.

Fig. 7—Wet testing is done first. A few drops of water are squirted on the trap. The water should start to vaporize immediately. If it does not, this indicates a cold trap.

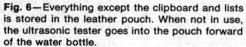

Fig. 8—A cold trap is marked by placing a colored clothespin on the trap's drain-valve handle. Wet-test each trap in the immediate area and mark those that are cold.

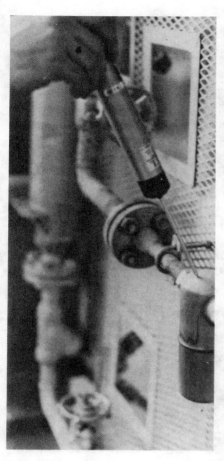

Fig. 9—Go back to the first trap and give it a sound test. When flow passes through a trap, an inaudible ultrasonic vibration is generated—the ultrasonic detector changes this to a frequency we can hear. A *bucket trap* (as shown here) should be relatively quiet, or cycle on and off at regular intervals (determined by condensate load and steam pressure). A ringing sound made by the bucket hitting against the trap wall tells that the trap is blowing. This sound is audible, but the ultrasonic detector will help pinpoint the noisy one in a bank of traps.

Fig. 10—*Disk traps* are checked by touching the detector directly to the top of the trap. A good trap cycles every 6 to 10 seconds. Cycling faster than once per second is called "machine-gunning," and indicates sufficient steam loss to merit replacement. A hot disk trap that does not cycle at all may have failed in the open position.

Fig. 11—On *bellows traps*, listen to the outlet piping. These traps should cycle or throttle depending on the condensate load. Ones that never close are either under a very heavy load or are blowing steam.

Fig. 12—For *level chambers and control automatics*, there is no way to use the ultrasonic detector, because a properly functioning system is always open, throttling the condensate flow. Such traps are usually maintained by the instrument repair group, separate from the normal trap-maintenance program. Record the percentage level as shown by the level indicator.

Fig. 13—Mark any "noisy" trap with an uncolored clothespin. Continue the ultra-sonic testing of all traps in the immediate area.

Fig. 14—Go back to the first trap in the group, and check it by observation to determine what is actually passing through it. To do this, close the condensate valve and open the trap drain-valve. This bucket trap in 35-psig steam-tracing service is functioning properly.

Fig. 15—This bucket trap, also in tracing service with 35-psig steam, is blowing. Note the clothespin that shows it had been found to be defective with the ultrasound test. Often, small bucket traps that are blowing steam can be corrected by repriming. To do this, close the drain valve and leave the condensate valve closed for about 10 minutes. Condensate collecting during this time will sometimes reprime the trap.

Fig. 16—This is normal discharge for a 550-psig drip-leg disk trap. A large number of high-velocity water droplets are surrounded by flashing steam. With 550-psi condensate, about one third will flash into steam—do not confuse this with live steam blowing through.

Fig. 17—Here is a worn disk trap that is leaking steam during the off-cycle. Note the clear vapor next to the outlet pipe. *Caution:* With high-pressure steam traps, the condensate and steam exit at very high velocities when the trap is opened ("decked"). Maintenance workers should not deck traps larger than ½ in. without special permission, because of the large volumes of condensate that might be present.

Fig. 18—After all traps in a group have been given the water test and ultrasonic test and the noisy traps have been tested by observation, they are ready to be marked for maintenance. For each trap marked with a colored clothespin (cold traps), make out a "Maintenance required" tag marked "C" (for "cold") on the back. The production department should check to see why the trap is cold (supply cut off, plugged or crimped tracer, plugged trap, check valve stuck closed, etc.) and add this information to the tag, to guide the Maintenance Dept. in making repairs. Remove the clothespin after the tag is in place.

Fig. 19—Plain clothespins mark traps that are either blowing or wasting steam. The observation (decking) test determines whether the high sound level is from blowing steam or from a very heavy load. (Normally, traps in tracer or drip-leg service will not be subjected to such loads.) When the trap has been confirmed as bad, attach a "maintenance required" tag marked "B" (for "bad"). The Maintenance Dept. will replace these with new assemblies. Remove the clothespin after tagging the trap.

Traps may also be marked "D" for external trap damage (cracks, corrosion, etc.) or "L" for leaks at welds, fittings and the like.

Fig. 20—Now enter the appropriate code letter for each trap on the trap listing sheet:

B **Bad** trap, blowing live steam.
C **Cold** trap.
D Externally **damaged** trap.
G **Good,** properly working trap.
L Externally **leaking** trap.
W Trap is **wasting** steam, but not yet blowing.

Now recheck to see that all valves are back in the "as found" condition.

When the trap listing is completed, copies should be sent to a predetermined group. We circulate them to the area production supervisor, the appropriate mechanical foreman, the plant energy coordinator and the appropriate division energy coordinator.

Setting up a steam-trap standard

Your plant will save money if you standardize on a type and size of trap for each of several pressures and condensate loads.

☐ The previous two articles have discussed how to set up an energy conservation program, to limit energy losses in steam traps by getting people involved at all levels. This article is directed at the hardware involved—"the right trap for the job." This basically means establishing a trap standard for all replacements and new installations.

Standardization is a vital part of any steam-trap program, to prevent premature failure caused by misapplication. Unfortunately, steam-trap misapplication is very common. Oversizing is probably the most usual malady. Far too often, steam traps are designed for heavy (>1,000 pph) condensate loads and then placed in services where the condensate load is very light (>50 pph). This keeps steam adjacent to a loosely fitting internal-trap-valve, resulting in live-steam loss.

To prevent such loss, a simple, effective steam-trap standard should be adopted and closely followed. The best way to gain acceptance and use of this standard is to: (1) keep it simple, (2) make it accessible, and (3) put it in the form of a drawing. This standard must be clear enough to be used by the field mechanic or foreman, the production supervisor, and the plant technical engineer.

A clear drawing is ideal because it can contain the most information in the simplest form.

Table I breaks down the various-sized steam traps by condensate loads for each pressure service: 94% light, 5% moderate, and 1% heavy. From this chart, you can see that the ideal trap standard would be to have three standard traps based on condensate loads alone. But the number of standard traps would not exceed nine if both pressure and condensate-load required separate traps.

Selecting traps for a standard

The problem here is which type of steam trap will give the best performance for each of the three condensate-load services—light, moderate and heavy. Trap testing and analysis was the method chosen for determining the best trap. Both existing process loads and trap test-stands were used in the evaluation process.

Fig. 1 shows one of the test stands used for evaluating various light-load steam traps. This test stand consisted of seven natural-convection steam condensers that provided a 6-lb/h condensate flow to the individual test steam-traps. Three-way valves were located downstream of each trap so that individual steam losses could be diverted for measurement. (This particular test installation was for 600-psig steam service.) A separate, smaller, stand was used for pressures under 100 psi. All testing of moderate- and heavy-load traps was done in the plant because of the large volumes of condensate involved.

The three main causes of steam-trap failures during testing were: (1) materials of construction, (2) weakness in steam-trap design, and (3) trash or solids buildup in the trap body.

The test work done by the author sifted the 38 different traps being used down to a standard of only four. These selected four trap systems are not ideal by any means, but are much superior to the multiplicity of types and sizes commonly used before the standard was adopted.

Selecting trap systems

Here is the basic way the four steam-trap systems were selected. All nine basic trap types were tested using the six general categories discussed below:

Steam Loss—Steam loss over the life of a trap was considered the most important criterion. Here, basic trap design, materials of construction, and size orifices played

Distribution of steam traps in a petrochemical plant	Table I

Total traps, by pressure, %	Total traps by pressure/condensate-load, %
9% 400 psig and up (High-pressure)	<1% Large condensate loads (>3,000 lb/h) 1% Medium condensate loads (500-3,000 lb/h) 8% Small condensate loads (<500 lb/h)
13% 100-400 psig (Moderate-pressure)	~1% Large condensate loads (>2,000 lb/h) 2% Medium condensate loads (200-2,000 lb/h) 10% Small condensate loads (<200 lb/h)
78% 5-100 psig (Low-pressure)	<1% Large condensate loads (>1,000 lb/h) 2% Medium condensate loads (100-1,000 lb/h) 76% Small condensate loads (<100 lb/h)

Originally published February 9, 1981

very important roles in steam loss. A trap valve that is erosion-resistant, just large enough to carry the required condensate load, designed to be tolerant of steam-line trash, and that has a steam loss, when new, of <1 pph was rated highest in this category. (See Tables II, III and IV for ratings.)

Materials of construction in steam traps have a dramatic effect on steam loss. The material around the flashing-condensate part of a trap should be stainless steel. It is common knowledge that flashing condensate is very erosive. If it is also slightly corrosive (and most condensate is), attack on carbon steel is certain. The inspection of several hundred steam traps that were removed from plant service (and those from the trap test stands), vividly points out the problems of carbon steel in steam traps. Fig. 2 shows how steam and condensate have bypassed the steam-trap valve by eroding the threaded portion of the carbon-steel trap cap. Fig. 3 shows examples of a safety problem—the carbon-steel threads were eroded away sufficiently to weaken the top of the trap to the extent that it would blow off while in service. In each case, the stainless-steel parts of the trap showed no signs of wear.

One carbon-steel trap on a 600-psig steam-header drip-leg was passing 4,200 lb/h of steam into a low-pressure collection header, which was vented to the atmosphere. This high volume of steam also contributed to additional steam loss by overpressuring the low-pressure header and caused traps that are sensitive to backpressure (disk and impulse types) to open prematurely and waste steam. Although a carbon-steel trap is the least expensive to buy, it normally is the most expensive to own!

Life—The life of a trap is governed by how long the valve system lasts, how long the body stays leakfree (both internally and externally), and how tolerant the trap is to steam-line trash. Tests show that all steam traps wear because they do not have a positive on/off action. Example: A disk trap appears to have an on/off valve actuation, but it actually does not. During condensate discharge, the disk does not fully retract in the valve cap, but flutters or wobbles very rapidly. All condensate has small suspended particles of scale and rust that the disk and seat constantly close on; this accounts for the faster wear rate on the moderate-load traps compared with the very-light-load traps.

Reliability—Trap reliability was determined by how well the steam trap responded when the condensate load or pressure changed. Occasionally, a trap would stay closed and back up condensate, or lock open. But most traps passed this test satisfactorily.

Size—This was a measure of the physical size and weight of the trap. Orientation of the trap was considered, but to a lesser degree.

Noncondensable Venting—All steam traps must have some way of discharging noncondensables, such as air. Most traps use a small leak somewhere in the valve design to do this (which results in some steam loss). The liquid-expansion design uses no direct gas venting but subcools the condensate enough so that most noncondensables will go back into solution. This will cause corrosion problems if the noncondensables include oxygen or carbon dioxide.

Cost—This was the last evaluation parameter, and is *least* important. When a ½-in. steam trap costs $25 but

Test stand used to evaluate light-load traps   Fig. 1

How steam eroded a carbon-steel trap cap   Fig. 2

Eroded carbon-steel threads pose a danger   Fig. 3

Criterion	Importance, range	Disk	Bucket	Impulse (piston)	Bellows	Float	Bimetallic	Orifice	Expansion	Instrumented
Steam loss	(0-10)	6	8	4	5	7	3	3	9	NA
Life	(0-8)	6	7	5	4	5	3	5	4	NA
Reliability	(0-6)	5	4	3	3	3	2	3	2	NA
Size	(0-3)	3	2	3	2	1	1	3	2	NA
Noncondensible venting	(0-2)	1	1	1	2	2	1	1	0	NA
Cost	(0-1)	1	1	1	0	0	0	1	0	NA
Small-trap types, total	**(30)**	**22**	**23**	**17**	**16**	**18**	**10**	**16**	**17**	**NA**

Steam traps for small condensate loads — Table II
Trap type

wastes an average of 100 lb/h of steam ($3,000/yr), the initial cost of the trap becomes insignificant.

The ratings for each trap type are shown in Tables II, III and IV for the light, moderate and heavy condensate-load trap types. There was not a perfect "30" among any of the tested types. All steam-trap systems have their weaknesses, with the ones selected being judged best by the author. There are traps in each category that rated lower that undoubtedly would also make a workable choice. The important thing to remember here is to take a position and make a selection of the trap that appears to do the job well. Standardize on this trap for all applicable services and see that it gets used.

Adoption of the steam-trap standard will improve overall plant performance. It will:

- Reduce energy loss through steam traps by minimizing misapplication.
- Provide higher plant utility and product quality by yielding more-temperature-stable processes.
- Lower trap installation costs by providing standardized trap stations.
- Lower company stores cost through the stocking of 80% fewer steam traps.
- Provide a nicer looking and safer plant by reducing steam plumes and wet spots.

Arranging traps in "banks" makes maintenance easier Fig. 4

Specifics of the steam-trap standard

As stated before, a drawing is the most useful form for a steam-trap standard. This single document can contain most of the necessary information for sizing, purchasing and installing steam traps. It is also easy to update and—since it is easily reproducible—it is simple and inexpensive for all plant areas to have copies.

Items that should be covered in the standard drawing:

- Sizing-chart and instructions for use.
- Drip- and tracer-trap-station layout.
- Small-process-load trap layout.
- Medium-process-load trap layout.
- Large-process-load trap layout.
- Steam-header drip-leg general design.
- Condensate-header general design.
- General notes.

The easiest way to describe a recommended "steam-trap standard" drawing is to show the one adopted for our local plant. The steam-pressure range of this standard covers 5 psig to 600 psig, and steam condensate flows from 2 to 80,000 lb/h. The piping schemes illustrated are a demonstrated minimum requirement for a safe, cost-effective installation. Arranging the traps in "banks" as shown in Fig. 4 is also recommended. The steam-trap models and sizes shown as standards for our petrochemicals plant may not necessarily be the best ones for the services specified, but they have provided satisfactory performance if monitored regularly on a 6-mo basis. Other trap types rating above "17" in the evaluation charts should also provide satisfactory service. The key is standardization and routine trap checks.

Trap-standard drawing

Steam-trap sizing chart—The sizing chart shown on the standard drawing should be made as simple and workable as possible. The sizing and selection chart is laid out so that only two pieces of information are needed: maximum operating pressure (psig), and maximum condensate load (lb/h). With only these two numbers, the chart tells you the trap size and type, and directs you to a recommended layout detail. This simple chart goes a long way toward solving the second greatest cause of steam loss—misapplication from oversizing. When setting up the sizing curve, be careful not to oversize the trap by more than 25%. It is very common for the field engineer to overestimate maximum condensate load for a given service. If, then, the

Steam traps for medium condensate loads

Table III

Criterion	Importance, range	Disk	Bucket	Impulse (piston)	Bellows	Float	Bimetallic	Orifice	Expansion	Instrumented
						Trap type				
Steam loss	(0-10)	6	7	5	8	7	4	2	NA	NA
Life	(0-8)	3	7	5	5	4	2	6	NA	NA
Reliability	(0-6)	4	4	4	4	4	3	3	NA	NA
Size	(0-3)	3	2	3	3	3	2	3	NA	NA
Noncondensible venting	(0-2)	1	1	1	2	2	1	1	NA	NA
Cost	(0-1)	1	1	1	1	0	0	1	NA	NA
Medium-trap types, total	**(30)**	**18**	**22**	**19**	**23**	**20**	**12**	**16**	**NA**	**NA**

sizing chart is also significantly oversized, a trap could be selected that may cause excessive steam loss.

Specifying each condensate-load group to use a different pipe size will make installing the wrong trap less likely. Example:

- Light loads—1/2-in. piping.
- Moderate loads—3/4-in. piping.
- Heavy loads—1-in.-and-larger piping.

Tracing and steam-main drip-pocket trap design details (Detail No. 1)—In Detail No. 1, the standard for a typical trap installation starts to take shape. A 15-in. standard trap assembly is established for easy removal and installation. The test valve is located on the removable part of the assembly, on the downstream side of the trap, because it is the valve most likely to leak and require replacing.

Normal trap maintenance is accomplished by removing the old trap assembly (piping, trap, drain valve, and flanges) and installing a complete replacement assembly. Then, back in the repair shop, the reusable parts of the removed assembly (piping, drain valve, and flanges) are used to make up a new assembly that is stored for later use. Use of standard piping sizes and dimensions makes trap changeout a 15-min, one-man assignment.

For many petrochemicals plants, the light-load traps—which includes tracing and drip-pockets—account for over 80% of all steam traps. These small devices in pressures above 100 psig should be very carefully monitored, with very close attention given to proper trap selection. It is strongly recommended that stainless-steel construction be used. Any leakage past the trap valve in these lightly loaded traps results in steam loss. Because steam conden-

sate can be both erosive and corrosive, a leaking trap nearly always gets worse.

Small-load-trap design details (Detail No. 2)—This system is very similar to the tracing and drip-pocket trap design above. Note that the design includes a second trap, in the form of an installed spare, for critical services. It should be emphasized that this is a spare trap and that only *one* trap should be "valved in" at a time. Running with both traps online invites early trap failure and steam loss.

Moderate-load-trap design details (Detail No. 3)—This covers about 15% of a petrochemical plant's traps. These are found on small unit-heaters, reboilers, and jacketed vessels where the steam condensate load is above 200 lb/h. Fast response, high reliability and dual traps are common installations. Because of the higher condensate loads, it is sometimes difficult to detect a steam loss of 100 lb/h when the condensate load is 500 lb/h. So, a high-reliability trap is also recommended for this service.

It is good engineering design to have a spare trap permanently installed on moderate condensate loads. This will greatly reduce equipment shutdown from steam-trap failures. Again, this is a *spare* trap assembly and should never be operated or sized as a dual-operation trap. If extra capacity is needed above the capacity of the standard single trap, the instrumented trap system as shown in Detail No. 4 should be used. Note that in all trap details the check valves are optional and should only be used when backflow of condensate is intolerable. An example of such a case would be a process that uses both steam and water alternately for heating and cooling. Check valves in

Steam traps for large condensate loads

Table IV

Criterion	Importance, range	Disk	Bucket	Impulse (piston)	Bellows	Float	Bimetallic	Orifice	Expansion	Instrumented
						Trap type				
Steam loss	(0-10)	4	5	6	NA	7	4	2	NA	9
Life	(0-8)	3	5	5	NA	4	3	4	NA	8
Reliability	(0-6)	3	4	4	NA	5	4	5	NA	5
Size	(0-3)	3	1	3	NA	2	2	3	NA	0
Noncondensible venting	(0-2)	1	1	1	NA	1	1	1	NA	1
Cost	(0-1)	1	1	1	NA	0	1	1	NA	0
Large-trap types, total	**(30)**	**15**	**17**	**20**	**NA**	**19**	**15**	**16**	**NA**	**23**

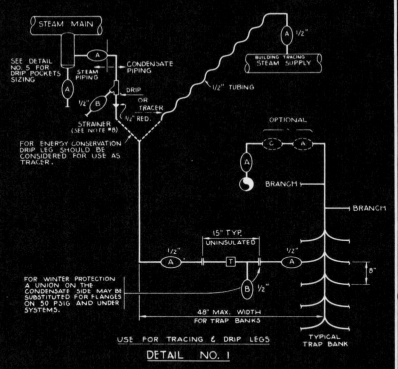

STEAM MAIN

SEE DETAIL NO. 5 FOR DRIP POCKETS SIZING

STEAM PIPING

CONDENSATE PIPING

DRIP OR TRACER

1/2" RED.

1/2" B

STRAINER (SEE NOTE #8)

FOR ENERGY CONSERVATION DRIP LEG SHOULD BE CONSIDERED FOR USE AS TRACER.

BUILDING TRACING STEAM SUPPLY

1/2" A

1/2" TUBING

OPTIONAL

C   A

A

BRANCH

BRANCH

FOR WINTER PROTECTION A UNION ON THE CONDENSATE SIDE MAY BE SUBSTITUTED FOR FLANGES ON 50 PSIG AND UNDER SYSTEMS.

15" TYP. UNINSULATED

1/2" A   T   1/2" A

B 1/2"

8"

48" MAX. WIDTH FOR TRAP BANKS

TYPICAL TRAP BANK

USE FOR TRACING & DRIP LEGS

DETAIL NO. 1

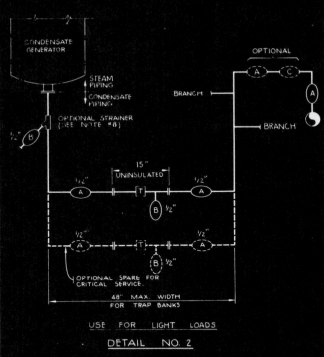

CONDENSATE GENERATOR

STEAM PIPING

CONDENSATE PIPING

OPTIONAL STRAINER (SEE NOTE #8)

1/2" B

OPTIONAL

A   C

A

BRANCH

BRANCH

15" UNINSULATED

1/2" A   T   1/2" A

B 1/2"

1/2" A   T   1/2" A

B 1/2"

OPTIONAL SPARE FOR CRITICAL SERVICE.

48" MAX. WIDTH FOR TRAP BANKS

USE FOR LIGHT LOADS

DETAIL NO. 2

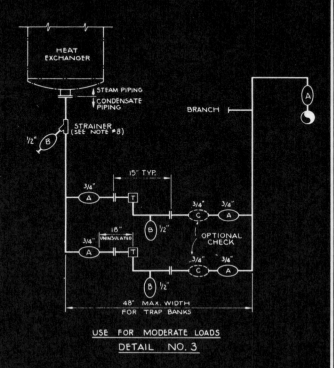

HEAT EXCHANGER

STEAM PIPING

CONDENSATE PIPING

1/2" B

STRAINER (SEE NOTE #8)

BRANCH

A

15" TYP.

3/4" A   T

B 1/2"

3/4" 3/4"
C   A

18" UNINSULATED

3/4" A   T

OPTIONAL CHECK

3/4" 3/4"
C   A

48" MAX. WIDTH FOR TRAP BANKS

USE FOR MODERATE LOADS

DETAIL NO. 3

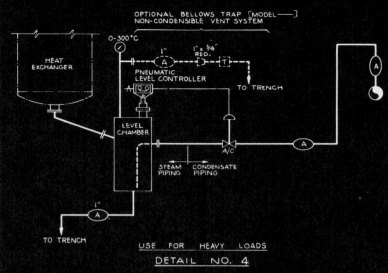

HEAT EXCHANGER

OPTIONAL BELLOWS TRAP [MODEL——] NON-CONDENSIBLE VENT SYSTEM

0-300°C

1"   A   1" x 3/4" RED.   T

PNEUMATIC LEVEL CONTROLLER

TO TRENCH

A

LEVEL CHAMBER

STEAM PIPING   CONDENSATE PIPING

A/C

A

1" A

TO TRENCH

USE FOR HEAVY LOADS

DETAIL NO. 4

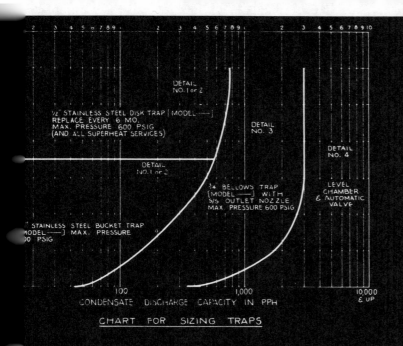

½" STAINLESS STEEL DISK TRAP [MODEL ——]
REPLACE EVERY 6 MO.
MAX. PRESSURE 600 PSIG
(AND ALL SUPERHEAT SERVICES)

DETAIL NO. 1 or 2

DETAIL NO.1 or 2

DETAIL NO. 3

DETAIL NO. 4

¾" BELLOWS TRAP [MODEL ——] WITH S/S OUTLET NOZZLE MAX. PRESSURE 600 PSIG

LEVEL CHAMBER & AUTOMATIC VALVE

" STAINLESS STEEL BUCKET TRAP [MODEL ——] MAX. PRESSURE
00 PSIG

CONDENSATE DISCHARGE CAPACITY IN PPH   10,000 & UP

CHART FOR SIZING TRAPS

TRAP SIZING:
1.) DETERMINE MAXIMUM QUANTITY OF CONDENSATE TO BE HANDLED (FROM ENGR'G FLOW SHEETS) AND MULTIPLY BY A FACTOR OF 1.25.
2.) SELECT TRAP SYSTEM WITH FLOW EQUAL TO, OR SLIGHTLY GREATER THAN REQUIRED FLOW.

STEAM MAIN

TO GRADE OR OTHER SAFE LOCATION

TO TRAP

STEAM MAIN	DRIP POCKET	GUARD VALVE	STEAM PSIG	PIPE CODE	INSULATION CODE
1½"	1½"	1"	0-70	STEAM PIPING CODES	INSULATION CODES
2"	2"		70-240		
3" & 4"	3"		240-600		
6" & 8"	4"				
10"	6"				
12" & 14"	8"	1½"			
12" & 18"	10"				
20"-24"	12"				
30"	16"				

STEAM
DETAIL NO. 5

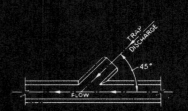

CONDENSATE HEADER INLET
TYPICAL DETAIL

TRAP DISCHARGE
45°
FLOW

VALVES	½" TO 2"		3" & LARGER				STEAM PSIG	INSULATION CODE
STM. PSIG	ALL PRESSURES TO 600 PSIG		0-70 PSIG	70-300 PSIG	300-600 PSIG		0-70	INSULATION CODES
PIPE CODES	SMALL STEAM PIPING CODES		LARGER STEAM PIPING CODES				70-600	
A	GATE	VALVE	CODES					
B								
C	CHECK	VALVE	CODES					

CONDENSATE
DETAIL NO. 6

GENERAL NOTES:
1.) SELECTION OF MATERIALS, VALVES, FABRICATION AND INSTALLATION OF ALL PIPING TO BE IN ACCORDANCE WITH COMPANY CODE INDICATED ON DIAGRAMS.
2.) FOR DESCRIPTION OF VALVES SEE CO. VALVE CODE. ALL PIPING ASSEMBLIES SHOWN ON THIS DRAWING ARE DIAGRAMMATIC ONLY AND ARE NOT TO BE USED AS PIPING ARRANGEMENTS.
3.) WHERE MORE THAN ONE TRAP SERVES A SINGLE PIECE OF EQUIPMENT, TRAPS SHOULD BE INSTALLED IN SAME HORIZONTAL PLANE AND NOT STACKED ONE ABOVE ANOTHER.
4.) WHERE "STANDARD" DRAWINGS DIFFER FROM PROJECT SERVICE DIAGRAMS, PROJECT SERVICE DIAGRAMS ARE TO TAKE PRECEDENCE.
5.) 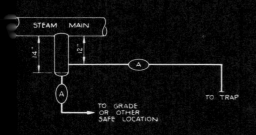 ON SERVICE DIAGRAMS INDICATES ASSEMBLY SHOWN IN DETAILS 1 THRU 4. SERVICE DIAGRAMS TO STATE LINE SIZE (L.S.) AND PIPE CODE TO AND FROM TRAP.
6.) CONDENSATE PIPE CODE IS TO COMMENCE AT VESSEL AND NOT DOWNSTREAM OF TRAP. IN CASE OF TRACING IT IS TO COMMENCE WHERE THE TUBING ENDS AND PIPING STARTS.
7.) OPEN END OF 304 S.ST. SCH. 40 ½" BLOWDOWN VALVE NIPPLE TO BE THREADED.
8.) INLINE STRAINER - SOCKET WELD ENDS.

SIZE	TYPE	MAT.	DESIGN PRESSURE	DESIGN TEMP.	BLOW-OFF CONN.	STD. SCREEN
½" & 1"	PLANT STD.	C.S.	600	850	½" NPT.	.033 PERF. S/s
1½"	PLANT STD.	C.S.	600	850	½" NPT.	.033 PERF. S/s

9.) USE ½" STAINLESS STEEL DISK TRAP [MODEL ——] FOR ALL SUPERHEATED STEAM SERVICES.

SAFETY NOTES:
1.) EXCEPT WHERE PROHIBITED IN ASSEMBLY ABOVE, TRAPS AND PIPING ARE TO BE INSULATED FOR PROTECTION OF PERSONNEL. WHERE INSULATION IS NOT PERMITTED, FIELD IS TO PROVIDE GUARDS AROUND TRAPS AND PIPING.

GENERAL (PLANT)
STEAM TRAPPING STANDARD
INSTALLATION & SIZING
DETAILS

DWG. 8993

this instance may be required to prevent water or steam from backflowing into the wrong header.

Also note that all steam traps return their condensate to a collection header. It is good energy-conservation practice to collect this condensate in a storage tank and pump it to the powerhouse or a hot-water process user. Not to reuse the condensate represents a sizable cost penalty in the loss of the condensate's heat content and the original boiler-feedwater treatment expense. Steam condensate is normally good-quality hot water and should not be wasted.

Heavy-load-trap design details (Detail No. 4) — The very-large-quantity steam users require steam traps of very high capacities. Determining steam wasting in these large traps is difficult, if not impossible. These traps, with capacities in the thousands of pounds of condensate per hour, are normally constructed of carbon steel. As earlier mentioned, failures at the internal sealing or gasketing surfaces can result in large steam losses that are almost impossible to detect if the steam loss is less than the condensate flow.

As a result, the instrumented condensate-level chamber was designed to overcome the two major problems with commercially available large steam-traps—frequent failure, and no way to test for steam loss. These two problems can be solved by installing a water seal between the steam and the condensate systems. The design shown in Detail No. 4 and Fig. 12 of the "steam trap testing" article (p. 89) does just this by using a level chamber (or catch pot), a level-sensing controller, and a control valve. The condensate chamber has a workable 30-in. level range and is sized for a condensate velocity through the vessel of about 1 in./s or less.

Our standard now consists of only two basic vessel sizes: 18 and 24-in. dia. The level instrumentation, which has proven very successful, is a compact pneumatic controller. This self-contained pneumatic instrument provides a positive level-readout, a level-sensor, a control point, and a pneumatic-output control signal all in one small housing. The level-output signal goes directly to the stainless-steel control valve (where the actual level-control takes place). For most flashing-condensate services, a standard Type 316 stainless-steel control valve will provide long reliable life. Special trims and seats that have been coated with hard-facing materials are available for very-high-pressure hot-condensate services.

Always size the control valve for flashing service. In many instances, this will require using a large valve-body with reduced trim to maintain valve-position control, because the flash steam causes high backpressure in the small valve-bodies. Using the level chamber to replace large steel steam-traps will result in a steam consumption reduction of 2-17%, depending on the steam pressure and condition of the trap being replaced.

General steam-header drip-pocket design details (Detail No. 5) — Most companies have a standard established for steam-main drip pockets. It should be reproduced—for clarity and easy access—on the steam-trap-standard drawing. Other company specifications are incorporated into this drawing in the form of codes for piping, insulation, and guard valves. Only the code numbers need be duplicated on the steam-trap standard. Details of these codes are generally found in company documents that are available elsewhere.

General condensate-header design details (Detail No. 6) — This detail incorporates the same general information as the steam-header details of No. 5, above. The general company codes and specifications are again duplicated here for the piping, valves and insulation.

General notes — These notes reflect good engineering practices. They call attention to special piping codes and other specifics not usually covered on the individual trap-layout detail.

Summary

Generally, a good steam trap is one that does its job by passing condensate and noncondensables, but without passing more than 1 lb/h of steam. Long trap life, high reliability, ease of installation, and low steam loss all should take precedence over initial trap cost.

In conclusion, here are the three main things to do for establishing an effective steam-trap standard:

1) Divide the plant steam traps into each of three condensate-load ranges—small, medium and large.

2) Select a steam trap or traps (if the steam pressure requirements dictate) for each of the three condensate ranges, using your own experience, plant tests, or the recommendations of a reliable source.

3) Combine items 1 and 2 above with applicable company standards for valves, piping insulation, etc., into a single plant-standard drawing.

If the above three steps are followed, and then put into practice, you will find that the second of the two greatest causes of steam loss from steam traps—misapplication—will be eliminated.

The author

Stafford J. Vallery is a Senior Engineer at the E. I. du Pont de Nemours & Co. Victoria Plant, P. O. Box 2626, Victoria, TX 77901. During the past 15 years, he has held various positions with Du Pont—development engineer at the Du Pont operated AEC facility at Aiken, S.C., and at the company's engineering department near Wilmington, Del. As part of his assignment at the Victoria plant, he has over the last five years worked to improve the plant's energy systems, especially in selection, testing and maintaining of steam traps. He holds a B.S.M.E. from Louisiana Tech University and is a member of the American Soc. of Mechanical Engineers.

STEAM-TRAP CAPACITIES

In selecting steam traps from published data, it is essential to check the conditions under which the trap capacities were measured. Some frequently-used procedures yield misleading results.

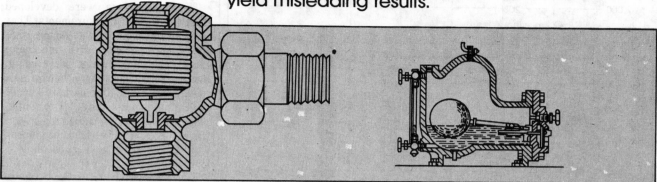

Elmer S. Monroe, *E. I. du Pont de Nemours & Co.* *

Despite the existence of a U.S. national standard for the capacity testing of steam traps, capacity data published by some steam-trap vendors continue to be misleading. Costly production and maintenance failures can occur if steam traps are not performing as expected. It is the purpose of this article to discuss the ways that the steam-trap user can avoid such problems.

Factors that affect steam-trap capacity

The following factors affect steam-trap capacity:
- Trap inlet pressure.
- Trap discharge pressure.
- Subcooling of inlet condensate.
- Static head of condensate on trap.
- Design of the trap.
- Design of the piping.

What a user wants from a steam trap determines the relative importance of these items. Since it is not possible for a vendor to include sufficient data in its catalogs to cover all variations, the user is often left in the dark. Some of the things that a user might want:
- Low installed cost.
- Low maintenance cost.
- Low cost of owning.
- Maximum performance from the trapped equipment.

Overemphasis on low *installed* cost explains the sweeping success of the thermodynamic disc trap in North America, when it was first introduced. Europeans have favored low *maintenance* cost, and have made the temperature-responsive bimetallic trap a favorite.

The cost of owning a trap, and the performance of the equipment trapped, are interrelated. A steam trap that wears out and wastes steam represents an average annual cost of owning, due to wasted steam, on the order of about $1,000 per year.

It is the performance of the equipment using the trap that is affected most by trap capacities and that is to be addressed. While improper capacities affect the cost of own-

Since writing this article, the author has retired. See "The author" on p. 75.

ing, it is difficult to assign cost penalties to lost production. Equipment performance is easier to consider.

It is no coincidence that heating and ventilating engineers tend to use float-and-thermostatic types of traps in large numbers, or that vendors of steam-heated equipment often specify inverted-bucket traps for use on their equipment. Both of these types are mechanical designs that work by sensing a level and, if properly sized, may keep the equipment drained and operating at maximum capacity.

There are, of course, other traps that can do an equal or better job of draining equipment, but the unfortunate fact is that many cannot. Simply stated, if condensate from a heat exchanger must be subcooled in order to pass through a trap, the exchanger's performance must suffer.

It suffers in three ways: A backing up of condensate into the heat exchanger reduces the overall heat-transfer coefficient, and thus, the throughput of product. Secondly, performance can suffer through inequality of heating. A coil half-flooded with water will not heat a product stream equally; to obtain a required average product-temperature, some product must be overheated. A traced pipeline might freeze in a segment that is in contact only with a tracer full of condensate. Lastly, noncondensable gases cannot find their way adequately to the steam trap through a water seal. Their cumulative effect is to further reduce heat transfer and, in some cases, to cause corrosion failure of the heat exchanger.

Obviously, the heating and ventilating engineers and original-equipment vendors are basically right. Not necessarily in their selection of traps, but in the basic principle that the equipment must be drained of condensate to provide optimum performance. Capacity under this condition is of paramount importance to the engineer who emphasizes such maximum performance from equipment.

Formulas for steam-trap capacity

It has previously been shown [1] that steam traps can be characterized, for the flow of saturated condensate, by a discharge coefficient:

$$W = K_t P_1^{1/3} [(P_1 - P_2)/P_1]^{1/2} \qquad (1)$$

where: W = lb/h of saturated condensate; K_t = trap dis-

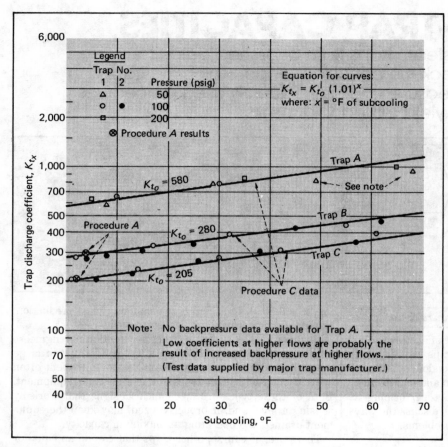

Figure 1—Variation of discharge coeffifient subcooling

charge coefficient; P_1 = inlet trap pressure, psig;

and: $P_1 \geq 64$ psig; P_2 = outlet trap pressure, psig, or:

$$W = \tfrac{1}{2} K_t (P_1 - P_2)^{1/2} \qquad (2)$$

where: $P_1 \leq 64$ psig.

These equations were developed from orifice flow measurements. They account for most of the factors that affect trap capacity, with one major exception — condensate subcooling. Further study has shown that for subcooling up to 70°F, the discharge coefficient can be adjusted by this formula:

$$K_{t_x} = K_t (1.01)^x$$

where: K_{t_x} = discharge coefficient, with subcooling x = °F subcooling $\leq$ 70°F.

Fig. 1 shows data for three typical traps calculated by using Eq. (1), (2) and (3). It should be noted that at 30°F subcooling, the capacity of a trap has increased 35%, and at 70°F, 100%.

Applying formulas for traps

Test data are required to experimentally determine the value of K_t in the formulas. Catalog data are not recommended for this purpose. While some vendors have published data at 30°F subcooling, others publish data "near

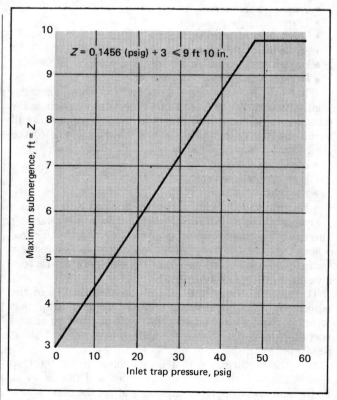

Figure 2—Static head limitation at low pressure

Figure 3—Effect of submergence on trap discharge coefficients

steam temperature" or "for hot discharge." This is ambiguous as well as misleading. A trap may easily have a good capacity at 30°F subcooling, and have no capacity at saturation temperature. The formulas apply only to traps that do.

Fortunately, in 1980, the American Soc. of Mechanical Engineers published a Code for Performance Testing of Steam Traps [2]. This code contains four acceptable test procedures. Three of these permit submergence of the traps under a static head of condensate during testing. At low pressures, this can introduce subcooling, even though the condensate is at the saturated temperature of the steam above the condensate. Fig. 2 shows a limiting curve for submergence, which is incorporated in a proposed International Standards Organization (ISO) Standard [3]. Unless this curve is followed, traps tested under Procedures B, C, and D of the ASME code can show abnormally high discharge coefficients at low pressures, as in Fig. 3. Procedure A of the ASME Code requires that the submergence of the trap not exceed 18 in., and that the subcooling not exceed 5°F.

This does two things. It tests the traps more nearly under the conditions that the wise user will want them to operate under and, most importantly, it eliminates those traps that the sophisticated trap user would not wish to use.

It is not enough to say that the procedures produce equal results when treated by the K_t method. Fig. 1 shows that Procedures A and C produce compatible results for the traps tested. It is only Procedure A that forces the trap to demonstrate wha* it can do under conditions that will adequately drain equipment.

For this reason, the engineering department of the author's company has adopted Procedure A of the ASME Code as the basis for rating trap capacities, in its own steam-trap

---

standard. Procedure A of the forthcoming ISO Standard is compatible with Procedure A of the ASME Code.

Once the trap discharge coefficient is obtained from Procedure A, it can be used to predict capacities at other trap operating conditions, such as part load (throttled-steam inlet), higher backpressures, and startup when subcooling is temporarily permissible.

Other trap tests

Capacity testing is only one aspect of proper steam-trap testing. Other tests include steam loss, wear, and air-handling capability. Steam-loss tests are covered by the ASME [2], and are to be covered by the ISO [4]. It is this author's opinion that such tests have their greatest value when applied to used traps, not to new ones. The accuracy of such tests is limited by the slight losses most new traps have. It is the large losses of worn traps that should concern the user.

Wear rate or failure rate can only be determined by field observation of specific traps in specific plants. Keeping adequate trained maintenance crews in the field is easily justified when one realizes that field inspection by the author revealed an average of 35% failed traps (range = 5–95%). Each failed trap represents about $1,000/yr in fuel costs.

Of course, each plant is different, and records will produce facts. The traps will speak for themselves if given a chance.

Air-handling-capacity tests are often important [5]. Development of suitable codes is under study by ASME PTC 39.1. A test procedure developed by the author's company shows promise in rating traps for air-handling capability, as does a similar procedure developed by a major trap company. Until such procedures are generally available, the user should remember that flooded traps with their water seals will have difficulty in handling noncondensable gases.

Conclusions

- Standardized procedures are available for capacity-testing of steam traps.
- Procedure A of Refs. 2 and 3 tests steam traps under optimum equipment operating conditions.
- Formulas are available for applying test capacity data to any operating condition.

References

1. Monroe, E. S., How to Size and Rate Steam Traps, *Chem. Eng.*, pp. 119–123, Apr. 12, 1976.
2. "Condensate Removal Devices for Steam Systems," ANSI/ASME PTC 39.1 — 19BO, American Soc. of Mechanical Engineers, 345 East 47th St., New York, N.Y. 10017.
3. "Discharge Capacity Tests for Steam Traps," ISO/DP 7842, Association Française de Normalisation (AFNOR), Tour Europe, Cedex 7, 92080, Paris La Défense, France.
4. "Steam Loss Tests for Steam Traps," ISO/DP 7841, AFNOR, Tour Europe, Cedex 7, 92080, Paris La Défense, France.
5. Monroe, E. S., Effects of CO$_2$ in Steam Systems, *Chem. Eng.*, pp. 209–212, Mar. 23, 1981.

The author

E. S. Monroe is now retired. At the time of writing this article, he was a principal consultant in the Engineering Dept. of E. I. du Pont de Nemours & Co., where he specialized in combustion and steam traps. He is chairman of ASME Performance Test Code 39 committee, and a U.S. delegate to an International Organization for Standardization (ISO) task force that deals with steam traps. Most of his articles on steam trapping have previously appeared in this magazine.

ISO steam trap standards

The ISO standards that are currently in draft form and have been circulated to member bodies worldwide include the following:

Draft Proposal ISO 7842
Automatic Steam Traps—Discharge Capacity Test

This is the trap-capacity draft referred to in the text. As now written, it specifies two alternative test methods that manufacturers will be permitted to use to determine the discharge capacity of traps.

In addition to the capacity standard, there are other steam trap standards under considerations. One is:

Draft proposal ISO 7841
Automatic Steam Traps—Steam Loss Tests

This standard specifies two alternative tests for determining steam loss in traps. One of the two is very similar to the American Soc. of Mechanical Engineers standard: ANSI/ASME PCT 39.1—1980. The other is based on a procedure developed by the U.K. National Engineering Laboratory.

Only the above two are at the Draft Standard stage at the present time. However, ISO is now also working on a standard that will cover methods of determining heat losses (by radiation and convection) from steam traps, and on a guide concerning materials of construction for traps.

SPECIFYING AND OPERATING DESUPERHEATERS

George E. Bowie, Humphreys & Glasgow Ltd.

The author describes the types of inline units available, and offers guidelines for picking the correct one for a particular service. Installation and maintenance are simple, provided a few rules are followed.

Efficient generation and use of steam systems means putting a lot of energy into a unit mass of steam. This is done by increasing the pressure and temperature — the latter by superheating, which avoids the production of wet steam during expansion in equipment at high pressures. (Here, we will not consider steam in the critical condition.)

Still, in process plants there is a significant demand for lower-pressure steam for heating, typically, at medium pressure. Although this demand can be satisfied by separately generating such steam, this is usually uneconomical, unless there is abundant low-grade waste heat available at a suitable temperature.

Thus, lower-pressure steam is commonly obtained by taking some of the high-pressure steam and reducing its pressure by a throttling device, such as a pressure-reducing valve. Also, some "pass-out" steam may be taken from the main-turbine drives at an appropriate pressure. A throttling device may or may not be used here. A typical system using both methods is shown in Fig. 1.

The letdown steam from either source is still superheated. Although it is now at the required pressure, it may be unsuitable for use owing to its temperature. This may be because either a more expensive design is needed for the higher temperature or, more generally, the high heat-transfer coefficient of condensing vapor is unavailable (it will not be until the superheat is removed). Thus, desuperheating needs to be considered in such cases.

Condition of the steam

In considering the degree of desuperheating necessary, should all of the superheat be removed to give dry, saturated steam or should a residual amount be left? The decision, of course, influences the method of desuperheating to be used.

It is infrequent that dry, saturated steam at medium to low pressure will be required, along with high-pressure superheated steam. Generally, a degree of initial superheat is needed to allow for heat losses, which otherwise would result in wet steam. The complexity and size of the steam piping system has to be considered, too. For simple, small systems, dry, saturated steam is generally acceptable. There are three methods for desuperheating:

• Direct-contact saturation with water.
• Indirect cooling.
• Controlled cooling by direct contact with water.

The first alternative is used when dry, saturated steam is required. This is the classic situation of producing such steam by use of a saturator. A simple setup is shown in Fig. 2.

Indirect cooling is used with an attemperator — this is a device through which superheated vapor or fluid is passed to reduce and control the temperature. Fig. 3 illustrates indirect cooling along with moderate-pressure desuperheated-steam production.

The third method is the one most commonly used: direct contact with water carried out in the steam piping. This method will now be covered in more detail.

Note that true saturated steam has a certain temperature that corresponds to its pressure. Thus, a change in pressure affects saturation temperature, making temperature control at or near the saturation temperature difficult.

Originally published May 27, 1985

There is a risk of over- or under-injecting water, both of which can upset operating conditions and result in equipment damage.

Therefore, makers of this type of equipment do not recommend its use when the controlled temperature must be close to the steam-saturation temperature. Further, the first and third alternatives increase steam mass-flow, since they cool by the evaporation of water. The second alternative does not.

Desuperheater requirements

Inline desuperheaters use direct injection of water, which involves some considerations:

1. Available water quality — The water quality affects the desuperheated steam quality, which may have limitations imposed on it by equipment. Also, since desuperheating is an evaporative process, any solids present in the water may result in the formation of scale in the desuperheater or piping.

2. Available water pressure — Generally, the water supply pressure must be a little above the steam pressure, although venturi-type desuperheaters can accept a pressure slightly below the steam pressure. If the main high-pressure boiler-feed pumps act as the water source, then pressure-reducing equipment may be needed because of the differen-

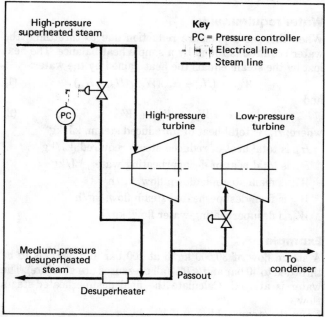

Figure 1 — Two typical ways to obtain lower-pressure steam, which must be desuperheated

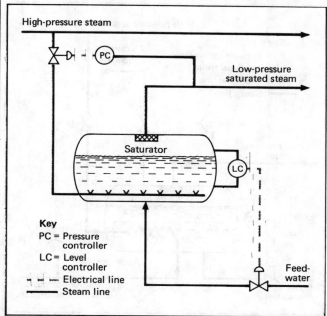

Figure 2 — Dry, saturated steam is produced by passing superheated steam through a saturator

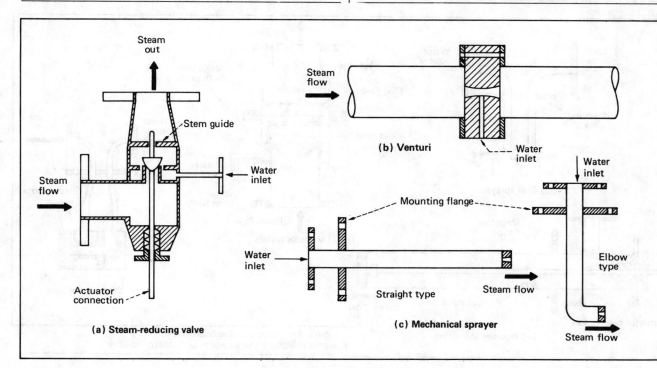

tial pressure between the pumps and desuperheated steam. If the deaerator feed-pumps are the water source, a booster pump may be needed.

3. Allowable steam-side pressure drop — If the allowable pressure drop of the desuperheated-steam system is small, some types of desuperheaters may account for a significant part of this drop.

4. Steam cooling range and required variation — Installed as single units, some types of desuperheaters have a temperature-range limitation. Variations in the initial and final steam temperatures may require special controls. The desuperheater spray angle may be affected by water pressure, which is varied by the temperature controller. Thus, the spray characteristic may be limiting — the angle may be too wide or narrow to yield effective mixing at the ends of the steam temperature range.

5. Sensitivity of controls — If the temperature of the desuperheated steam must be held within close limits, normal control-system responses may be too slow and controls with quick responses may be needed.

6. Steam-flow variation — If, during normal operation, the steam flow varies, it must be accounted for in the desuperheater design. If there is a turndown requirement, it needs to be specified. In extreme cases, parallel units may be needed. (The effect here is similar to that in Point 4.)

Water requirements

When there is a pressure reduction due to throttling, the water required is found by a simple heat balance. The heat lost by the steam equals the heat gained by the water:

$$W_w = (H_{s_1} - H_{s_2})(W_{s_1})/(H_{s_2} - H_w) \tag{1}$$

and

$$W_{s_2} = W_{s_1} + W_w \tag{2}$$

where: H_{s_1} is total heat of unreduced steam, kJ/kg

H_{s_2} is total heat of reduced steam required, kJ/kg

H_w is total heat of desuperheating water, kJ/kg

W_{s_1} is main steam-letdown flow, kg/h

W_{s_2} is final desuperheated-steam flow, kg/h

W_w is desuperheating-water flow, kg/h

Example

A steam flow of 30,000 kg/h at 100 bar and 550°C is to be throttled to 40 bar and 350°C (all pressures are absolute). The water is at 70°C. Calculate the final desuperheated-steam flow.

From the steam tables:

H_{s_1} = 3,499.8 kJ/kg; H_{s_2} = 3,095.1 kJ/kg; H_w = 293 kJ/kg.

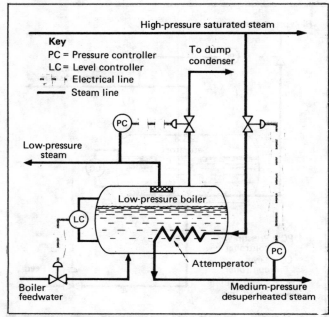

Figure 3 — An attemperator is used to indirectly cool steam to desuperheat it

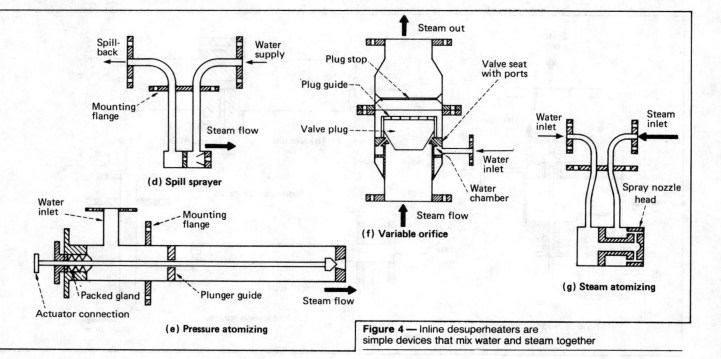

Figure 4 — Inline desuperheaters are simple devices that mix water and steam together

Applying Eq. (1):

$W_w = (3,499.8 - 3,095.1)(30,000)/(3,095.1 - 293) = 4,333$ kg/h

From Eq. (2):

$$W_{s_2} = 30,000 + 4,333 \cong 34,300 \text{ kg/h}$$

If the initial steam condition had been 100 bar at 400°C (H_{s_1} = 3,099.9 kJ/kg), the reduced-pressure steam would have arrived at the same final condition without desuperheating. If the steam had been dry and saturated at 100 bar (H_{s_1} = 2,727.7 kJ/kg), then at 40 bar it would be slightly wet (H_{s_2} *saturated* = 2,800.3 kJ/kg). However, for dry, saturated steam at 30 bar (H_{s_1} = 2,802.3 kJ/kg) throttled to 5 bar, then for H_{s_2} = 2,802.3 kJ/kg, the steam would have about 24°C of superheat. Thus, the steam tables (or a Mollier diagram) will establish the need for desuperheating.

Inline desuperheaters

There are several basic types (see Fig. 4). For purposes of reference, these will be numbered:

Type 1 — Steam-reducing valve (Fig. 4a). This is, essentially, a pressure-reducing valve in which water can be injected into the lower-pressure side to reduce the steam temperature. It is used where temperature control is needed in addition to pressure reduction.

Type 2 — Venturi (Fig. 4b). The desuperheating water enters a high-velocity nozzle for efficient heat transfer. The venturi requires a lower water pressure (i.e., one below the steam pressure) than do the other types, but its steam-side pressure drop is greater than that of some other models. However, this can be used to advantage if the venturi is placed in series with a steam reducing valve. The overall pressure drop can be shared in an economical way, especially if it is large.

Type 3 — Mechanical sprayer (Fig. 4c; two types are shown). This is a nozzle-type sprayer that is inserted into the

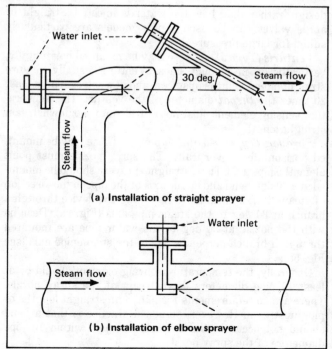

Figure 5 — Here is how to mount a sprayer-type desuperheater for proper performance

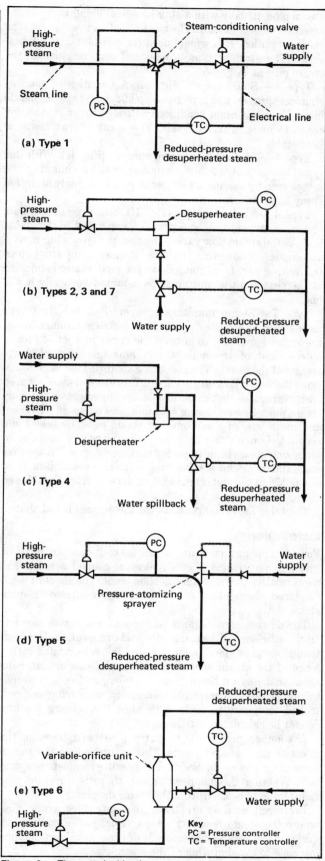

Figure 6 — These typical backups prevent damage to the system in case the desuperheater fails

steam pipe, usually with a stainless-steel lining in the spray area to minimize erosion. The mechanical sprayer is often the cheapest option. It is widely used for constant-duty applications. (The straight type is mounted perpendicular to the steam pipe as in Fig. 5.)

Type 4 — Spill sprayer (Fig. 4d). A variation of the mechanical sprayer, this type has a spillback and special nozzle internals to maintain conditions at the sprayer reasonably constant under varying loads. The spray characteristic is kept stable.

Type 5 — Pressure-atomizing sprayer (Fig. 4e). With this type, load variations are accommodated by changing the orifice area by moving a plunger or piston. A finely atomized spray is produced.

Type 6 — Variable orifice (Fig. 4f). This type uses a ball, piston or valve plug that floats on a valve seat in the steam line. As the steam flow varies, the ball, piston or plug moves, automatically adjusting the flow. A cage (plug stop) stops the moving part from going down the pipe. Water is injected at the seat and the unit must be mounted vertically in the line.

Type 7 — Steam-atomizing sprayer (Fig. 4g). This type uses steam at a higher pressure to increase atomization of the water spray and to entrain the spray in a jet of steam, independent of the main steam flow immediately downstream of the nozzle. The atomizing steam is kept at constant flow; the water flowrate is varied. This method avoids water-spray variations that cause droplet impingement on the main steam pipe. In considering this unit, the effect of total heat and mass flow of the atomizing steam must be taken into account. Consult the manufacturer, since the amount of atomizing steam is different for various designs. An approximate starting point is to assume atomizing steam flow to be about 25% of the water required, as derived from Eqs. (1) and (2).

The table lists requirements for the devices listed above.

Installation

For all inline desuperheaters, careful consideration must be given to positioning the temperature-control sensing element relative to the water-injection point. The element must be placed where all water has been evaporated into the main steam flow.

The downstream distance of the water injection depends on the efficiency of heat transfer and evaporation. Evaporation depends on the droplet size, and the velocity and turbulence of the steam at the mixing point. Desuperheater designs that rely on downstream pipe-length to effect proper mixing may be susceptible to error in measuring the controlled temperature. This is because the sensing position cannot be calculated with accuracy.

A common practice is to put a pipe bend between the injection and sensing points. This assumes that the turbulence at the change in direction will ensure complete evaporation. Whether this happens or not, there is a risk of pipe failure due to erosion from the water droplets.

Preferred designs are those that introduce a fine mist where the steam flow has a high velocity. Generally, such designs employ special materials of construction in the mixing zone to avoid erosion of the steam pipe.

General installation requirements are, for:
Valves (Type 1) — These are mounted to suit the valve

Table — How requirements vary for the basic types of inline desuperheating devices

Type No.	Description	Typical steam-side pressure drop	Typical water pressure above steam pressure, bars	Notes
1	Steam-reducing valve	Main pressure-reduction duty	2+ at valve	
2	Venturi	As required by duty	As required by duty but always less than steam pressure	
3	Mechanical sprayer	Negligible	9 at sprayer	Usually the cheapest
4	Spill type	Negligible	9–20 at sprayer	See notes below
5	Pressure-atomizing	Negligible	9–20 at inlet	See notes below
6	Variable orifice	0.25 bar	2+ at inlet	See notes below
7	Steam-atomizing	Negligible	4 at inlet	See notes below

Notes
Atomizing steam pressure should be approx. 110% of line steam pressure; 5 bar for Type 7. Types 3, 4, and 5 may be used in conjunction with a venturi in the steam pipe, where significant steam flow turndown is required. However, Type 2 may be a better alternative as the water pressure required is below steam pressure. In all cases, a significant steam-side pressure drop may have to be allowed for. Where either Type 4, 5 or 6 can be used, and appears to have equal technical merit, it may be necessary to make preliminary inquiries to establish the most economic selection. Types 4 and 5 can be used in place of Type 6 if it is less costly to increase water supply pressure for a particular duty. Type 2 may be less costly overall than Type 6.

design, rather than how the water is injected. Straight or angle valves can be used. Angle models are particularly suited for large pressure drops.

Venturis (Type 2) — These may be mounted horizontally, vertically or in sloping lines. Venturis usually require straight piping lengths upstream and downstream of at least 20 times the *throat* diameter of the venturi. The temperature-sensing element must not intrude into the downstream straight length.

Sprayers (Types 3–5 and Type 7) — These may be mounted horizontally or vertically. The spray nozzle must point along the pipe axis. Thus, straight sprayers should be mounted at a 90-deg bend and on the axis of the pipe in the direction of steam flow; a variation is mounting the device through a branch at 30 deg to the steam-pipe axis (Fig. 5a). Designs with the nozzle at 90 deg to the water pipe are mounted through a branch perpendicular to the steam-pipe axis (see Fig. 5b).

Generally, the temperature-sensing element should be at least 10 pipe diameters downstream of the spray nozzle. There are no requirements for minimum straight lengths of pipe upstream. However, standard practice is that a bend should not occur in the downstream pipe within 10 pipe diameters of the spray nozzle.

Variable orifices (Type 6) — These are mounted in a vertical line with steam flow upward. There is no require-

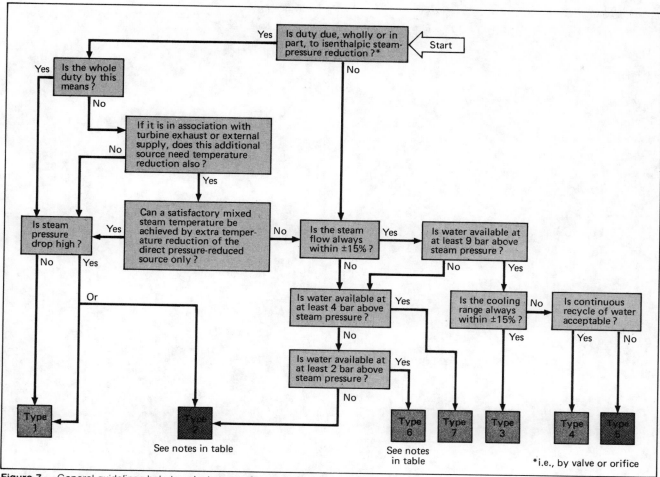

Figure 7 — General guidelines help to select proper desuperheater for a particular application

ment for any specific number of straight lengths upstream or downstream of these desuperheaters. However, as noted before, a close-downstream bend may suffer erosion. A temperature-sensing device should be placed 10–20 pipe diameters downstream.

Operation

For proper operation, the prime requirement is that there always be a supply of clean water at adequate pressure. However, to avoid erratic operation, the water temperature should not be high enough to cause flashing.

Although normal maintenance of controls is sufficient, attention must be paid to installations that are not normally in service, e.g., in steam-turbine trip-bypasses. This is to ensure that such desuperheaters will operate properly when they are needed. In desuperheater selection, variable duties must be considered and specified. If the unit is not selected for the full range of flowrates, no amount of fiddling will right the situation. Turn-up or turndown ratios are important to proper selection. This applies to temperature and pressure variations of the steam, not just its flowrate.

The effect of equipment failure must be considered. If the desuperheating water supply fails, how will it affect the steam system and associated equipment? Will fully superheated steam damage piping, piping supports, and equipment? What happens if the main steam flow stops or drops

sharply — will the water control system react quickly enough to avoid water slugs passing through the system? Is there a danger of thermal shock?

Answers to such questions will determine the necessary instrumentation and standby equipment. A desuperheater that fails "safely" for a particular hazard may avoid expensive additional equipment and maintenance. Fig. 6 shows typical backups for the types of desuperheaters covered before.

Selection

Fig. 7 shows the most likely choices to be made in specifying a desuperheater. It can be used to assist in comparing alternative offers, based on a general inquiry. It should not be taken as definitive.

The author

George E. Bowie is manager of the Engineering Development Group in the Process Dept. of Humphreys & Glasgow Ltd., Chestergate House, 253 Vauxhall Bridge Rd., London SW1V 1HD, U.K. His group is responsible for the engineering of piping and instrument diagrams from process flowsheets. Bowie originally trained and served as a sea-going marine engineer. During his 30-year service with his firm, he has been involved with equipment design, project management and plant development.

Section VI
Cost

Estimating costs of shell-and-tube heat exchangers

Costs of most exchangers can be estimated via this method.
Among key parameters taken into account are exchanger geometry
and metallurgy, and shellside and tubeside design pressures.

G. P. Purohit, Fluor Corp.

☐ This cost-estimating method covers the fixed-tube-sheet, U-tube, split-ring floating-head, and pull-through floating-head designs. For each of these types, it considers various configurations, as classified by the Tubular Exchanger Manufacturers Assn. (TEMA) (Fig. 1) [1].

Labor and material costs are identified as percentages of total exchanger cost, based on actual cost data. Material cost is divided according to the major cost contributors: tubes, shell, channels and tubesheets.

All the parameters are modeled as a function of shell inside diameter. The cost equations include the effects of shell and tube diameter, and tube construction (welded or seamless), wall gage (average or minimum tolerance), pitch, layout angle and length. Separate curves for shell-and tubeside take into account the effect of design pressures. Relative costs of various metals and alloys are extensively tabulated.

In addition to the various TEMA configurations, the method covers shell diameters of from 12 to 148 in., tube lengths of from 8 to 36 ft, tube diameters of from ¾ to 2 in., tube-wall thicknesses of from 10 to 20 Birmingham Wire Gage (BWG), tube passes of from 1 to 8, shell design pressures of from 100 to 2,800 psig, tubeside design pressures of from 100 to 2,500 psig, construction of a wide range of alloys (also clad construction), and specialty exchangers with high-flux tubing. A flowchart of the estimating procedure is presented in Fig. 2.

The results of the cost-estimating method are compared against actual purchased prices, and found to be accurate to within ±10% in 60% of the cases, and to within ±15% in the remaining 40%. However, the overall accuracy for all the exchangers is within 2%.

Fixed-tubesheet design

Straight tubes in the fixed-tubesheet exchanger are secured at both ends in tubesheets welded to the shell. TEMA front-end stationary-head types **A**, **B** or **N** and rear-end head types **L**, **M** or **N** can be used with this type

of exchanger (Fig. 1). An example of a TEMA-type configuration is **AEL**. These exchangers are generally inexpensive. Besides having fewer parts, they are slightly smaller and of simpler construction than other types.

The tubes can be easily cleaned mechanically. However, because the tube bundle cannot be removed, the shellside cannot be cleaned in this way. Chemical cleaning is possible but sometimes unsatisfactory. Because of the fixed tubesheet, an expansion joint in the shell may be necessary to prevent high stresses resulting from differential thermal expansion between the tubes and shell. The cost of the expansion joint may offset the price advantage of this type of exchanger over the others.

If corrosion occurs, fixed-tubesheet exchangers are difficult and expensive to maintain. Retubing may be expensive and time-consuming, and even impractical. Although bonnet or flat covers may be fitted at either end, flat covers are generally specified when frequent tube-cleaning is expected.

U-tube design

Both ends of U-shaped tubes are fastened to a single tubesheet in the U-tube design. This eliminates the problem of differential thermal expansion, because the tubes are free to expand and contract. This design may be fitted with any of the TEMA front-end stationary-head types (Fig. 1). An example of a TEMA configuration is **AEU**. U-tube exchangers are moderate in cost.

The tube bundle can be removed, making manual cleaning of the shellside easy. However, because cleaning the inside of the tubes is difficult, tubeside service is restricted to clean fluids. High-pressure fluids are generally placed on the tubeside.

Floating-head designs

Floating-head rear-end configurations (Fig. 1) may be split-ring (TEMA type **S**), pull-through (TEMA type **T**), externally-sealed-tubesheet (TEMA type **W**), or outside-

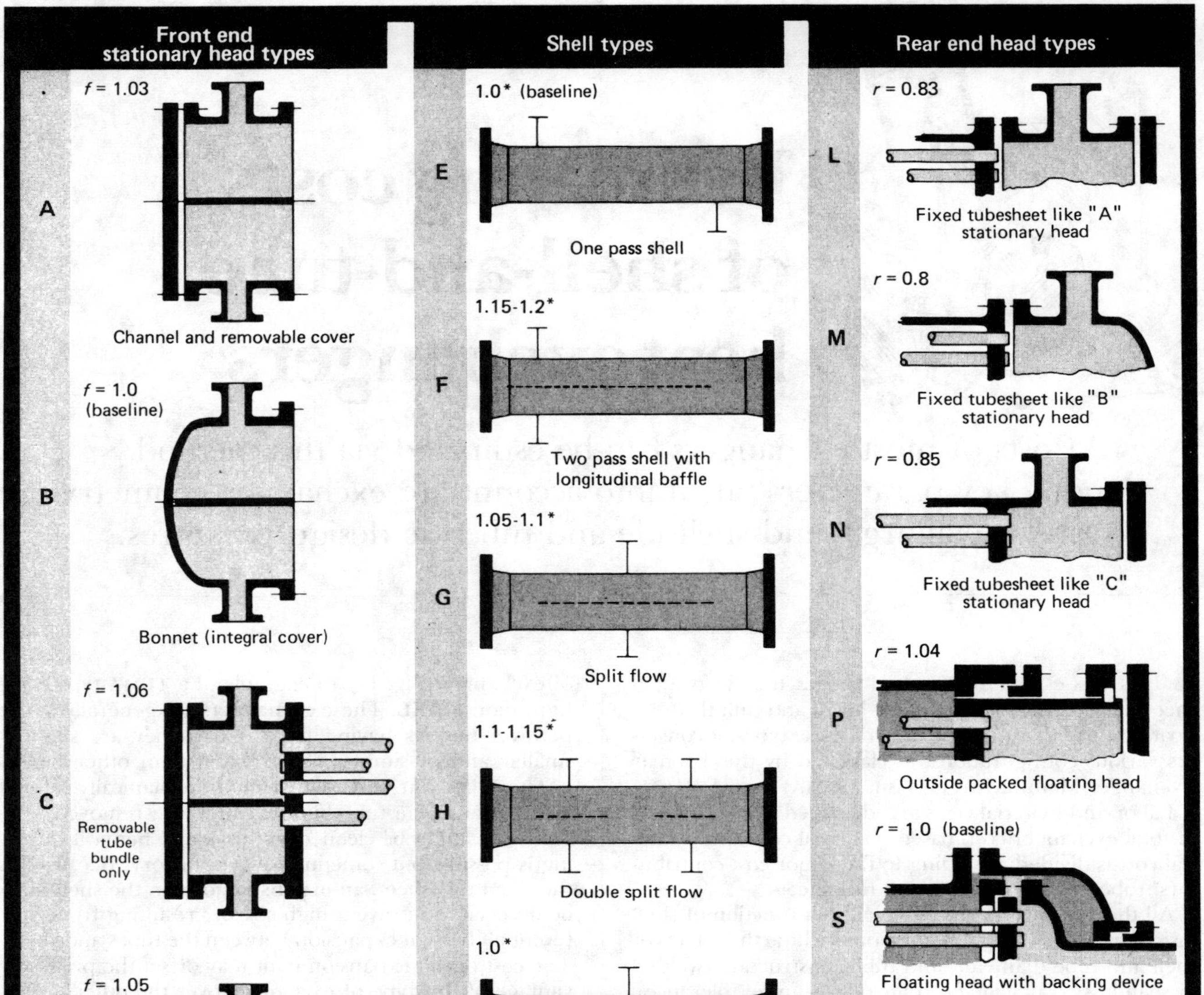

Front end stationary head types	Shell types	Rear end head types

Front end stationary head types

A — $f = 1.03$ — Channel and removable cover

B — $f = 1.0$ (baseline) — Bonnet (integral cover)

C — $f = 1.06$ — Removable tube bundle only

N — $f = 1.05$ — Fixed tubesheet only — Channel integral with tubesheet and removable cover

D — $f = 1.5-1.7$ — Special high pressure closure

Shell types

E — $1.0*$ (baseline) — One pass shell

F — $1.15-1.2*$ — Two pass shell with longitudinal baffle

G — $1.05-1.1*$ — Split flow

H — $1.1-1.15*$ — Double split flow

J — $1.0*$ — Divided flow

K — $1.25-1.35*$ — Kettle type reboiler

X — $1.0*$ — Cross flow

*Cost factor relative to baseline

Rear end head types

L — $r = 0.83$ — Fixed tubesheet like "A" stationary head

M — $r = 0.8$ — Fixed tubesheet like "B" stationary head

N — $r = 0.85$ — Fixed tubesheet like "C" stationary head

P — $r = 1.04$ — Outside packed floating head

S — $r = 1.0$ (baseline) — Floating head with backing device

T — $r = 1.05$ — Pull through floating head

U — $r = 0.9$ — U-tube bundle

W — $r = 1.02$ — Packed floating tubesheet with lantern ring

TEMA heat exchanger nomenclature, with cost factors for shell types, and front-end and rear-end head types relative to their respective designated baseline type

Fig. 1

Tubular Exchanger Manufacturers Assn.

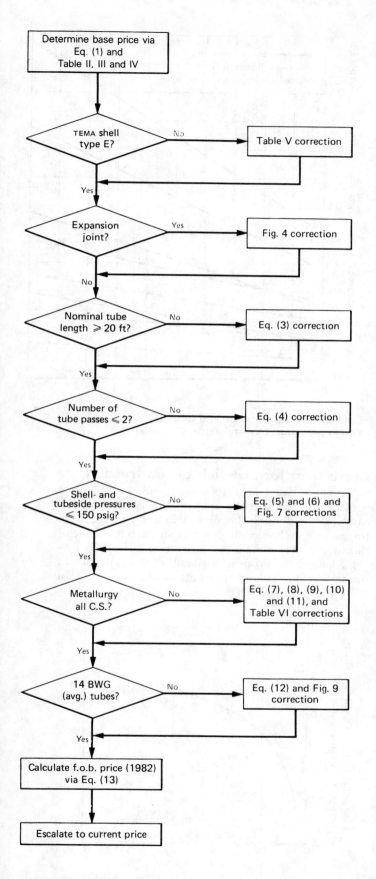

**Flowchart for estimating costs of
shell-and-tube heat exchangers**   **Fig. 2**

Nomenclature

A	Surface area, ft²/shell
a	Cost multiplier for layout angle, Eq. (2) and (7)
b	Base price of exchanger (defined in Table 1), $/ft²
C_g	Cost correction for tube gage, Eq. (12)
C_L	Cost correction for tube lengths shorter than 20 ft
C_{mc}	Cost correction for channel material
C_{ms}	Cost correction for shell material
C_{mt}	Cost correction for tube material
C_{mts}	Cost correction for tubesheet material
$C_{N_{TP}}$	Cost correction for number of tube passes
C_{PS}	Cost correction for shellside design pressure >150 psig
C_{PT}	Cost correction for tubeside design pressure >150 psig
C_S	Cost correction for TEMA shell type
C_T	Sum of all cost corrections
C_x	Correction for expansion joint
D_i	Shell I.D. (bundle dia. for kettle reboiler), in.
d_o	Tube outside diameter
E_b	Estimated price of exchanger (January 1982), $
f	Cost multiplier for type of TEMA front-end stationary head
g	Cost multiplier for tube gage, Eq. (11)
L	Nominal tube length, ft
M_1	Relative cost of alloy tubing to welded carbon-steel tubing, Eq. (8) and Table VI
M_2	Relative cost of plate materials to carbon-steel plate material of same dimensions for exchanger shells, channels and tubesheets (Table VI)
N	Total number of shells per heat-exchange unit
N_{TP}	Number of tube passes
p	Cost multiplier for tube outside dia. x pitch x layout angle
P_{DS}	Shellside design pressure, psig
P_{DT}	Tubeside design pressure, psig
p_i	Tube pitch, Eq. (2)
r	Cost multiplier for type of TEMA rear-end head
X	Adjustment parameter for shellside design pressures greater than 2,000 psig, Eq. (5) and Fig. 7
y	Cost of carbon-steel tubes as a fraction of total price of base carbon-steel exchanger

packed (TEMA type **P**). Types **S** and **T** are more common than types **W** and **P**. An example designation is **BES**.

The floating head eliminates the need for an expansion joint. Because the tube bundle can be removed, both the shell- and tubeside can be cleaned mechanically. Floating-head exchangers are more expensive than the fixed-tube and U-tube types.

The split-ring exchanger's shell-cover diameter is larger than the shell diameter, usually by about 3–4 in. In the pull-through design, the shell cover and shell are of the same diameter. Whereas the shell cover of the split-ring exchanger can be removed, this need not be the case with the pull-through design.

Split-ring construction permits tubes to be close to the inside of the shell. In the pull-through design, space must be provided around the periphery of the floating tubesheet to accommodate bolts. For the same number

Cost corrections for TEMA shell type   **Table V**

	TEMA shell type	Correction, C_S
	Open-tube exchanger (no shell) (special case)	−0.2
E	One-pass (base)	0
J	Divided-flow	0
X	Cross-flow	0
G	Split-flow	0.05 - 0.1
H	Double split-flow	0.1 - 0.15
F	Two-pass with longitudinal baffle	0.15 - 0.2
K	Kettle-type reboiler	0.25 - 0.35

($/ft²) is less than for an exchanger having 20-ft tubes; however, no credit should be taken for tube lengths longer than 20 ft. If $L > 20$ ft, $C_L = 0$. Eq. (3) is plotted in Fig. 5. Although the curves only extend to shell diameters of 60 in., they are valid beyond this limit.

For an exchanger with a 35-in. shell diameter and 12-ft-long tubes, $C_L = 0.55$, via Eq. (3). If the exchanger base price is $10/ft², the shorter tubes would boost the price by $5.5/ft².

Correction for tube passes

Larger numbers of tube passes increase exchanger cost by hiking labor costs and reducing the available heat-exchange surface. Exchangers can have as many as 20 tube passes. Although based on limited data, Eq. (4) is suggested for estimating the extra cost for more than two tube passes:

$$C_{N_{TP}} = (N_{TP} - 1)/100 \qquad (4)$$

Here, N_{TP} = actual number of tube passes. For single or double tube passes, $C_{N_{TP}} = 0$.

For a $10/ft² base-price exchanger with 8 tube passes, the cost of the additional passes is $0.7/ft².

Corrections for design pressures

Exchanger price rises with increasing design pressure. For shellside design pressures higher than 150 psi, calculate the cost fraction to be added via Eq. (5):

$$C_{PS} = [(P_{DS}/150) - 1][0.07 + 0.0016(D_i - 12)] + X \quad (5)$$

For tubeside pressures greater than 150 psi, the cost fraction of base price to be added is calculated via:

$$C_{PT} = [(P_{DT}/150 - 1)[0.035 + 0.00056(D_i - 12)] \quad (6)$$

Eq. (5) (with $X = 0$) and Eq. (6) are plotted in Fig. 6. Although the curves extend to only 60-in. shell diameter, they can be extrapolated beyond this limit. As Fig. 6 shows, the cost effect of shellside pressure is twice that of tubeside, up to 2,000 psig, and even greater beyond 2,000 psig. The X term in Eq. (5) is an adjustment factor for shellside design pressures higher than 2,000 psig. Values of X (based on limited data) are plotted in Fig. 7 as a function of shell diameter. For shellside pressures equal to, and less than, 2,000 psig, $X = 0$.

For an exchanger with a 35-in. shell diameter and designed for a shellside pressure of 450 psi and a tubeside pressure of 600 psi, $C_{PS} = 0.2136$ and $C_{PT} = 0.1436$. Therefore, the higher shellside pressure adds $2.14/ft², and the tubeside pressure $1.44/ft², to the $10/ft² base cost of the exchanger.

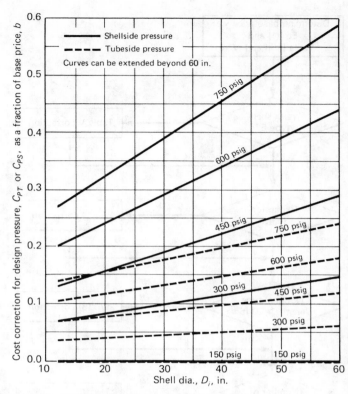

Cost corrections for shell- and tubeside design pressures; C_{PT}—tubeside, C_{PS}—shellside   Fig. 6

Corrections for materials of construction

Eq. (1) is based on carbon-steel construction. Alloy construction can alter the price of an exchanger significantly, depending on the alloy's cost and its quantity in the exchanger. Sometimes, exchanger parts are made of different alloys, making cost estimating difficult.

Exchanger costs consist essentially of labor, profit and materials. The costs of materials can be broken down

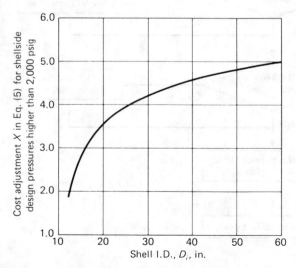

Additional cost correction for shellside design pressures higher than 2,000 psig   Fig. 7

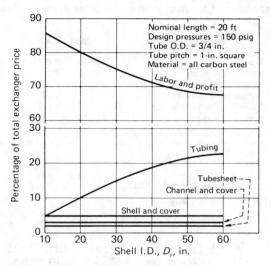

Typical price breakdown for low-pressure carbon-steel exchangers   **Fig. 8**

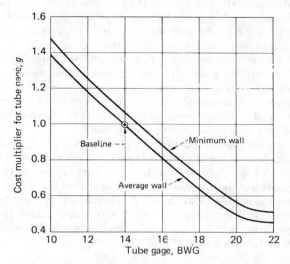

Tube-gage cost multiplier depends on specification of average or minimum wall tolerance   **Fig. 9**

into those for shell, tubes, tubesheets and channels. These costs are related to shell diameter for a low-pressure (shell- and tubeside design pressures up to 150 psig), carbon-steel exchanger with a ¾-in. tube O.D. × 1-in. tube pitch × 45- or 90-deg. layout in Fig. 8.

Fig. 8 shows that, with increasing shell diameter, unit labor cost declines, whereas the costs of the shell, channel and tubesheet remain fairly constant –at approximately 10%, 6% and 4%, respectively—relative to the total cost of the base carbon-steel exchanger. The unit tube cost, however, increases with shell diameter. The cost of carbon-steel tubes, y, as a fraction of the total price of the base carbon-steel exchanger can be calculated via Eq. (7):

$$y = 0.129 + 0.0016(D_i - 12)[d_o/0.75(p_i)^2 a] \qquad (7)$$

In Eq. (7), $a = 0.85$ for a triangular tube-pitch pattern (30 or 60 deg.), and $a = 1.0$ for a square tube-pitch pattern (45 or 90 deg.).

For a carbon-steel exchanger having a 35-in. shell diameter and tubes of ¾-in. O.D. in a 1-in. square-pitch pattern, $y = 0.166$. In the case of a $10/ft^2 base-price exchanger, this means that the cost contribution of the tubing is $10/ft^2 × 0.166, or $1.66/ft^2. Similarly, the cost contribution of the shell is $10/ft^2 × 0.1, or $1/ft^2; that of the channels is $10/ft^2 × 0.06, or $0.6/ft^2; and that of the tubesheets is $10/ft^2 × 0.04, or $0.4/ft^2. The total material cost is, therefore: $1.66/ft^2 + $1/ft^2 + $0.6/ft^2 + $0.4/ft^2 = $3.66/ft^2. The balance of $6.34/ft^2 is attributed to labor costs and profit.

Cost corrections for tube material calculated from Eq. (7) run higher—from 4 to 10%—in the shell diameter range of 20–12 in. The purpose of this is to provide conservative estimates for exchangers of small diameters with alloy tubes.

Knowing the cost relationships of the carbon-steel tubes, shell, channels and tubesheets to the base carbon-steel exchanger, one can calculate the higher cost due to other-than-carbon-steel construction for these exchanger components by multiplying the relative cost of the other material to that of carbon steel.

Calculate the cost correction for tube material via:

$$C_{mt} = y(M_1 - 1) \qquad (8)$$

The cost correction for shell material by:

$$C_{ms} = 0.1(M_2 - 1) \qquad (9)$$

The cost correction for channel material from:

$$C_{mc} = 0.06(M_2 - 1) \qquad (10)$$

And the cost correction for tubesheet material by:

$$C_{mts} = 0.04(M_2 - 1) \qquad (11)$$

The value of y in Eq. (8) is calculated via Eq. (7). In Eq. (8), M_1 is the ratio of tubing cost (welded or seamless) relative to welded carbon-steel tubes of the same diameter, wall thickness and cut length (i.e., for welded carbon-steel tubes, $M_1 = 1$ and $C_{mt} = 0$). In Eq. (9) through (11), M_2 is the relative cost of plate materials to carbon-steel plate of the same dimensions. The cost ratios M_1 and M_2 for various materials used in heat exchangers listed in Table VI are calculated from early 1982 prices.

Tubing prices depend on market conditions and such factors as required cut lengths and quantity of purchase. Price differences among manufacturers can differ by as much as 40%. Prices of low-alloy tubing vary significantly, depending on whether construction is welded or seamless. As can be seen in Table VI, the price of seamless carbon-steel tubing is 2.5 times that of welded carbon-steel tubing. As the material price increases, however, the cost difference between seamless and welded construction diminishes to as low as 10%. Values for M_1 chosen from Table VI will depend on whether construction is seamless or welded.

The M_2 ratios in Table VI for shell, channel and tubesheet materials are based on plate prices. The accuracy of M_1 and M_2 values is expected to be within ±20%. The ratios should be updated periodically.

Exchanger components constructed of carbon steel or low alloys are sometimes clad with a more-expensive material, rather than making them entirely out of the

of tubes, therefore, the shell diameter of the pull-through exchanger is about 2 in. larger than that of the split-ring type. Because the seal is not visible externally in either type, a leak is not easily detected.

In the externally sealed design (TEMA type **W**), the shell- and tubeside streams are individually sealed and separated by a lantern ring. This type is the least expensive of the floating-head designs, and can be fitted with an **A**, **B** or **C** type front head. It is limited to a maximum of two tube passes, 300 psi and 375°F.

In the outside-packed design (TEMA type **P**), a skirt attached to the floating tubesheet extends through the back of the shell. Several layers of packing in a packing gland seal the space between the skirt and shell. A packing failure is readily detected. This design permits high tubeside pressures and more than two tube passes. However, the packing tends to limit shellside fluids to less than 150 psi and 300°F.

The base-cost equation

Costs of shell-and-tube exchangers vary according to: TEMA type; shell diameter; tube length, diameter, construction (welded or seamless), gage, pitch and layout; number of tube passes; shell- and tubeside design pressures; and materials of construction. Each of these factors is related to a baseline exchanger, which is defined in Table I.

Exchangers having small-diameter tubes (¾- or 1-in. O.D.) are more common, and generally more economical, than those having large-diameter tubes. Because tube pitch is at least 1.25 × tube O.D., large-diameter tubes provide less surface area (for a given shell diameter) than small-diameter tubes, and increase cost almost proportionally. Nevertheless, tube diameters, in a few cases, range up to 2 in. Triangular tube patterns (30 or 60 deg.) are more economical than square-pitch patterns (45 or 90 deg.), because they generally accommodate small surfaces in a given shell better.

Unit labor costs are higher than unit material costs for small-diameter exchangers. As shell diameter increases, however, the unit material cost rises and the unit labor cost declines. For this reason, exchangers of large shell diameters are more economical on a $/ft² basis. This makes shell diameter a convenient parameter on which to base estimates of shell-and-tube-exchanger costs.

Analysis of the purchase prices of many shell-and-tube exchangers received by means of competitive bids during the early part of 1982 resulted in the development of Eq. (1) for estimating the cost of the baseline exchanger defined in Table I:

$$b = \left[\frac{6.6}{1 - e^{[(7-D_i)/27]}}\right]pfr \qquad (1)$$

Here, D_i = shell I.D. or bundle dia. of a kettle reboiler, in.; p = cost multiplier for tube O.D., pitch and layout angle (Table II); f = cost multiplier for TEMA-type front head (Table III); and r = cost multiplier for TEMA-type rear head (Table IV).

Eq. (1) is plotted in Fig. 3 as a function of shell diameter for TEMA **BES** exchangers having ¾-in. O.D. tubes × 1-in. pitch × 45-deg. layout. Eq. (1) can theoretically be used for shell diameters as small as 8 in. However, at shell diameters of less than 12 in., double-pipe and

Definition of baseline shell-and-tube exchanger   Table I

Parameter	Base designation
Tubes	Welded c.s., 14 BWG, avg. wall
Nominal tube length	20 ft
Number of tube passes	1 or 2
Shellside design pressure, psig	≤150
Tubeside design pressure, psig	≤150
Material of construction	All carbon steel

Base price, which includes cost of ASME Sec. VIII, Div. 1 code stamp and of exterior paint, was developed via evaluation of bids by U.S. manufacturers.

Cost multiplier for tube O.D., pitch and layout angle   Table II

	Multiplier, p	
Tube O.D. × pitch	Triangular pitch (30- or 60-deg.)	Square pitch (45- or 90-deg.)
5/8 in. × 25/32 in.	0.62	Not common
3/4 in. × 15/16 in.	0.80	Not common
3/4 in. × 1 in.	0.85	1.0 (base)
7/8 in. × 1 3/32 in.	0.87	Not common
1 in. × 1 1/4 in.	0.98	1.16
1 1/4 in. × 1 9/16 in.	1.23	1.45
1 1/4 in. × 1 37/64 in.	1.29	1.49
1 1/2 in. × 1 7/8 in.	1.47	1.73
1 1/2 in. × 1 57/64 in.	1.56	1.80
1 3/4 in. × 1 3/16 in.	1.72	2.03
1 3/4 in. × 1 13/64 in.	1.81	2.13
2 in. × 2 1/2 in.	1.97	2.32
2 in. × 2 17/32 in.	2.08	2.45

Cost multiplier for TEMA front-end stationary-head type   Table III

	TEMA type	Multiplier, f
	No front end (special case)	0.95
B	Bonnet (integral cover) (base)	1.0
A	Channel and removable cover	1.02 - 1.03
N	Channel integral with tubesheet, and removable cover	1.05
C	Channel integral with tubesheet, and removable cover	1.06 - 1.07
D	Special high-pressure closure	1.5 - 1.7

Cost multiplier for TEMA rear-end head type   Table IV

	TEMA type	Multiplier, r
S	Floating head with backing device (base)	1.0
M	Fixed tubesheet, like B stationary head	0.8
L	Fixed tubesheet, like A stationary head	0.83
N	Fixed tubesheet, like N stationary head	0.85
U	U-tube bundle	0.9
T	Pull-through floating head	1.05
P	Outside-packed floating head	1.04
W	Externally sealed floating tubesheet	1.02

multitube exchangers are more economical [2]. Eq. (1) has been checked against purchase prices of exchangers ranging in diameter from 12 to 148 in. and found to be satisfactory.

Values for cost multiplier p in Eq. (1) are listed in Table II for commonly used tube diameters, pitches and layouts. Values of p not given in Table II can be approximated by means of Eq. (2):

$$p = 0.75(p_i)^2(a/d_o) \qquad (2)$$

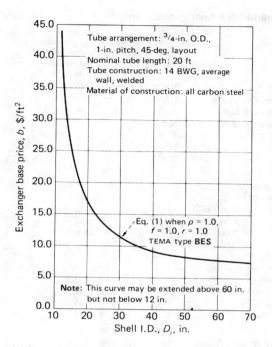

Base price of TEMA BES exchangers (1982)   **Fig. 3**

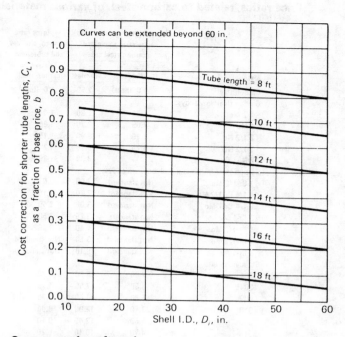

**Cost corrections for tube
lengths shorter than 20 ft**   **Fig. 5**

In Eq. (2), p_i = tube pitch, in., and a is a constant. For 45- or 90-deg. layouts, $a = 1$; for 30- or 60-deg. layouts, $a = 0.85$.

For a U-tube exchanger having a 35-in. shell diameter, ¾-in. tube O.D., 1-in. tube pitch, 45-deg. tube layout, and a TEMA type **A** front end, $p = 1.0$ (Table II), $f = 1.02$ (Table III), and $r = 0.9$ (Table IV). Via Eq. (1), $b = \$10/\text{ft}^2$. To arrive at a cost estimate for this exchanger, this base price must be corrected by means of other parameters, which are now discussed.

Correction for shell type

Eq. (1) is based on TEMA shell type **E**. For shell types other than **E,** the base price calculated via Eq. (1) must be adjusted by a cost-correction factor, C_S, from Table V. For example, for a kettle reboiler (TEMA type **K**), $C_S =$ 0.3. Therefore, the estimated price of a kettle reboiler is $\$13/\text{ft}^2$—i.e., $\$10/\text{ft}^2 + (\$10/\text{ft}^2 \times 0.3)$.

Correction for expansion joint

If differential thermal expansion is expected, an expansion joint may be incorporated into the shell of a fixed-tubesheet exchanger. This, of course, hikes the cost of an exchanger.

Two types of expansion joints are common: the flanged-and-flued and the bellows. The first is cheaper but limited to ³⁄₁₆-in. expansion. For greater thermal expansion, the bellows type is usually selected.

Cost corrections for expansion joint as a fraction of base price are plotted in Fig. 4 as a function of shell diameter. For example, consider a fixed-tubesheet exchanger with a 30-in. shell diameter at a base price of $\$10/\text{ft}^2$. A flanged-and-flued-type joint would add $\$1.4/\text{ft}^2$ to the base price, and a bellows type would boost the price by $\$2.9/\text{ft}^2$.

Correction for tube length

On a $/\text{ft}^2$ basis, long, skinny exchangers are more economical than short, fat ones, because of the labor cost and material scrapped in cutting tubes to required lengths. However, the labor cost decreases somewhat as shell diameter increases, because of the larger number of tubes involved.

For the baseline exchanger, tubes are of 20-ft nominal length (Table I). Cost-correction factors for tube lengths less than 20 ft can be calculated via Eq. (3):

$$C_L = \left[1 - \frac{L}{20}\right]\left[1.5 - \frac{0.002083(D_i - 12)}{1 - (L/20)}\right] \quad (3)$$

If tube length, L, is 20 ft, $C_L = 0$. If $L > 20$ ft, C_L is negative, indicating that the unit cost of the exchanger

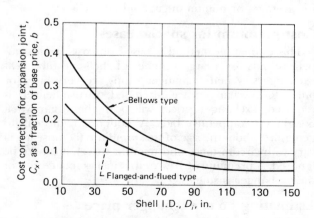

Cost corrections for shell expansion joints   **Fig. 4**

Price ratios, related to carbon steel, of various materials for shell-and-tube heat exchangers   Table VI

Material	M_1, tubing-price ratio relative to welded carbon-steel tubes		M_2, price ratio for shell, channel and tubesheet relative to carbon steel	Material	M_1, tubing-price ratio relative to welded carbon-steel tubes		M_2, price ratio for shell, channel and tubesheet relative to carbon steel
	Welded	Seamless			Welded	Seamless	
Carbon steel	1.0 (base)	2.50	1.0 (base)	Ferralium (Alloy 255)	12.00	23.90	14.00
Carbon steel, low alloys:				Carpenter 20 Cb-3	15.10	––	16.00
1/2 Mo	1.04	2.60	1.04	Carpenter 20 Mo-6	18.90	––	––
1 Mo	1.05	2.70	1.05	AL-6-X	12.20	––	––
2 1/2 Ni	1.15	2.90	1.15	AL-29-4	12.00	––	––
3 1/2 Ni	1.20	3.10	1.20	AL-29-4-2	11.80	––	––
2 Ni-1 Cu	––	3.30	1.30	AL-29-4-C	5.0	––	––
Carbon steel, chromium-molybdenum alloys:				Nickel 200	––	20.90	18.40
1 Cr-1/2 Mo	Not standard	2.60	2.00	Monel 400 (Alloy 400)	––	15.50	14.50
1 1/4 Cr-1/2 Mo	Not standard	2.70	2.10	Inconel 600 (Alloy 600)	19.40	––	15.30
2 1/4 Cr-1 Mo	Not standard	3.00	2.40	Inconel 625 (Alloy 625)	––	32.70	27.40
3 Cr-1 Mo	Not standard	3.20	2.50	Incoloy 800 (Alloy 800)	11.00	21.80	9.00
5 Cr-1/2 Mo	Not standard	4.40	3.50	Incoloy 800H (Alloy 800H)	––	18.00	––
7 Cr-1/2 Mo	Not standard	5.50	Not standard	Incoloy 825	––	23.50	––
9 Cr-1 Mo	Not standard	6.10	Not standard	Hastelloy B-2	34.90	48.60	38.40
Stainless steels:				Hastelloy C-4	28.70	40.00	31.30
304	2.80	6.50	3.70	Hastelloy C-276	29.10	38.10	31.00
304L	3.00	7.50	4.70	Hastelloy G	15.30	24.70	18.10
309	5.80	14.50	7.70	Hastelloy X	16.70	27.10	21.30
310	7.40	12.00	9.80				
310L	7.60	12.40	10.10	Titanium (Grade 2)	11.00	22.00	11.00
316	4.70	10.10	6.20	Titanium (Grade 7)	21.00	42.00	––
316L	4.80	11.00	6.40	Titanium (Grade 12)	14.00	28.00	––
317	8.10	13.30	8.10				
317L	8.30	13.60	8.30	Zirconium 702	35.00	43.70	36.80
321	4.20	9.50	5.60	Zirconium 705	39.00	48.70	40.00
329 (Carpenter 7 Mo)	10.50	17.20	10.50	Aluminum	Not standard	1.60	1.60
330	7.90	12.90	9.50	Naval rolled brass	Not standard	3.50	3.50
347	5.50	13.70	7.30	Admiralty	Not standard	3.60	3.60
405	6.00	15.00	6.90	Aluminum brass	Not standard	3.70	3.70
410	6.90	17.20	7.90	Aluminum bronze (5%)	Not standard	4.10	4.10
430	5.40	10.60	6.20	Copper (arsenical or			
439	5.00	11.20	5.80	deoxidized)	Not standard	4.20	4.20
444 (Alloy 18-2)	7.80	8.80	9.00	90-10 cupro-nickel	3.50	4.60	4.60
446	4.70	10.00	5.40	70-30 cupro-nickel	4.20	5.50	5.50
904L (Sandvik 2RK-65)	15.30	19.20	17.00	Union Carbide high-flux tubing:			
Sandvik 2RE-69	––	14.50	––	SA-214, welded	4.40	––	––
Sandvik 3RE-60	––	10.10	––	SA-334-1, welded	4.70	––	––
Sandvik 253 MA	––	12.70	––	SA-334-3, seamless	––	9.00	––
Sandvik SAF 2205	––	11.80	––	SA-214, fluted, welded	7.00	––	––
Sanicro 28	16.10	20.20	18.20	SA-334-1, fluted, welded	7.40	––	––
E-brite-26-1 (XM-27)	9.00	––	10.00	SA-210, fluted, seamless	––	8.40	––

more-expensive material. In such cases, the average of the material ratios of the base and clad materials should be used for cost-estimating purposes. If, for example, a carbon-steel ($M_2 = 1.0$) tubesheet is clad with Monel ($M_2 = 14.50$), the average $M_2 = 7.75$.

Correction for tube gage

The price of tubing depends on the tube-wall gage (because cost is proportional to the weight of the material), and on whether an average or minimum gage is specified. Because only positive tolerances are allowed, minimum-wall tubing will cost more than average-wall tubing by about 5–10%.

This cost-estimating method is based on 14 Birmingham Wire Gage (BWG), average-gage wall, welded tubing. For gages other than this, the cost correction as a fraction of base price (C_g) is calculated from Eq. (12):

$$C_g = y(g - 1) \qquad (12)$$

Values for y are calculated via Eq. (7). Values for g, the cost multiplier for tube gage, are obtained from Fig. 9, for average- or minimum-gage tubing walls.

Cost premium for special cases

This estimating method is based on costs from competitive bids for a large number of shell-and-tube heat exchangers. When exchanger shops are very busy, or only a few exchangers are purchased, or rush delivery is required, exchanger prices will be higher than when bought via competitive bids. Although it is difficult to generalize about the size of this premium (because it will vary with circumstance and manufacturer), it cannot be ignored. A premium of 10% of the final estimated cost of the exchanger is suggested.

Estimating f.o.b. exchanger price

After the exchanger base price and all the corrections for alternatives have been calculated, the free-on-board

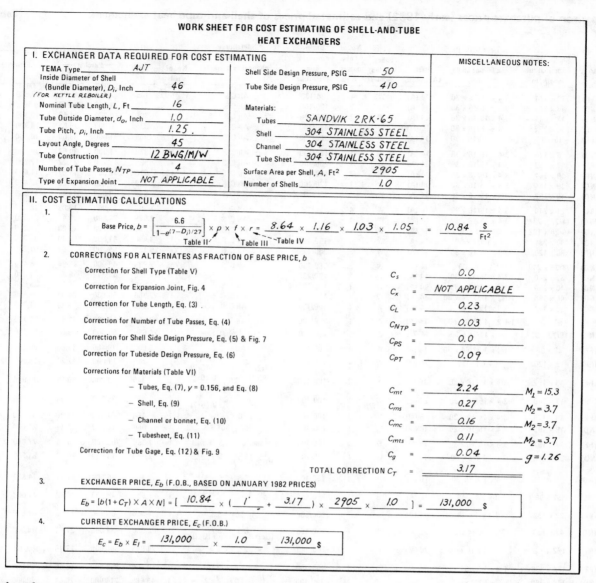

Worksheet for cost estimating of shell-and-tube heat exchangers

Fig. 10

(f.o.b.) January 1982 price of the particular exchanger can be estimated via Eq. (13):

$$E_b = [b(1 + C_T)A]N \qquad (13)$$

In Eq. (13), C_T represents the sum of all the base-price cost corrections for alternatives, as calculated via Eq. (3) through (12) and obtained from Table V and Fig. 4.

Because equipment costs generally change from the date that an order is placed to when it is shipped, estimators usually incorporate an escalation factor into cost estimates. When manufacturers submit bids, they usually include formulas for adjusting prices.

An exchanger price estimated by this method can be escalated to a price for a date beyond January 1982 by means of an escalation index. One such index that applies to process heat exchangers is the Fabricated Equipment component of the Chemical Engineering Plant Cost Index. This component is updated monthly.

Other possibilities are the Marshall & Swift Equipment Cost Index, the Nelson Refinery Cost Index (or the materials-cost component), and the U.S. Bureau of Labor Statistics General Purpose Machinery and Equipment Cost Index.

An estimated corrected base-price of an exchanger, E_b, is escalated from the January 1982 date to another date by multiplying it by the ratio of the values of the Fabricated Equipment component on the two dates.

When exchanger price quotations more-current than those of the base date are readily available, an escalation index specific to exchangers can be determined. For comparable exchangers, the sum of the average of the lowest and second-lowest current quotations, divided by the sum of cost estimates for an equal number of exchangers, provides a highly reliable escalation index. An index determined in this manner should remain valid for about six months. Taking the average of lowest and

Comparison of purchase prices (early 1982) and estimated prices of shell-and-tube heat exchangers   Table VII

Exchanger no.	TEMA type	Shell I.D., in. x nominal tube length, in.	Tube O.D., in. x pitch, x layout angle, deg.	BWG (Avg./Min.)	Tubes, Welded, Seamless	No. of Tube passes, N_{TP}	Design Pressures, psig shell/tube	Metallurgy tube/shell/ channel or bonnet/ tubesheet	No. of shells	Surface area per shell, ft²	Purchase $	Purchase $/ft²	Estimated $	Estimated $/ft²	Accuracy, %
E - 101	BEM (NO)*	12 x 144	3/4 x 15/16 x 30	14 (A)	W	4	180/130	All C.S.	1	185	7,700	41.62	7,600	41.08	− 1.3
E - 102	BEM (NO)	13 x 144	3/4 x 15/16 x 30	14 (A)	W	2	100/100	All C.S.	1	310	9,600	30.97	10,500	33.87	+ 9.4
E - 103	BEM (NO)	13 x 144	3/4 x 15/16 x 30	14 (A)	W	2	150/130	All C.S.	1	305	10,500	34.43	10,300	33.77	− 1.9
E - 104	NEN (NO)	13 x 192	3/4 x 15/16 x 30	14 (M)	W	2	480/100	All C.S.	1	335	10,000	29.85	10,900	32.54	+ 9.0
E - 105	BEM (FF)	15 x 96	3/4 x 15/16 x 30	14 (A)	W	1	200/100	All C. S.	1	305	11,600	38.03	10,700	35.08	− 7.8
E - 106	BEM (NO)	15 x 120	3/4 x 15/16 x 30	14 (A)	W	1	130/100	All C. S.	1	370	11,300	30.54	10,600	28.65	− 6.2
E - 107	BEM (FF)	17 x 192	3/4 x 15/16 x 30	14 (A)	W	4	130/100	All C. S.	1	590	13,700	23.22	12,200	20.74	−11.0
E - 108	BEM (NO)	19 x 144	3/4 x 15/16 x 30	14 (A)	W	4	100/150	All C. S.	1	610	12,000	19.67	11,600	19.02	− 3.3
E - 109	BEM (NO)	19 x 144	3/4 x 15/16 x 30	14 (A)	W	1	130/100	All C. S.	1	720	14,600	20.28	13,400	18.61	− 8.2
E - 110	BEM (NO)	19 x 192	3/4 x 15/16 x 30	14 (A)	W	1	130/100	All C. S.	1	965	13,900	14.40	14,600	15.13	+ 5.0
E - 111	BEM (BL)	21 x 240	3/4 x 15/16 x 30	14 (A)	W	4	200/100	All C. S.	1	1,300	17,000	13.08	18,100	13.92	+ 6.5
E - 112	NKN (NO)	41/27 x 288	3/4 x 15/16 x 30	14 (M)	W	4	350/180	All C. S.	1	2,575	34,600	13.44	35,300	13.71	+ 2.0
E - 113	AKL (NO)	46/28 x 192	3/4 x 1 x 30	14 (A)	W	1	230/230	All C. S.	1	1,660	23,200	13.98	24,200	14.58	+ 4.3
E - 114	AEL (NO)	30 x 384	1 x 1 1/4 x 30	14 (A)	W	4	150/410	All C. S.	1	3,490	41,500	11.89	37,300	10.69	−10.1
E - 115	BGM (NO)	30 x 420	3/4 x 15/16 x 30	14 (A)	W	2	300/1,000	All C. S.	1	5,125	65,700	12.82	57,200	11.16	−12.9
E - 116	AKL (NO)	78/35 x 240	3/4 x 1 x 30	14 (A)	W	2	360/230	All C. S.	1	2,935	40,900	13.94	38,400	13.08	− 6.1
E - 117	AFL (NO)	37 x 240	3/4 x 15/16 x 30	14 (A)	W	8	160/230	All C. S.	1	4,240	41,300	9.74	37,200	8.77	− 9.9
E - 118	BEM (NO)	40 x 240	3/4 x 15/16 x 30	14 (A)	W	1	100/100	All C. S.	1	6,010	36,700	6.11	36,000	5.99	− 1.9
E - 119	CEM (NO)	45 x 168	1 1/2 x 1 57/64 x 30	14 (A)	W	1	1,100/100	All C. S.	1	2,360	56,800	24.07	59,700	25.30	+ 5.1
E - 120	BEM (NO)	58 x 144	3/4 x 15/16 x 30	14 (A)	W	1	130/100	All C. S.	1	7,325	52,600	7.18	54,700	7.47	+ 4.0
E - 121	NEN (FF)	59 x 96	1 1/2 x 1 57/64 x 30	14 (A)	W	1	200/100	All C. S.	1	2,420	52,400	21.65	50,700	20.95	− 3.2
E - 122	BKM (NO)	118/72 x 324	3/4 x 15/16 x 30	14 (M)	S	4	220/330	High-flux /C.S. → C.S. tube /	1	24,650	469,000	19.03	454,100	18.42	− 3.2
E - 123	NEN (FF)	83 x 96	1 1/4 x 1 37/64 x 30	14 (A)	W	1	100/100	All C. S.	1	5,800	79,400	13.69	85,400	14.72	+ 7.6
E - 124	BEM (FF)	12 x 120	3/4 x 1 x 30	20 (M)	W	4	100/100	Ti (Gr2) /C.S./C.S./ Ti (Gr2)	1	145	17,000	117.24	16,800	115.86	− 1.2
E - 125	BEM (FF)	13 x 192	3/4 x 15/16 x 30	16 (A)	S	2	100/120	90-10 Cu-Ni /C.S./ 90-10 → Cu-Ni	1	340	20,700	60.88	17,900	52.65	−13.5
E - 126	BEM (FF)	17 x 240	3/4 x 15/16 x 30	20 (M)	W	1	100/100	Ti (Gr2) /C.S./C.S./ Ti (Gr2)	1	850	35,900	42.24	37,100	43.65	+ 3.3
E - 127	BEM (FF)	18 x 120	3/4 x 15/16 x 30	16 (M)	S	2	100/150	Ni/C.S./Ni →	1	420	47,000	111.90	40,100	95.48	−14.7
E - 128	BEM (NO)	19 x 192	3/4 x 15/16 x 30	16 (M)	W	1	130/100	304 S.S. /C.S./ 304 S.S. →	1	960	23,200	24.17	21,000	21.88	− 9.5
E - 129	BEM (FF)	27 x 192	3/4 x 15/16 x 30	16 (M)	S	1	100/100	Monel/C.S./C.S./Monel	1	1,985	69,700	35.11	77,800	39.19	+11.6
E - 130	AJL (NO)	34 x 240	3/4 x 1 1/8 x 30	14 (M)	W	1	100/100	Ti (Gr2) /C.S./C.S./ Ti (Gr2)	1	2,740	73,900	26.97	81,200	29.64	+ 9.9
E - 131	−EM (NO)	36 x 144	3/4 x 1 1/4 x 30	20 (A)	W	1	130/100	Ti (Gr2) /C.S./C.S./ Ti (Gr2)	1	2,400	65,000	27.08	74,800	31.17	+15.0
E - 132	−EM (BL)	37 x 96	3/4 x 1 x 45	14 (A)	S	1	300/450	304-L S.S. / C.S./ 304-L → S.S.	1	1,385	48,200	34.80	41,000	29.60	−14.9
E - 133	BEN (BL)	37 x 132	3/4 x 1 x 30	14 (A)	W	1	300/300	All 304-L S.S.	1	2,350	68,700	29.23	67,600	28.77	− 1.6
E - 134	−EM (NO)	42 x 120	3/4 x 1 1/4 x 30	20 (M)	W	1	130/100	Ti (Gr2) /C.S./C.S./ Ti (Gr2)	1	2,650	72,000	27.51	78,300	29.55	+ 8.7
E - 135	BXM (NO)	122 x 432	3/4 x 15/16 x 30	20 (M)	W	2	100/140	Ti (Gr2) /C.S.→	1	74,845	1,300,000	17.18	1,489,000	19.89	+14.5
E - 136	NEN (FF)	128 x 336	1 1/2 x 1 57/64 x 30	14 (M)	S	1	100/100	All 1 Cr-1/2 Mo	1	40,550	669,000	16.50	600,000	14.80	−10.3
E - 137	BEM (NO)	148 x 360	2 x 2 1/2 x 30	14 (M)	S	1	100/100	70-30 Cu-Ni /C.S./ 70-30 → Cu-Ni	1	45,565	1,214,000	26.64	1,070,000	23.48	−11.9
E - 138	BEU	12 x 192	3/4 x 1 x 90	14 (A)	W	2	170/130	All C. S.	1	200	9,400	47.0	9,200	46.0	− 2.1
E - 139	BEU	12 x 192	3/4 x 1 x 45	14 (A)	W	4	100/100	All C. S.	1	200	8,600	43.0	9,300	46.50	+ 8.1
E - 140	BEU	13 x 96	1 x 1 1/4 x 30	14 (A)	W	2	100/120	All C. S.	1	140	8,700	62.14	7,800	55.71	−10.3
E - 141	BEU	13 x 192	3/4 x 15/16 x 30	14 (A)	W	4	160/180	All C. S.	1	355	10,400	29.30	11,300	31.83	+ 8.6
E - 142	AEU	15 x 240	1 x 1 1/4 x 30	10 (M)	W	2	2,520/200	All C. S.	1	305	31,500	103.28	36,200	118.69	+14.9
E - 143	BEU	18 x 144	3/4 x 15/16 x 30	14 (A)	W	2	180/130	All C. S.	1	565	14,400	25.49	12,900	22.83	−10.4
E - 144	BJU	23 x 192	3/4 x 1 x 45	14 (A)	S	2	100/100	All C. S.	1	900	16,200	18.0	18,100	20.11	+11.7
E - 145	DEU	26 x 288	1 x 1 1/4 x 45	10 (M)	S	2	150/2,530	All C. S.	1	4,830	171,000	35.40	191,000	39.54	+11.7
E - 146	BKU	49/29 x 192	3/4 x 1 x 90	14 (A)	W	2	650/740	All C. S.	1	1,780	47,100	26.46	40,000	22.47	−15.0
E - 147	BKU	100/45 x 240	1 x 1 1/4 x 30	14 (A)	W	2	380/300	All C. S.	1	7,965	103,300	12.97	97,500	12.24	− 5.6
E - 148	NFU	49 x 421	3/4 x 15/16 x 30	14 (A)	W	2	300/130	All C. S.	1	14,425	107,000	7.42	116,600	8.08	+ 9.0
E - 149	BEU	52 x 192	3/4 x 1 1/4 x 30	14 (A)	W	2	350/650	All C. S.	1	3,040	55,300	18.21	47,100	15.49	−14.8
E - 150	DEU	19 x 168	1 x 1 1/4 x 30	14 (M)	W	2	2,540/2,310	430 S.S. /1¼ Cr-½ Mo→/ 430 S.S.	1	515	112,500	218.45	96,400	187.26	−14.3
E - 151	BHU	21 x 144	1 x 1 1/4 x 45	13 (A)	W	2	300/650	E-brite- 26-1 / 304-L S.S. / 304 S.S. clad on C.S. →	1	515	39,200	76.12	33,400	64.85	−14.8
E - 152	BKU	42/21 x 192	3/4 x 1 x 90	16 (A)	S	2	150/150	Sandvik SAF 2205 / C.S./ 304 S.S. clad on C.S. →	1	925	41,700	45.08	46,900	50.70	+12.5
E - 153	BKU	33/22 x 144	3/4 x 1 x 90	16 (A)	S	2	300/650	Al- 26-1/ C.S. → / 304 S.S.	1	585	31,100	53.16	29,800	50.94	− 4.2
E - 154	AJU	23 x 288	1 x 1 1/4 x 30	16 (A)	S	8	350/100	304 S.S. / 304 S.S. / C.S./ 304 S.S.	1	1,100	34,500	31.08	35,100	31.62	+ 1.7
E - 155	BEU	24 x 120	3/4 x 15/16 x 30	16 (A)	W	4	210/100	All 304 S.S.	1	885	24,100	27.23	24,000	27.12	0.0

*FF = Flanged-and-flued expansion joint, BL = bellows-type expansion joint, NO = No expansion joint

(continued next page)

Comparison of purchase prices (early 1982) and estimated prices of shell-and-tube heat exchangers   (cont'd) Table VII

Exchanger no.	TEMA type	Shell I.D., in. x nominal tube length, in.	Tube O.D., in. x pitch, x layout angle, deg.	BWG (Avg./Min.)	Tubes, Welded, Seamless	No. of Tube passes N_{TP}	Design Pressures, psig shell/tube	Metallurgy tube/shell/ channel or bonnet/ tubesheet	No. of shells	Surface area per shell, ft²	Purchase $	Purchase $/ft²	Estimated $	Estimated $/ft²	Accuracy, %
E - 156	BEU	26 x 192	1 x 1 1/4 x 30	16 (A)	S	4	100/130	Sandvik 3 RE 60 / 304L S.S. → Sandvik 3 RE-60	1	1,125	47,900	42.58	46,800	41.60	− 2.3
E - 157 A, B	DEU	28 x 360	1 x 1 1/4 x 45	16 (A)	S	4	2,340/2,500	304 S.S. / 309 S.S. clad on 2¼ Cr-½ Mo →	2	1,950	753,000	193.08	649,800	166.62	−13.7
E - 158	DEU	32 x 132	1 x 1 1/4 x 45	12 (M)	S	2	2,770/2,540	321 S.S. / 304 S.S. clad on C.S. / 309 S.S. clad on 2¼ Cr-½ Mo →	2	1,290	499,000	193.41	436,400	169.15	−12.6
E - 159 A, B	AEU	32 x 240	1 x 1 1/4 x 45	16 (A)	W	4	670/900	430 S.S. / 304 S.S. clad on C.S.	1	1,815	60,700	33.44	51,700	28.48	−14.8
E - 160	AEU	32 x 240	1 x 1 1/4 x 45	16 (A)	S	4	670/900	316L S.S. /304 S.S. clad on C.S. → 316L S.S.	1	1,815	85,000	46.83	72,300	39.83	−14.9
E - 161	BFU	34 x 180	3/4 x 15/16 x 30	16 (A)	W	2	720/350	304 S.S. /C.S. / 304 S.S. →	1	3,090	63,100	20.42	61,800	20.0	− 2.1
E - 162	DEU	35 x 168	1 x 1 1/4 x 45	12 (M)	S	2	2,750/2,540	321 S.S. / 304 S.S. clad on C.S. / 309 S.S. clad on 2¼ Cr-½ Mo →	1	1,915	316,300	165.17	306,300	159.95	− 3.2
E - 163 A, B	DEU	38 x 144	1 x 1 1/4 x 45	12 (M)	S	2	2,750/2,540	321 S.S. / 304 S.S. clad on C.S. / 309 S.S. clad on 2¼ Cr-½ Mo →	2	1,995	547,800	137.29	629,100	157.67	+14.8
E - 164	AES	12 x 240	3/4 x 1 x 45	14 (M)	S	2	100/130	All C. S.	1	235	13,300	56.60	11,400	48.51	−14.3
E - 165	AES	12 x 240	3/4 x 1 x 45	14 (M)	W	4	100/300	All C. S.	1	255	11,500	45.10	10,600	41.57	− 7.8
E - 166	AES	15 x 192	3/4 x 1 x 45	14 (M)	W	4	230/100	All C. S.	1	400	12,100	30.25	13,900	35.25	+14.2
E - 167	AES	15 x 240	3/4 x 1 x 45	14 (M)	W	4	150/130	All C. S.	1	470	15,200	32.34	13,000	27.66	−14.5
E - 168	AES	17 x 192	3/4 x 1 x 45	14 (M)	W	4	290/100	All C. S.	1	465	14,000	30.11	14,200	30.54	+ 1.4
E - 169 A, B	AES	19 x 144	3/4 x 1 x 45	12 (M)	S	4	100/100	All C. S.	2	415	34,500	41.57	29,400	35.42	−14.8
E - 170 A, B	AES	19 x 192	3/4 x 1 x 45	14 (M)	W	2	150/130	All C. S.	2	645	28,400	22.02	31,500	24.42	+10.9
E - 171A, B, C	AES	19 x 192	3/4 x 1 x 45	14 (M)	W	2	230/170	All C. S.	3	660	44,800	22.63	50,200	25.35	+12.0
E - 172	AES	19 x 240	3/4 x 1 x 45	14 (M)	W	4	300/130	All C. S.	1	735	17,800	24.22	15,500	21.09	−12.9
E - 173	AES	21 x 144	3/4 x 1 x 45	14 (M)	W	2	150/300	All C. S.	1	650	15,700	24.15	17,800	27.38	+13.4
E - 174	AES	21 x 192	3/4 x 1 x 45	14 (M)	W	2	150/420	All C. S.	1	855	18,300	21.40	19,600	22.92	+ 7.1
E - 175	AES	21 x 192	3/4 x 1 x 45	14 (M)	S	6	250/130	Ali C. S.	1	740	21,700	29.32	20,000	27.03	− 7.8
E - 176	AES	21 x 240	3/4 x 1 x 45	14 (M)	W	4	180/100	All C. S.	1	1,000	17,400	17.40	17,600	17.60	+ 1.1
E - 177	AES	23 x 192	3/4 x 1 x 45	14 (M)	W	4	250/100	All C. S.	1	770	17,600	22.86	16,300	21.17	− 7.4
E - 178	AES	23 x 192	1 x 1 1/4 x 45	12 (M)	W	2	180/500	All C. S.	1	830	20,200	24.34	21,000	25.30	+ 4.0
E - 179 A, B	AES	23 x 192	3/4 x 1 x 45	14 (M)	W	2	180/100	All C. S.	2	910	34,400	18.90	35,900	19.73	+ 4.4
E - 180	AES	23 x 192	3/4 x 1 x 45	14 (M)	S	6	250/130	All C. S.	1	920	23,800	25.87	22,600	24.57	− 5.0
E - 181A, B, C	AES	25 x 192	3/4 x 1 x 45	14 (M)	W	4	180/100	All C. S.	3	780	36,800	15.73	42,200	18.03	+14.7
E - 182	AES	30 x 144	3/4 x 1 x 45	14 (M)	W	6	250/200	All C. S.	1	1,200	23,700	19.75	24,200	20.17	+ 2.1
E - 183	AES	34 x 192	3/4 x 1 x 45	14 (M)	S	6	100/100	All C. S.	1	2,130	37,600	17.65	35,600	16.71	− 5.3
E - 184	AJS	34 x 240	3/4 x 1 x 45	14 (M)	W	4	200/180	All C. S.	1	2,640	33,100	12.54	31,100	11.78	− 6.0
E - 185	AES	37 x 240	3/4 x 1 x 45	14 (M)	S	4	200/130	All C. S.	1	3,350	48,100	14.36	45,200	13.49	− 6.0
E - 186	BES	17 x 240	3/4 x 1 x 45	20 (M)	W	2	130/100	Ti (Gr2) /C.S. → Ti (Gr2)	1	605	33,900	56.03	37,800	62.48	+11.5
E - 187	BES	23 x 240	3/4 x 1 x 45	16 (M)	S	8	170/170	Monel / C.S. →/ Monel	1	1,145	65,900	57.55	66,100	57.73	0.0
E - 188	AET	30 x 240	3/4 x 1 x 45	14 (M)	W	2	150/100	All C. S.	1	1,760	25,900	14.72	22,200	12.61	−14.3
E - 189	AJT	31 x 240	3/4 x 15/16 x 30	14 (M)	S	2	300/150	All C. S.	2	2,290	31,600	13.80	31,800	13.89	− 1.0
E - 190	AET	33 x 192	3/4 x 1 x 45	12 (M)	W	2	180/500	All C. S.	1	1,935	34,900	18.04	34,000	17.57	− 2.6
E - 191	AET	33 x 192	3/4 x 1 x 45	14 (M)	W	4	230/100	All C. S.	1	2,095	26,100	12.46	30,000	14.65	+14.9
E - 192	AET	37 x 192	1 x 1 1/4 x 45	12 (M)	W	2	180/500	All C. S.	1	1,935	34,900	18.04	34,000	17.57	− 2.6
E - 193	AJT	41 x 192	3/4 x 1 x 90	14 (M)	W	6	130/140	All C. S.	1	2,955	34,700	11.74	38,000	12.86	+ 9.5
E - 194 A, B	AET	44 x 192	1 x 1 1/4 x 45	14 (M)	W	2	180/410	All C. S.	2	2,830	83,000	14.66	85,000	15.02	+ 2.4
E - 195	AET	44 x 192	1 x 1 1/4 x 45	12 (M)	W	2	290/410	All C. S.	1	2,830	44,000	15.55	47,600	16.82	+ 8.2
E - 196	AJT	47 x 192	3/4 x 1 x 90	14 (M)	W	8	100/140	All C. S.	1	4,000	42,500	10.63	48,100	12.03	+13.2
E - 197A, B	AET	47 x 240	3/4 x 1 x 45	14 (M)	W	4	290/100	All C. S.	2	5,425	94,100	8.67	108,000	9.95	+14.8
E - 198 A, B	AET	49 x 192	3/4 x 1 x 45	14 (M)	W	6	130/100	All C. S.	2	4,475	84,200	9.41	96,500	10.78	+14.6
E - 199	AET	52 x 240	1 x 1 1/4 x 45	12 (M)	W	4	180/410	All C. S.	1	4,985	57,200	11.47	61,000	12.24	+ 6.6
E - 200A-H	AJT	54 x 240	1 x 1 1/4 x 45	12 (M)	W	4	100/410	All C. S.	8	4,950	522,000	13.18	495,000	12.50	− 5.2
E - 201A, B	AET	54 x 240	1 x 1 1/4 x 45	12 (M)	W	4	290/500	All C. S.	2	5,195	124,300	11.96	141,000	13.60	+13.4
E - 202	AET	57 x 192	3/4 x 1 x 45	14 (M)	W	4	100/420	All C. S.	1	5,675	63,000	11.10	64,800	11.42	+ 2.9
E - 203	AET	57 x 240	3/4 x 1 x 45	14 (A)	W	4	250/100	All C. S.	1	7,350	67,700	9.21	62,200	8.46	− 8.1
E - 204	AET	59 x 240	3/4 x 1 x 45	14 (M)	W	4	150/420	All C. S.	1	7,905	152,600	9.65	150,700	9.53	− 1.2
E - 205	AET	15 x 240	3/4 x 1 x 90	16 (A)	W	4	100/140	304L S.S. / C.S. →/ 304L S.S.	1	970	36,200	37.32	37,800	38.97	+ 4.4
E - 206	AET	32 x 240	1 x 1 1/4 x 45	14 (A)	S	2	290/500	All 5 Cr-½ Mo	1	1,730	62,000	37.05	52,700	30.46	−15
E - 207	AET	46 x 240	1 x 1 1/4 x 45	16 (M)	S	4	100/410	Sandvik 2 RK-65 /C.S. →	1	3,630	166,000	45.73	167,100	46.03	+ 1.0
E - 208	AET	53 x 240	1 x 1 1/4 x 90	14 (A)	S	4	140/500	5 Cr-½ Mo/C.S. →	1	5,005	99,500	19.88	88,500	17.68	−11.1

second-lowest bids is recommended, because lowest bids are sometimes not realistic.

Estimating delivered cost

This method estimates exchanger cost f.o.b. the manufacturer's shop. The cost of the exchanger delivered to the plantsite depends on the distance between the shop and the site, the means of transportation, and the size and weight of the exchanger. Without knowledge of these details, the delivered cost cannot be accurately estimated. However, increasing the f.o.b. cost by 1–4% provides an approximate delivered cost.

A worksheet example

A simple example that illustrates the method is presented in the worksheet shown in Fig. 10. The problem is to estimate the f.o.b. cost of a pull-through floating-head, TEMA-type **AJT**, exchanger purchased in December 1982.

The single shell's diameter is 46 in. Nominal tube length is 16 ft, and surface area is 2,905 ft^2. The tubes are 1-in.-O.D., welded construction with a minimum wall gage of 12 BWG, arranged in a 1¼-in. rotated-square pitch and in four passes. Shell- and tubeside design pressures are 50 and 450 psig, respectively. The shell, channels and tubesheets are constructed of 304 stainless steel, and the tubes of Sandvik 2RK-65.

These data are entered into Part I of the worksheet.

In Part II, the base price, b, is first calculated via Eq. (1) and Tables II, III and IV. Next, in order, are calculated: shell-type correction, C_S (Table V); expansion-joint correction, C_x (Fig. 4); tube-length correction, C_L via Eq. (3); tube-pass correction, $C_{N_{Tp}}$, via Eq. 4; shellside-pressure correction, C_{PS}, via Eq. (5) and Fig. 7; tubeside-pressure correction, C_{PT}, via Eq. (6); tube-material correction, C_{mt}, via Eq. (8), with $y = 0.156$, from Eq. (7); shell-material correction, C_{ms}, via Eq. (9); channel-material correction, C_{mc}, via Eq. (10); tubesheet-material correction, C_{mts}, via Eq. (11); and tube-gage correction from Eq. (12). All the foregoing corrections are then added to arrive at C_T.

Next, the estimate of the f.o.b. price, based on January 1982 prices, E_b, is calculated via Eq. (13). Because the December value of the Fabricated Equipment component of the CE Plant Cost Index was 324.5—essentially its January 1982 value—the escalation index, E_I, is taken to be 1. Therefore, the December 1982 current-price estimate, E_c, is $131,000.

Accuracy of estimates

An attempt was made to include in this estimating method as many cost variables as possible, to make it as accurate as possible, but yet keep it simple and quick. However, regardless of the analytical approach taken, and the level of details incorporated into a method, exchanger price estimates based on it may vary widely, because of the many factors that affect exchanger costs. Indeed, an evaluation of a large number of competitive bids received from five manufacturers revealed differences between lowest and highest quoted prices of from 15% to as much as 60% (and, in some cases, even higher). Thus, it is important to always be aware that cost estimates are only that—estimates.

Estimates derived by means of the method presented are compared against actual prices of 108 exchangers purchased during the first half of 1982. Included were exchangers of different TEMA types, geometry and metallurgy, and for various design conditions. The results, summarized in Table VII, indicate that the estimates of all the individual exchangers fall within ±15%, with 60% within ±10%, of actual purchase prices. The average deviation is ±9.6%.

Accuracy is defined as: %-accuracy = [(estimated price − purchase price)/purchase price]100.

When the accuracies of the individual cost estimates for all of the exchangers fall within ±15%, the overall cost estimate for all of the exchangers purchased for a project will be more accurate, because positive and negative deviations will cancel each other out when added together. Thus, the total purchase price of all 108 exchangers in Table VII is $10.615 million, and the estimated price of the total is $10.469 million—an overall accuracy of −1.4%.

References

1. "Standards of Tubular Exchanger Manufacturers Assn.," Tubular Exchanger Manufacturers Assn., Tarrytown, N.Y.
2. Purohit, G. P., Thermal and hydraulic design of hairpin and finned-bundle exchangers, *Chem. Eng.*, May 16, 1983, p. 62.

Acknowledgement

Deep appreciation is extended by the author to R. B. Ritter of Fluor Engineers and Constructors, Inc. (So. Calif. Div.) for his critical evaluation of this article.

The author

G. P. Purohit, a consultant in the design and analysis of energy systems (3801 Parkview Lane, No. 6C, Irvine, CA 92715; telephone 714–857-4762), has until recently been a heat-transfer engineer with Fluor Corp. Previously, he had been a systems development engineer with the Jet Propulsion Laboratory of the California Institute of Technology. He holds an M.S. degree in engineering from the University of California at Los Angeles (UCLA) and a B.E. degree in mechanical engineering from Birla Vishvakarma Mahavidyalaya of Sardar Patel University, Vallabh Vidyanagar, India.

Costs of double-pipe and multitube heat exchangers — Part 1

A method for estimating costs that takes into account numerous parameters is presented in this, the first of two articles. In Part 2, the method will be illustrated, and estimated results compared with actual prices.

G. P. Purohit, Fluor Corp.

A cost-estimating method and cost data are provided for double-pipe and multitube exchangers having bare tubes and external-longitudinally-finned tubes. These units (which are depicted in Thermal and Hydraulic Design of Hairpin and Finned-Bundle Exchangers, page 129) are usually available off-the-shelf in a range of dimensions (Table I).

The effects of shell diameter, tube length, number of tubes, fin geometry and density, and metallurgy are taken into account in this method. Cost factors are provided for major components: tubes, shell, and shell-to-tube closure. Relative costs of various metals and alloys are extensively tabulated. A flowchart of the method is shown in Fig. 1.

The results predicted by this method will be compared with actual purchase prices in the Apr. 1, 1985 Cost File. These will be shown to be accurate within ±10% up to shell diameters of 8 in. for carbon-steel exchangers. (Because of limited data availability, the accuracy of the method could not be tested for shell diameters greater than 8 in.)

The base-cost equation

The costs of double-pipe and multitube exchangers are comprised of three major components--tubes, shell and closure. Shell and tube costs increase with exchanger length. Closure costs, which contribute significantly to total exchanger costs, remain constant over the entire range of exchanger lengths. This makes longer units more economical on a $/ft² basis, and provides a convenient primary parameter on which to base cost estimates.

From the analysis of purchase prices of many double-pipe

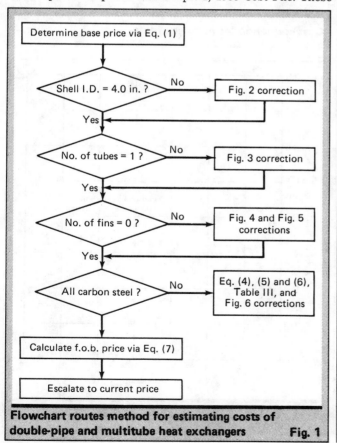

Flowchart routes method for estimating costs of double-pipe and multitube heat exchangers   **Fig. 1**

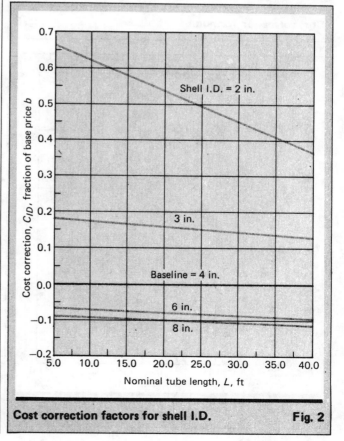

Cost correction factors for shell I.D.   **Fig. 2**

Originally published March 4, 1985

and multitube exchangers submitted via competitive bids during the period 1980-1982, Eq. (1) was developed for estimating exchanger base cost in $/ft²:

$$b = (485/L^{0.705})(A_o/A_T) \qquad (1)$$

Here, L is the nominal length of the exchanger in feet; A_o is the surface area based on bare tubes; and A_T is the total external surface (bare surface + finned surface).

The bare-tube surface area, A_o, is based on tube O.D., and is calculated in ft² via Eq. (2):

$$A_o = (2N_TL/12)(\pi d_o - N_FX) \qquad (2)$$

The total external surface area (tube plus fins), A_T, is calculated in ft² by means of Eq. (3):

$$A_T = (2N_TL/12)(\pi d_o + 2N_FY) \qquad (3)$$

For a bare-tube exchanger, $A_o = A_T$.

Consider a 20-ft-long finned multitube exchanger having 19 ¾-in.-O.D. tubes, with 20 fins/tube that are 0.21 in. high and 0.035 in. thick. With $A_o/A_T = 6.5$, the base price via Eq. (1) comes to $9.03/ft².

However, to arrive at the actual cost-estimate of such an exchanger, the base cost must be corrected by means of the following other parameters.

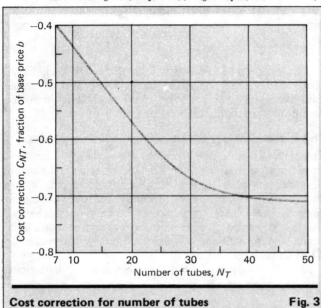

Cost correction for number of tubes   Fig. 3

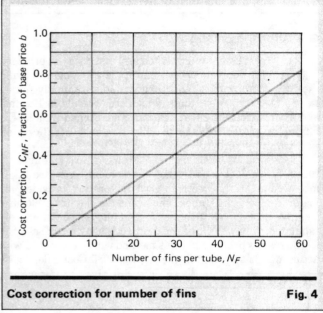

Cost correction for number of fins   Fig. 4

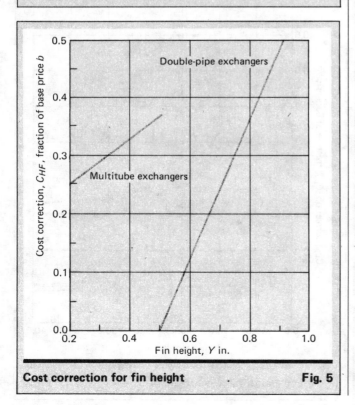

Cost correction for fin height   Fig. 5

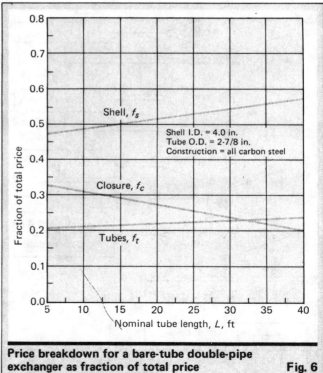

Price breakdown for a bare-tube double-pipe exchanger as fraction of total price   Fig. 6

Correction for shell diameter

Cost corrections for shell I.D. as fractions of base price are plotted in Fig. 2 in terms of exchanger length. The base diameter is the 4-in.-I.D. shell (i.e., $C_{ID} = 0$). Cost estimates for other-size shells must be adjusted by the factor C_{ID}.

For example, for a 6-in.-I.D. shell and a 20-ft-long exchanger, $C_{ID} = -0.07$. Therefore, $9.03/ft^2 + (-0.07 \times $9.03/ft^2) = $8.4/ft^2$.

Correction for number of tubes

As the number of tubes increases, the exchanger unit cost decreases. For this cost-estimating method, the double-pipe exchanger ($N_T = 1.0$) was selected as the base unit. Therefore, for $N_T = 1.0$, $C_{NT} = 0$. For multitube units, $C_{NT} < 0$. Values for C_{NT} are plotted in Fig. 3.

For a multitube exchanger having 19 tubes, $C_{NT} = -0.55$. Therefore, its estimated cost is $9.03/ft^2 + (-0.55 \times $9.03/ft^2) = $4.06/ft^2$.

Correction for number of fins

Exchanger unit cost rises with increasing fin density. The cost correction factors, C_{NF}, as fractions of the base price, b, are plotted in Fig. 4. For bare tubes, $N_F = 0$, and $C_{NF} = 0$.

For example, $C_{NF} = 0.265$ for an exchanger having 20 fins/tube. Therefore, the estimated cost = $9.03/ft^2 + (0.265 \times $9.03/ft^2) = $11.42/ft^2$.

Correction for fin height

Cost corrections due to fin height are plotted in Fig. 5 for double-pipe and finned multitube exchangers. For bare tubes, $C_{HF} = 0$.

For example, $C_{HF} = 0.265$ for the latter type of exchanger having fins 0.21-in. high. Therefore, the estimated cost = $9.03/ft^2 + (0.265 \times $9.03/ft^2) = $11.42/ft^2$.

Corrections for materials of construction

Eq. (1) is based on carbon-steel construction. Alloy construction affects the price of an exchanger significantly, depending on the cost of the alloy and its quantity in the exchanger. To incorporate these effects on cost, the cost contributions of the major exchanger components (tubes, shell and closure) to the base carbon-steel exchanger must first be known.

Analysis produced the representative cost breakdown shown in Fig. 6 for a 4-in.-dia. bare-tube double-pipe exchanger having a 2⅜-in.-O.D. tube. The Fig. 6 cost fractions will vary, depending on whether the exchanger is a double-pipe or multitube unit, and on whether the tubes are bare or finned. Tube cost fractions for multitube units are higher than those for double-pipe units, and are also higher for exchangers having finned tubes than for those having bare tubes (Table II).

Having the cost relationships of carbon-steel tubes, shell and closure to the base carbon-steel exchanger, one can calculate the higher cost due to other-than-carbon-steel construction of these components by multiplying by the relative-cost differential for the other material.

Calculate the tube-material cost correction by:

$$C_{mt} = f_t (M_1 - 1) \qquad (4)$$

Calculate the shell-material cost correction by:

$$C_{ms} = f_s (M_2 - 1) \qquad (5)$$

Dimensions of double-pipe and multitube heat exchangers   Table I

Parameter	Dimension range
Inside dia. of shell (outer pipe)	2-16 in.
Nominal length of section	5-40 ft
Outside dia. of tube (inner pipe)	0.75-4 in.
Number of tubes	4-208
Number of fins/tube	0-72
Fin height	0.21-1 in.
Fin thickness	0.02-0.035 in.
Surface area/section	3-1,600 ft²

Cost breakdown of double-pipe and multitube heat exchangers   Table II

Component	Percent of total exchanger cost			
	Double-pipe		Multitube	
	Bare	Finned	Bare	Finned
Tubes	15-25	20-40	30-40	35-60
Shell	40-55	40-50	35-50	25-40
Closure	20-35	20-35	15-30	15-30

Price ratios, related to carbon steel, of various materials for shell-and-tube heat exchangers   Table III

Material	M_1, tubing-price ratio relative to welded carbon-steel tubes		M_2, price ratio for shell and closure relative to carbon steel	Material	M_1, tubing-price ratio relative to welded carbon-steel tubes		M_2, price ratio for shell and closure relative to carbon steel
	Welded	Seamless			Welded	Seamless	
Carbon steel	1.0 (base)	2.50	1.0 (base)	Ferralium (Alloy 255)	12.00	23.90	14.00
Carbon steel, low alloys:				Carpenter 20 Cb-3	15.10	—	16.00
1/2 Mo	1.04	2.60	1.04	Carpenter 20 Mo-6	18.90	—	—
1 Mo	1.05	2.70	1.05	AL-6-X	12.20	—	—
2 1/2 Ni	1.15	2.90	1.15	AL-29-4	12.00	—	—
3 1/2 Ni	1.20	3.10	1.20	AL-29-4-2	11.80	—	—
2 Ni-1 Cu	—	3.30	1.30	AL-29-4-C	5.0	—	—
Carbon steel, chromium-molybdenum alloys:				Nickel 200	—	20.90	18.40
1 Cr-1/2 Mo	Not standard	2.60	2.00	Monel 400 (Alloy 400)	—	15.50	14.50
1 1/4 Cr-1/2 Mo	Not standard	2.70	2.10	Inconel 600 (Alloy 600)	19.40	—	15.30
2 1/4 Cr-1 Mo	Not standard	3.00	2.40	Inconel 625 (Alloy 625)	—	32.70	27.40
3 Cr-1 Mo	Not standard	3.20	2.50	Incoloy 800 (Alloy 800)	11.00	21.80	9.00
5 Cr-1/2 Mo	Not standard	4.40	3.50	Incoloy 800H (Alloy 800H)	—	18.00	—
7 Cr-1/2 Mo	Not standard	5.50	Not standard	Incoloy 825	—	23.50	—
9 Cr-1 Mo	Not standard	6.10	Not standard	Hastelloy B-2	34.90	48.60	38.40
Stainless steels:				Hastelloy C-4	28.70	40.00	31.30
304	2.80	6.50	3.70	Hastelloy C-276	29.10	38.10	31.00
304L	3.00	7.50	4.70	Hastelloy G	15.30	24.70	18.10
309	5.80	14.50	7.70	Hastelloy X	16.70	27.10	21.30
310	7.40	12.00	9.80	Titanium (Grade 2)	11.00	22.00	11.00
310L	7.60	12.40	10.10	Titanium (Grade 7)	21.00	42.00	—
316	4.70	10.10	6.20	Titanium (Grade 12)	14.00	28.00	—
316L	4.80	11.00	6.40	Zirconium 702	35.00	43.70	36.80
317	8.10	13.30	8.10	Zirconium 705	39.00	48.70	40.00
317L	8.30	13.60	8.30	Aluminum	Not standard	1.60	1.60
321	4.20	9.50	5.60	Naval rolled brass	Not standard	3.50	3.50
329 (Carpenter 7 Mo)	10.50	17.20	10.50	Admiralty	Not standard	3.60	3.60
330	7.90	12.90	9.50	Aluminum brass	Not standard	3.70	3.70
347	5.50	13.70	7.30	Aluminum bronze (5%)	Not standard	4.10	4.10
405	6.00	15.00	6.90	Copper (arsenical or deoxidized)	Not standard	4.20	4.20
410	6.90	17.20	7.90	90-10 cupro-nickel	3.50	4.60	4.60
430	5.40	10.60	6.20	70-30 cupro-nickel	4.20	5.50	5.50
439	5.00	11.20	5.80	Union Carbide high-flux tubing:			
444 (Alloy 18-2)	7.80	8.80	9.00	SA-214, welded	4.40	—	—
446	4.70	10.00	5.40	SA-334-1, welded	4.70	—	—
904L (Sandvik 2RK-65)	15.30	19.20	17.00	SA-334-3, seamless	—	9.00	—
Sandvik 2RE-69	—	14.50	—	SA-214, fluted, welded	7.00	—	—
Sandvik 3RE-60	—	10.10	—	SA-334-1, fluted, welded	7.40	—	—
Sandvik 253 MA	—	12.70	—	SA-210, fluted, seamless	—	8.40	—
Sandvik SAF 2205	—	11.80	—				
Sanicro 28	16.10	20.20	18.20				
E-brite-26-1 (XM-27)	9.00	—	10.00				

Calculate the closure-material cost correction by:

$$C_{mc} = f_c (M_2 - 1) \qquad (6)$$

Approximate values for f_t, f_s and f_c are provided by Fig. 6. The cost ratios M_1 (tubing cost, welded or seamless, relative to welded carbon-steel tubing of same diameter, wall thickness and length) and M_2 (plate-material cost relative to carbon-steel plate of same dimensions) listed in Table III were calculated from early-1982 prices. In some cases, prices differences among manufacturers were as high as ±40%. As Table III shows, the cost of seamless carbon-steel tubing is 2½ times as high as that of welded carbon-steel tubing. Thus, prices of carbon-steel and low-alloy tubing vary significantly, depending on whether construction is welded or seamless. As material prices increase, however, this cost difference declines to as low as 10%.

References

1. Purohit, G. P., Thermal and Hydraulic Design of Hairpin and Finned-Bundle Exchangers, *Chem. Eng.*, May 16, 1983, p. 62.
2. Purohit, G. P., Estimating Costs of Shell-and-Tube Heat Exchangers, *Chem. Eng.*, Aug. 22, 1983, p. 56.

The author

G. P. Purohit, a consultant in the design and analysis of energy systems (3801 Parkview Lane, No. 6C, Irvine, CA 92715; telephone 714/786-4762), has until recently been a heat-transfer engineer with Fluor Corp. This article is based on his work at Fluor. Previously, he had been a systems development engineer with the Jet Propulsion Laboratory of the California Institute of Technology. He holds an M.S. degree in engineering from the University of California at Los Angeles (UCLA) and a B.E. degree in mechanical engineering from Birla Vishvakarma Mahavidyalaya of Sardar Patel University, Vallabh Vidyanagar, India.

Costs of double-pipe and multitube heat exchangers — PART 2

A method for estimating costs that takes into account numerous parameters was presented in Part 1. In this part, the method is illustrated, and estimated results are compared with actual prices.

G. P. Purohit, Fluor Corp.

After the exchanger base price and the corrections for all the alternatives have been calculated, the 1982 f.o.b. price of the exchanger can be estimated by means of Eq. (7), in which C_T represents the sum of all the base-price cost corrections for the alternatives, as determined via Eqs. (2) through (4), Figs. 2 through 6, and Table III in Part 1 (p. 331):

$$E_b = [b(1 + C_T)A]N \qquad (7)$$

Because costs usually change from the date an order is placed to when the equipment is delivered, an escalation factor is generally incorporated into estimates. Manufacturers usually include price-adjustment formulas with bids.

Escalating via an index

An exchanger price estimated by this method can be escalated for a date beyond 1982 by means of an escalation index. One such index that applies to process heat exchangers is the "Fabricated equipment" component of the Chemical Engineering Plant Cost Index. Others that are applicable include the Marshall & Swift Equipment Cost Index, the Nelson Refinery Cost Index (or its "Materials cost" component), and the U.S. Bureau of Labor Statistics' General Purpose Machinery and Equipment Cost Index.

An escalation index that is specific to heat exchangers can be formulated when price quotations more current than those of this method's base date are readily available. For comparable exchangers, the average of the lowest and second-lowest current quotations, divided by the sum of the cost estimates (i.e., the E_b total), provides a highly reliable escalation index that should remain valid for at least six months. Taking the average of the low-

est and second-lowest quotations is recommended because lowest bids are sometimes unrealistic.

Costs estimated via this method are f.o.b. manufacturer's shop. The cost delivered to the plantsite depends on the distance involved, the means of transportation, and the size

Exchanger Data:

Double-pipe/Multitube _Multitube_

Bare/Finned _Finned_

Surface area/section, A, ft² _681_

No. of sections, N _2_

Ratio bare tubes/total surface area, A_o/A_T _1/6.5_

Materials: Tubes _5 Cr - ½ Mo._

Shell _C. S._

Closure _C. S._

Shell I.D., D_i, in. _6.0_

Tube O.D., d_o, in. _3/4_

Nominal tube length, L, ft _20_

No. of tubes, N_T _19_

No. of fins/tube, N_F _20_

Fin height, Y, in. _0.21_

Fin thickness, X, in. _0.035_

Calculations:

1. Base price, $b = (485)(L^{-0.705})(A_o/A_T) = 485\,(20)^{-0.705}\,(1/6.5) = \underline{9.03}$ $/ft²

2. Corrections:

For shell I.D. (Fig. 2) $C_{ID} = \underline{-0.07}$

For number of tubes (Fig. 3) $C_{NT} = \underline{-0.55}$

For number of fins (Fig. 4) $C_{NF} = \underline{0.265}$

For fin height (Fig. 5) $C_{HF} = \underline{0.265}$

For materials (Table III):

Tubes, Eq. (4): $f_t = \underline{0.241}$ (Fig. 6), $M_1 = \underline{4.4}$ (Table III) $C_{mt} = \underline{0.819}$

Shell, Eq. (5): $f_s = \underline{0.503}$ (Fig. 6), $M_2 = \underline{1.0}$ (Table III) $C_{ms} = \underline{0.0}$

Closure, Eq. (6): $f_c = \underline{0.256}$ (Fig. 6), $M_2 = \underline{1.0}$ (Table III) $C_{mc} = \underline{0.0}$

Total correction, $C_T = \underline{0.729}$

3. Estimated price, E_b (f.o.b., based on January 1982 prices)

$E_b = b(1 + C_T)\,A\,N = \underline{9.03}\,(1 + \underline{0.729})(\underline{681})(\underline{2}) = \$\,\underline{21,260}$

4. Current exchanger price, E_c (f.o.b.):

$E_c = E_b \times E_I = \underline{21,260}\ (\underline{335.9/324.5}) = \$\ \underline{22,000}$

Figure 7 — Worksheet for estimating costs of double-pipe and multitube heat exchangers

Table IV — Estimated exchanger prices compare well with quoted and purchased prices for late 1980 to early 1982

Exchanger	Shell I.D. × Nominal tube length, in.	Tube O.D., in.	Number of tubes	Number of fins/tube	Fin height, in.	Fin thickness, in.	Total surface to bare tube surface ratio	Number of sections	Surface area per section, ft²	Quoted or purchased	Estimated	Accuracy, %
Bare double-pipe												
E-101 A/B/C	2 × 60	1	1	0	0	0	1	3	3.1	2,500	2,400	−4.0
E-102	2 × 180	1	1	0	0	0	1	1	8.5	900	960	6.7
E-103	3 × 120	1 29/32	1	0	0	0	1	1	11.0	1,200	1,250	3.3
E-104	3 × 360	1 29/32	1	0	0	0	1	1	31.0	1,550	1,550	0.0
E-105 A/D	4 × 60	2 7/8	1	0	0	0	1	4	9.3	5,800	5,800	0.0
E-106	4 × 360	2 7/8	1	0	0	0	1	1	47	2,100	2,100	0.0
Finned double-pipe												
E-107 A/B	4 × 60	2 7/8	1	40	0.5	0.035	6.5	2	43	3,300	3,200	−3.0
E-108	4 × 120	2 7/8	1	40	0.5	0.035	6.5	1	84	1,800	1,900	5.5
E-109 A/B	4 × 120	2 7/8	1	20	0.5	0.035	3.5	2	50	3,300	3,450	4.5
E-110	4 × 120	2 3/8	1	40	0.75	0.035	11.2	1	114	1,700	1,800	5.9
E-111	4 × 180	2 3/8	1	40	0.75	0.035	11.2	1	170	1,850	2,000	8.1
E-112	4 × 240	2 7/8	1	20	0.5	0.035	3.5	1	100	2,000	2,100	5.0
E-113	4 × 240	2 3/8	1	20	0.75	0.035	5.6	1	126	1,930	2,070	7.2
E-114	4 × 240	1 29/32	1	20	1.0	0.035	8.8	1	154	1,900	1,900	0.0
E-115	4 × 240	1 29/32	1	40	1.0	0.035	18.9	1	287	2,000	1,900	−5.0
E-116	4 × 300	1 29/32	1	20	1.0	0.035	8.8	1	192	2,100	2,050	−2.3
E-117	4 × 300	1 29/32	1	40	1.0	0.035	18.9	1	359	2,200	2,050	−6.8
E-118	4 × 360	2 3/8	1	20	0.75	0.035	5.6	1	188	2,250	2,300	2.2
Bare multitube												
E-119	4 × 60	3/4	7	0	0	0	1	1	17	1,600	1,600	0.0
E-120	4 × 120	7/8	7	0	0	0	1	1	35	2,000	2,000	0.0
E-121	4 × 420	3/4	7	0	0	0	1	1	120	2,800	2,850	1.8
E-122	4 × 480	7/8	7	0	0	0	1	1	140	3,000	3,000	0.0
E-123	6 × 120	3/4	19	0	0	0	1	1	86	3,000	3,100	3.3
E-124	6 × 180	3/4	26	0	0	0	1	1	165	3,400	3,850	13.2
E-125	6 × 240	3/4	26	0	0	0	1	1	216	3,600	4,050	12.5
E-126	6 × 240	3/4	19	0	0	0	1	1	161	3,400	3,500	2.9
Finned multitube												
E-127	4 × 60	3/4	7	20	0.21	0.035	5.8	1	56	1,800	1,700	−5.5
E-128	4 × 60	7/8	7	20	0.21	0.035	5.0	1	59	2,050	2,100	2.4
E-129	4 × 180	7/8	7	20	0.21	0.035	5.0	1	172	2,700	2,800	3.7
E-130	4 × 180	3/4	7	20	0.21	0.035	5.8	1	191	2,400	2,650	10.4
E-131	6 × 120	1	7	20	0.5	0.035	9.5	1	284	3,200	3,300	3.1
E-132	6 × 360	1	7	20	0.5	0.035	9.5	1	826	4,600	4,400	−4.3
E-133	6 × 360	3/4	19	20	0.21	0.035	5.8	1	1,030	6,800	7,000	2.9

and weight of the exchanger. If these details are not known, delivered cost can be approximated by increasing the f.o.b. cost by 1–4%.

An example worksheet

The Fig. 7 worksheet illustrates the method in the estimation of the f.o.b. cost of a finned-multitube exchanger to be purchased in September 1984. Except for Eq. (7) and Table IV, the equations, figures and tables referred to in the worksheet are to be found in Part 1.

Accuracy of estimates

To make this estimating method as accurate as posssible, many cost variables were included, but not so many as to keep it from being simple and quick in application.

Estimates derived via the proposed method, and actual purchased or lowest-quoted prices collected from 1980 to 1982, are tabulated and compared in Table IV. The tubes, shells and closures of all the exchangers listed in Table IV are constructed of carbon steel. Also, tube construction in all cases is welded.

Except in three instances, estimated prices fall within ±10% of actual or quoted prices. Accuracy % = [(estimated $ − purchased $) ÷ purchased $] × 100. For all the exchangers, the average deviation is ±4.1%; with the three most divergent estimates omitted, ±3.0%. The overall cost estimate for all the exchangers purchased for a project is likely to be more accurate, as plus and minus deviations will cancel each other when added. Thus, the total purchase, or quoted, price for all the exchangers in Table IV is $86,730, and the total estimated price is $88,680; overall accuracy = +2.2%.

Current costs of process equipment

Capital costs for key process equipment are represented in over 50 graphs.* Among the equipment included are tanks, heat exchangers, columns, pumps, compressors, centrifuges and filters. These are mostly f.o.b. costs, up to date as of January 1982.

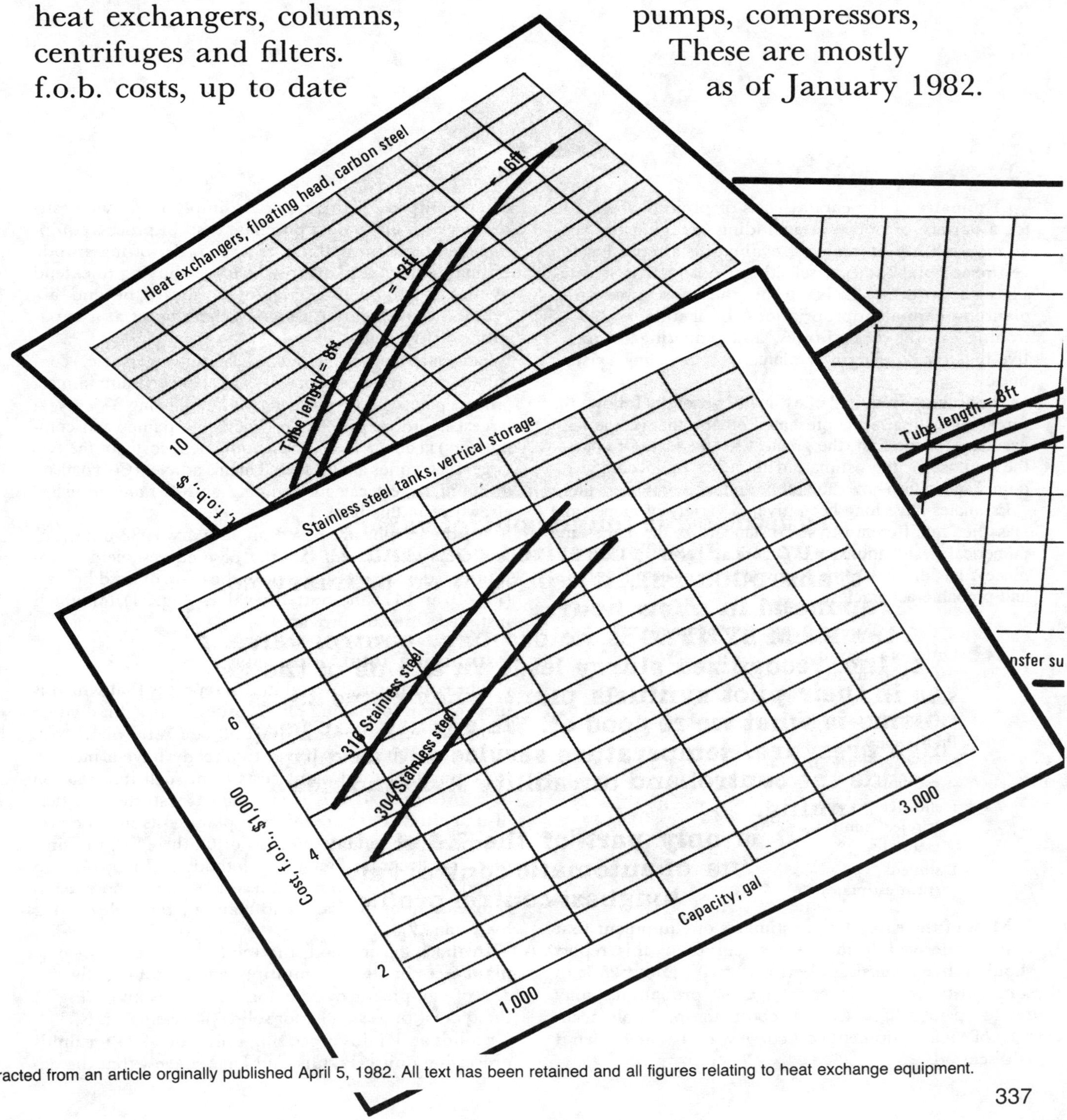

Heat exchangers, floating head, carbon steel

Tube length = 16ft

Tube length = 12ft

Tube length = 8ft

Cost, f.o.b. $

10

Stainless steel tanks, vertical storage

316 Stainless steel

304 Stainless steel

Cost, f.o.b. $1,000

6

4

2

Capacity, gal

1,000

3,000

Tube length = 8ft

nsfer su

*Extracted from an article orginally published April 5, 1982. All text has been retained and all figures relating to heat exchange equipment.

Richard S. Hall, Richard S. Hall and Associates,
Jay Matley and *Kenneth J. McNaughton,* Chemical Engineering

☐ Estimates of the capital costs of projects are made for a variety of purposes, including: gauging the economic viability of projects, evaluating alternative investment opportunities, selecting from alternative designs the process likely to be the most profitable, planning capital appropriations, budgeting and controlling capital expenditures, and tendering competitive bids for building new plants or remodeling existing ones.

The accuracy required of estimates generally follows the foregoing listing in the presented order—that is, the least accuracy for the first, the greatest for the last. Of course, the purpose of the estimate determines the accuracy required and in turn how much time and money is spent on it.

Estimates have long been given a variety of names and classified in different ways. To standardize the names and numerical designations, the American Assn. of Cost Engineers has drawn up the following list of estimate types and probable accuracies:

Type	Accuracy, ±%
Order-of-magnitude (ratio estimate)	40
Study (factored estimate)	25
Preliminary (budget authorization estimate)	12
Definitive (project control estimate)	6
Detailed (firm estimate)	3

Most of the capital cost estimates of equipment that may be derived from the cost curves in this report should range in accuracy between ±10% and ±25%. In some instances, however, the error probability may range up to ±35%. In this report, the probable accuracy of each equipment cost curve will be noted when it is discussed.

The purpose of this report is simply to provide estimators with up-to-date purchase costs of process equipment, suitable for making study or order-of-magnitude estimates of total plant costs, without having to extend available plant-construction-cost indexes beyond acceptable extrapolation ranges (generally set at a maximum of five years).

Fabricated equipment (tanks, heat exchangers, etc.) constitutes the largest category of capital expenditure in most process plants — a representative figure being 37%. Next largest is process machinery (including pumps and compressors) at about 21%. Thus, equipment costs for the two largest categories are presented in this article. (Cost data on the third largest category, piping at 20%, are furnished elsewhere in this book.)

All the graphs are based on January 1982 data, unless otherwise stated. All the cost curves yield shop-fabricated f.o.b. costs, except in the cases of fired heaters (Fig. 46 and 47) and butterfly valves (Fig. 54), for which installed costs are provided.

Estimating total plant cost

The equipment covered in this report is that which is most critical to making plant cost estimates, that which generally accounts for at least 50% of total plant cost.

The estimator is assumed to already have a method for estimating total plant costs from such information as is presented. If this is not the case, study or order-of-magnitude estimates of total plant costs may be prepared via the module method of Guthrie [3], the simpler, less accurate factoring methods of Lang [6] and Hand [4], or the more complex, and likely more accurate, variant of the Lang and Hand methods developed by Viola [9].

In the Lang method, the total investment cost of a plant is estimated by multiplying the total delivered cost of equipment by a factor that varies according to the type of process—3.1 for solids processing units, 3.63 for solids and fluids processing units, and 4.74 for fluids processing units. (Pikulik and Diaz suggest that the delivered cost of equipment may be approximated by increasing the purchased cost of equipment, f.o.b. manufacturer's shop, by 3% [8].)

In the Hand method, installation factors (multipliers) for each type of major equipment relate total battery-limit costs to equipment costs. The factors are different for each type of equipment. Some typical factors are 4 for distillation columns and pressure vessels, $3\frac{1}{2}$ for heat exchangers, $2\frac{1}{2}$ for compressors and 2 for fired heaters [2]. Estimated equipment costs are multiplied by the factors to arrive at total installed costs, and the sum of these products represents the estimated total inside-battery-limit cost of the complete installed plant.

In the Viola method, plant complexity factors are correlated against estimated capital costs. This correlation depends on a base curve that is prepared by determining the costs of individual pieces of equipment for several plants. The complexity factor is a function of: the number of major operating steps in the process, a correction factor for pressure level and materials of construction, the ratio of raw material to product, an average-throughput correction factor, and the fraction of major operating steps handling solid-fluid mixtures.

Both the Lang and Hand methods are only suitable for order-of-magnitude and, at best, study estimates. The accuracy of the Viola method should be higher, and that of the Guthrie method still higher. However, the latter two methods require more work.

Equipment cost data

Most of the cost information in this report has been supplied by Richard S. Hall and Associates, a manufacturers-representative and cost-consulting firm. Considerable data have also been contributed by equipment manufacturers, by operating companies that systematically track equipment and construction costs, and by PDQ$ Inc., a cost-estimating service firm that furnishes computer-calculated designs and detail-grade cost estimates of equipment when provided with the key process-design variables. The major contributors of cost information are acknowledged at the end of this report.

Costs are, of course, always changing, so equipment costs derived from the following graphs should not be considered unassailable, but rather be adjusted in light of cost data from other sources, according to one's judgment and experience.

Storage tanks and process vessels

Storage tanks often represent the largest single expense of process plants. It has been said of them that operating personnel never have enough and accountants never too few.

Atmospheric tanks usually contain liquids whose vapor pressure at storage conditions remains at about 15 psia.

Fig. 1 yields f.o.b. purchase costs of vertical fiberglass-reinforced-plastic tanks suitable for storing liquids (including such corrosive ones as 50% sulfuric acid) having atmospheric vapor pressures. The tanks have dished heads and flat bottoms. Included in the indicated costs are hold-down lugs, a manway, two nozzles and a vent connection.

Fig. 2 also gives costs for vertical atmospheric storage tanks, of stainless steel construction, Types 304 and 316. Shells and flanged-and-dished heads are of 12-gage material. Costs cover a manway, four nozzles and steel support lugs.

Fig. 3 and 4 present costs of light-gage vertical stainless-steel storage tanks of capacities to 10,000 gal. The first is cone-bottomed and supported by steel legs. The second is flat-bottomed and must be supported by a concrete pad or other such foundation. Both are flat-topped. Tank shells are reinforced with angle or channel rings of steel. In both cases, costs include a manway and four nozzles. Material gages are based on requirements for weight of water, with no allowance for corrosion. Gages vary in the shell section, heavier in the lower and lighter in the upper.

Fig. 5 takes the costs of light-gage vertical stainless-steel storage tanks of flat top and bottom construction to capacities up to 30,000 gal. Again, costs include a manway and four nozzles, and such tanks must be fully supported. As before, gages are based on weight of water, with no allowance for corrosion, and vary similarly with shell section. Shells are also reinforced with steel angle or channel rings.

Fig. 6 gives costs of vertical atmospheric-pressure storage tanks of stainless steel to 8,000-gal capacity. These are cone-shaped, top and bottom, therefore supported on legs, and can contain liquids weighing up to $9\frac{1}{2}$ lb/gal. Included are a manway and three half-coupling connections for inlet, outlet and vent.

Fig. 7 presents costs of vertical atmospheric-pressure stainless-steel storage tanks built to API 650 Appendix J requirements. Tops are cone-shaped and bottoms flat. Tops and upper shell sections are $\frac{3}{16}$ in. thick, and bottoms and lower shell sections $\frac{1}{4}$ in. Included are an 18-in. manway and three 3-in. flanged nozzles.

Fig. 8 provides costs of horizontal atmospheric stainless-steel storage tanks. Shell walls and flanged-and-dished heads are $\frac{3}{16}$ in. thick. Included are four nozzles and two support saddles.

Fig. 9 gives costs of horizontal stainless-steel tanks for full vacuum and 50 psi at 350°F, ASME construction and stamp. Costs include steel saddles, an 18-in. manway, four flanged nozzles and three half-couplings.

Fig. 10 yields costs of vertical stainless-steel liquid receivers for full vacuum and 25 psi at 350°F to 350-gal capacities, inspected for ASME standards and stamped. Costs include shell supports, three flanged nozzles and three half-couplings.

Fig. 11 extends the range of Fig. 10 receivers to 2,000-gal capacities and includes an 18-in. manway.

Fig. 12 takes the capacity range of vertical stainless-steel receivers to 12,000 gal. Design is for full vacuum and 50 psi at 300°F, and vessels ASME stamped. These vessels are mounted on legs that provide a 24-in. clearance from bottom nozzle to floor. Costs include a manway, four nozzles and six half-couplings.

Cost estimates derived from Fig. 1 through 12 should be accurate to about ±10% as of January 1982.

Heat exchangers

The most versatile equipment for process heat transfer is the shell-and-tube heat exchanger. Curves are presented for the Tubular Exchanger Manufacturers Assn. (TEMA) Class C, as well as the ASME code for unfired pressure vessels, Section VIII, Div. 1. TEMA classifies this equipment as "unfired shell-and-tube heat exchangers for the generally moderate requirements of commercial and general process applications."

The curves all represent f.o.b. costs for carbon-steel single-pass shells, two-pass carbon steel and stainless steel tubeside exchangers. Tubes are $3/4$ in. O.D. and design temperature and pressure are 400°F and 75 psi, respectively.

Three categories are dealt with:

The *fixed-tubesheet* design has straight tubes secured at both ends in tubesheets welded to the shell. This type is moderate in cost and cleanable on the tubeside.

In the *U-tube* design, both ends of the U-shaped tubes are fastened to a single tubesheet, thus eliminating the problem of differential thermal expansion because the tubes are free to expand and contract. This exchanger is low in cost and the removable bundle makes it easy to clean the shellside manually. However, it is difficult to clean the tubes.

With the *floating tubesheet,* straight tubes are secured at both ends in tubesheets, but one tubesheet is free to move, thereby providing for differential thermal expansion between the tube bundle and the shell. This type is the highest in cost, but both the tubes and the shell are easily cleaned.

Fig. 13 through 24 show f.o.b. cost versus surface area of the heat exchanger, and are presented in two ranges, 0 to 240 ft² and 300 to 1,500 ft², in tube lengths of 8, 12 and 16 ft.

Example: Let us estimate the f.o.b. cost of a TEMA Class C shell-and-tube exchanger with stainless steel tubes for use at 75 psi and 400°F. The shell is to be single-pass and made of carbon steel. The tubes will be two-pass, $3/4$ in. O.D. Calculations indicate that a heat-exchange surface area of 200 ft² will be required.

Space is limited and the exchanger should be as short as possible. The nature of the process dictates a fixed-tubesheet design, to minimize the possibility of leaks.

Fig. 15 shows that this exchanger would cost about $6,050 in Type 304 stainless steel, using an 8-ft tube-length. Using the 1.1 multiplier for 316 stainless steel gives $6,655.

Cost estimates derived from Fig. 13 through 24 should be accurate to ±10% as of January 1982.

Tank-vent condensers

Coolant is circulated through the tubeside of the vent condenser, and vapors that would normally be vented to atmosphere are passed over the extended fin surface. This causes the vapors to condense and drip back into the tank. If more surface is required than that provided by a single unit, additional condensers can be installed in parallel.

The housing can be obtained in aluminum, galvanized steel, and stainless steel. The fins and tubes may be of aluminum, copper or stainless steel. Vent diameters of 3–12 in. may be fitted with ASA standard flanges.

The shellside is designed for 15 psi at 350°F; the tubeside is designed for 150 psi at 350°F. Fig. 25 is for a stainless-steel finned tube enclosed in stainless steel housing. Fig. 26 is for an aluminum-finned copper coil, with the housing as shown.

Example: Prices are required for tank-vent condensers with a 1,200-ft² area and a Type 304 stainless steel housing. Fig. 25 and 26 indicate that these conditions could be met with a condenser containing a stainless-steel finned tube for $12,200, or one with an aluminum-finned copper coil costing $5,800.

Cost estimates derived from Fig. 25 and 26 should be accurate to about ±10% as of January 1982.

Dimple-jacketed reactors

In terms of control efficiency and product quality, jacketing provides the optimum method of heating or cooling process vessels. Most liquids, as well as steam and other high-temperature vapors, can be used as the heat transfer fluid. The jacket allows accurate control of both circulation temperature and velocity of the heat transfer media. In many cases, the jacket may be fabricated from a material less expensive than the vessel material.

A dimple jacket is made by pressing a pattern of depressions, or dimples, into sheets of lightgage metal. When these sheets are wrapped around the vessel, the dimples are welded to the vessel wall to give protection against pressure and vacuum. This design is considerably cheaper than other jackets.

The curves in Fig. 27 are for stainless-steel baffled reactors as shown, with one 18-in. manway, one vapor nozzle, one agitator nozzle, four flanged nozzles and one flanged drain nozzle. The reactors are designed for 75 psi and 350°F.

The stainless-steel dimple jacket is designed for 125 psi and 350°F. Curves are shown for 316 and 304 stainless steel.

Example: Estimates are required on a 6,000-gal reactor of this description, for both Type 316 and Type 304 stainless steel. Fig. 27 shows prices of $43,000 (Type 304) and $48,500 (Type 316) for these two vessels.

Costs should be accurate to ±10% as of January 1982.

Reactor heating system

This system is designed to heat a 50% ethylene glycol solution from 100 to 135°F using 40-psig steam.

It is a skid-mounted assembly ready for installation and operation. The assembly includes heat exchanger(s), tank, insulation, pumps, piping, valves and fittings, instrumentation, structural steel, painting, and engineering design—all as shown in Fig. 28.

The equipment is generally of all-steel construction. Heat exchangers are built to comply with TEMA Class C specifications, and the tank to Underwriters' Laboratories (U.L.) specifications. Pumps are of cast iron and include base plates, couplings, guards, totally-enclosed fan-cooled (TEFC) motors, and mechanical seals.

Jobsite requirements: unload unit, locate same on purchaser's foundation, install piping for glycol and

(text continues on p. 317)

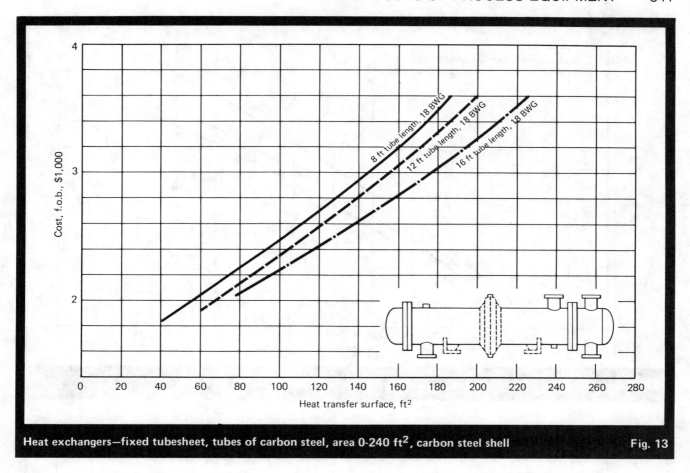

Heat exchangers—fixed tubesheet, tubes of carbon steel, area 0-240 ft², carbon steel shell   Fig. 13

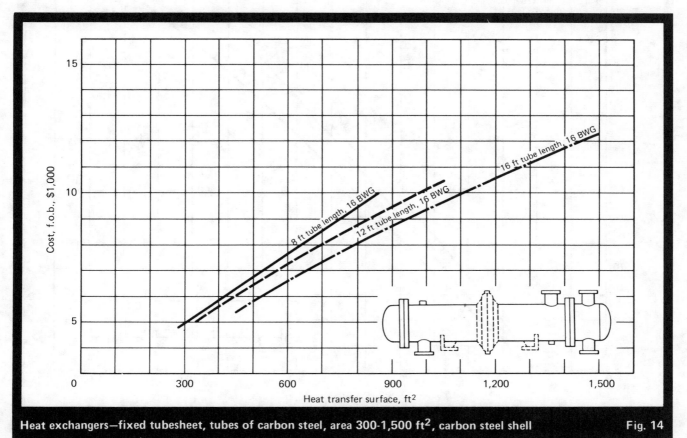

Heat exchangers—fixed tubesheet, tubes of carbon steel, area 300-1,500 ft², carbon steel shell   Fig. 14

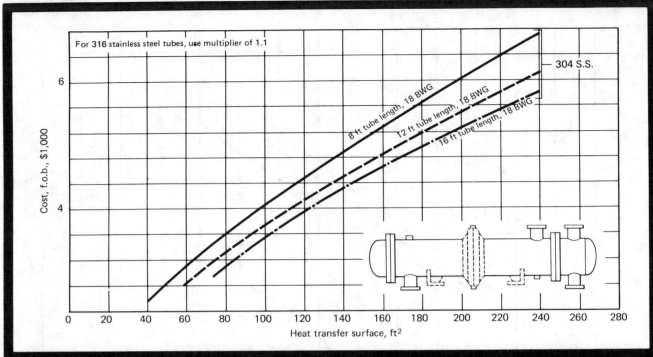

For 316 stainless steel tubes, use multiplier of 1.1

304 S.S.

8 ft tube length, 18 BWG

12 ft tube length, 18 BWG

16 ft tube length, 18 BWG

Cost, f.o.b., $1,000

Heat transfer surface, ft²

Heat exchangers—fixed tubesheet, tubes of stainless steel, area 0-240 ft², carbon steel shell   **Fig. 15**

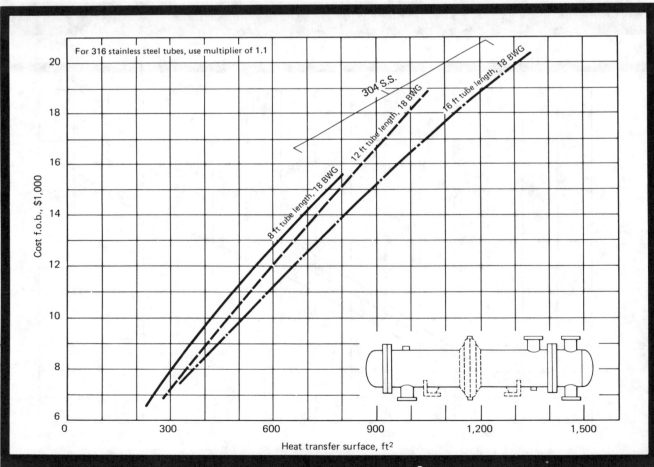

For 316 stainless steel tubes, use multiplier of 1.1

304 S.S.

8 ft tube length, 18 BWG

12 ft tube length, 18 BWG

16 ft tube length, 18 BWG

Cost f.o.b., $1,000

Heat transfer surface, ft²

Heat exchangers—fixed tubesheet, tubes of stainless steel, area 300-1,500 ft², carbon steel shell   **Fig. 16**

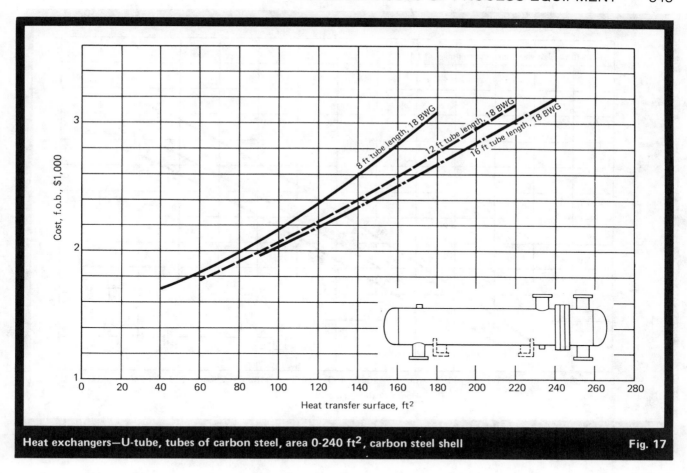

Heat exchangers—U-tube, tubes of carbon steel, area 0-240 ft², carbon steel shell   Fig. 17

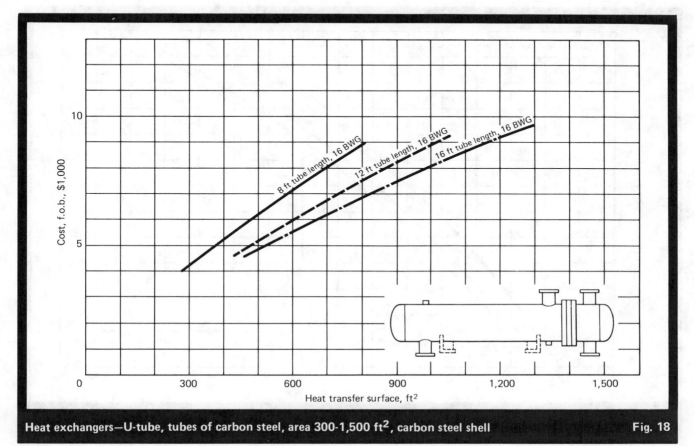

Heat exchangers—U-tube, tubes of carbon steel, area 300-1,500 ft², carbon steel shell   Fig. 18

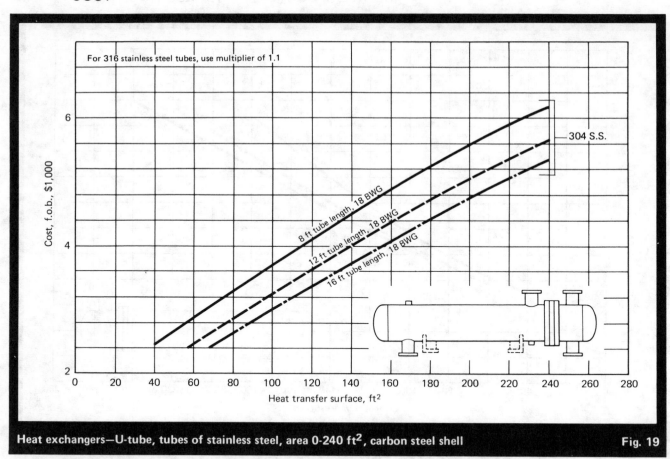

Heat exchangers—U-tube, tubes of stainless steel, area 0-240 ft², carbon steel shell

Fig. 19

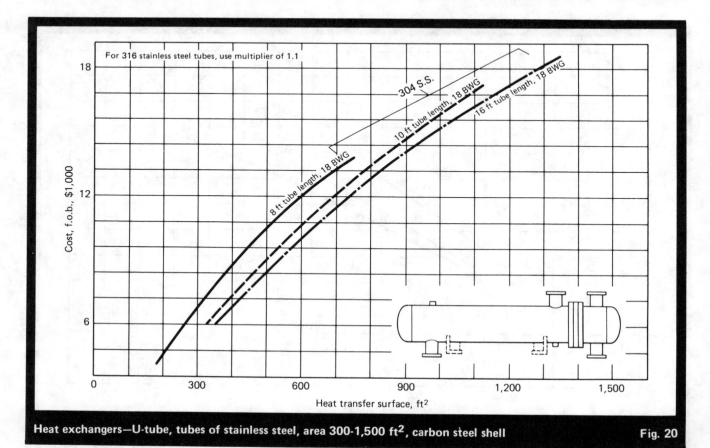

Heat exchangers—U-tube, tubes of stainless steel, area 300-1,500 ft², carbon steel shell

Fig. 20

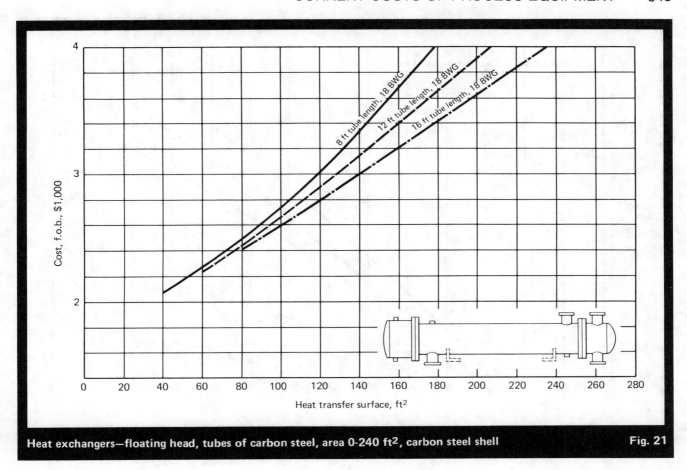

Heat exchangers—floating head, tubes of carbon steel, area 0-240 ft², carbon steel shell   Fig. 21

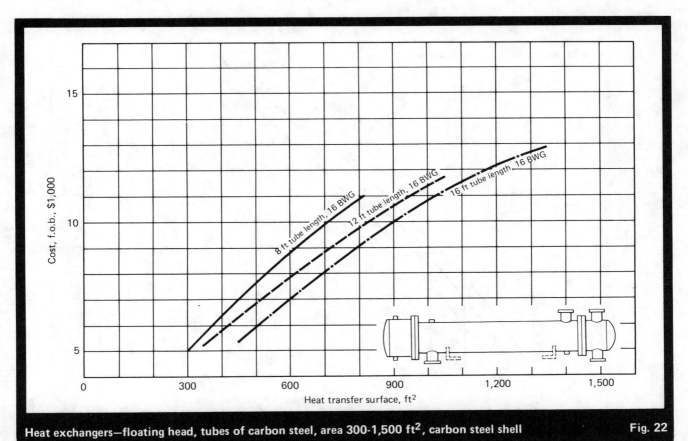

Heat exchangers—floating head, tubes of carbon steel, area 300-1,500 ft², carbon steel shell   Fig. 22

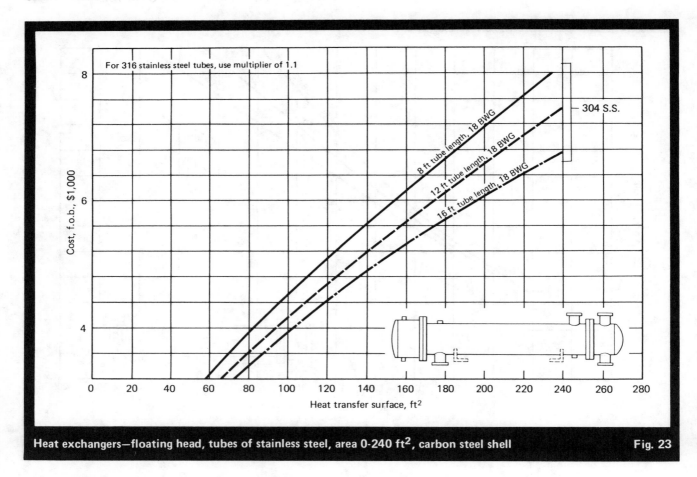

Heat exchangers—floating head, tubes of stainless steel, area 0-240 ft², carbon steel shell   **Fig. 23**

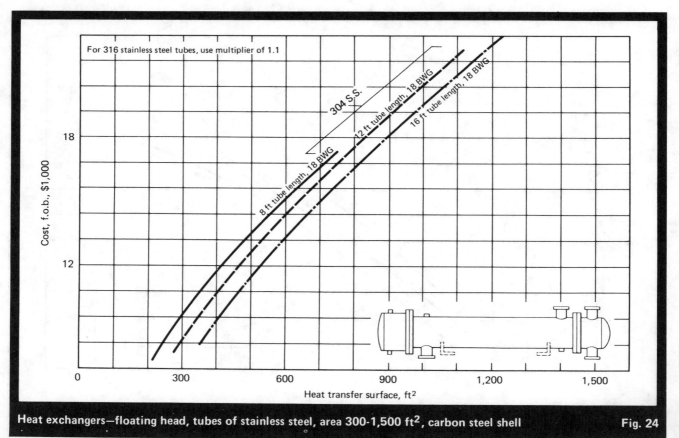

Heat exchangers—floating head, tubes of stainless steel, area 300-1,500 ft², carbon steel shell   **Fig. 24**

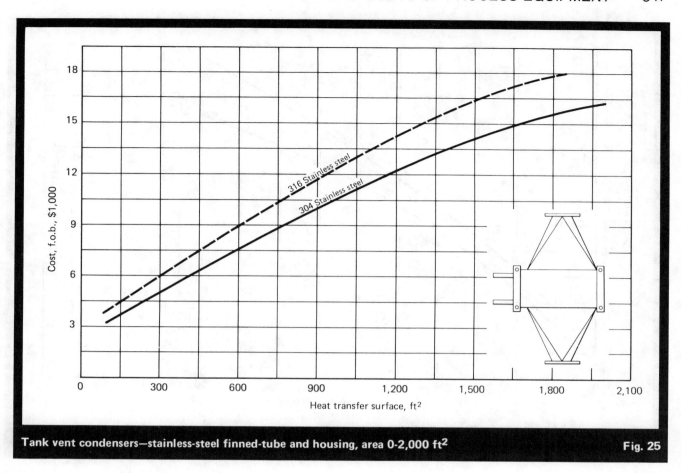

Tank vent condensers—stainless-steel finned-tube and housing, area 0-2,000 ft²   Fig. 25

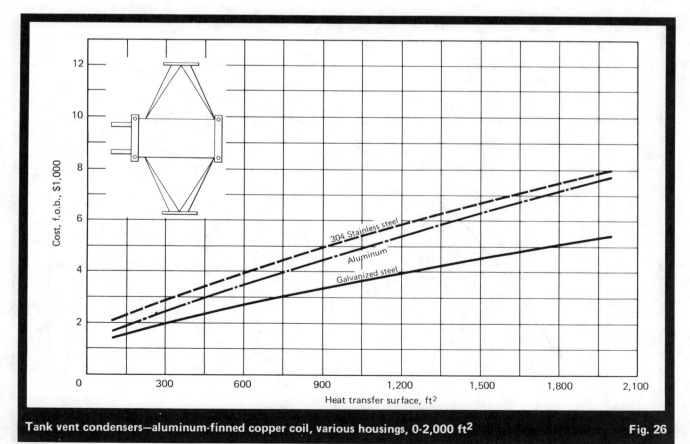

Tank vent condensers—aluminum-finned copper coil, various housings, 0-2,000 ft²   Fig. 26

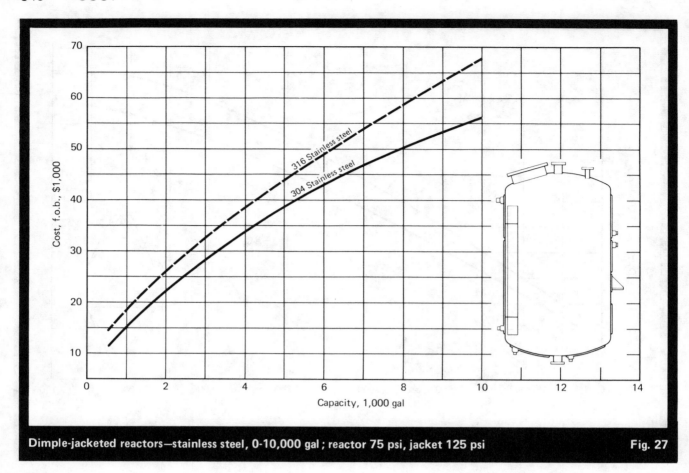

Dimple-jacketed reactors—stainless steel, 0-10,000 gal ; reactor 75 psi, jacket 125 psi   **Fig. 27**

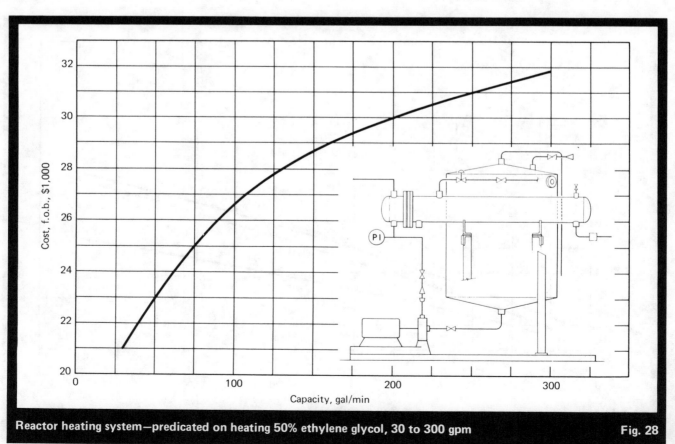

Reactor heating system—predicated on heating 50% ethylene glycol, 30 to 300 gpm   **Fig. 28**

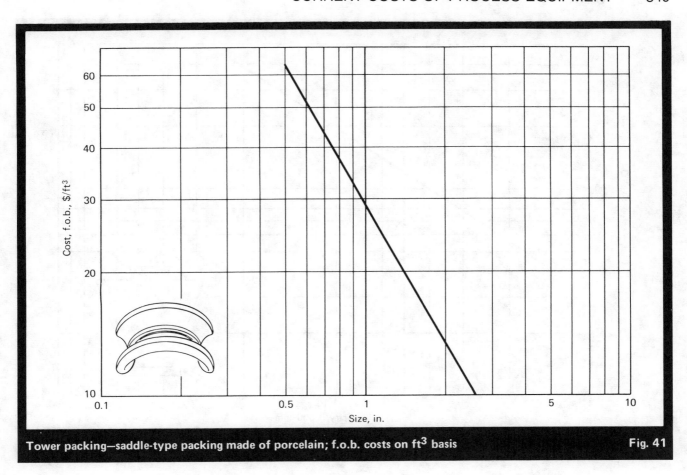

Tower packing—saddle-type packing made of porcelain; f.o.b. costs on ft³ basis   Fig. 41

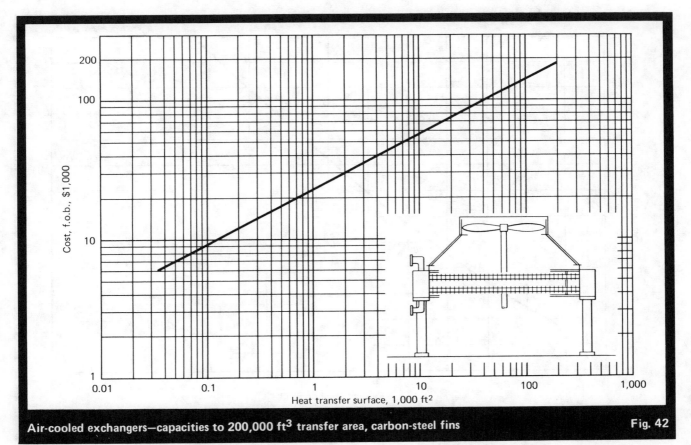

Air-cooled exchangers—capacities to 200,000 ft³ transfer area, carbon-steel fins   Fig. 42

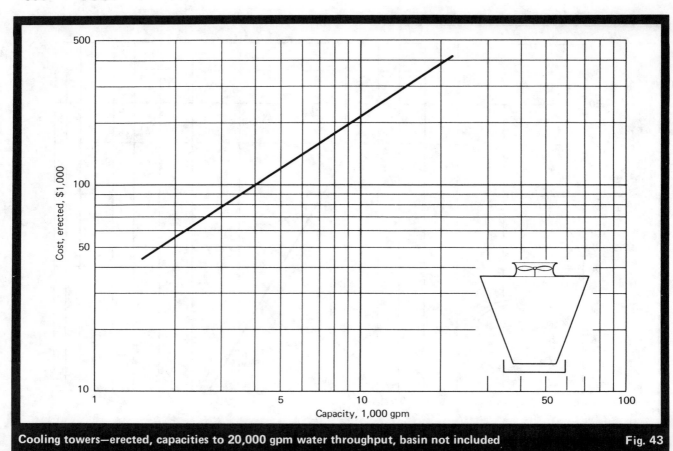

Cooling towers—erected, capacities to 20,000 gpm water throughput, basin not included   Fig. 43

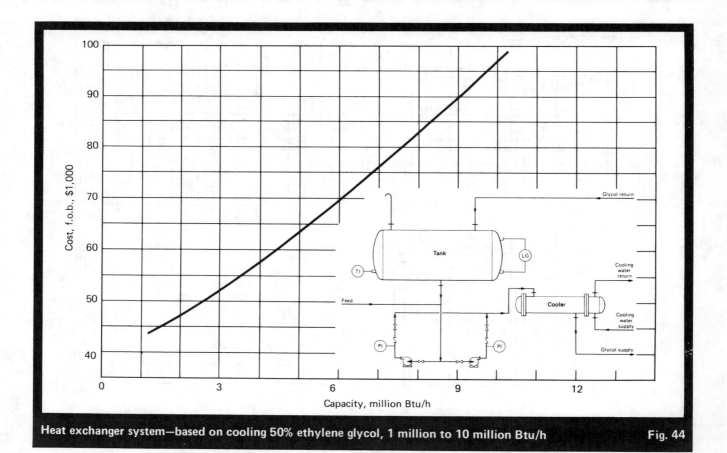

Heat exchanger system—based on cooling 50% ethylene glycol, 1 million to 10 million Btu/h   Fig. 44

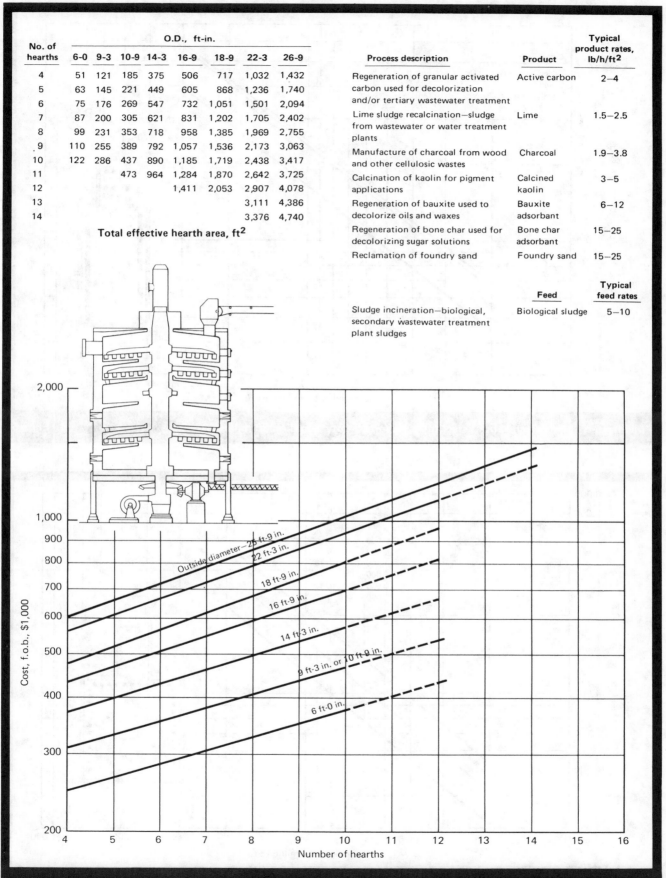

No. of hearths	O.D., ft-in.							
	6-0	9-3	10-9	14-3	16-9	18-9	22-3	26-9
4	51	121	185	375	506	717	1,032	1,432
5	63	145	221	449	605	868	1,236	1,740
6	75	176	269	547	732	1,051	1,501	2,094
7	87	200	305	621	831	1,202	1,705	2,402
8	99	231	353	718	958	1,385	1,969	2,755
9	110	255	389	792	1,057	1,536	2,173	3,063
10	122	286	437	890	1,185	1,719	2,438	3,417
11			473	964	1,284	1,870	2,642	3,725
12				1,411	2,053	2,907	4,078
13							3,111	4,386
14							3,376	4,740

Total effective hearth area, ft^2

Process description	Product	Typical product rates, lb/h/ft^2
Regeneration of granular activated carbon used for decolorization and/or tertiary wastewater treatment	Active carbon	2–4
Lime sludge recalcination—sludge from wastewater or water treatment plants	Lime	1.5–2.5
Manufacture of charcoal from wood and other cellulosic wastes	Charcoal	1.9–3.8
Calcination of kaolin for pigment applications	Calcined kaolin	3–5
Regeneration of bauxite used to decolorize oils and waxes	Bauxite adsorbant	6–12
Regeneration of bone char used for decolorizing sugar solutions	Bone char adsorbant	15–25
Reclamation of foundry sand	Foundry sand	15–25

	Feed	Typical feed rates
Sludge incineration—biological, secondary wastewater treatment plant sludges	Biological sludge	5–10

Multiple hearth furnaces—outside diameters 6 ft to 25 ft 9 in., number of hearths 4-14　　　Fig. 45

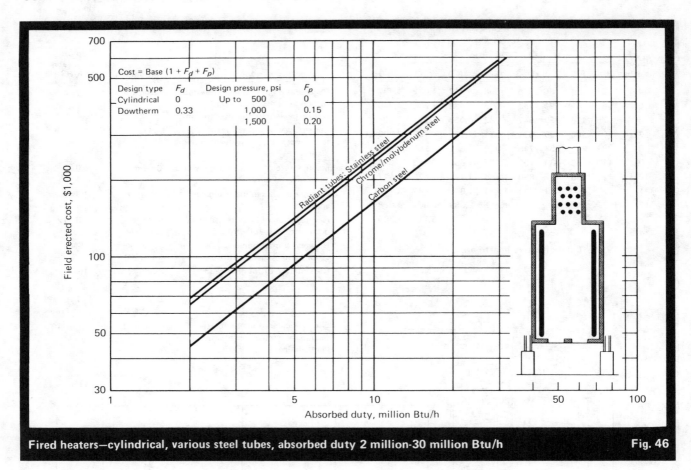

Fired heaters—cylindrical, various steel tubes, absorbed duty 2 million-30 million Btu/h   Fig. 46

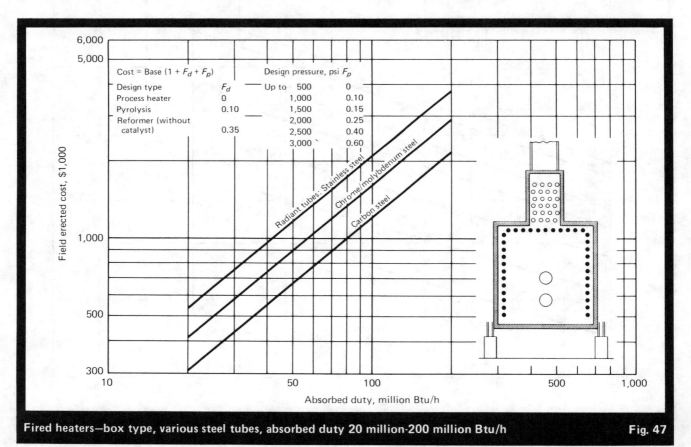

Fired heaters—box type, various steel tubes, absorbed duty 20 million-200 million Btu/h   Fig. 47

cooling water, reconnect piping if required, and provide utilities to the unit.

Example: How much would it cost to purchase a complete skid-mounted assembly as shown in Fig. 28 for heating a 180-gpm 50% ethylene glycol feed from 100 to 135°F? Fig. 28 indicates that this cost would be about $29,000.

Cost estimates derived from Fig. 28 should be accurate to about ±10% as of January 1982.

Centrifugal pumps

When specifying pumps for any service, it is necessary to know the liquid to be handled, total dynamic head, suction and discharge pressures, and the properties of the fluid—temperature at operation, viscosity, specific gravity and vapor pressure.

Pumps are specified at the early stages of a project when a plant layout is not completely defined and most of the equipment dimensions and elevations are sketchy. The easiest way to estimate the price of pumps is by using a current price-data book from one of the leading pump manufacturers, unless the pumping requirements are unique, and a special design or even a first-of-a-kind pump has to be created. In this case, the usual procedure is to send detailed specifications to several pump manufacturers for their bids.

Fig. 30 provides costs for centrifugal pumps designed in accordance with American National Standards Institute's (ANSI) B73.1—1977, which specifies horizontal, end-suction, single-stage centrifugal pumps with centerline discharge for chemical processes.

Temperatures are limited to 300°F by the standard, but in practice, temperatures between 500 and 900°F are not uncommon. The motors used are T-frame, National Electrical Manufacturers Assn. (NEMA) motors with a 1.0 service factor.

The input required to make a cost estimate is: capacity (gpm), differential pressure (psi), suction pressure (psi), system temperature (°F), and casing material.

The items included in the estimate are: ANSI pumping unit, TEFC T-frame NEMA motor, coupling, and base plate.

Centrifugal pump cost, $ = Base cost $\times F_m \times F_o$ where the adjustment factors are:

Material	F_m
Cast iron	1.00
Bronze	1.35
Cast steel	1.41
Stainless steel	1.94
Alloy 20	2.27
Monel	3.31
Nickel	3.47
Hastelloy C	3.78
Hastelloy B	3.93
Titanium	5.71

Operating limits	F_o
Suction pressure:	
—below 275 psig at 100°F	1.0
—below 350 psig at 100°F	1.5
System Temperature:	
0–500°F (at ANSI allowable pressures)	1.0
above 500°F (at ANSI allowable pressures)	1.5

Example: A centrifugal pump is required to deliver 1,430 gpm of a heavy organic mixture (specific gravity = 0.952) with a differential head of 77 ft. Suction pressure is below 275 psig at 100°F. The casing material is to be Alloy 20.

The head, in psi, is (77 × 0.952)/2.31 = 31.7 psi, where 2.31 ft/psi is a conversion factor.

$$\text{Capacity/head factor} = 1{,}430 \times 31.7$$
$$= 4.5 \times 10^4$$

From Fig. 30, base cost = $3,500
$$F_m = 1.2$$
$$F_o = 1.0$$
$$\text{Cost} = \$3{,}500 \times 1.2$$
$$= \$4{,}200$$

Cost estimates derived from Fig. 30 should be accurate to about ±25–35% as of January 1982.

Liquid and slurry blenders

Fig. 29 gives f.o.b. costs of light-gage atmospheric-pressure liquid and slurry blending tanks constructed of Type 304 and 316 stainless steel. The tanks are jacketed for heating or cooling, with transfer surfaces approximating 0.03 ft^2 per gallon of capacity. Dimpled or panelcoil jackets are designed for 125 psi at 350°F. The agitator is turbine type, with motor requirements of approximately 1 hp/1,000 gal of liquid. Electricals are explosionproof, Class I, Group D. Costs include a manway and drain nozzle, and should be accurate to ±10% as of January 1982.

Compressors

Process gases are required in a wide range of capacities, pressures and temperatures; process gas compression is a complex operation involving either centrifugal or reciprocating machines.

Commercial equipment has the following general characteristics: *centrifugal*—high-capacity, low-discharge pressures; *reciprocating*—lower capacity, high-discharge pressures.

Custom-designed equipment is usually necessary for applications exceeding the above capacities and pressures. Motor, turbine or gas engine drives can also be used.

For cost-estimating purposes, consider the many variables, such as volume or weight capacity (ft^3/min or lb/h), molecular weight, k-values (ratio of specific heat at constant pressure to that at constant volume), and compression ratios, as represented by the bhp parameter. These calculations are required before Fig. 31 and 32 can be used. Carbon steel construction is assumed.

Example: A turbine centrifugal compressor rated at 400 bhp may be priced from Fig. 31 at $143,000.

A reciprocating air compressor is required at 800 bhp and could be driven by steam or a gas engine. What would be the approximate capital investment for each type of compressor?

From Fig. 32, it may be seen that the steam-driven compressor would cost $252,000 and the gas-engine version would cost $412,000.

Cost estimates derived from Fig. 31 and 32 should be accurate to about ±25–35% as of January 1982.

Distillation columns

Expenditures for distillation columns and packed towers rank high among equipment costs for process plants. Columns have become even more expensive because of a trend, brought on by soaring fuel costs, to increase their heights so as to reduce the energy costs of refluxing. Packed towers—long widely used in gas absorption, distillation and liquid-liquid contacting—have found growing application in pollution control for removing dusts, mists and odors.

Fig. 33 presents costs, plotted against column diameter in feet, of distillation columns containing 25 trays. Fig. 34 does the same for columns having 50 trays, and Fig. 35 for columns with 100 trays. The parameters are operating pressures of 0, 100 and 300 psig. Costs presented are for October 1981.

Column heights, actually tangent-to-tangent shell heights, noted on the graphs show those of the 25-tray column lengthening from $57\frac{1}{2}$ to 68 ft as the diameter goes from 5 to 6 ft, those of the 50-tray column extending from 95 to 188 ft as column diameter enlarges from 5 to 6 ft, and those of the 100-tray column remaining constant at 170 ft.

Construction is of SA-285-C carbon steel, except for the 11 and 12-ft-dia. 25-tray columns and the 4-ft-dia. 100-tray columns, which are made of SA-515-65 low-alloy steel. Not included are tray costs, which are given in Fig. 37 and 38.

Costs decline with increasing diameter in the narrower column ranges (1–3 ft dia.) because the taller and slenderer the column, the heavier its base must be to withstand wind-load stresses. This is particularly noticeable in Fig. 34 and 35.

Probable accuracies for Fig. 33, 34 and 35 are about ±15%.

Fairly accurate costs can be derived for fabricated vessels, such as distillation columns, on the basis of their weights (see Corripio, et al. [1]). Although vessel weights cannot be readily obtained from process calculations, certain design standards can be used to translate volumetric data into vessel weight equivalents.

A method for determining installed costs of distillation columns is presented by Miller and Kapella [7].

Packaged distillation system

As an example of module estimating, Fig. 36 presents costs for a packaged continuous-operation distillation system for separating methanol from water.

In this system, the methanol content in the feed is 80% by weight, 99% in the product, and 1% in the effluent water. The utilities that must be brought to the battery limits of the unit consist of 100-psig steam, cooling water at about 85°F and a minimum pressure of 50 psig, instrument air at 50 psig, and 220/400-V three-phase 60-Hz electrical power.

The package, skid-mounted and ready for installation and operation, includes the distillation column, all column internals, reboiler, heat exchanger, condenser, pumps, piping, fittings and valves, instrumentation, structural steel, painting, insulation and all design engineering work—all as per the Fig. 36 schematic. The process side of the system is of stainless steel, with all other construction generally of carbon steel. Not in-

cluded are transportation, rigging, site preparation, foundations, lighting and fireproofing.

Probable accuracies of cost estimates from Fig. 36 are about ±10%. Costs of other systems requiring comparable fractionation and energy input may also be estimated from Fig. 36.

Fig. 37 and 38 give costs of sieve and valve trays, respectively. The graphs are based on 20 trays being purchased at a time, so factors from the quantity table included in the graphs must be used to adjust tray costs for quantities other than 20. The plots are only for single-flow trays. Probable accuracies are about ±10% as of November 1981.

Packed towers and packing

Fig. 39 presents costs of single-bed packed towers with process sides constructed of carbon steel and Types 304 and 316 stainless steel. The packed height section is 4 ft, 0 in. tall, and the tower is supported by a 6-ft carbon steel skirt. Towers are designed for full vacuum and 25 psi at 350°F. Internals are not included; costs should be accurate to about ±10%.

Fig. 40 yields costs for two-bed packed towers, with process-side construction of carbon steel and Types 304 and 316 stainless steel. The packed height of each section is 5 ft, 0 in., and the tower is supported by a 10-ft carbon steel skirt. The towers are designed for full vacuum and 25 psi at an operating temperature of 350°F. Internal structures and packings are not included in the costs. Probable accuracies are about ±10%.

Fig. 41 shows costs of saddle-type packings made of porcelain. Saddle packings are useful in absorption and regeneration operations because they provide good liquid redistribution, and porcelain gives corrosion resistance at low cost. Costs are of December 1981, and should be accurate to ±10%.

Other cooling equipment

Fig. 42 presents costs of air-cooled exchangers. Costs cover finned tubes of carbon steel, hot-fluid piping, fan with motor and speed-reducer, plenum and venturi. Draft may be forced or induced. Costs are accurate to ±10% as of December 1981.

Fig. 43 yields cost of erected cooling towers on the basis of water-handling capacity. Included in the costs are the tower, fans, pumps and motors and the expense of erection, but not the cost of the basin. These costs are accurate to ±10% as of December 1981.

Packaged heat exchanger

This system is designed to cool a 50% ethylene glycol solution from 120 to 105°F using cooling water at 90°F.

It is a skid-mounted assembly ready for installation and operation. The assembly includes heat exchanger(s), tank, pumps, piping, valves and fittings, instrumentation, structural steel, painting, and engineering design—all as shown in Fig. 44.

The equipment is generally of all-carbon-steel construction. Heat exchangers are built to comply with TEMA Class C specifications, and the tank to U.L. specifications. Costs include cast-iron pumps, baseplates, couplings, guards, TEFC motors, and mechanical seals.

Jobsite requirements: unload unit, locate same on

purchaser's foundation, install piping for glycol and cooling water, reconnect piping if required, and provide utilities to the unit.

Example: Find the price of a skid-mounted assembly as shown in Fig. 44, to be used for cooling a 50% ethylene glycol solution from 120 to 105°F. The duty of the heat exchanger has been calculated at 7.5 million Btu/h.

Fig. 44 indicates that the price would be about $80,000.

Cost estimates based on Fig. 44 should be accurate to about ±10% as of January 1982.

Multiple-hearth furnaces

The Nichols/Herreshoff furnace consists of a series of circular hearths, placed one above the other and enclosed in a refractory-lined steel shell. A vertical rotating shaft through the center of the furnace carries arms with rabble teeth that stir the charge and move it in a spiral path across each hearth. The material is then fed to the hearth below.

Discharge takes place through one or more ports at the bottom. Hot gases flow countercurrently, heating the charge to reaction temperature and carrying on the desired reaction.

In many operations, combustion of the volatiles in the charge provides the necessary heat. In other instances, heat is furnished by combustion of various fuels introduced through burners into certain hearths (direct firing), or in separate combustion chambers (external firing).

Some applications with process descriptions are given in the tables in Fig. 45.

Example: The decolorization of an organic acid solution produces 24,000 lb/d of spent granular activated-carbon. From Fig. 45, a typical product rate of 2 to 4 lb/(h)(ft^2) is obtained.

The required furnace effective-area is calculated as follows:

$$24,000\frac{\text{lb}}{\text{d}} \times \frac{\text{d}}{24 \text{ h}} \times \frac{\text{h-ft}^2}{3 \text{ lb}} = 333 \text{ ft}^2$$

From Fig. 45, a suitable furnace size is obtained—in this case, a 10-ft, 9-in.-O.D. by 8-hearth unit, and the f.o.b. price of such a unit is estimated at $410,000.

Cost estimates based on Fig. 45 should be accurate to about ±10% as of January 1982.

Fired heaters

A fired heater, for our purposes, will include a number of devices in which heat liberated by the combustion of fuel within an internally insulated enclosure is transferred to fluid contained in tubular coils. Typically, the tubular heating elements are installed along the walls and roof of the combustion chamber, where heat transfer occurs primarily by radiation, and, if economically justifiable, in a separate tube bank, where heat transfer is accomplished mainly by convection.

Industry identifies these heaters with such names as process heater, furnace, process furnace and direct-fired heater, which are all interchangeable.

The fundamental function of a fired heater is to supply a specified quantity of heat at elevated temperatures to the fluid being heated. It must be able to do so without localized overheating of the fluid or of the structural components.

Fired-heater size is defined in terms of the equipment's heat-absorption capability, or duty. Duties range from about a half-million Btu/h for small, specialty units to about 1 billion Btu/h for super-project facilities such as the mammoth steam hydrocarbon-reformer heaters. By and large, the vast majority of fired-heater installations fall within the 10 to 350-million-Btu/h range.

There are many variations in the layout, design, and detailed construction of fired heaters. Thus, virtually every fired heater is custom-engineered for its particular application.

The input required to use Fig. 46 and 47 is: absorbed heat duty (Btu/h), furnace type, design pressure (psig), and radiant-tube material.

The cost includes complete field erection and subcontractor indirects. Fig. 46 applies to cylindrical designs and Fig. 47 applies to the "box" or A-frame type of construction.

Example: Two fired heaters are required in a plant. One is to be a simple cylindrical type with chrome-molybdenum steel tubes, to be used at a pressure of 1,500 psi and a duty of 1.5 million Btu/h. Fig. 46 shows:

$$\begin{aligned}
\text{Base cost} &= \$340,000 \\
F_p &= 0.2 \\
\text{Cost} &= 1.2 \times \$340,000 \\
&= \$408,000
\end{aligned}$$

The other unit is a reformer (without a catalyst) to be designed for service at 2,000 psi and 66 million Btu/h. Radiant tubes should be of stainless steel.

Fig. 47 shows:

$$\begin{aligned}
\text{Base cost} &= \$1.95 \text{ million} \\
F_d &= 0.35 \\
F_p &= 0.25 \\
\text{Cost} &= 1.60 \times \$1.95 \text{ million} \\
&= \$3.12 \text{ million}
\end{aligned}$$

Cost estimates based on Fig. 46 and 47 should be accurate to about ±25% as of January 1982.

Major types of filters

Operations involving the separation of solids from liquids are so prevalent in process plants that many estimates of plant capital cost would be less reliable if they did not include costs of the equipment that perform this function, chiefly filters and centrifuges.

Fig. 48 through 51 shows costs per square foot of most of the filter types that represent major capital-cost expenditures in process plants.

Fig. 48 gives f.o.b. costs of batch pressure-leaf or tubular filters of carbon steel construction but with filtering elements of stainless steel, the most common material of construction. Costs are for filters only, of accuracies of about ±10% as of January 1982. Installations would involve feed pumps, compressed air, and usually feed storage and precoat tanks.

These filters can be arranged vertically or horizontally. Slurry is introduced under pressure and forced through the filter elements. Suspended solids are retained on the media. The clarified liquid flows to the interior of the leaf. The solids are removed from the elements and the unit is ready for another cycle. These filters are described in the "Kirk-Othmer Encyclopedia of Chemical Technology," pp. 298 and 303–309 [5].

Fig. 49 presents costs of four types of vacuum filters: the horizontal table filter in both carbon and 316 stainless steel, a corrosion-resistant horizontal belt filter, a 316 stainless-steel tilting pan filter, and a 316 stainless-steel single-compartment drum filter. These filters compete with each other in handling fast-filtering granular and crystalline materials. The belt unit also competes with drum filters in washing applications. Vacuum systems for these filters are similar to those for disk and drum filters, often with the addition of extra receivers and pumps for handling wash filtrates. Vacuum pumps often become a dominant cost consideration. Probable cost accuracies run about ±10%. These filters are described in "Kirk-Othmer," pp. 300–305 and 318–321.

Fig. 50 indicates costs of continuous rotary vacuum disk filters of two basic designs: a heavy-duty unit for general industrial applications (especially metallurgical, such as taconite, coal and flue dust), and a unit specifically designed for fiber recovery and concentration in the pulp and paper industry.

The curve for the general industrial filter yields costs only for carbon steel construction, because corrosion-resistant designs are not common. Costs cover only the filter with agitator. A typical filter station would include a vacuum system consisting of a vacuum receiver (to separate liquid from air), a filtrate pump, a vacuum pump, and possibly a moisture trap, perhaps with a scrubber or condenser, to protect the vacuum pump. Vacuum-system costs can vary so widely that they cannot be lumped into the filter cost.

The curve for the pulp and paper unit represents costs for 304 stainless-steel construction. To upgrade from 304 to 316 stainless-steel construction, use a cost multiplier of 1.19 for smaller units and 1.24 for larger units. Costs do not include an agitator but do include a feedbox, hood and repulper, which are normally provided. The vacuum system consists of a barometric leg rather than a pump (taking advantage of the very high liquid flows) and a vacuum receiver.

Probable cost accuracies are about ±10%. These types of filters are discussed in "Kirk-Othmer," pp. 300–301 and 313–314.

Fig. 51 shows costs of two basic types of continuous multicompartment rotary vacuum drum filters: one for chemical, metallurgical and general industrial services (the lower pair of curves), the other for pulp and paper applications (the upper pair). (One other type of drum filter is included in Fig. 49, with horizontal filters, with which it competes.)

The general filter has five major types of discharge device: scraper, belt, string, roll and precoat (two others, coils and wire, are more specialized). The two curves plot costs only for filters with belt and scraper discharge, because these bracket the costs for all types of dischargers. The costs of filters with roll and string discharge are intermediate between the belt and scraper, and the precoat is comparable to the belt.

Costs for the general filter cover only carbon-steel construction. Multipliers for converting costs from carbon to 316 stainless steel are 1.15 for small units, 1.4 for medium-sized units, and 1.7 for larger units. Probable accuracies are ±10%. This type of filter is described in "Kirk-Othmer," pp. 300–305 and 314–317.

The pulp and paper filters are totally different because of the high hydraulic capacities required (tens of thousands of gal/min) and the discharge of pulp sheets. Brownstock and bleach washing represent major applications, with deckering (pulp thickening) important also.

The curves give costs only for carbon-steel and 304 stainless-steel construction. To convert costs from 304 to 316 stainless, multiply by 1.08 for small units, 1.11 for medium-sized units, and 1.16 for large units. To convert from 304 to 317 stainless, multiply by 1.23 for small units, by 1.31 for medium-sized units, and 1.40 for large units. Included in costs are normal accessories—feedbox, wash showers and repulper. As with pulp and paper disk filters, barometric legs are used to generate vacuum. Cost accuracies are about ±10%, as of January 1982.

This filter's rotating cylindrical drum is made up of longitudinal sections, which form a drainage grid on which the filter medium is placed. Vacuum applied to the cylinder causes the formation of the sheet as the cylinder rotates and submerges into the vat containing pulp stock. The sheet, after passing through a spray washing, is continually discharged. The filtrate goes through the medium and out through an automatic valve to the barometric leg.

Not presented are two major types of filters—filter presses and filter-belt presses. The first are manufactured in so many materials of construction and with such varied features that their costs could not be represented via graphs. The second also come in so many different configurations as to preclude graphical representation of costs.

Continuous centrifuges

Fig. 52 and 53 give costs of centrifuges of two basic categories, the first for inorganic chemical applications, the second for organic chemical applications. Both categories are represented in single groups of curves by the solid-bowl, screen-bowl and pusher types of centrifuges.

In the inorganic chemicals applications, solids are suspended in a mother liquor of water, caustic soda, brine, or some other nonflammable liquid. Therefore, housings can be unsealed. Materials of construction can be as basic as carbon steel, sometimes as exotic as nickel, but rarely of Hastelloy.

In the organic chemicals applications, it is assumed that the unit must be sealed against oxygen inflow. Such sealings add from 10 to 20% to costs. Materials of construction usually involve nothing less costly than stainless steel and sometimes require Hastelloy C-276.

Costs represented by Fig. 52 and 53 are for the complete but basic centrifuge, f.o.b. point of shipment, January 1982. Not included are drive motor and special structures, as well as shipping and installation. As average costs of the three basic centrifuges are presented, probable accuracies are ±25%.

The solid-bowl centrifuge's two principal elements are a rotating bowl, which is the settling vessel, and a conveyor, which discharges the settled solids. The bowl has adjustable overflow weirs at its larger end for discharging clarified effluent, and solids ports on the opposite end for discharging dewatered solids. Feed enters through a supply pipe and passes through the conveyor hub into the bowl itself. As the solids settle out in the bowl, due to centrifugal force, they are picked up by the conveyor scroll and carried along continuously to the solids outlets. At the same time, effluent continuously overflows the weirs.

In the screen-bowl centrifuge, dilute feed is concentrated in a solid bowl. The solids, substantially drained of mother liquor, are then conveyed to the screen section for washing, if required, and final discharge.

Pusher centrifuges usually deliver a drier cake than other continuous centrifuges. Feed enters through an inlet pipe and accelerating cone, and is introduced on a first-stage basket, where the solids are retained. The first-stage basket, actuated by a hydraulic pushing mechanism, reciprocates under a static pusher plate to advance the cake from the first to the second stage on the back stroke. The forward stroke of the first basket pushes the cake off the second basket and into a cake chute.

Butterfly valves and pipe strainers

Fig. 54 presents installed costs of butterfly valves, ANSI classes 150, 300 and 600, plotted against valve size. The cost of the valve operator is not included. Probable accuracies are ±10% as of January 1982.

Fig. 55 gives f.o.b. costs of pipeline strainers. Outer cases are of carbon steel, the baskets of stainless steel. Baskets may be slope-top for inline flow, or flat-top for offset flow. Probable accuracies of costs are ±10% as of January 1982.

References

1. Corripio, A. B., Mulet, A., and Evans, L. B., Estimate Costs of Distillation and Absorption Towers via Correlations, *Chem. Eng.*, Dec. 29, 1981.
2. Desai, M. B., Preliminary Cost Estimating of Process Plants, *Chem. Eng.*, July 27, 1981.
3. Guthrie, K. M., Capital Cost Estimating, *Chem. Eng.*, Mar. 24, 1969.
4. Hand, W. E., From Flow Sheet to Cost Estimate, *Petroleum Ref.*, Sept. 1938.
5. "Kirk-Othmer Encyclopedia of Chemical Technology," Vol. 10, 3rd ed., John Wiley & Sons, Inc., 1980.
6. Lang., H. J., Simplified Approach to Preliminary Cost Estimates, *Chem. Eng.*, June 1948.
7. Miller, J. S., and Kapella, W. A., Installed Cost of a Distillation Column, *Chem. Eng.*, Apr. 11, 1977.
8. Pikulik, A., and Diaz, H. E., Cost Estimating for Major Process Equipment, *Chem. Eng.*, Oct. 10, 1977.
9. Viola, J. L., Jr., Estimate Capital Costs via a New Shortcut Method, *Chem. Eng.*, Apr. 6, 1981.

Acknowledgements

We wish to thank the following companies and individuals for contributing cost information to this report: Bird Machine Co. (John J. Campbell), Cortech Plastics, Inc., Chuck Gurdin, Inc., Dorr-Oliver Inc. (Robert M. Talcott), Duriron Co., Durco Pump Div. (Joseph Trendy), Foster-Wheeler Energy Corp., Glitsch Packaged Plants, Monsanto Co. (Harley C. Nelson), Nichols Engineering & Research Corp. (Stuart S. Spater), PDQ$ Inc. (Gustav Enyedy, Jr.), Pakbilt Co., Papeco, Inc., Posi-Seal Intl., Inc., Rubicon Industries, Inc., Taylor Tank Co., and Xchanger, Inc.

The authors

Richard S. Hall is president of Richard S. Hall and Associates, manufacturers' representatives, 145 Cortlandt St., Staten Island, NY 10302 (telephone 212-442-2460), as well as president of C-P Associates, a marketing and sales development company. Formerly, he was president of Chem-Pro Marketing Service for Walster Corp. and vice-president of marketing and sales for Doyle & Roth Manufacturing Co. See "Who's Who in Finance and Industry" for a complete biography.

Jay Matley is a senior associate editor with CHEMICAL ENGINEERING. Previously, he worked for Kaiser Aluminum & Chemical Corp., W. R. Grace & Co., Callery Chemical Co. and Monsanto Co. His industrial experience has largely been in operations supervision, process engineering and process development.

A graduate of Kansas State University, with a B.S. degree in chemical engineering, and State University of Iowa, with a B.A. in liberal arts, he is a member of AIChE, American Assn. of Cost Engineers and The Filtration Soc.

Kenneth J. McNaughton is an associate editor at *Chemical Engineering* and editor of the "You and Your Job" department. He received his B.Eng. (Chem.) at Melbourne University and an M.Eng. Sci. at Monash University in Australia. Mr. McNaughton did research at London University in biochemical engineering, and has worked with sugar and oil refineries in Australia and with Texaco Inc. in the Bahamas. He has published technical articles on fluid flow and has edited three books for McGraw-Hill on materials.

Cost update on specialty heat exchangers

Estimate costs via a correlation based on actual prices for two of the most common specialty heat exchangers: the plate-and-frame and spiral-plate designs.

Jimmy D. Kumana, Henningson, Durham and Richardson, Inc.

☐ Specialty heat exchangers are being used increasingly in the chemical process industries, especially when unfavorable temperature profiles would make conventional shell-and-tube prohibitively large and expensive. Among the specialty exchangers that have gained fairly widespread acceptance are the plate-and-frame and the

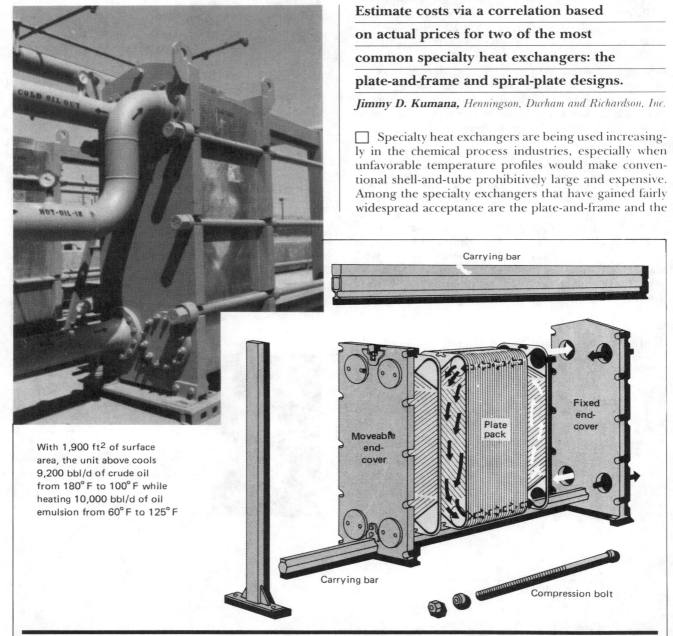

With 1,900 ft² of surface area, the unit above cools 9,200 bbl/d of crude oil from 180°F to 100°F while heating 10,000 bbl/d of oil emulsion from 60°F to 125°F

Carrying bar

Moveable end-cover

Plate pack

Fixed end-cover

Carrying bar

Compression bolt

Alfa-Laval, Inc.

Gasketing pattern around each port distributes hot and cold fluids to alternate plate flow channels   Fig. 1

Flowrates for these plate-and-frame exchangers range from 11,000 to 5 gal/min   Fig. 2

A.P.V. Equipment, Inc.

spiral-plate (which should not to be confused with the spiral-tubecoil).

Neither design equations nor cost data on these types of exchangers are generally available. Process design engineers must rely on the manufacturer to provide a workable design. Aware that they are frequently contacted only when a conventional shell-and-tube exchanger is unsuitable for a particular application, vendors of specialty exchangers are often able to command full market prices for these products.

Basis of correlation

Engineers at Henningson, Durham and Richardson, Inc. recently (2nd and 3rd quarters of 1983) completed the process engineering design and equipment bid evaluation for two alcohol plants for which large numbers of plate-and-frame and spiral-plate exchangers were purchased. Several shell-and-tube exchangers were also bought at this time.

It was found that the price quotations from all the manufacturers correlated well with exchanger surface area via the equation:

$$C = kA^n$$

Here, C = purchased cost, F.O.B. factory, \$, and A = exchanger surface area, ft². Values for k and A are given in the table.

Reliability of cost estimates

For both shell-and-tube and plate-and-frame heat exchangers (see Fig. 1 and 2 for illustrations of the latter

Coefficients for cost-estimation correlation

Exchanger type	Materials of construction	Applicable size range, ft²	Coefficients	
			k	n
Shell-and-tube	All Type 304 stainless steel	400-9,000	235	0.665
Plate-and-frame	Frame: carbon steel; plates: 304 stainless steel	100-5,000	100	0.778
Spiral-plate	All 304 stainless steel	100-1,500	660	0.590

Alfa-Laval, Inc.

Normal installation is horizontal (left), vertical if one fluid is gas; spacers help keep plates apart (right)   Fig. 3

These four spiral exchangers are cooling water containing fibers in solution   **Fig. 4**

A.P.V. Equipment, Inc.

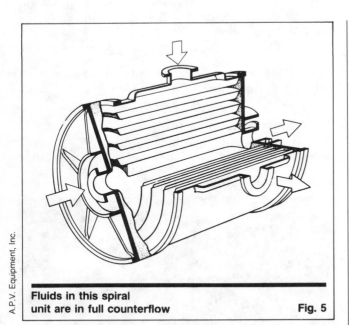

**Fluids in this spiral
unit are in full counterflow**   **Fig. 5**

A.P.V. Equipment, Inc.

type), the accuracy of the foregoing equation is ±10%.*
For spiral-plate exchangers (this type is shown in Fig. 3,
4 and 5), it is ±20%.

Part of the reason for the greater variation in the costs
of spiral-plate exchangers is that the cost of manufacturing this type is very sensitive to the thickness of the
plates, which in turn depends on the distance between
spacers supporting the plates. If, for example, the process fluid contains suspended fibrous solids that could
get caught on the spacers, and eventually block the
flowpath, the distance between the spacers should be at
least 3 inches, and preferably 4. This requirement could
make a difference of as much as 30% in the cost of a
spiral-plate exchanger.

The mechanical design and cost of shell-and-tube and
plate-and-frame exchangers, on the other hand, are
relatively insensitive to the physical properties of process
fluids.

It should be noted that, in all cases, the cost data
presented are for operating temperatures and pressures
of, respectively, less than 350°F and 200 psig.

The author

Jimmy D. Kumana is Chief Process
Engineer of HDR's Technical Services
Div. (P.O. Box 12744, 101 West
Garden St., Pensacola, FL 32575; tel:
904-432-2481), responsible for process
design work on chemical, petrochemical
and biomass energy (including alcohol)
projects. Prior to joining HDR, he was a
partner in a consulting firm that
specialized in alcohol fermentation and
distillation technology. Holder of an
M.S. degree in chemical engineering
from the University of Cincinnati, he is
a registered engineer in Ohio, Florida,
Louisiana and Nebraska, has published
several articles, and holds U.S. and
foreign patents on alcohol distillation.

*For shell-and-tube exchangers, a more detailed method of estimating costs
and extensive cost data are available in C. P. Purohit's "Estimating Costs of
Shell-and-tube Heat Exchangers," pp. 289–300.

INDEX